ATLANTIC STUDIES ON SOCIETY IN CHANGE

NO. 111

HUNGARIAN ARTS AND SCIENCES
1848-2000

"I have created another world out of nothing."
János Bolyai

Edited by
László Somlyódy and Nóra Somlyódy

Copyedited by Tim Wilkinson
Typeset by Andrea T. Kulcsár

SOCIAL SCIENCE MONOGRAPHS, BOULDER, COLORADO
ATLANTIC RESEARCH AND PUBLICATIONS. INC.
HIGHLAND LAKES, NEW JERSEY

DISTRIBUTED BY COLUMBIA UNIVERSITY PRESS, NEW YORK

2003

EAST EUROPEAN MONOGRAPHS NO. DCXXXV

ISBN 0-88033-533-5
Library of Congress Control Number 2003113232

Printed in the United States of America

Table of Contents

Preface to the Series and Acknowledgments

The present volume is a component of a series that is intended to present a comprehensive survey of the many aspects of East Central European society.

The books in this series deal with peoples whose homelands lie between the Germans to the west, the Russians, Ukrainians and Belorussians to the east, and the Mediterranean and Adriatic seas to the south. They constitute a particular civilisation, one that is at once an integral part of Europe, yet substantially different from the West. The area is characterised by a rich diversity of languages, religions and governments. The study of this complex area demands a multidisciplinary approach, and, accordingly, our contributors to the series represent several academic disciplines. They have been drawn from universities and other scholarly institutions in the United States and Western Europe, as well as East and Central Europe.

The editor-in-chief is responsible for ensuring the comprehensiveness, cohesion, internal balance, and scholarly quality of the series he has launched. He cheerfully accepts this responsibility and intends this work to be neither justification nor condemnation of the policies, attitudes, and activities of any person involved. At the same time, because the contributors represent so many different disciplines, interpretations, and schools of thought, our policy in this, as in the past and future volumes, is to present their contributions without major modifications.

We are thankful to the editors of this volume: László Somlyódy, a prominent professor at the Budapest University of Technology and Economics—the father—, and the most reliable partner in compiling the book—his daughter. The contributors are well-known fellows of various fields of natural sciences and humanities.

I also wish to express our gratitude to Andrea T. Kulcsár, Tim Wilkinson, Attila Tevely, Csaba Horváth, and Béla Nagy.

The volume was already completed when the happy news was received about the 2002 Nobel Prize for Literature having been won by Imre Kertész, a new glorious achievement of a distinguished Hungarian.

Budapest, October 23, 2002

Béla K. Király
Editor in Chief

László Somlyódy
Nóra Somlyódy

PREFACE

This is the latest addition to the 'Atlantic Studies on Society in Change' series, under which banner more than one hundred and twenty monographs have appeared to date. The majority of these deal with aspects of the history, evolution, economy, culture and foreign relations of Hungary and the surrounding region, with a particular focus on the period from the country's 1848–49 War of Independence to the 1989 change in political régime. The contributions to the present volume embrace the same century and half, although in most cases the purview has been slightly extended to bring it up to the turn of the millennium. In a departure from most of its predecessors, the book is not in the form of a monograph but has called on the assistance of a large group of specialists to review Hungarian achievements in the various fields of the natural sciences, technology, the arts, and sport.

From inception on, the enterprise has, by its very nature, thrown up issues in determining its scope. These start with the issue of defining who should be regarded as Hungarian, given that the twists of history have prompted many to leave the country and go on to produce achievements of undeniable significance in their new homeland—a phenomenon that is not limited to Hungarians, of course. Others were drawn abroad by the greater opportunities offered for their talents. Thus, the figures who are known to the wider world under the names John von Neumann, Michael Curtiz and Dennis Gabor were all three born in Budapest, where they were known as Neumann János, Kertész Mihály and Gábor Dénes. Equally it would have been wrong to turn a blind eye to instances where this process worked in the reverse direction, because

Hungary's period of intensive economic development in the latter third of the nineteenth century, its *Gründerzeit*, was also a time of substantial immigration to the country, as a result of which the likes of Swiss-born Ábrahám Ganz and András Mechwart were among the major founder figures of Hungarian industry. They all find mention in this volume, along with many others whose names will be less familiar because their contribution to the respective fields found resonance 'merely' within the frontiers of their native land.

The idea of bringing together an English-language conspectus of outstanding Hungarian achievements in different fields is not entirely new. Francis S. Wagner's compilation, *Hungarian Contributions to World Civilization*, which appeared in 1977, provided what still stands in any language, Hungarian included, as an unparalleled review of those achievements from the foundation of the state to the time of publication. The present volume is more modest in its temporal scope but instead aims to give a more detailed treatment of each field, with plentiful references to the historical and cultural context, up to the turn of the millennium. Alongside Wagner's book, George Marx's *The Voice of the Martians* represents a kindred attempt, through a series of interviews and essays, to sketch portraits of many innovators of Hungarian descent who were active outside the country, and we have chosen here to present a modified version of its chapter on physics.

As with any project, we have been forced to make certain compromises in putting this volume together, some of which may justifiably be perceived as shortcomings. The biggest lacuna is the decision to omit from the book any direct coverage of the social sciences such as economics, history, psychology and anthropology, which has meant that figures such as the sociologist Karl Mannheim or art historian Charles de Tolnay, whose reputations are well established in the United States as elsewhere, here gain at best just a passing mention. Much the same goes for a number of 'uncategorizable' innovators like László József Bíró, inventor of the ball-point pen that in the English-speaking world is still usually referred to as the 'biro'. Then again, the choice of which contemporary personalities to include was governed by the perspectives of the different contributors. In many areas consensus has not yet been reached on the exact place of contributions from those who are still active in the field, and for that reason we have not pushed authors too strenuously to cover the most recent developments in much detail. Likewise, in presenting the various individual fields the aim has not been to achieve completeness but rather to encourage our contributors to live with an element of personal subjectivity in drawing up what, we hope, will be found to be enjoyably readable accounts, full of intriguing details and curiosities.

Much thought went into choosing the book's motto. The prime consideration was that it should express the impact that the talents of even a relatively small nation like Hungary are capable of exerting in the wider world, without in any way wishing that to be seen as implying any superiority over the achievements of other small nations. In the end, the solution was provided courtesy of János Bolyai, who in a letter to his father, Farkas Bolyai, announcing his discovery of non-Euclidean geometry wrote: "I have created another world out of nothing."

Apart from this Preface, the book comprises nine chapters. Tibor Frank's 'Introduction' gives a very brief sketch of Hungary's history, geography, cultural influences and other factors that have shaped the creativity of her populace. Chapter 2, edited by Zsuzsa Szentgyörgyi, deals with the natural sciences and technology, with subsections on mathematics, physics, computer technology, biology, astronomy, chemistry, agriculture and engineering. This is followed, in order, by chapters on architecture by Ferenc Vámossy, the visual arts by László Beke, the cinema by Gergely Bikácsy, literature by Mihály Szegedy-Maszák, music by Péter Halász, and, finally, sport by Gyöngyi Földesi Szabó. The book is rounded off by a biographical dictionary and a map to show the most important geographical locations.

We would like finally to thank all those who have helped in putting this volume together. There are many who generously facilitated the editors' work with creative suggestions at various stages in preparatory discussions, in the course of the editing and translation, and in the selection of illustrations. We would like here to give particular thanks to three individuals for their sterling assistance. Pride of place, naturally, goes to Béla Király, who originally broached the idea of writing the book and has been a constant source of inspiration and reassurance throughout its preparation. Tibor Vámos provided much invaluable advice during the crucial early phase of the project's gestation. Tim Wilkinson was much more than our language editor: he provided a critical outside eye, offering many helpful suggestions to strengthen the book. The consistency of language that has been achieved between the various chapters is largely due to his unstinting efforts.

Budapest, June 2002

Tibor Frank

THE SECRET OF SURVIVAL: PROBLEM SOLVING AND CREATIVITY IN HUNGARIAN HISTORY

Introduction

Hungarians typically tend to suppose that their country's story is one of failure. They were beaten by many enemies, failed in repeated revolutions and wars of independence, and were on the losing side in two world wars. The country was overrun by Mongols in the thirteenth century, by Ottoman Turks in the sixteenth, by Germans and Soviet Russians in the twentieth, and in between it spent nearly half its entire history under foreign domination as part of the Habsburg Empire. When it finally regained sovereignty in 1920, after almost 400 years, that was at the price of losing over two-thirds of its historical territory and some three and a half million kinsmen. Many national political leaders died in exile, including Prince Ferenc Rákóczi II, Governor–President Lajos Kossuth, President Count Mihály Károlyi, Regent Adm. Miklós Horthy, and the Communist dictators Béla Kun and Mátyás Rákosi. A distressingly long list of the country's best minds were also lost to the West, including all twelve Nobel laureates of Hungarian origin—among them, Georg von Békésy, János Harsányi, George de Hevesy, György Oláh, Albert Szent-Györgyi, and Eugene Wigner—as well as composers Franz Liszt, Béla Bartók, and Ernő (Ernst von) Dohnányi; conductors Fritz Reiner, George Szell, Eugene Ormandy, Sir Georg Solti, Antal Doráti, Ferenc Fricsay, and István Kertész; scientists Leo Szilárd, Edward Teller, Theodore von Kármán, and Michael Polanyi; mathematicians John von Neumann, George Pólya, and John Kemény; film directors and producers

Alexander Korda, Michael Curtiz, and Joe Pasternak; photographers Brassaï, Robert Capa, and André Kertész; artists/designers Laszlo Moholy-Nagy and Marcel Breuer; social scientists and scholars such as Karl Mannheim, Arnold Hauser, Charles de Tolnay, Arthur Koestler, and Karl Polanyi, and a host of others.

For all the centuries of political failure, then, Hungarians at the individual level have evidently thrived, developing an art of survival and a readiness to restart their lives against all odds. They have mastered a special kind of tenacity to prevail under highly adverse circumstances in the stormy heart of Europe into which their forefathers led them, late in the late ninth century A.D., on what was no doubt originally seen as a temporarily forced stop-over that has now lasted more than eleven hundred years. They have endured in those 'Lands Between,' to use the telling phrase of the historian Alan Palmer, at the meeting-point of North with South and West with East, at the confluence of some of the largest and most powerful empires in world history. The Roman, Ottoman, Russian, Holy Roman, Habsburg, Napoleonic French, Nazi German, and Soviet empires all stretched as far as Hungary and at some time laid claim to part or all of her territory, treating it as war booty, a border area, a *cordon sanitaire,* or a defensive line.

The Hungarians have never been able to ward off such powerful empires for long, yet they have always been able to perpetuate themselves as a nation, even if that was sometimes merely a virtual entity, and as an indigenous culture. Located in a region of the world that has an almost in-built geopolitical menace as a honeypot for hungry neighbours, Hungary and the Hungarians have proved indefatigable and unwavering in their determination to outwit, outmanoeuvre and outlast their adversaries, and continually resurrect their nation: predestined losers, perhaps, yet shrewd survivors all the same.

Yet is it right to regard Hungary's history as one of either abysmal failure or unparalleled success? More likely it is both these things at once. The outcome is a national mentality that is split between a sense of inferiority and an exaggerated sense of self-worth: insecurity and self-pity on the one hand, overconfidence and inflated national ego on the other—both extremes equally unjustified and unsound in their hyperbole.

The Setting: Climate, Geography, Anthropology

Many nations claim to lie at the centre of Europe, and Hungary is no exception. At the same time, however, it has an isolated language and culture. Its location in the Carpathian basin means it is surrounded by some of Europe's highest mountain ranges, set around the River Danube and its many tributaries, big and small. With its large expanse of arable land and continental climate, it has become one of the bread- and fruit-baskets of Continental Europe.

Until World War I, Hungary included large regions which today make up foreign states in their own right, as in the case of Slovakia and Croatia, or now form part of Romania, the Ukraine, Serbia, and Austria. Once almost the size of the state of Arizona or modern Italy, Hungary was reduced to a small (35,907 square mile), landlocked country similar in size to Indiana or Portugal. The national coat of arms indeed still depicts the four major rivers and three large mountains that the country once boasted. That one-time kingdom of the millennium from 1000 A.D. is still commemorated in the much vaunted, and often profaned, royal crown of St. Stephen, the first king of Hungary (1000–38), whose long and successful reign put the country on the map of Christian Europe through the alignment to the religious values of the medieval West that came with its embracing of Roman Catholicism.

The Hungarians brought their isolated language—a branch of the Finno-Ugrian family of languages, which includes Finnish, Estonian and a scattering of minor, near-extinct relatives—from the Ural region of Russia. On account of the singularities of their language, they have never been able to shake off a sense of isolation throughout the 1,100 years that they have spent surrounded by Slavs, Germans, and Latins. That has endowed the nation with an aura of apartness and loneliness, which has helped foster a sense of exceptionalism and pride that are prevalent to the present day. Unrelated to any neighbour's language, the Hungarian vernacular nevertheless preserves elements of all the tongues its speakers came into contact with in the course of their long migrations to their present homeland, incorporating extensive vocabularies from Germans, northern and southern Slavs, Romanians, and Turks. Most tellingly, Hungarians continued to use medieval Latin as a literary language up until the fifteenth century, and as the official language of administration right up to 1844, making Hungary one of the few countries in the world where Latin survived as a living language well into the modern era. In addition, German increasingly became the lingua franca by which Hungary communicated with other areas of the Habsburg Empire, to which it was officially yoked to as a result of a marriage contract with the Habsburg dynasty that took effect after the death, in 1526, of Lajos (Louis) II, independent medieval Hungary's last king in the battle of Mohács against the invading Ottoman army.

Isolation thus became both a destiny and a virtue, making it imperative for Hungarians to interact with other cultures, use other languages and borrow from the vernaculars of all the neighbouring ethnic groups of the region. That readiness to express themselves in diverse tongues and take over, or share, alien idioms and other cultural paradigms has had a big hand in keeping them relatively open to the outside world, encouraging an engagement with other cultures and the evolution of a cultural pluralism which has been a hallmark of the Hungarian tradition.

As many villages in old Hungary's far-flung border areas were ethnically mixed, multi-culturalism, cross-fertilisation and assimilatory drift have all been standard features of Hungarian society and culture since the Middle Ages. In receiving a wide variety of immigrants over the centuries, Hungary has absorbed and been enriched by an astonishing range of influences, showing considerable ability to adapt many of the features of what constitutes 'Hungarianness' to the ebb and flow of those impacts.

'The Lands Between'

The geography and geopolitics of the Danubian basin have obliged the Hungarians to struggle for survival during most of their national history. There have been few opportunities for them even to contemplate, let alone embark upon, expansionist adventures. An attempt was made, albeit fruitlessly, in the tenth century, and although Lajos I (Louis the Great) was more successful on behalf of the house of Anjou in the fourteenth century, and Mátyás (Matthias) Hunyadi managed to enlarge an already unwieldy realm at the expense of its neighbours in the fifteenth, the country was more often itself the target of foreign imperial encroachments. Thus, the thirteenth century saw invasion by the Mongols in 1241, the fifteenth and sixteenth centuries the Ottoman push first to Belgrade, in 1456, and then to Mohács, where a bloody battle in 1526 put an end to the independent kingdom of Hungary, initiating a prolonged struggle against the Ottoman and, later, the Habsburg Empire, a period that ultimately saw Hungarian sovereignty vanish altogether.

The anti-Ottoman and anti-Habsburg struggles are central to any understanding of the Hungarian mind in the early modern and the modern era, generating both a fiercely nationalistic sentiment and a pervasive sense of victimisation, threat, and fear. The dominant national mood in the late nineteenth century was still, at bottom, defensive in nature, even though Hungarian nationalism had by then turned to aggressive intolerance of neighbouring ethnic groups. It is a tragic irony of history that when the long-cherished desire to regain national independence was finally met after World War I, the country lost over two-thirds of its original territory and a goodly proportion of its native speakers.

Hungary's early modern history was so much taken up with strife that it was granted little respite, few of the sustained periods of tranquillity and security that a nation needs in order to nurture and develop its non-martial talents. Yet its cultural achievements in the eighteenth and nineteenth centuries were impressive, notably in the fields of mathematics (János Bolyai), medicine (Ignác Semmelweis), mining (at Selmecbánya), theology (Cardinal Pázmány), to give just a few shining examples. The renewal of the national language during the early decades of the nineteenth century stimulated the emergence of a great generation of innovative poets, novelists, essayists, and writers in general—Mihály Vörösmarty, Sándor Petőfi, and János Arany foremost amongst them—who helped to shape the Hungarian language and its literature into a more flexible vehicle to meet the needs of a by then rapidly changing world. In parallel, the emergence on the political stage of reform-minded figures such as Count István Széchenyi, Lajos Kossuth, Ferenc Deák, Baron József Eötvös and Count Lajos Batthyány mediated a transformation to national politics in the modern sense of the word.

Under these illustrious leaders, Hungary fought a bitter fight against the Habsburgs in 1848–49, the longest revolution and war of independence of that turbulent time in Europe. The war gave birth to a short-lived republic in 1849; more importantly, it put Hungary on the political map, thanks in no small measure to the oratorical genius of Kossuth in exile. This bore fruit when international political events in the 1860s forced Austria to regroup and abandon the monolithic structure of a Habsburg Empire, firmly centred on Vienna, for the Austro–Hungarian ('Dual') Monarchy. The Compromise *(Ausgleich)* of 1867 created unparalleled economic opportunities for Hungary, opening up a 'golden age' of rapid economic and social advance and Great Power status that is still looked back on with nostalgia and fondness.

Whilst Hungary continued to be the breadbasket of Central Europe, that half-century of prosperity, with no internal war, gave her a modern industrial base, up-to-date communications, and a thriving international commerce and banking sector. In 1873 the hitherto separate municipalities of Buda, Old Buda and Pest were unified as a newly inaugurated capital city of Budapest, vying with Vienna in ambition and splendour. By the turn of the century, that metropolis was the very symbol of the era with its magnificent avenues and boulevards, its imposing public buildings, its subway (the first in continental Europe), its opera house, theatres, museums, thermal baths, hotels and great parks. But this was also a highly productive period for the sciences, the humanities, and the arts. The University of Budapest and its prestigious faculty was blossoming; Franz Liszt had founded the Academy of Music that bears his name to this day; new medical clinics approached the highest standards in Europe; and achievements at these higher levels were sustained by the general excellence of the country's secondary schooling.

This boom was cut short by World War I; the ensuing revolutions of 1918–19 and the Treaty of Trianon in particular provided a final blow. Grand Trianon, the bigger of the two small châteaux in the park of the great palace of Versailles, where Hungary signed a humiliating peace treaty with the Allies on 4 June 1920, became synonymous with defeat, disgrace, and despair—with little doubt, the single most important event in modern Hungarian history. The territorial and populational concessions dictated by the treaty turned Hungary overnight into an independent but small, landlocked and vulnerable state, and one that was equipped with a full-blown small-country complex. The entire history of Hungary in the inter-war period and World War II can best be understood as a quest to regain that lost territory. That quest led the country to support, and enter the war on the side of, Hitler's Germany, albeit as a somewhat unwilling ally, though the support included contributing to the Holocaust by officially ordering the ghettoisation of its large Jewish population and facilitating the deportation of the great bulk of that community to perish in Hitler's gas chambers.

With Germany's defeat the outcome for Hungary was inevitable. In 1945 the country again found itself at the mercy of the Great Powers, and the Soviet Union above all, the power she had chosen to go to war with, alongside Hitler, in 1941. The Red Army occupied Hungary, just as Russian imperial troops had done back in 1849, when the tsar had rushed to the support of the Habsburgs. That military occupation was not dislodged during the ensuing periods of Cold War and détente, though with the revolution of October 1956 Hungary mounted the only full-blown armed challenge that was to shake Soviet domination of East-Central Europe over the next 45 years and so salvage for her people a proud page in world history.

Hungarian Creativity: A Social History

The standard joke about Hungarians is that they are the ones who can enter a revolving door behind you but leave ahead of you. A back-handed compliment, to be sure, but if there is such a thing as national character, then it can be taken as a sign of sneaking respect for a certain shrewdness, ingenuity, originality, and an uncommon approach to problem-solving.

These were aptitudes bred by the vicissitudes of Hungary's history, which put a premium on cunning, bold determination, and unexpected reactions to contend with the challenges of the day. Inventiveness was the weapon by which adversaries were deceived and outfoxed, wrong-footed and trapped, riposted and refuted. The social history of the Hungarian cast of mind—indeed, the way of thinking across much of East-Central Europe—is deeply rooted in war and conflict, abetted by a foe of an entirely different nature: poverty. Naturally well-endowed as Hungary is, most of its inhabitants down the ages have faced hardship, and often outright privation, which called for considerable resourcefulness just to survive. Survival strategies had to be developed as a matter of course to obtain a bare minimum of food, clothing, shelter, and protection. A keen sense of the unpredictability of the future engendered in many a certain ironic, flippant, devil-may-care attitude whereby they were ready, almost without a thought, to sacrifice themselves for king and country. Equally, many have been prone to feelings of pessimism, hopelessness, disillusionment, as reflected in the appallingly high suicide rates that typify the region. Lamentation over the nation's lost glory permeates Hungarian poetry, painting, opera, indeed the words of the national anthem itself, as penned by the early 19th century poet Ferenc Kölcsey: although there is always a sense that individual courage may count, the nation as a whole is doomed.

A *sine qua non* of this blend of national quirks has been a propensity for problem-solving which permeates all aspects of life, from the most mundane to the highly abstruse. Hungarians have been ready to accept whatever solutions they could find or devise, even if that has meant flying in the face of received yardsticks or devising new, unorthodox approaches—provided, of course, that these served their purpose. One may conjecture that this call for problem solvers goes part way to explaining the country's longstanding abundance of brilliant mathematicians, to mention János Bolyai, Lipót Fejér, John von Neumann, George Pólya, and Paul Erdős as merely the best-known in a seemingly unending succession of outstanding talents which spills over, beyond the realm of pure mathematics, into physics, chemistry, engineering, and many other fields.

A potent factor in maintaining that record of achievement lay in the manner secondary schooling was reorganised after the

Compromise of 1867. In 1870–72, the educationist Mór Kármán, father of aviation pioneer Theodore von Kármán, was commissioned by the minister of education, Baron József Eötvös, to undertake a first-hand study of Germany's acclaimed high-school system. This laid the ground for ensuring that the best Hungarian schools consistently had access to first-class teaching resources capable of encouraging students to standards of attainment that compare favourably with many lower-tier colleges in the United States today. On the German model, the high school, or *gimnázium*, placed heavy emphasis on the Classics, Hungarian language and literature, and universal culture, without neglecting mathematics and the natural sciences. These were unashamedly élitist institutions, with a student intake drawn typically from a rather narrow upper-middle section of Hungary's then still relatively conservative, even feudalistic society. However, they could attract teaching staff of a very high calibre, many of them recognised scholars in their own fields, as reflected in their subsequent membership of the Hungarian Academy of Sciences and appointment to university professorships. As a result, the country's top schools, such as the Lutheran Gimnázium in Buda or the 'Model' Gimnázium of the University of Budapest, succeeded for several decades in cultivating an astonishingly consistent succession of brilliant young minds, of whom John von Neumann, Theodore von Kármán, and physicists Eugene Wigner and Edward Teller, were only a few of the more prominent.

The German influence during this era also reached more widely, with Hungary in many ways constituting itself an outpost of German culture, whose icons—from writers Goethe and Schiller or philosophers Kant and Schopenhauer, through composers Beethoven, Brahms and Wagner, or painters Kaulbach and Piloty, to scientists Gauss, Haeckel and Brehm—were held in unparalleled esteem. Even news of the wider world outside the German universe usually reached Hungarian aristocratic libraries or the coffee-houses and salons of Budapest's middle classes refracted through the medium of the German language and, inevitably, cultural paradigms.

Tradition and Innovation

Although Hungary was in several ways an ideal creative spawning ground, many of its achievements were made in the face of official Austrian and Hungarian disapproval. For the greater part of the nineteenth century the national tradition was conservative and the mentality hostile to innovation not least due to the obverse side of German and Austrian culture, with its authoritarian insistence on strict and often antiquated rules and standards, established patterns of thinking, and unalterable methods. Creativity and innovation were often ill received. The general ambience favoured preserving the *status quo* rather than supporting new ideas, and accordingly the ruling conservative forces of the pre-1867 period ignored or spurned many reform-minded Hungarians, which more often than not led to exile, the lunatic asylum or to suicide.

After the Compromise, however, Francis Joseph, Emperor of Austria and Apostolic King of Hungary, whether he liked it or not, presided over a tide of change during the half a century of his 'dual' Monarchy. Innovative spirits flourished in many walks of life; big industrial firms sprang up and in their search for competitive edge founded product-oriented experimental laboratories in fields as diverse as telephony, lighting, pharmaceuticals, armaments, and electric locomotion, to name just a few. Later generations were to look back on these as the 'good old days' of peace and prosperity.

At the same time, the likes of Sigmund Freud, Franz Kafka, Robert Musil, Lajos Kossuth, and Endre Ady, in their very different ways, were all highly critical of the Habsburg monarchy and exposed its troubled nature. In many ways, as the poet Endre Ady said, official Hungary was a 'cemetery of souls', where new ideas were still doomed to failure. Unfortunately, the country was unable to sustain either full employment or a bare modicum of social services. Between the 1880s and World War I, one and a half to two million people left Hungary, mainly for the United States, as part of an unparalleled international migration. Their exodus, in some ways a natural and recurrent historical pattern, was essentially economic, a search for work by, in many cases, uneducated, illiterate peasants that would enable them to remit part of their pay packet to their families back home or, having 'made it', return with whatever earnings they had managed to save. Indeed, the majority of these masses are more properly considered 'birds of passage' rather than as true 'emigrants', in the sense that many of them never intended to become American citizens and longed to return to their native land. Many were stranded there by World War I and its aftermath, however, with the collapse of the Monarchy and Hungary's subsequent dismemberment and dire economic plight leaving nothing to which they might return.

The economic advance that occurred under the Austro–Hungarian Monarchy did, however, spur further development of the Hungarian language to furnish it with an adequate technical vocabulary to serve as a continued vehicle for professional communication and understanding—again reflecting the willingness of the culture to adapt itself to the modern world. This was particularly notable at the universities, where Hungarian gradually displaced German as the main language of tuition. One example symbolic of that shift is that the generation of professors who were teaching at the Budapest Academy of Music around the turn of the century—Robert Volkmann, Hans Koessler, Victor von Herzfeld, David Popper, or Ferenc Xavér Szabó—in most cases did so in German, whereas their students—the young and modernist generation of Hungarian composers and performing artists, such as Béla Bartók, Zoltán Kodály, and Ernő (Ernst von) Dohnányi, and Leo Weiner—were gripped by the Hungarian national idea and went on to teach exclusively in Hungarian. Bartók and Kodály, in particular, set out to collect the Hungarian and East-Central European folk music heritage and in the process rediscovered an ancient musical paradigm which they incorporated into mainstream European classical music—a radically bold move at the time, though one validated by the important place their works still occupy in the concert repertory of the present day.

Defeat in World War I ended all pretensions Hungary had to Great Power status. The ensuing revolution of 1918, and particularly that of 1919, played a big part in the blame for Hungary's wartime misfortune and subsequent foreign occupation being shifted onto the country's Jewish population. As in Austria and Germany, the Jews were perceived as a monolithic, alien, leftist, disruptive element in society. Though this was a gross travesty of the realities of the situation, it gained credence from the fact that many of the leaders of the 1919 Commune, a short-lived offshoot of Lenin's 1917 Bolshevik Revolution, were of Jewish descent. In the poisonous atmosphere engendered by being on the losing side in the war and seeing the country stripped of so much of its historical lands, the new régime that installed itself under Admiral Horthy in the fall of 1919 produced the continent's first anti-Semitic legislation of that century, the 'Numerus Clausus' (Law XXV of 1920), which effectively barred most students of Jewish origin from entering higher education. Though the system largely lapsed, for a while, by the end of the decade, it was a powerful spur for many bright young Hungarian intellectuals of Jewish background to leave the country and complete their education or seek employment elsewhere, especially (prior to Hitler's rise to power) in the German-speaking parts of Europe, which then included the German universities of Prague and Brno in Czechoslovakia. Many no doubt would have preferred to emigrate to the United States, but entry was made extremely difficult by the introduction of the quota system, starting with the emergency immigration restriction law of

1921 and finalised by the even more restrictive national origins scheme of the Reed–Johnson Act of 1924, under which a mere 869 (later only 473) persons annually were to be admitted from Hungary.

As events were to prove, Hungary thereby lost large numbers of budding natural scientists, physicians, engineers, architects, musicians, artists, film-makers, sociologists, and other thinkers who would go on to make distinguished names for themselves years and sometimes decades later in the West. That was compounded by further rounds of increasingly drastic anti-Jewish legislation in 1938–41, now modelled on that introduced in Nazi Germany. Émigrés (temporary or otherwise) were not just of Jewish origin but included the likes of authors Gyula Illyés, Lajos Kassák, and Sándor Márai, the painter Aurél Bernáth, later composer Béla Bartók, and, after World War II, Nobel laureate Albert Szent-Györgyi.

The still predominantly élitist and highly conservative school system nevertheless managed to maintain such high academic standards that, despite the loss of some of the very brightest students, Hungarian universities held their place, continuing to produce a stream of graduates who went on to demonstrate excellence in many fields, from mathematics and chemistry to medicine, music, and engineering, the positive effects of which were still evident after World War II. Founded by Cardinal Péter Pázmány at Nagyszombat (Trnava, Slovakia) in 1635 and now named after the reputable physicist and former professor Loránd Eötvös, the University of Budapest was the country's first and, for a long time, only university. During the nineteenth century its staff had boasted scientists such as Eötvös and scholars such as the orientalist Ignác Goldziher who were full equals to (and indeed became accepted as such by) their colleagues in the West. Hierarchical though the system may have been, the intellectual training it provided was rigorous and laid the foundation for solid future research. The chemical laboratories of Professors Béla Lengyel and Géza Zemplén, the medical clinics of Professors Sándor Korányi and János Bókay, the history seminars of Professors Henrik Marczali, Gyula Szekfű and Elemér Mályusz, or, for that matter, the violin school of Professor Jenő Hubay, were just some of the high spots of Hungarian higher education in the otherwise dour decades of the interwar period.

World War II saw the demise of this productive and rich culture. Hungary managed to steer clear of active involvement until 1941 but was then caught up on the horns of the dilemma of whether to fear the Nazis or the Soviets most. In truth, the die had been cast long before, given the long tradition of pro-German orientation, including alliance during World War I, and the adverse experience of the country's experiment with Bolshevism in 1919. Disastrously, an army corps was committed to Hitler's invasion of Russia. It is a measure of the reluctance of Hungary's support, however, that Hitler deemed it necessary to order an army of occupation into the country in March 1944 in an effort to keep her on his side. With Germany deciding to make a stand in Hungary against the advancing Red Army by the end of that year, the outcome was virtually inevitable. Having ousted the Germans, the Soviet Union sponsored a fledgling but aggressive local Communist Party under Mátyás Rákosi, which rapidly proceeded to seize full control over the now nominally democratic Hungary. By 1947–48 the country had effectively lost whatever slight independence it possessed and was thence absorbed into the Soviet bloc.

One singularly important episode made the subsequent Cold War history of Hungary very different from that of all the other Soviet satellites: the Revolution of 1956. Though this failed to achieve its aim of gaining Hungary her sovereignty, it showed the world that the USSR was vulnerable, and that Hungarians were ready to die rather than accept a humiliating defeat. This had important legacies. First, the appalling bloodshed and recriminations served as a warning to hold the post-revolutionary régime, during the 32 years under János Kádár, to a more liberal and tolerable course than anywhere else in East-Central Europe. Second, it made the 1989–90 transition to democracy, independence, and capitalism smoother, speedier, and more civilised than in most neighbouring countries. Significantly, too, Hungary still retained a capacity to produce excellence in various fields, particularly in the areas of scientific endeavour that could largely escape direct contamination by political ideology, though it is unfortunate that even here slavish imitation of Soviet models resulted in a division of efforts and resources between the universities and the Hungarian Academy of Sciences that weakened the universities. Despite the pressures and adverse circumstances, the educational system managed to preserve a large measure of its high professional standards in training the intellectual and professional élite who are now set to move the country into Europe.

How Far Is Europe?

Today Hungary is *ante portas*—before the gates of Europe. In the dozen or so years since it emerged from her period of more than six decades of political, economic, and social isolation, the country has become a member of NATO and is well on her way toward entry into the European Union. This prospect has galvanised Hungarian education, which no longer lives under the baneful influence of misguided Soviet examples and has been quickly transformed into a diverse, open, and ambitious system of choice and freedom of thought. State education is intended to be non-ideological, and the many competing programs now available at high-school level offer a wide range of individual choice. At her universities Hungary has received European and American visiting faculty and students since the early 1980s. In the reverse direction, an attractive system of scholarships enables a sizeable cohort of students, at both undergraduate and graduate level, to attend some of the best universities and colleges internationally.

If there is a problem on this 'road to Europe' then perhaps it lies with the linguistic isolation of most students, whose knowledge of foreign languages is still imperfect. In addition, the more competitive structure and pace of studies elsewhere in Europe and the U.S. can cause difficulties for Hungarian students on initial encounter. Against that, most Hungarian schools are equipped with computers and access to the Internet, and it must be hoped that this will help to erode any slight disadvantages on such fronts.

So are Hungarian students, on balance, ahead of or behind their peers in other countries? Clearly, it depends what you look at, but the country's outstanding legacy of creative thinking should augur well in the long run. It is to be hoped that the rich heritage surveyed in this volume will serve as a lasting inspiration and model. Hungary is a country where the mind has had a long tradition of excellence, and that tradition is alive and comparable to the best in the world.

Zsuzsa Szentgyörgyi (ed.)

NATURAL SCIENCES AND ENGINEERING: A PROCESSION OF GIANTS

Writing for an English-speaking public about the outstanding scientists and engineers—or, for that matter, economists and media or film personalities—who were (or are) of Hungarian birth is both easy yet problematic. Easy in the sense that names like John von Neumann, Theodore von Kármán, Joseph Pulitzer or the Warner brothers have a familiar ring; yet problematic in that as many or more who possessed talents stayed in Hungary and remain little known to the wider world—excluded by the twists of history, language and economics that often obscure the significance of those living and working in smaller countries.

This chapter attempts to throw some light on a wide range of Hungarians who brought about substantial and permanent achievements in the various fields of the natural sciences and engineering. In so doing, particular attention will be drawn to the important role that education and the cultural climate played in nourishing their talents. This is the reason why the names of great teachers will be seen to appear again and again, as well as the titles of Hungarian journals, scientific associations and competitions, some of them still living on, which form part of the process of selecting the best and ensuring that high standards are maintained. They have been reliable and treasured resources both for scientists who remained in Hungary and often also for many who chose to spend some or all of their working lives outside the country.

We present these surveys in the order of disciplines, starting from the fundamental sciences of mathematics and physics and building towards the applied sciences. However, life makes a mockery of all orderly categorisations. As a result, a person trained as a chemist, for example, may create something special in mathematics or economics. Equally, it is impossible, and even futile, to separate theory and practice. The emphasis here is, indeed, on the importance of such overlappings and interactions, and for that reason alone the names of certain individuals will be found to crop up in more than one of the following sections.

1. Mathematics: Under Lucky Stars (László Filep)

During the early decades of the nineteenth century, a broad movement got under way in Hungary to promote national revival, first and foremost through political and linguistic reforms, but also by building up a modern institutional base for broader scholastic and scientific activities. A key step in this process was the establishment of the Hungarian Academy of Sciences (Magyar Tudományos Akadémia, or MTA for short) in 1825 on the initiative of Count István Széchenyi. In line with one of the Academy's primary aims, much of the early efforts of the mathematicians amongst its founding members went into extending the Magyar language to fit it for the role of replacing the Latin and German, in which discourse on the subject had been conducted hitherto. On the whole, then, this was not a period of original investigation and discovery, but it did produce a sound basic stock of secondary and higher education textbooks in the native language. With all of its first academicians taking an active part in the country's War of Independence in 1848–49, mathematics for a while suffered very directly from the cultural reprisals that came in the wake of defeat in that struggle.

The Two Bolyais

The shining exceptions to the above characterisation of Hungarian mathematics in the Reform Era were the activities of the two Bolyais—both with claims to a place in even the briefest survey of mathematics history. The father, Farkas Bolyai (1775–1856), a teacher, was a versatile, excellent mathematician in his own right, but his real merit lay in providing the foundation for the brilliant discovery made by his son, János. Farkas Bolyai had studied at the University of Göttingen, gaining the friendship of the renowned Karl Friedrich Gauss, and on returning home to Transylvania joined the staff of the Calvinist College in Marosvásárhely (Tîrgu Mureş,

Romania), where he continued to teach for 47 years. His main work, the two-volume *Tentamen... in Elementa Mathesios Purae* (1832–33), contains many significant mathematical findings. Farkas Bolyai (like Gauss himself) was interested in the theory of parallels, specifically in the eleventh axiom in Book XII of Euclid's *Elements* and the premise that only a single non-intersecting (parallel) straight line can be drawn through a point outside of a given line in the same plane. Fruitless attempts had already been made in ancient times to prove the postulate directly by deriving it from the others, and those were later supplanted by the indirect method of taking denial to prove that it would contradict the other axioms. Neither approach had fully cracked the problem, though they had yielded a host of partial results and substitute axioms—one of the most suggestive being produced by Farkas Bolyai: any three points on a plane must be located either on the same straight line or on the circumference of a circle.

János Bolyai (1802–60), having studied under his father at Marosvásárhely, was sent to the Vienna Military Academy before embarking on a military career as a sub-lieutenant in Temesvár (Timişoara, Romania). Here he worked out the non-Euclidean theory that prompted him to write to his father, in 1823, "I have created another world out of nothing." He introduced a radical change in approach to the problem of the parallels by supposing that the lack of contradiction found on denying the axiom might arise because there was no contradiction in the first place; in other words, the axiom was independent of the others and, therefore, a contradiction-free geometry could be constructed whether the axiom was accepted as true or denied—yielding the familiar Euclidean geometry in the former case, and a new, non-Euclidean one in the latter case.

János's new geometry was published as an Appendix to the first volume of his father's *Tentamen* (1832), with the subheading 'Exhibiting the absolutely true science of spaces, independent of

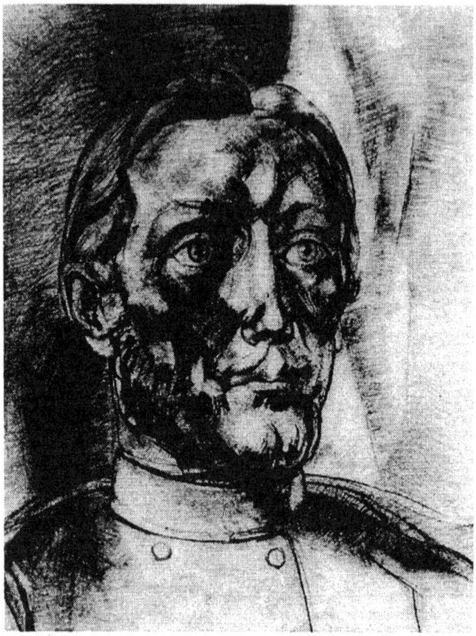

János Bolyai

Axiom XI of Euclid (not to be decided *a priori*) with the geometrical quadrative of a circle in the case of its falsity.' As that suggests, the most surprising result of this non-Euclidean geometry was that any circle can be quadrated, that is, yield a square of exactly the same area. Little wonder that contemporary mathematicians received the idea coolly, as few understood the new geometry. Gauss (to whom Farkas Bolyai sent a copy) did understand it, but feared for his own reputation and did nothing to propagate it (any more than he did for Russian N. Lobachevsky, the other innovator in this area). Recognition came only in 1860, after Bolyai's death, when another German mathematician, R. Baltzer, published a book praising the Appendix's importance, whilst the American G. Halsted also did much to have the two Bolyais recognised. The first Congress on Mathematics in 1896 ruled on the competing claims for priority by calling the new theory Bolyai–Lobachevsky geometry. Later generations expressed their gratitude by naming the independent Hungarian University of Kolozsvár (Cluj, Romania) for János Bolyai in 1945 (it was later closed down), and the Hungarian Mathematics Society likewise in 1949.

The discovery of non-Euclidean geometry radically transformed mathematical thinking. By shaking faith in the geometric approach, it opened the way for new algebraical and analytical methods, and is comparable in importance to Newton's and Leibniz's work on the foundations of infinitesimal calculus. In the words of M. Kline, the American historian of mathematics: "The creation of non-Euclidean geometry was the most consequential and revolutionary step in mathematics since Greek times."

The 1867 Austro–Hungarian Compromise gave Hungarian society a huge impetus, resulting in improved education laws and the establishment of new universities, new associations, new specialist journals. Two important research centres for mathematics came into being in the early 1870s: one in Budapest, at the Technical University,[1] the other at the then new University of Kolozsvár. The best-known mathematics professors at the Technical University then were Gyula Kőnig (1849–1913) and József Kürschák (1864–1933).

Gyula Kőnig's most important discovery was the Kőnig inequality among the numerous infinite sets, which is still a basic concept in this theory, and his name is also perpetuated in number theory by the Kőnig–Rados theorem.[2] John von Neumann was later to use Kőnig's lectures on the concept in integration that is now known as the Stieltjes integral (this was before the work done by its name-giver) as the basis for one of his axiomatic systems on set theory.

József Kürschák's work was important primarily in laying the foundation for the theory of evaluation—first published in Hungarian in 1912 then in German the next year—which enabled the number fields (in general, the abstract fields) to be completed. For example, although the set of rational numbers can be considered arithmetically complete, it is incomplete in two aspects: first, equations can be written in the field which have no solution within it (e.g., $x^2-2=0$, since $\sqrt{2}$ is not a rational number); second, not all (convergent) series of rational numbers have a rational number as the limit value. With Kürschák's notion of evaluation, the resulting evaluated fields could be completed, and thereby algebraic and analytical concepts and methods extended to cover them. The concept of evaluation has since been generalized to certain abstract fields as well.

Kolozsvár University was established in 1872 with a separate faculty for natural sciences, which was quite an innovation for those times. Gyula Farkas (1847–1930) and Lajos Schlesinger (1864–1933), its first professors of mathematics, may not have been of the highest rank but they did give the faculty a European standing. (It was here, incidentally, that Gyula Vályi submitted his work on the mathematical theory of a propeller in 1881 as the first Hungarian doctoral dissertation to present an independent discovery.)

Working on a problem in physics, Gyula Farkas proved the theorem of homogeneous linear inequalities (later generalised by Alfréd Haar to inhomogeneous cases). The Farkas theorem subsequently provided one of the bases for linear programming of computers, its use allowing derivation of the duality theorem, whilst in game theory it contributed to the so-called minimax theorem—both theorems first proven by John von Neumann. Schlesinger, on the other hand, taught and did research into the theory of differential equations, his monographs on which were widely used at German universities. His major achievements were made in continuing Gauss' researches in the field of ordinary differential equations with complex variables.

Lipót Fejér (1880–1959) and Frigyes Riesz (1880–1956), two of Hungary's great mathematicians and both professors at Kolozsvár University, were both born in 1880. The university's reputation in this subject was further raised when they were joined by Alfréd Haar, though Fejér later went to Budapest University in 1911 to build up a new centre there.

The Mathematics and Physics Society was founded in Budapest in 1891, with the physicist Loránd Eötvös (1848–1919) as its first president and Gyula Kőnig as his deputy. Three years later, the society instituted an annual mathematics competition for secondary school pupils, the first such in the world, which was held conjointly in Budapest and Kolozsvár. Long named after Loránd Eötvös (and later József Kürschák), the list of the competition's winners and runners-up is virtually a roll-call of the country's great mathematicians and physicists. Another pioneering initiative, originally conceived as a means of helping aspirants prepare for this competition, came from Dániel Arany (1863–1945), a teacher in Győr, who started the *Középiskolai Matematikai Lapok* (Secondary School Mathematics Papers). The editing work soon passed to László Rátz and later on was shared between him and Márk Antal, both secondary school teachers, who together arguably should take the real credit for spotting and nourishing many of the young talents. Rátz spent his entire career teaching at the renowned Lutheran Gymnasium in Városligeti fasor (Alley), just south of Budapest's City Park. Antal also started teaching in Budapest but later became head master of a school in Kolozsvár. After Transyvania's annexation to Romania following World War I, he also held private seminars at home in an attempt to compensate for the closure of the city's Hungarian university. Among others, Ábrahám Wald (see below) was one of his 'graduates'.

Secrets of the Twentieth-Century Success Story

World War I and its aftermath brought another caesura to the promising development of mathematics in Hungary. Quite apart from the losses of those killed in action, the most painful being that of Zoárd Geőcze in 1916,[3] the trickle of significant practitioners leaving for posts abroad that had occurred before the war (e.g. Lajos Schlesinger, Marcell Riesz) turned into flood with the country's defeat, the unrest of the ensuing 'Red' and 'White' terrors, and above all the territorial and population losses that Hungary suffered under the 1920 Treaty of Trianon. Many university graduates fled from the successor states to Hungary, among them the teachers from the closed Hungarian universities at Kolozsvár and Pozsony (Bratislava, Slovakia), though the activities of the institutions themselves were transferred to form new universities at Szeged and Pécs, respectively. The role of the Kolozsvár mathematics centre was thus taken up by Szeged, led by Frigyes Riesz and Alfréd Haar, who in 1922 started *Szeged Acta*, the first Hungarian mathematics journal of international standing. Other outstanding figures of the Szeged school at this time were Béla Kerékjártó, Tibor Radó, László Kalmár, László Rédei, Gyula Szőkefalvi-Nagy and István Lipka.

Alfréd Haar (1885–1933) in 1933 generalised the concepts of measure and the integral into certain topological groups,[4] which in turn links the concepts of topological space and group. The interpretation of the Haar measure and the integral within those concepts was to develop into a major branch of abstract harmonic analysis (John von Neumann later used the Haar measure to solve Hilbert's fifth problem for a special case that is too complex even to summarise here). Béla Kerékjártó (1898–1946) meanwhile was doing research into the new branch of topology, which was then emerging out of geometry, writing one of the first monographs on the topic in 1923. In H. Weyl's judgement, one strength of this book was that it amalgamated concept and approach in its methods of proof. Kerékjártó went on to write another important monograph on the foundations of geometry. Prompted by Riesz, Tibor Radó generalised Geőcze's results on the measurement of surfaces by applying Riesz's work on the theory of functions.

Dénes Kőnig (1884–1944), son of Gyula, was briefly one of Lipót Fejér's students but a rare exception in that he was influenced more by József Kürschák. He introduced the new field of graph theory into Hungary and wrote the first monograph on the subject in 1936 (in German), which remained the sole work of its kind for over twenty years. One of Dénes Kőnig's theorems, as later generalised by Jenő Egerváry, was found by H. W. Kuhn in 1955 to lend itself perfectly, with minor adjustment, to the solution of the Marriage Dilemma and other linear programming problems. Kuhn called his approach the Hungarian Method, and this is the name under which it became more widely known.

Riesz and Fejér—A New Binary Star

Frigyes Riesz and Lipót Fejér, though not bound by familial ties like the Bolyais, were linked by many threads in addition to the accidental near-coincidence of being born just a few days apart, on 22 January and 9 February 1880, respectively. They largely overlapped in their places of university study, and both started their professional careers at Kolozsvár University, though Fejér then went on to Budapest, whereas Riesz spent a period at Szeged before rejoining his friend and rival as a colleague at Budapest towards the end of their working lives. Similar in lifestyles and mentality, neither of them married, both living for their beloved science and for teaching, and striving to remain aloof from politics. The pair of them played an invaluable role in elevating Hungarian mathematics to an international level, and were key figures—Fejér especially—in bringing on young talents scouted in the country's secondary schools. Many of Fejér's grateful students—John von Neumann ('Hungary's greatest Johnny', as Fejér dubbed him) notable amongst them—kept in close contact with him, mostly by letter, even after they had moved abroad.

Frigyes Riesz's best known achievement, published in the *Comptes Rendus* in 1907, was what is known today as the Riesz–Fischer theorem,[5] which is based on the recognition that geometric characteristics could be introduced into certain sets of functions to construct a sort of space. M. Fréchet, a Frenchman, and S. Banach, a Pole, independently came to similar findings, and thus the three are jointly credited with laying the foundations of the theory of function space in functional analysis, which was subsequently applied by John von Neumann to provide a mathematical foundation of quantum mechanics. Riesz and his pupil Béla Szőkefalvi-Nagy co-authored the definitive monograph (*Functional Analysis*, 1955) on the topic, which appeared in French in 1952. Riesz also lay the basis for the theory of subharmonic functions with his generalisation of the theory of potential in 1925. He and Lipót Fejér jointly offered the most elegant of proofs for the basic theorem on conformal mapping (an approach with wide application in aerodynamics and hydrodynamics, amongst other fields), whilst with brother Marcell Riesz he generalised one of Fatou's theorems on analytical functions. It was also Frigyes Riesz who introduced the concept of topological space at the 1908 International Congress of Mathematics in Rome, whilst at the 1928 Congress he defined the concept of the vector lattice (known as Riesz spaces ever since).

Though he continued to produce enduring work on the theory of the Fourier series, functional analysis and interpolation theory, Lipót Fejér's most significant individual discovery, made when he was still only twenty, was of a new method for summing Fourier series and the related Fejér theorem. At Budapest University, Fejér also built up a wide-ranging school of mathematics, which included Mihály Fekete, György Pólya and Gábor Szegő among the more notable members of his team, though a host of other outstanding Hungarian mathematicians who did not subsequently enter his own field of research—Jenő Egerváry, Pál Erdős, György Hajós, Dénes Kőnig and Pál Turán, to name but a few—were also amongst his students.

Pólya and Szegő—Another Stellar Pairing

Though not even temporal 'twins' like Riesz and Fejér (Pólya (1887–1985) was eight years older than Szegő (1895–1985)), this pair nevertheless also shared much in common in both being Fejér students, co-authoring two books together, and both ending highly successful careers at Stanford University in the USA before they both retired in 1953, even dying in the same year (1985). Of their two books, the first—a collection of analytical exercises, first published in German in 1925 ('Pólya–Szegő', as it is generally referred to)—still counts as a basic textbook to this day, whilst the second, written in 1951, was a summary of their own work in the field of mathematical physics.

György Pólya had the wider range of interests, venturing into combinatorial analysis and graph theory (Pólya theorem, 1937), probability calculus (random walks), as well as issues of mathematics education, where he laid the foundations for heuristics, the method of teaching through letting pupils learn things for themselves. His famous book (*How to Solve It?*, 1945) has been translated into 16 languages.

Gábor Szegő, one of the greatest analytical scientists of the twentieth century, attained his most significant results in the field of orthogonal polynomials, in part on Haar-type orthogonal systems, publishing a monograph on the subject (*Orthogonal Polynomials*, 1939) which has since gone into four editions and countless reprints. Besides his seven books, several theorems bear his name.

The pioneer of Hungarian research in mathematical statistics and probability calculus, Károly Jordán (1871–1959), a professor at Budapest Technical University, was interested primarily in potential applications, which led him to the method of finite differences. His English-language textbook on this subject (*Method of Finite Differences*, 1947) was published several times in the USA. A further important work was a book on mathematical statistics (*Mathematical Statistics*, 1927), published at Paris.

The excellence of mathematics instruction at Hungary's secondary schools, coupled with the competitions started at the end of the nineteenth century, bore fruit in a stream of talented mathematicians (and physicists) entering university and going on to achieve prominence. Whatever the explanation (and it is undoubtedly complex) for this sudden invasion of 'Martians', as they have been jokingly termed, the country, now shrunk to one-third of its former size, was no longer able to absorb them all, so many were forced to look abroad for employment. Adding to this as a major cause of the mass efflux of scientists was the growing officially backed anti-Semitism. The most immediate manifestation of this was a 'Numerus Clausus', imposed as early as 1920, which drastically cut university intakes of students of Jewish descent, later compounded by even harsher legislation, restricting virtually all employment opportunities for Jews, as the régime became nazified in the late 1930s and during World War II. Many emigrants initially went to Germany, but after Hitler's rise to power in 1934 that option too was increasingly unattractive, with the result that the USA quickly became the preferred destination. The contribution that Hungarians of this generation made to the development of American mathematics was recognised by the Mathematics Association of America in its Bicentennial Tribute to American Mathematics, 1776–1976. This particularly highlights the role played by John von Neumann across various fields but also mentions that of György Pólya, Tibor Radó, Ottó Szász, Gábor Szegő and Aurél Wintner, all in the field of analysis; Theodore von Kármán and Eugene Wigner in applied mathematics; Pál Erdős in number theory; and Ábrahám Wald in statistics). The fact that several of these names (e.g. Kármán and Wigner as well as Neumann) recur in subsequent sections of the present chapter is a further demonstration that the various disciplines of science are broad, interrelated and open fields of endeavour.

Neumann and Harsányi—Two Johns and Game Theory

John von Neumann (1903–57), one of the greatest and most versatile mathematicians of the twentieth century, lived and worked in the USA from the age of 26 until his death. The roots of much of his work, however, some aspects of which have already been touched on (set theory, duality theorem, minimax theorem, Hilbert's fifth problem, the mathematical bases of quantum mechanics)—went back to Hungary. He was the founder of game theory and worked out the basic principles of computer operation; contributed to operational calculus, ergonomic theory and the study of automata; and, through the considerable attention he paid to economic applications of mathematics, he pioneered many of the basic tenets of operations research. Tracing Neumann's intellectual 'descendants' would be a fascinating subject in its own right, so integral were his contributions to the very beginnings of the whole field of computer science, research into artificial intelligence (and by that token, indirectly, into 'natural' intelligence, the link between the human brain and mind), operations research, theory of games, and modern economics. The complex interlinkages of the multiplicity of 'schools' arising from his work do not permit the full story to be told here, but a sketch of the steps by which one 'disciple', John Harsányi, came to apply game theory to economic processes offers a good illustration of the reach of Neumann's heritage.

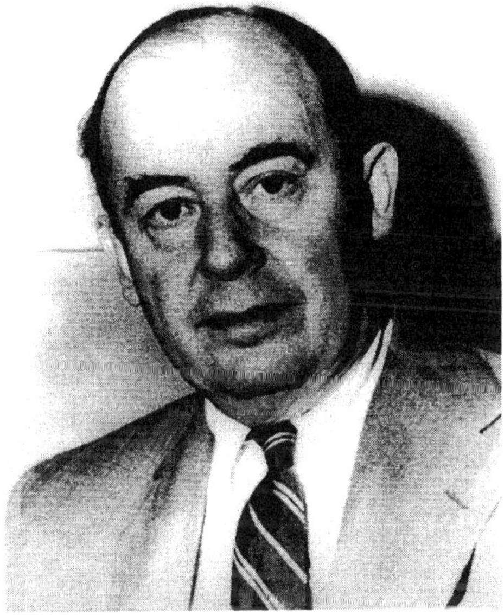

John von Neumann

The theory of games is concerned with the mathematical analysis of conflicts whose outcome can be influenced by the behaviour of the participants and thus is not dependent on chance alone. Such conflicts are called strategic games, their participants—players, and their behaviours—strategies. The task is to identify the optimal strategies that represent the best possible outcome of the conflict for any particular player. A game can be said to have been solved if a strategy can be defined which is optimum for all the players, that is, creates an equilibrium position in the game. Strategic games occur not only in formal games, like chess or bridge, but in many areas of life (e.g. markets, wars and other human conflicts, competition for resources) and so have wide applications in economics, environmental science, military and political science, psychology, sociology, etc. Neumann proved the existence of a solution (the minimax theorem) for the simplest case, that of so-called matrix games, in which two players of equal intelligence but conflicting interests are opposed. By recourse to his own duality principle, Neumann also discovered a link between matrix games and linear programming. If one of the players in a matrix game is not intelligent, and its interests are not necessarily conflicting (e.g. the weather), then, following Ábrahám Wald's subsequent work, it is called a game against nature. This first extension of game theory led Wald into statistical decision theory, a field that he helped to found, and his name is also linked with the elaboration of statistical methodologies for continuous sampling and sequential analysis.

The introduction to *The Theory of Games and Economic Behaviour* (1944), the first monograph on the subject, which Neumann wrote jointly with the Austrian economist Oskar Morgenstern, mentions the earlier research that Neumann carried out with Dénes Kőnig. Definitions of new types of games started to evolve in the 1950s: many players, conflicting and non-conflicting interests, complete and incomplete information, etc. John Harsányi (1920–2000) worked out the theory of games with conflicting participants holding incomplete information, developing a mathematical model that is still used in various economic activities (e.g.

designing auctions) and which played a role even during the strategic arms limitation talks between the USA the Soviet Union that led to the 'S.A.L.T.' agreements of the 1970s. The ultimate recognition of game theory was the Nobel Prize for economics awarded to John Harsányi in 1994, jointly with J. Nash and R. Selten. The citation for the award highlights the pioneering work done by Harsányi in the area of equilibrium analysis within the theory of non-co-operative games.

Pushing Ahead Elsewhere

Even though the USA was the main goal, some very considerable Hungarian mathematicians chose to settle in other countries: Pál Dienes (Great Britain), Mihály Fekete (Israel), Gyula Pál (Denmark), Marcell Grossmann (Switzerland), and the already mentioned Marcell Riesz (Sweden). Dienes's major achievements lay in elucidating Taylor's series, on which he published a significant monograph, and in being one of the first to investigate infinite matrices. Fekete did research in algebra and into the theory of complex functions, most notably in developing the theory of set tools. Pál's name is generally associated with the solution of the so-called Kakeya problem for convex domains, whilst Grossmann helped to put Einstein's theory of relativity into mathematical form by using absolute differential geometry.

Despite this huge loss of talent, mathematics within Hungary still retained many figures with considerable international reputations. The brilliant Alfréd Rényi (1921–70), who died at a sadly young age, built up a school at Debrecen after World War II through his work on probability calculus, in which he redefined Kolmogoroff's axiomatic system and generalised the Borel theorem, though his first major result was in number theory with a proof relating to the Goldbach conjecture that every even number greater than 2 is the sum of two primes.

At Budapest University, the most outstanding results achieved were in discrete geometry (László Fejes Tóth), number theory and graph theory (both Pál Turán), geometry (György Hajós), and mathematical logic (Rózsa Péter).

Discrete geometry deals with problems of tiling planes and packing spaces. This is an area that Farkas Bolyai had been amongst the first to investigate, with the Bolyai–Gerwien equidecomposability theorem[6] still often quoted to this day. The first famous problem[7] was solved by the Norwegian A. Thue, then quite independently by László Fejes Tóth. Since the publication of his monograph on the subject in 1953, however, the latter is considered to be the founder of discrete geometry as a branch.

Pál Turán (1910–76) is remembered particularly for developing the sum of powers method in 1949 in connection with his efforts to prove the famous Riemann hypothesis of prime number theory. His method turned out to be useful for solving other problems in number theory, and Turán himself published some fifty articles on it. Prior to that, in 1941, he had raised the following graph theory problem: What is the largest number of edges a (simple) graph can have without including any full graph segments? This prompted an investigation into so-called extremal problems within graph theory. The Turán graph, which provided the solution, can be looked on as a kind of generalisation of the case of the paired graph earlier investigated by Dénes Kőnig.

Using the tools of group theory, György Hajós (1912–72) in 1942 proved the Minkowski conjecture in geometric number theory, as a result of which it hardened into the Minkowski–Hajós theorem. Rózsa Péter (1905–77) takes the credit for introducing the concept of recursive functions into the study of problems of mathematical logic and working out the theory behind them. Through her 1951 monograph summarising these results she is justly regarded as the 'founding mother' of recursive functions theory.

At Szeged, Béla Szőkefalvi-Nagy (1922–98) and his students carried on the centre's tradition in analytical number theory. A school of abstract geometry also developed here, under László Rédei (1900–80), whilst László Kalmár (1905–76) was both productive in the field of mathematical logic and also responsible for establishing computer science as a discipline in Hungary.

Szőkefalvi-Nagy's most important results were achieved in functional analysis; the monograph that he co-wrote with Frigyes Riesz (*Functional Analysis*, 1955) was translated into six languages. His own work was focused primarily on studying the linear operators of Hilbert space that John von Neumann had used in laying the mathematical foundations for quantum mechanics.

Though László Kalmár had an unusually wide range of interests, his most significant achievements were in mathematical logic, to which he had been led after reading one of Neumann's articles, and specifically in connection with the decidability problem of number theory as it related to his exploration of the relationship of Church's theorem ('There is no infallible method for telling theorems of typographical number theory from non-theorems') to Gödel's 1931 theorem ('All consistent axiomatic formulations of number theory include undecidable propositions').

László Rédei published five books and numerous papers in various areas of abstract algebra, producing his main results in the theory of finite Abel groups. His monograph *Algebra I*, which has been translated into German and English, is the first work on algebra which points beyond Van der Waerden's 'structural' approach by treating algebra as a general theory of sets with real operations.

Pál Erdős—Roving Ambassador of Mathematics

Pál Erdős (1913–96) occupies a very special place within twentieth-century mathematics, as he did not fit neatly into any single research group or category, any single field, or even any single country. He personally embodied the globalisation of the discipline and the necessity for international co-operation among researchers.

Though he grew up in Hungary and was domiciled in the USA for much of his career, his real home might better be said to have been the international mathematics conference circuit. Whenever he was in Hungary, he systematically tried to seek out young Hungarian talents; more than a few members of the Academy of Sciences have him to thank for launching them on their scientific careers, and it is largely due to Erdős that combinatorics (including graph theory) is the most intensively studied field of mathematics in Hungary today.

Erdős contributed particularly to combinatorics and to number theory. In 1949, he and the Norwegian Atle Selberg gave an elementary proof of the prime number theorem regarding the distribution of primes. Many of his later great achievements—the Erdős–Kac, Erdős–Szekeres, Erdős–Wintner and Erdős–Selfridge theorems might, perhaps, be highlighted—were likewise reached jointly with other mathematicians. In 1983, at the age of 70, he received the Wolf Prize for his lifetime work.[8] In 1999 that prize was given to László Lovász, himself an Erdős 'discovery', with the awarding committee emphasising in its citation his work not only in combinatorial analysis but also in computer science. The only other mathematicians of Hungarian origin to receive this recognition, considered to be the Nobel Prize of mathematics, have been Péter Lax (1987) and Raoul Bott (2000), both resident in the US. Lax was outstanding in analysis and applied mathematics, while Bott has made his mark in topology and differential geometry.

Lax and Bott, though Hungarian by birth, left Hungary at a relatively young age and thus their fields of research are not rooted in Hungarian mathematics. Amongst others who have risen to distinction outside their native land, John G. Kemeny (1926–92), a professor at Dartmouth College, is perhaps the best known as one of the developers of the BASIC computer programming language and for his work on finite mathematics, which plays an important role in computer science. Cornelius Lánczos (1893–1974) was for a while Einstein's assistant in Berlin before becoming a professor of mathematical physics in Dublin, publishing more than 120 articles and books on different areas of theoretical physics and applied mathematics. Jenő Lukács (1906–87), a professor at several American universities, was one of the founders of modern mathematical statistics, and the first to apply differential equations in the theory of eigenfunctions. He had a particularly fruitful collaboration with Ábrahám Wald.

Mathematics in Hungary Today

Given the lack of historical perspective, it would be premature to attempt an assessment of the still ongoing research and results of Hungarian mathematicians of today. It would be difficult, moreover, to separate research done in Hungary from that done elsewhere in the world. The leading research institutions remain the Mathematics Research Institute (now named for Alfréd Rényi) in Budapest and the universities of Budapest, Debrecen and Szeged, but it is not so easy to define main lines of research due to the enormous complexity of modern mathematics. During the first half of the twentieth century the overwhelming bulk of Hungary's mathematical research effort went into analytical mathematics, and particularly Lipót Fejér's explorations of series. After World War II, however, under the influence of Erdős and Turán, combinatorics (graph theory) and number theory came to the fore—a shift in emphasis that has been reflected in a perceptible widening of the range of research topics undertaken by the country's academicians. The current picture may be summarised under the following broad headings.

Research into set theory and mathematical logic is concentrated in Budapest and is particularly linked with the team around András Hajnal. The legacy of Erdős and Géza Fodor has been continued in the strong lead that the Hungarian schools of combinatorial set theory and combinatorial graph theory have maintained under László Babai, Béla Bollobás, László Lovász and Endre Szemerédi, which has permitted the emergence of unified concepts and methodology across the field of combinatorics. Similarly, extremal graph theory, initiated by Turán's work, has been a largely Hungarian preserve, its 'Bible' being Bollobás's monograph on the subject, as has been the field of random graph theory founded by Erdős and Rényi.

In number theory there are four main areas of current research. Imre Kátai is successfully applying the 'big sieve' methodology devised by Rényi for locating new primes. Gábor Halász has had great success in the study of multiplicative number theory functions. Combinatorial number theory, which links number theory with combinatorics, has entered a new era under the stimulus of the work of Endre Szemerédi and András Sárközi, aided in part by Erdős. Meanwhile at Debrecen, work on Diophantine equations under Kálmán Győry's leadership has given rise in recent years to a veritable 'Debrecen school'.

Algebraic research is intensively pursued at Szeged, mainly in the field of universal algebras and also, led by Ferenc Gécseg and István Peák, into the mathematical foundations of automation, now one of the key areas in all of applied algebra and one in which they wrote the basic monograph.

Analytical research is very strong in all three of Hungary's university departments of mathematics. At Szeged, Károly Tandori and László Leindner continue a line of research into the Fourier and other orthogonal series that stretches back to Fejér and Haar. At Debrecen Zoltán Daróczy leads research into functional equations, keeping in close touch with János Aczél, the founder of this theory, who now lives in Canada. It would be fair to say, however, that the widest international acclaim has been that accorded to Miklós Laczkovich, working in Budapest, following his solution in 1988 of the Banach–Tarski paradox, the modern equivalent of the classic problem of quadrating the circle. Taking one of Farkas Bolyai's theorems on the latter as his starting-point, Laczkovich demonstrated that a circle (or virtually any other planar domain) is divisible into a finite number of pieces in such a way that the pieces can be rearranged into a square of surface area identical to that of the circle. For this achievement, he received the MTA's Prize in Hungary and the Ostrowsky Prize in the USA.

Along with discrete geometry, one of the fields in which Hungarians have specialised is that of differential geometry, started up by Ottó Varga at Debrecen and still largely concentrated there. In the field of general topology, the elaboration by Ákos Császár of a theory of syntopogenic spaces has turned out to be of considerable importance because it enables various types of spaces to be studied with a unified approach.

A note should also be made of two scholars whose work is directed towards the history of mathematics. Árpád Szabó has cast new light on the origins of ancient Greek mathematics, whilst Elemér Kiss has managed to show that János Bolyai worked out the theory of divisibility of complex whole numbers (nowadays called Gaussian integers) essentially simultaneously with, and totally independently from, Gauss himself, making Bolyai a co-founder of the new branch of algebraic number theory.

To turn to some final reflections, there are solid grounds to be hopeful for the future of mathematics in Hungary and new generations of researchers. Teaching of the subject in schools remains of an enviably high standard, even a potential model, judging by recent delegations from other countries (Finland, Great Britain) to study Hungarian maths teaching methods. The now century-old tradition of actively spotting and nurturing exceptional talent through such means as the *Secondary School Mathematical Papers* continues unbroken, and its success is perhaps best placed in an international perspective by the continuing success that Hungarian teams enjoy at the Mathematics Student Olympics. The first of these problem-solving competitions, aimed at children of secondary-school age, was organised at Brassó (Braşov, Romania), in 1959, with participants coming from just seven Socialist-bloc countries; nowadays, however, teams from 70–80 countries across the world come together at these annual events. A Hungarian team was entered in all but one of the first 40 Olympics held up to 1998 (1978 was the sole exception), and at 22 of these a Hungarian was ranked in the top three places in the individual point scores, thus aiding the Hungarian team as a whole to a total of 67 first and 106 second places over that period. Undoubtedly the high spot has to be the unprecedented hat trick of first places that was pulled off by the team of László Babai, Béla Bollobás, László Lovász and József Pelikán. Pelikán went on to become a coach to the Hungarian team, and in that capacity in 1992 he won the Pál Erdős Prize of the International Association of National Mathematics Competitions. The best performance by a girl to date has been the feat of Marianna Csörnyei, from Budapest's Mihály Fazekas Model Gymnasium, in garnering a first place in 1993 and second place in 1994.

On any sober assessment—whether of prestigious prizes won, recognition for significant pieces of work, standing in international competitions and conferences, or proliferation of publications—all the signs are that Hungary ranks highly in the world as a centre of mathematical research. This was clearly one of the factors behind the decisions to choose Hungary as the host country for the 6th International Congress on Mathematics Education (1988) and the 2nd European Mathematics Congress (1998).

2. Hungarian Physicists at Home and Abroad: the Great Generations (György Marx)[9]

Deep Roots

In the sixteenth century, the western part of Hungary, seeking help from Europe in its struggle against the Ottoman power that had conquered most of central and southern Hungary, elected a Habsburg king. The eastern parts of the country, including Transylvania, meanwhile retained a large measure of autonomy, allowing the Reformation to gain a hold there. They therefore looked particularly to the Protestant strongholds of the West—Calvinists to Switzerland and the Netherlands, Lutherans to Germany—over the ensuing two centuries, as the Catholic Habsburgs and the Islamic Porte fought for supremacy. In turn, this helped implant the generally more liberal ideas of the West on Hungarian soil. It was Hungarian Calvinists, for instance, who offered astronomer Johannes Kepler a temporary refuge in 1598, when he became embroiled in a dispute with religious zealots at the imperial court of Rudolf II at Prague.

The Calvinist *gymnasia* (colleges) of eastern Hungary, founded in the sixteenth century, were the most progressive centres in the whole region through their propagation of Erasmian humanism and, later, Cartesian rationalism. Magnets and other experimental tools were listed in the inventory of Debrecen's Calvinist gymnasium as early as 1703. Johannes Andreas Segner (1704–77), who invented the 'Segner wheel', a water reaction turbine (working on the same principle as the modern garden sprinkler) in order to demonstrate the phenomenon that would now be called rocket propulsion, taught at Debrecen before obtaining a chair in Göttingen. István Hatvani (1718–86), who had studied in Basle and Leyden, likewise taught in Debrecen (1748–86), and fabricated galvanic cells within a year after Galvani's discovery.

Electricity also provides a direct link to the work of Reverend Ányos Jedlik (1800–95), who can be regarded as the first great Hungarian physicist. A Benedictine monk, Jedlik taught at the gymnasium in Győr and sought ways of improving the existing galvanic batteries in order to produce the higher voltages needed to make his classroom demonstrations more impressive. This led to his discovering the concept of the cascade electrostatic generator: by first charging a group of capacitors connected in parallel then discharging them when reconnected in series, a huge spark is produced. Poggendorff, the editor of *Annalen der Physik und Chemie*, refused to publish Jedlik's account, and although that put the latter off ever again seeking publication of his work, he continued to show off his invention to students. The ageing teacher's demonstration of 90-cm sparks at the 1873 Vienna World Exposition prompted Ernst Siemens to recommend Jedlik for a Gold Medal award. As Jedlik failed to turn up for award ceremony, however, the Viennese E. Marx is nowadays generally credited as the discoverer of the condenser cascade as he was the first to publish the concept (in 1876), but in doing so he failed to make any reference to the priority of Jedlik's work. That had already been the fate of a hand-cranked electromagnetic dynamo that Jedlik constructed in 1861 to produce a bigger sustained electrical current than could be obtained from the chemical batteries of the day (again just for classroom demonstration purposes)—a full six years before Siemens independently patented his own design. Thus, whilst the modest monk used his dynamo as a teaching aid, the Siemens Company built up an entire electrical industry in Germany on the strength of their invention.

A university had operated intermittently in Buda since the late fourteenth century, until the Ottoman seizure of the city in 1541, and it was not until a new institution was founded in 1635—a year before Harvard—at Nagyszombat (Trnava, Slovakia) that Hungary regained a university. That institution was later transferred briefly to Buda (1777) and then to its present site in Pest (1783). Already the first physicist member of the Hungarian Academy of Sciences (in which position he proposed for regular membership scientists of the calibre of Faraday (1858), Clausius (1872) and Boltzmann (1888) as well as several more honorary members), Jedlik was eventually appointed professor of physics at this Hungarian University in 1840.

The Victory of Alternating Current

It was some time before Jedlik's legacy bore practical fruit. The dynamo concept was taken a step further by Károly Zipernowsky (1853–1942) in constructing electric generators for the Ganz Company in Budapest. The alternating current thus obtained had the great advantage that its voltage could be easily transformed up or down; this enabled energy losses to be cut by employing a high voltage (low intensity) current for long-distance transmission and transforming it down to a low voltage at the consumers' end to avoid accidents. Zipernowsky, working with Miksa Déri (1854–1938) and Titusz Ottó Bláthy (1860–1939), constructed and (in 1885) patented an iron-core transformer for this purpose. At a time when the rest of the world was still relying on direct-current electricity supplies, a thousand light bulbs were lit by means of this AC device at the 1885 National Exhibition in Budapest. Siemens and Edison were reluctant to accept the new technique, perhaps due to their commercial investments in DC ventures, with Edison even resorting to scare-tactics over the alleged dangers of alternating currents.[10] The system was first employed on a large scale by the Ganz Company in the city of Rome (1885–92). Kálmán Kandó (1869–1931) constructed the first efficient engine for AC-powered electric trains by changing the single-phase voltage to multiphase voltage, and this too first went into operation on the Italian railway system (1930). As a result of Zipernowsky's insight, AC eventually prevailed over DC as the standard form of electricity supply throughout the world.

Loránd Eötvös—Great Scientist and Organiser

The founding father of professional physics research in Hungary was Baron Loránd Eötvös (1848–1919), son of the liberal statesman and gifted writer József Eötvös (1813–71), who can claim the credit for introducing compulsory elementary schooling in Hungary, during his second term as minister of education, after 1867. During his studies in Heidelberg, Loránd fell under the spell of the exactitude of classical mechanics. On becoming professor at Budapest University (called the Loránd Eötvös University since 1950), at the age of 30, he set out to prove Newton's law of gravity by showing that the proportionality of gravitational mass to inertial mass was valid to a far greater degree of accuracy than the four places of decimals previously attained. Over two decades of careful, patient work in his laboratory, using his own design of double-armed torsion balance, he increased this accuracy to eight decimal places and thereby supplied the most reliable empirical foundation for Einstein's theory of general relativity. Due to its sensitivity to minute changes in gravitational acceleration, the Eötvös balance was also a popular tool for early oil prospectors in Texas and elsewhere.

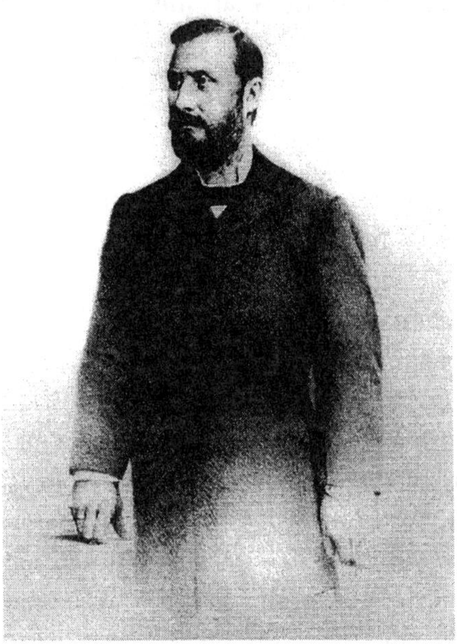
Loránd Eötvös

Eötvös was a professional experimentalist and a consummate politician, becoming rector of Budapest University, president of the Academy of Sciences (1889–1905), and (in 1894) briefly even minister of education. Nevertheless, he stuck to the Newtonian principle of 'action at a distance' throughout his life. He had first-class assistants like Philipp von Lenard (1862–1947)[11] and Győző Zemplén (1879–1916), the latter of whom was asked to write the chapter on shock waves for the *Encyclopaedia Physica*, forerunner of the *Handbuch der Physik*. Eötvös preferred observers like himself, who were ready to record the torsion balance readings without preconception or partiality. The only problem was that some of these diligent assistants in turn became professors and, during the first two decades of the twentieth century were still Newtonian mechanical physicists at heart, suspicious of the ideas of Maxwell, Planck and Einstein, and still transmitting that conservatism to their students.

The most progressive of those students was Károly Tangl (1869–1939). When he took over Eötvös's chair after the great man retired, it was Tangl who became the powerhouse for experimental physics in Hungary. It was one of Tangl's students, Zoltán Gyulai (1887–1968), who created a centre for crystal physics at the University of Debrecen. When Gyulai subsequently moved to the Technical University in Budapest, it was another Tangl student, Sándor Szalay (1909–87) who after a spell at Rutherford's laboratory in Cambridge, set up the centre for nuclear physics that still flourishes at Debrecen to the present day. In a country where physics has been dominated by theoretical approaches, both Gyulai and Szalay brought on a generation of genuine experimentalists.

It was Szeged that carried the torch of progress during the 1920s and '30s, after the University of Kolozsvár was relocated to there within the redrawn borders of post-Trianon Hungary. Centres were built up there by Frigyes Riesz for modern mathematics, by Rudolf Ortvay (1885–1945) for quantum mechanics, by Zoltán Bay

(1900–92) for nuclear physics, by Albert Szent-Györgyi for molecular biology, and by László Kalmár for computer science. Amongst Ortvay's postgraduate students at Szeged was the theoretical physicist Cornelius Lánczos.

The clock of physics in Budapest only truly caught up with the present when strong centres for modern theoretical approaches were established after Ortvay accepted a chair at Loránd Eötvös University and Bay moved to the Technical University. Ortvay, who had studied with Arnold Sommerfeld at Munich, in 1929, launched the Ortvay Colloquia, the regular Thursday afternoon meetings that have continued to the present day. In his own day the top priority was given to quantum mechanics, with lecturers like himself, Lánczos, John von Neumann, Edward Teller, László Tisza (another of Ortvay's postgraduate students, who, working in the USA, was later to earn recognition for putting forward the two-liquids model of superfluid helium), and Eugene Wigner, along with leading Hungarian scientists who were working abroad and the likes of Debye, Dirac, Heisenberg, Planck, and Sommerfeld. Secondary school teachers could also attend the colloquia and were given the opportunity to speak about their own research work in electronics, materials science and quantum electrodynamics. Ortvay's regular correspondence with George von Hevesy, Neumann, Wigner and others after they left Hungary helped maintain an important intellectual bridge to scientific progress elsewhere—a tradition that others continued even in the Cold War period. Even though Ortvay took his own life during the war, his school of modern theoretical physics survived under his successor, Károly Novobátzky (1884–1967), who taught and researched in quantum field theory and relativity, despite attacks by Marxist hard-liners, turning the Loránd Eötvös University into a leading centre for basic theory. Pál Gombás (1900–71), a former Ortvay assistant, meanwhile established a highly successful centre for theoretical nuclear and solid-state physics at the Technical University.

Foundation of the Mathematical and Physical Society

The Thursday afternoon timing for physics colloquia was in fact deliberately fostering a tradition that had been established by Loránd Eötvös in 1885. Recognising the importance of actively encouraging scientific discovery, he started to organise regular informal meetings with mathematicians and physicists over a lunch table so that they could swap news and learn about one another's research work. The step of formalising those arrangements came with a letter that Eötvös sent out in 1890 to invite attendance at a lecture he was due to give on 'Terrestrial gravitation and its measurement', for the letter went on to announce the formation of a new society: "If we carry out the task of self-education with dedication and seriousness, this will also have the result that in future the researchers and developers of science will come from amongst our ranks. I hereby declare the Mathematical and Physical Society founded."

With Ányos Jedlik as Member No. 1, the Society had some 400 members in its early years (the No. 1 card later passed to Eugene Wigner during the inter-war years). Some idea of its importance, though, may be conveyed by the fact that, for example, X-rays were already being demonstrated to members on 16th January 1896, shortly after the discovery was made (9th November 1895) and even prior to Röntgen's public announcement in Würzburg (23rd January 1896). Likewise, Hevesy reported on several of his key discoveries (radioactive tracing, discovery of the element hafnium) within weeks of making them. The Society (which was renamed the Loránd Eötvös Mathematical and Physical Society in 1919, after the death of its founder) immediately started publishing its own journal, with a print run of 2,000 copies. Apart from a gap of several years following World War II, this has continued to the present day as *Fizikai Szemle* (Physics Review), the title under which it was restarted in 1950.

Sub pondere crescit palma, runs a Latinised saw—Trees grow under weight. Accordingly it is not so surprising that the Society was at its most active during World War II and again during the early decades of Cold War, when personal contacts with people abroad were limited and governments had different priorities. It was restructured in 1949 by splitting into a separate Loránd Eötvös Physical Society and the János Bolyai Mathematical Society. The Eötvös Society rapidly extended its reach, organising local groups throughout Hungary and, from 1951, rotating its sessions around those locations. A lively scientific forum was built up, encouraging a results-oriented ethos, and quantum tunnelling through the Iron Curtain even during the chilliest decades of the Cold War.

Industrial research reached world standards in northern Budapest, where the United Incandescent Lamp & Electrical Company (Egyesült Izzólámpa és Villamossági Rt.) was established (1896). They manufactured Tungsram light bulbs, which got their name from the heat-resistant tungsten (or wolfram) that Sándor Just and Ferenc Hanaman patented (1903) for use as the glowing filament in bulbs instead of Edison's fragile carbon fibres. Tungsten gave a much more durable and hence cheaper product, enabling the light bulb to become an everyday commodity. (When the General Electric Co. started to use tungsten, United Incandescent took them to court for patent infringement and won, though through the ironic twists of fate the Hungarian company is now owned by them and is now known as GE Lighting.) Lipót Aschner (1872–1953), the firm's managing director, recognised the commercial importance of supporting a strong R & D activity, even in fairly pure science. Thus, whilst Imre Bródy (1891–1944), Egon Orowan (1901–89) and Michael Polanyi (1891–1976) worked as a team to develop a krypton-filled bulb (1934) as the most economic incandescent lamp, and to work out how to mass produce the krypton gas that this required, Polanyi and Orowan were also researching their theory of crystal dislocations as the carriers of plasticity (1934), whilst pioneering experiments were being undertaken by Zoltán Bay in radio astronomy and by Pál Selényi (1884–1954) on photocopying. Both Dennis Gabor and Edward Teller were also, for a short time in their early careers, amongst those who worked in the United Incandescent laboratories.

Excellent Schools and a Crop of Geniuses

"In the early twentieth-century, the Hungarian education system was the most brilliant one the world had seen until its close imitation in post-1945 Japan. The use of the word 'gymnasium' was bor rowed by most of German-speaking Europe, and by any country that looked to Germany for educational leadership... The advantage of the gymnasium system is that, in the best gymnasia, students can be pushed to the limit of their capacities. They are exposed to an intellectual rigor that is not usually reached in more democratic countries' secondary schools. In particular, the gymnasium system gives dignity to those who provide instruction in top secondary schools... The most elite and successful of these gymnasium systems in history was definitely that of little Hungary from about 1890 to nearly the 1930s... Hungarian schools entered their brightest 18-year-olds each year for the Eötvös Competition in mathematics and in physics," explains Norman Macrae, a former editor of *The Economist*, in his biography of John von Neumann.[12]

Much of the credit for those high standards lies with the churches. On the Catholic side there were the schools run by the Benedictine and Piarist teaching orders, of which the Piarist Gymnasium in Budapest—numbering Loránd Eötvös and Nobel laureate chemists George von Hevesy and George Olah amongst its ex-pupils—is particularly noteworthy, excelled only by the Lutheran Gymnasium near the City Park and its alumni: Nobel laureates Neumann, Wigner and economist–mathematician John Harsányi. Eugene Wigner himself cited the merits of its legendary teachers: László Rátz in mathematics and Sándor Mikola in physics. Though Budapest has named no street after Neumann or Wigner, it does have a Rátz Street (albeit in the industrial 11th District of Kelenföld). Today, the János Bolyai Mathematical Society awards a Rátz Prize annually for the country's best mathematics teacher, and the Eötvös Loránd Physical Society bestows a Mikola Prize on the best physics teacher. Increasingly, though, the purely secular schools that were established as a direct result of the educational reform package introduced by József Eötvös began to attain similar standards through the efforts of their teaching staffs—a prime example being the Berzsenyi Gymnasium in Budapest, whose students included Marcel Grossmann, John G. Kemeny, George Klein and George Soros and in post-war years has been a veritable breeding-ground for high-performers at the Student Olympics.

The excellence of Hungary's undergraduate intake has, in turn, allowed its universities to further reinforce the exacting requirements that are demanded for scientific success. Many of the country's top scientists have indeed been conspicuously active in higher educational institutions as well. A fine example is provided by physicist Károly Simonyi (1916–2001). As a young researcher, he joined Zoltán Bay's team at United Incandescent, which was engaged on detecting the echoes of radar signals bounced off the moon's surface. Simonyi spent most of his working life after the war as a professor at Budapest Technical University, teaching theoretical electricity, and his textbook on the subject, which has been translated into a number of languages, has given tens of thousands of students an insight into this important science. Another of his books, *A Cultural History of Physics*, is a unique approach to showing that the discipline, despite its very diverse manifestations, can indeed claim to have a unified overall culture. Simonyi was the first recipient of the Hungarian Association of Scientific Writers' award for the popularisation of science, and even has a star named after him.

Instructive in a different way is the career of George von Hevesy (1885–1966). Having studied chemistry in Budapest and Berlin, his father's wealth gave him the private means to travel widely, as a result of which he spent some time in Manchester as an assistant to Rutherford, who was by then famous for providing the first direct proof of the atomic nature of matter. Meeting Hevesy one day in the corridor, Rutherford is reported to have told him: "My boy, if you are worth your salt, you will separate the active Radium-D from that nuisance of lead!" Hevesy tried hard, using all the tricks of chemistry, and though he failed to achieve that specific goal, he made a crucial observation that atoms of the selfsame chemical element (e.g. lead) could still differ from each other, with one variant being stable whilst another variant might decay spontaneously, with the emission of radioactivity. This was a breakthrough that led directly to Frederick Soddy's elaboration of the theory of isotopes (1912) as well as to Hevesy's own discovery of radioactive tracing back in Budapest (1913). He served in the Austro–Hungarian army during World War I, and on demobilisation resumed research work at the Veterinary College in Budapest whilst, on Loránd Eötvös's recommendation, his appointment as professor of applied physics at the University came through. However, Eötvös died soon afterwards and, due to the intervening Communist Republic the promotion was officially signed by Theodore von Kármán, who had become state secretary for university education, but later that year, with the collapse of the Commune, was compelled to flee the country for Germany. In 1920, the professorship was withdrawn from Hevesy by the new right-wing university administration, under pressure from ultraconservative Eötvös epigones. After staying on in Budapest for a short while, later that year he accepted an invitation from Niels Bohr to become professor at the University of Copenhagen, never to return to his own country.

Leo Szilard, Georg von Békésy, Dennis Gabor, Eugene Wigner, John von Neumann, Peter C. Goldmark, Miklós Kürti and

Edward Teller all belong to a generation born some 15–20 years after Hevesy, and thus attended their schools during World War I and its immediate aftermath, when political systems and 'ultimate truths' followed one another thick and fast. During the short-lived Communist Republic, Szilard was in fact amongst the many students who wrote and distributed propaganda leaflets and was consequently expelled from the Technical University after the régime fell in the summer of 1919. (Arthur Koestler was another of the same generation whose leftist sympathies led him to leave Hungary at this point.) Szilard, fortunately, was able to resume his training abroad, eventually defending his Ph.D. thesis on statistical mechanics at Berlin. The drain continued as the growing anti-Semitism in Hungary persuaded the families of several of those listed above (Kürti, Neumann, Teller, Wigner), along with many others, to seek opportunities for their sons to gain their university training outside Hungary.

Within a generation the haemorrhaging of talent from Hungarian science began again in the turmoil of World War II and its aftermath. Imre Bródy, who had done fruitful work with Max Born at Göttingen before joining Polanyi's team at United Incandescent, was a victim of the Holocaust; Rudolf Ortvay committed suicide; Rezső (Rudolf) Schmid (1904–43), who had been a pioneer in molecular spectroscopy, died in a Russian prison camp. The laboratory teams that had been built up by Zoltán Bay (at United Incandescent), Georg von Békésy (at the Loránd Eötvös University) and Albert Szent-Györgyi (at Szeged) survived, and they enthusiastically began to rebuild scientific life in the new democratic republic. However, the Communist take-over of power in 1949 soon forced them and many others into exile, an efflux that was further swollen by the suppression of the 1956 Revolution, which prompted the likes of Andrew Grove and George Olah to leave the country.

3. Computer Technology: Late Starters Come Good

(Zsuzsa Szentgyörgyi)

Hungary was late in joining the computer club, as compared to developed countries, especially the United States and Great Britain. Nevertheless, in speaking about computer technology in Hungary, two factors need to be borne in mind. Firstly, although several internationally renowned figures in electronics and the closely associated field of computer technology were born and educated in Hungary (John von Neumann, John G. Kemeny and Andrew Grove, to take just the best-known examples), their impact on Hungarian scientific life and technology was only indirect and no stronger than anywhere else. None established a school in Hungary, or indeed even visit Hungary after World War II. This survey is restricted largely to figures who were active in Hungary, though it is hardly possible to avoid mentioning the names and activities of such widely respected individuals who, willingly or not, spent most of their working lives outside their native land. Secondly, with post-war Hungary having been abandoned by the Allies to the Soviet sphere of interest, the negative impact of that isolation from the most developed technology definitely left its mark. The official Soviet line into the early 1950s was that cybernetics—and consequently semi-conductor and computer technology—was a 'bourgeois pseudo-science'. By the time this crippling ideology could be laid to rest, after Stalin's death, the knowledge gap between the Soviet bloc and the West was already insurmountable. With the Cold War had come the embargo that Western governments placed on technology transfer, including scientists, engineers, manufacturers and users—the so-called COCOM list, which cut off the region, Hungary included, from many huge technological advances, with microelectronics, telecommunications and computer equipment heading the restricted areas. Despite these major drawbacks, computer technology in Hungary still contrived to make some significant progress, thanks to the undoubted excellence of basic training in mathematics and science.

The Launch Pad of the 'Golden Age'

In the final phase of what many refer to as the 'long' nineteenth century, running from the 1880s until the outbreak of World War I, Hungary's economy boomed as never before, or since. This *Gründerzeit*—the period when many of the country's institutions were founded—was marked by a general proliferation of new factories and uninterrupted growth (employment in manufacturing industry increased fourfold between 1884 and 1913, whilst manufacturing output rose tenfold): "The *Gründerzeit* poured out an enormous number of business talents and geniuses. Purely for the sake of example, one could cite such entrepreneurial types as a [Károly] Haggenmacher, [Ábrahám] Ganz, [László] Láng, a Brüll, [Manfréd] Weiss, a [József] Hatvany-Deutsch, the Goldberger family, a [Zsigmond] Kornfeld, a [Leó] Lánczy or a [Ferenc] Chorin... In the *Gründerzeit*, the economy was like a magnet in attracting geniuses, who at any other time... would have become scientists, artists, writers. The great founders were lured not just by money, the chance to get rich quick, but also by a creative urge... The intellectual energies that were released, the broad vistas opened up, by the emancipation and assimilation of the middle classes, especially the Jewish community, were such that talented individuals could freely forge ahead and be consummated in major creative ventures—first and foremost in the business sphere, but equally in science and later also in many cultural areas."[13]

This period also saw the advent of electric and telecommunications industries—the early forerunners of computer technology—capable of responding almost instantaneously to modern developments across the globe. Thus, the first telephone factory was established in Budapest as early as 1882, just six years after the patent was filed by Graham Bell, and phone calls could be made within Budapest already in 1883, with the other larger towns rapidly following. The country's first telecommunications institute was founded in 1891, and in 1896 the United Incandescent Lamp & Electrical Co. (Egyesült Izzólámpa és Villamossági Rt.) set up its operations in Budapest to manufacture electric products, especially its modern Tungsram light bulbs (by the early '20s it was effectively controlled by the American International Standard Electric Corp. and hence, ultimately, by ITT, which resulted in the spinning-off of its wireless equipment division as the separate Hungarian Standard Company). United Incandescent's managing director, Lipót Aschner, stands out among the great founders, though it was actually Ignác Pfeifer (who joined the company at Aschner's invitation in 1912, having been forced into early retirement from his post at Budapest Technical University for political reasons) who set up the research laboratory and recruited many of the talented young researchers who won international recognition during the 1930s.

The Golden Age came to a close with World War I and its losses of life, then the ensuing wholesale dismemberment of Hungary's territory, infrastructure and industry under the 1920 Trianon Treaty. The intelligentsia was also greatly diminished by a large-scale efflux of young talent, particularly amongst Jews seeking to escape the anti-Semitic restrictions of the 1920s and '30s.

Amongst those emigrants was the young engineer, László Kozma (1905–76), who joined Bell Telephone Laboratories at Antwerp in 1930 and was soon displaying his promise, with his name featuring as inventor or co-inventor on 25 of the patents registered by the company between 1934 and 1938. Kozma was primarily interested in computation, demonstrating his first prototype electric calculator in the autumn of 1938. He returned home in 1942 to become manager of the Hungarian Standard Electric factory (which, though still American-owned, was nominally assigned to the German affiliate), staying on at the end of World War II only to become one of those arrested and sentenced (narrowly escaping a death penalty) on fabricated spying charges in the notorious Standard 'show trial', staged by the Rákosi régime in 1950, being eventually amnestied in 1954 (see below).

Naturally, there were also many creative individuals who stayed, among them István Juhász (1894–1981), who in 1930 patented an automatic, ground-radar-based target selector for anti-aircraft defence batteries and three subsequent refinements (the last in 1939), of which more than a thousand were manufactured, mostly for export sale. Only in the early '50s was this superseded by a fully electronic version, which was still being manufactured by the Gamma Works into the '70s.

As already indicated, Hungary largely missed out on the rapidly accelerating developments in electronic computer technology of the '50s and '60s, though the legacy of certain pre-war initiatives in government data-processing systems was carried forward. The 1949–50 census, the data from which were worked up by a relatively large array of punched-card sorters, for instance, constitutes an important landmark in domestic computer technology as it provided the Central Statistical Office with a core competence that made it one of the important centres for computer applications in the country.

A Strange Story

An odd side-light on the desperate plight of Hungarian society at the height of the Rákosi era in the late '40s and early '50s, typical of its perverse contradictions, is that László Kozma, far from being the only engineer or physicist to fall victim to trumped-up charges, a number of other computer scientists, including József Hatvany (1926–87; a scion of the aforementioned Hatvany-Deutsch family), Rezső Tarján (1908–78) and László Edelényi, landed up in prison—often working in closed workshops, as in the 'special camps' of Stalinist Russia, in some cases carrying out work of high scientific quality. In December 1953, the Hungarian Academy of Sciences (MTA) received a letter from the director of one such workshop offering to build an analogue or digital electronic computer, should they receive the right kind of support—indeed, following this up with a second letter listing (on behalf of the anonymous prisoners) the equipment and components that would be needed and their costs. The po-faced Academy rejected the plan, however, on the grounds that there was no budget for it.

Computer R & D in Hungary received a considerable boost in the 'Thaw' of the post-Stalinist period (albeit suffering a further major setback from the flood of talent that left the country in the wake of the 1956 Revolution) as wrongfully imprisoned scientists were released, a new generation of youngsters receptive to its challenges emerged from the universities, and a string of new research facilities were set up.

The MTA's Institute for Metrology and Instrumentation (Mérés- és Műszertechnikai Intézet), established in 1955 with the rehabilitated Rezső Tarján as its director, developed various magnetic memories of modern design and also tackled cybernetic problems. Most of that latter team in 1957 transferred to a new Cybernetics Research Group, which can today be considered the real cradle of modern computer science in Hungary, with many of its staffers—most still alive—now the 'old pioneers' of the field. The creation of this group was not without its own incongruities, however. Whether out of conservatism or political considerations, many of the MTA's leaders, with a few notable exceptions, were unwilling to support either computer science or cybernetics; moreover, when it was decided that a 'modern' electronic digital computer should be built, there were heated debates even amongst experts within the field as to whether to develop the computer with domestic resources alone or to rely on Soviet documentation. Director Sándor Varga, in the end, went for the latter option, and so the Soviet M3, along with all necessary components and peripherals, was used as the model. The M3, with its 4-kilobit magnetic drum memory and its snail-like speed of 30 instructions per second, served as the basis for a whole host of computer applications in scientific, technical and economic fields and in R & D up till 1963–64.

The universities also became involved in computer science. László Kozma, by then a professor at Budapest Technical University, together with a team of young assistants developed and built the MESZ-I electronic calculator, a programmable digital machine assembled from cheap binary relays of Hungarian manufacture. More significant for future progress, however, was the work of Prof. László Kalmár at the University of Szeged, who, like most leading figures of the day was cut off from contacts with, and due international recognition from, his colleagues in the West. As a mathematical logician, Kalmár not only significantly simplified certain aspects of the Gödel and Church theorems but also contributed greatly to the solution of a number of cybernetic problems. Had the Socialist bloc been able to offer a more adequate technological base, his design of a formula-controlled computer might have been one of the big breakthroughs of that era. Even more grievous, though, was the lack of acknowledgement accorded to Rózsa Péter. During the 1920s and '30s she had been excluded even from working as a school teacher by the pre-war anti-Jewish laws yet her work on recursive functions after the war formed a similarly key contribution to the theoretical foundations of computer science. A conference on automation theory that she and Kalmár organised in 1956 gave a big push-start to computer science in Hungary, and the two of them were also the instigators of, and driving-force behind, the country's first university courses in programming.

By the late 1950s the emphasis had begun to shift to computer-construction efforts, for the simple reason that the commercially available machines of the time were notoriously unreliable. The M3 and Kozma's relay calculator have already been mentioned, but in the EMG factory László Edelényi and electrical engineer László Ladó designed a new machine, the EDLA, in which the calculations were done on circuit boards and the operations involving peripherals were handled by relays. By 1963 they had made a fully operational prototype but it proved to be unviable despite several then highly original features such as its use of multiplexing, a foil memory (an early ancestor of the floppy disk, as it were), and a form of hard disc memory.

In the mid-'60s the Hungarian economy was starting to grow, largely in response to the global boom of that period. Several R & D facilities were established or expanded which were to play a major part in the subsequent wide-scale adoption of a computer culture in the country. The Automation Research Institute (SZTAKI) was founded in 1964, under the direction of Ottó Benedikt (who had fled to the Soviet Union after the collapse of Hungary's own Communist Republic in 1919 but not been allowed to return home until 1955). In 1965, the Central Statistical Office established an Information Processing Laboratory (INFELOR), headed by the late Ferenc Rabár (who was to become the first minister of finance in the Antall cabinet after the change of political régime), with the primary aim of developing and spreading the application of computers.

The end of the decade was marked by the realisation within parts of the Soviet leadership, especially circles close to Premier Kosygin, that the yawning technological gap with the West, especially in computer sciences, was compromising their defence capabilities. Experts were agreed on the need to catch up, but there were fierce debates over whether to pursue original developments or simply to copy an already existing system. The latter camp (which had Kosygin's backing) eventually won the argument, and as a result the Hungarian government set up an Inter-Government Committee on Computer Technology, chaired by János Sebestyén, then deputy chairman of the National Technical Development

Committee (OMFB), the 'godfather' of computer technology. The next year, the government launched a Central Development Program for Computer Technology to assist the development of computer technology in Hungary, from basic research all the way through to industrial manufacture, taking in training, education, and promoting its applications. Within this program a new Institute for the Co-ordination of Computer Technology (SZKI) was set up under Zsolt Náray. This initially tiny office rapidly grew into a significant institution, carrying out not just R & D but also production, marketing, purchasing and servicing activities, or in other words, covering essentially the whole cycle of innovation. Recognita, a still popular optical character-recognition program, and the MProlog language for expert systems utilising artificial intelligence were both developed at SZKI.

Socialist Clones

Before the political change of 1989–90, the Socialist bloc countries were rigidly denied access to most of the West's advanced technologies—particularly in the wake of critical events like the Soviet invasion of Czechoslovakia in 1968 or the occupation of Afghanistan in 1979. The pressures to keep their economies expanding impelled a reverse-engineering approach to circumvent the technological blockade and to ensure that vital equipment was still made available. Far from being a simple copying job, this requires a high level of expertise and a thorough knowledge of the product or system, often including research into and innovation of tricky new processes. During the 1960s, a team of young researches at the Central Research Institute for Physics (KFKI) in Budapest realised that control of the strictly programmed ('hard-wired') multi-channel analysers that were collecting and processing the masses of data generated in the Institute's experiments could be replaced by a software-controlled computer. They quickly recognised, however, that much more rapid progress could be made if the applications were compatible with the world's leading computers; in other words, if they were to adopt a specific 'model'. Their choice fell on the PDP series manufactured by DEC in the USA. Produced under the name of TPA, the KFKI computers were relatively modern units with an up-to-date architecture, running on software that met Western industrial standards of the day. Due to the import restrictions on Western equipment, these machines were in great demand in the Socialist bloc, and it is little wonder that by the late '80s roughly 600 systems, catering for a very wide range of applications, had been installed.

By the early '80s, then, computer technology had grown deep and strong roots in Hungary spurred in part by the interest shown by large enterprises but equally by state support, particularly in the sphere of training and education. This was the dawn of the personal computer era but most of the activity was centred on the host of mainframe machines that were in operation by then, principally at universities and research institutions, government computer centres, and of course business enterprises, mostly carrying out vital data-processing and office administration or management tasks. As far as the future of the field is concerned, however, the work being done in the research institutes was much more important. In a number of areas the results bore international comparison, scientifically speaking, and moreover bore practical fruit.

At SZTAKI the significance of computer-aided design and manufacture (CAD/CAM) was recognised early on, and R & D programs were mounted into its applications in mechanical engineering (under József Hatvany) and electronic systems (under Miklós Uzsoky). Hatvany's group gained international recognition for their work on spatial modelling, especially of sculptured surfaces, whilst the team under Uzsoky, one of the cleverest engineers of those days, achieved significant results in the design and production of printed circuits.

Much valuable earlier research had already been conducted in Hungary in the field of artificial intelligence, with its exacting requirements for mathematical skills and advanced software, by the time the Prolog language for handling AI problems was published in the USA in 1975. By the end of the decade, a team of young researchers working on logical programming at SZKI had developed a modular version, the MProlog that was capable of supporting a broad spectrum of practical problems (e.g. three-dimensional graphics, windows- and menu-driven operations, etc.). This had considerable commercial success, with more than 1,500 MProlog-based systems being installed in 25 countries by the late '80s.

Researchers at Budapest Technical University managed to improve voice recognition to the point of being able to distinguish between identical twins reading the same text. They also demonstrated in field trials that another system developed there, employing near-natural speech synthesis of a fixed vocabulary, was more accurate than traditional methods for the diagnosis of hearing impairments (especially in young children)—a system that had been refined considerably further by the late '80s. Meanwhile a team at the MTA's Institute of Linguistics were developing a text-to-speech system, and SZKI researchers achieved outstanding results in optical character recognition with Recognita, which allows printed and (within limits) hand-written texts to be entered directly into a computer without any keying. The versions of Recognita developed for other languages enjoyed considerable commercial success abroad, allowing its developers to set up their own company in the nineties, which has since passed into American ownership.

Through work that went back over a decade, several research institute and university teams were active in the area of image-processing by the early '80s. Initially this was focused on its purely scientific uses (e.g. in biology, molecular chemistry, physics, astronomy), but this was subsequently extended to industrial and other tasks (e.g. the evaluation of computer graphics, photographs, radiographs, satellite pictures, and so on). Further important applications were developed for texture analysis, shape recognition by industrial robots, and the evaluation of micrographs. Several items of equipment developed at KFKI for the Soviet VEGA spacecraft to monitor the passage of Halley's comet in 1984 gained international attention, leading to their subsequent use in some of NASA's space research programs.

The social changes that Hungary underwent during the 1990s have naturally had a big impact on computer technology. With the lifting of most remaining restraints on the exchange of ideas and the COCOM restrictions, state-of-the-art equipment became readily accessible and the scope for collaborative projects widened. Large multinational companies (IBM, Compaq, Hewlett–Packard, Oracle, Cisco, Ericsson, Nokia, etc.) set up Hungarian affiliates, and not just to exploit the country's marketing potential or to tap a skilled, relatively low-cost labour force for assembly work, but also to take advantage of the intellectual resources that are to be found there. Hungarian computer experts, researchers as well as engineers, take an active role in international scientific programmes and manufacturing arrangements. To give just a few examples: within the framework of the Cellular Neuron Network, partly funded by the US Navy Research Office, a team of Hungarian researchers under Tamás Roska (1940) has been collaborating with Californian and Spanish scientists; the Recognita OCR system developed by Emőke Kovács and colleagues is being developed further by its American owners; the domestic company which markets the ArchiCad architectural CAD suite, developed by Gábor Bojár (1949), has built up international sales. The list could be carried on with examples of the international companies that have been taking advantage of the considerable experience in successful development and manufacture that has been built up over the past three decades in Hungary's 'Silicon Valley' in and around Székesfehérvár in western Hungary.

A Respectable Position

Young Hungarians continue to win prizes at the Student Olympics for mathematics and physics, but even the average standard of performance attained in the country's school remains good. The International Mathematics and Science Study (TIMSS) that has been carried out since 1996 by the Organisation for Economic Co-operation and Development (OECD) surveys results obtained in standardised interviews and ability tests in reading, mathematics and science administered to large groups of 7-, 11- and 15-year-old students in 38 countries. In the June 2000 report, the rankings for science showed Hungary in third place—ahead of, for instance, Japan (4th), Great Britain (9th) and the USA (14th), whilst in mathematics Hungarians were placed sixth—again ahead of Japan (4th), the USA (13th) and Great Britain (14th).

4. Biology: from Anatomy to Ecology (István Matskási)

The surge with which biological research got under way in Hungary, in the second half of the nineteenth century, was closely bound up, at first, with the burgeoning medical sciences, as witness the title of *Proceedings of the General Assembly of Hungarian Physicians and Natural Scientists* (Magyar Orvosok és Természetvizsgálók Nagygyűlésének Közleményei) that was given to one of the chief vehicles for publishing work at the time. For a considerable time two distinct broad lines of research are discernible, one ranging over anatomy, histology and cytology, the other exploring zoology and botany, with both lines producing scientists of international repute. Of course, that picture grew more complex over the course of the twentieth century, as the whole field of biological studies expanded and new disciplines were spawned, but the country's scientists have continued to make significant contributions to their chosen fields of study.

Building on an Early Controversy in Neurology

Anatomy, the study of the physical structure of living organisms, was greatly assisted by the huge strides that were made in microscopy during the latter half of the nineteenth century. This was the field in which István Apáthy (1863–1922) became the first Hungarian zoologist to achieve international prominence. The pioneering work that he conducted in the field of microscopic anatomy led him, at the turn of the century, to advocate the principle of neuronal continuity in the transmission of the nerve impulse, claiming (mistakenly as it turned out) that at the synaptic junctions between two nerve cells the end plates of the pre-synaptic nerve fibre and the dendrites of the post-synaptic fibre are fused, and thus the two neurons form a system that is continuously inter-linked at the cytoplasmic level. This was at odds with the contiguity principle being expounded by the great Spanish neurologist and histologist S. Ramón y Cajal, who held that the cells, albeit in close contact, are separated by an inter-neuronal gap and thus remain physically independent. Apáthy may have lost the argument, but the controversy received much attention and thus gave a big fillip to advancing the then still rudimentary understanding of the nervous system.

Apáthy's chief sparring partner in this debate was his contemporary and compatriot, Mihály Lenhossék (1863–1937), who had worked abroad for several years before returning home to become director of the Institute of Anatomy in Budapest—an appointment he held until his retirement. Not only was Lenhossék the first to provide convincing microscopic evidence in support of Ramón y Cajal's view, he also clarified numerous details in regard to the structure of the cerebrum and medulla oblongata and first identified many of the nervous system's typical cell types. Particularly through the German translation of his seminal book, *Der feinere Bau des Nervensystems* (1895), he lay the foundations for modern research into the nervous system and its functioning.

Nerve physiology was just one of the areas cultivated by Endre Hőgyes (1847–1906), the third member of this outstanding 'triumvirate'. He was well ahead of his contemporaries in noting the part that vestibular reflexes and eye-movements play in the sense of balance—work that only became more widely known through the influence it exerted on the research of Róbert Bárány. The best-known field of Hőgyes' activity was bacteriology, largely because he was the first person in Hungary to undertake systematic research in the area. He founded a Pasteur Institute in Budapest (1890), and in introducing the rabies vaccine into the country, he made various improvements in Pasteur's original process that were still in use until fairly recently.

Of Hungarian parentage but born in Vienna (and as he never actually worked in Hungary ranked as Austrian), Róbert Bárány (1876–1936) dedicated his entire working life to the border area between otology and neurology, being awarded the 1914 Nobel Prize for Medicine “for his work on the physiology and pathology of the vestibular apparatus (the organ of balance).” His attention was originally drawn to this organ by a simple clinical observation that his patients often became markedly dizzy when their external ears were syringed. On closer investigation, he found that the dizziness did not occur if the liquid used to flush the ear had been heated to body temperature (37°C). He reasoned that dizziness is caused when a temperature difference between the syringe fluid and the otic sac (utricle, saccule and semicircular canals) of the inner ear causes the perilymph flowing in the latter to expand or contract, thus altering its flow in the canals and so activating the receptors that sense movement, giving rise to the sensation of dizziness—a physiological reflex mechanism that in Hungary is still referred to as the Bárány caloric response. This led him to the discovery that this pathway for providing information on body posture is also linked to the involuntary movements of the eye that refresh the visual field by his demonstration that the eye disorder of nystagmus (in which the involuntary movements are continual) is due to a vestibular disturbance. After completing his clinical work in Vienna, Bárány took up a post in the medical school at Uppsala, where he died.

Unlike Bárány, Georg von Békésy (1899–1972) is a curiosity amongst winners of the Nobel Prize for Medicine in having been a physicist by training and won his honour through physics-based research. On gaining his Ph.D. from Budapest University, Békésy started his career at the Hungarian Post’s Research Institute working on ways of improving the quality of voice transmission by telephone. His comparative study of the telephone membrane and the human eardrum aroused his interest in acoustics, and for some twenty years, alongside work as an acoustic designer and as a teacher of practical physics, he spent much of his time researching what happens physically in the human hearing process. Apart from constructing the first mechanical model that gave a reasonably accurate representation of how the inner ear functions, he made use of sound generation to study the mechanical processes on which hearing ultimately depends. He went on to make preparations of inner ears from various mammals and birds, and observations on these specimens under the microscope led him to note how a pressure (from a sound wave) applied to the inner ear sets into motion the fluid within the cochlear duct, which vibrates the spiral lamina and its basilar membrane carrying the sensory hair cells, with this vibration wave passing along the entire length of the cochlea. The hair cells have tiny cilia which are imbedded in an overhanging gelatinous tectorial membrane, so that as the spiral lamina vibrates in response to the sound wave, the hair cells are bent and stretched, generating nerve impulses in the auditory nerve-fibres of the basilar membrane. By this time Békésy had left Hungary (1946) to work in the USA, first at Harvard then at the University of Honolulu (where he remained until his death), continually adding further details of the precise role that the cochlea plays in the perception of sound. Thus, he showed that different sound frequencies are perceived by hair cells placed at different points along the basilar membrane, with high-pitched sounds being sensed by hair cells at the base (nearest the middle ear) and low pitches by cells at the far tip. He then went on to demonstrate that the basilar membrane only carries out a rough analysis of frequencies, the task of pitch discrimination being handled largely by the nervous system. The theory was further refined by Békésy's calculations of the energy consumption involved in these processes.

A Giant of Brain Research

For all those illustrious predecessors, the individual who had the greatest impact on neurology in Hungary, in a remarkably successful career covering practically all areas of the subject, was János

Szentágothai (1912–94), who, through his mother, was related to Hőgyes and had Lenhossék as a university teacher. Though the great debate on the fine structure of the neuron had been essentially settled by the time he began to work, it was Szentágothai who provided the first physiological evidence corroborating the principle of contiguity—several decades before electron microscopes were available to come up with the conclusive proof. He also made important contributions to the study of the vestibulo-ocular system, by then almost a Hungarian preserve, and was the first in Europe to utilise stereotaxic surgery as a tool of brain research.

Szentágothai's research and teaching activities began in the Institute of Anatomy at Pécs University, where he was appointed professor in 1946, after he had returned from a period as a prisoner-of-war. His early findings on the innervation of eye-movements and the role of the receptor organs in the membranous labyrinth of the inner ear were soon textbook data. His 1948 discovery of the monosynaptic masseter reflex was likewise rapidly acknowledged, already being cited by the eminent neurophysiologist J. Fulton in his definitive 1949 monograph *Physiology of the Nervous System*. Szentágothai's contributions to the functional anatomy of the cerebellum and the medulla oblongata, the functioning of certain units of the nervous system, reflex arcs, synaptic junctions and the transmission of nerve impulses are too numerous to list here. However, with one of his groups of research students, he was one of the early pioneers in the then-nascent area of neuroendocrinology. This team was the first in the world to show that hormones of the peripheral endocrine glands regulate hormone production by the anterior lobes of the pituitary gland indirectly via the hypothalamo-hypophyseal tract (i.e. through receptor neurons located in the hypothalamus) rather than by direct action via the bloodstream. Their English-language monograph on these findings went into four editions.

The thrust of the third main area of Szentágothai's research was directed at elucidating what controls the orderly establishment of neuronal connections during embryonic development to ensure that the nervous system is functionally co-ordinated by the time a baby is born. He was the first to apply cybernetic approaches to research of nerve centres and reflex mechanisms, and some of his results are still used today in the special architectures that are employed for computer modelling of biological structures and processes (one example is the joint research being carried out in Hungary by computer scientist Tamás Roska and brain researcher József Hámori into the mechanism of vision).

By 1960, Szentágothai was heading the Institute of Anatomy at the Semmelweis Medical University, Budapest. It was here that he demonstrated the existence of special inhibitory nerves and their functioning via specific pathways; he and his colleagues were also the first to spot that the cerebral cortex has a columnar modular structure. With J.C. Eccles and M. Ito, he co-authored a classic monograph, *The Cerebellum as a Neuronal Machine* (1967), which presented a rational model of cerebellar function. However, many of those who have surveyed Szentágothai's career would suggest that his major achievement, from a theoretical standpoint, was the discovery that the human brain is capable of creating information *sui generis*; in other words, the nervous system does not merely process and analyse sensory data but also makes use of information that it has actively generated itself.

An outstanding research scientist, Szentágothai was also a great teacher. The *Atlas of Human Anatomy* that he co-wrote with Ferenc Kiss, and for which he personally made the preparations and drawings, has gone into more than 50 editions in many languages since it first appeared (in Hungarian) in 1946, and is a basic text for medical training in many countries.

The Founding Fathers of Biochemistry

Biochemistry was somewhat late to get under way in Hungary, but when it did so, it had the good luck to be spearheaded from the outset by Albert Szent-Györgyi (1893–1986). This extraordinary scientist, who was born in Budapest and died at Woods Hole, USA, was the first Hungarian Nobel laureate to win his award for work actually done in Hungary. He began his career in England, at the Cambridge laboratory of Frederick Hopkins, the world's first professor of biochemistry, working initially with German-born Hans Krebs on cell respiration—the complex chain of reactions by which carbohydrates are broken down within the mitochondria of every living cell to release energy (stored in hydrogen radicals) for use in the multitude of other life processes that go on within the cell. In the early 1930s, he and Krebs elucidated the essential details of how this oxidative pathway—now usually referred to simply as the Krebs cycle—feeds into a key final cycle of catalytic transformations of dicarboxylic acids (succinic, fumaric, malic, oxalic, etc. acids). Szent-Györgyi's most celebrated discovery, however— made after his return home to take a chair at the University of Szeged in 1931—was of the anti-scorbutic properties of a substance called ascorbic acid that he isolated from the adrenal cortex, thereby identifying the chemical composition of vitamin C.

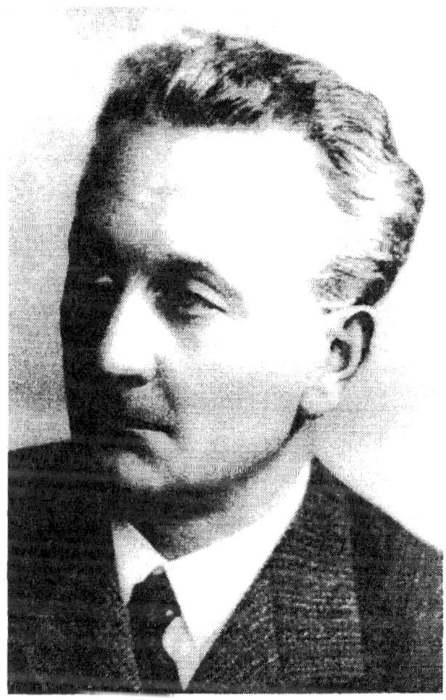

Albert Szent-Györgyi

At Szeged, he worked out a process for large-scale manufacture of ascorbic acid by extraction from paprika. The Nobel Prize for Medicine that Szent-Györgyi received in 1937 was both for his work on the pathway of biological oxidation and for the discovery and production of vitamin C.

Before leaving Hungary in 1948, Szent-Györgyi also isolated and described the biological activities of another water-soluble substance, citrin, to which he ascribed vitamin-like properties in what was thought to be its role in controlling the permeability of blood capillaries (for a while the substance was called vitamin P, though it is now known not to be a vitamin, nor indeed does it have the role then attributed to it). More crucially, however, whilst studying muscle function, he demonstrated how the molecular filaments of actin and myosin, the two proteins of muscle fibres, achieve the contractile function of muscle. On being activated by the energy-donating adenosine triphosphate (ATP) molecule formed in the Krebs cycle, the two molecules temporarily cross-bond to form an actomyosin complex, which is accompanied by a coiling (and hence shortening) of actin molecules amongst the surrounding rigid rods of myosin. This muscle biology research was continued by Szent-Györgyi's disciples after his departure, forming the foundations for the study of enzymology and subsequently also molecular biology in Hungary. During his long career at Woods Hole Szent-Györgyi opened up a new line of biophysical research by continuing his investigations of electron transfer processes in biologically active substances and proteins, finally devoting the last period of his life to trying to elucidate the role of these and other phenomena in the formation and growth of cancerous cells.

Szent-Györgyi's mantle as the head of what had been built up into a renowned Hungarian school of biochemistry research was inherited by Brunó F. Straub (1914–96). Straub had worked on cell respiration with Szent-Györgyi at Szeged in the 1930s and then for several years at Cambridge before returning to Szeged, where he isolated and purified malate and lactate dehydrogenases, two enzymes that play a vital role in biological oxidation, and discovered a diaphorase, which bears his name. On switching to muscle research, along with his master, it was Straub who identified actin as the key contractile protein, which alone secures his place in the history of biochemistry. Though much of his time after World War II was taken up by duties as a teacher, administrator and author of major textbooks, his research activities did not diminish. Of particular note was his elucidation of the role that ATP plays in pumping potassium ions out of red blood cells, and his discovery of the enzyme disulphide isomerase.

Pursuing the Classical Field of Zoology

Botanical and zoological research, the classical fields for exploring the diversity of the living world, formed the other major branch of biological investigation in Hungary. For obvious reasons, this work has concentrated primarily on the domestic flora and fauna of the Carpathian basin, but that has not precluded findings that are of wider significance or, indeed, the frequent extension of work to countries beyond the country's borders.

As far as Hungary's fauna are concerned, the single biggest contribution to knowledge during the latter half of the nineteenth century was that made by János Frivaldszky (1822–95). Having learned about collecting and systematic classification of plants and animals from his naturalist uncle, Imre Frivaldszky, he embarked on a pioneering survey of the cave life of the Carpathian basin, which generated considerable attention, and no small number of emulators, outside Hungary. His area of special interest was beetles (Coleoptera), though he also studied other insect orders (Orthoptera), the respect that he garnered internationally in these fields being reflected in the 32 species and varieties that bear his name. However he was also an expert on birds, writing the monograph

Aves Hungariae, which for many years was the definitive reference work on the subject. The collecting work that Frivaldszky and other Hungarian naturalists carried out not only won international recognition but resulted, during a re-organisation of the National Museum in 1870, in setting up a separate Botanical and Zoological Collections, of which he was appointed to direct. Under Frivaldszky's leadership, the activities of the section were extended to cover the collection and study of the animal world further afield.

Ottó Herman (1835–1914) was one of Hungary's polymaths, making contributions of enduring value not only to natural history, zoology in particular, but also with his wide-ranging studies of various areas of ethnography, linguistics, anthropology and cultural history—alongside a career as a political activist (and member of parliament from 1879). Born in what is now the town of Brezno, Slovakia, he became interested in entomology whilst living in Vienna as a young man, developing this further after obtaining a post with the Transylvanian Museum Society in Kolozsvár (Cluj). Having moved to Budapest in 1876, he was engaged by the Natural History Society to study the spider fauna of Hungary, producing a major three-volume treatise on the subject (1876–79), which included descriptions of 36 species new to science and formed the basis for a whole new area of study. This was followed by a switch to investigations of the native fish species and the ethnography of fishing, documented in *A magyar halászat könyve* (A Handbook of Fishing in Hungary, 1887–88). In parallel with further ethnographic work, he also pursued an interest in ornithology, primarily in relation to the migration of birds, recording much information that remains valid to the present day in his book *A madarak káráról és hasznáról* (On the Damage and Use of Birds, 1901), which reached an international readership through the several editions of its English and German translations—an unprecedented achievement at the time for a popularising scientific work from Hungary. On the strength of this success, Herman founded the Hungarian

Ornithology Centre, of which he was director until his death, and in 1894 launched the journal *Aquila*, which is still published today as one of the internationally recognised organs for publishing ornithological research papers. He had a big hand in drawing up an International Convention on the Protection of Birds, historically the first such agreement on nature conservation, which was adopted in 1902.

The adoption of new zoo-geographical and taxonomic approaches by Endre Dudich (1895–1971) signalled the start of a new phase in the application of modern scientific methodology in Hungarian zoology. Dudich produced significant work on the morphology and classification of arthropods, though as a university lecturer his contributions ranged over virtually all areas of zoology, and he personally considered his main field of research to be cave biology, in which indeed he was a leading international figure. Besides writing several general studies on the subject, he carried out a detailed survey of the fauna of the Baradla cave in the Aggtelek complex of north-eastern Hungary, publishing his findings in German. He made noteworthy contributions to population biology and colour ecology, although these branches of science truly started to develop only after his death, and was similarly decades ahead of his time in initiating an international collaborative research on the River Danube and establishing the Hungarian Danube Research Station.

The most outstanding researcher of the Dudich school was Zoltán Kaszab (1915–86). A remarkably prolific taxonomist, over the half century of his working career he published 397 papers, totalling some 10,000 printed pages, in the scientific journals of 25 countries, including descriptions of 3,700 new insect taxa. He was one of the world's best-known entomologists, as witness the 435 species and 52 other taxa that have been named in his honour. More than a hundred of the papers were devoted to establishing the phylogenetic system of the Meloidae beetle family, but his monographs and key reviews cover the entire world of insects of five continents. Special mention should be made of his work on the darkling beetles (Tenebrioidae) of the world's arid habitats, for his studies of this family in Oceania yielded evidence to support the plate-tectonic theory of continental drift—just one of several basic zoo-geological connections between continents and islands of different geographical zones that his wide-ranging knowledge allowed him to demonstrate. Kaszab gained special international attention with his one-man expeditions to Mongolia, in the course of which he collected some half a million insects. More important than that, however, he enlisted the help of practically every entomologist in the world to work up this huge collection of specimens scientifically.

The major advances in the study of cell structures made possible by improved microscopes by the turn of the nineteenth into the twentieth century were obviously of particular significance for the development of protozoology, the study of the simplest forms of animal life. József Gelei (1885–1952) was one of the first in Hungary to enter the field and enjoyed considerable success with the new histological techniques that he developed to study the morphology, physiology and taxonomy of such unicellular creatures. He made his name internationally known by identifying the 'neuronema' system of the Ciliata (i.e. organisms like *Paramecium*, bearing cilia over the whole or part of their body) and proving that the neuroplasmatic network and the cilia of such organisms are linked. He did original research into the mechanism of protozoan cell division and the structure and function of the sensory and other organelles, as well as discerning many new species of free-living ciliates. Towards the end of his career, he turned his attention towards protozoan morphogenesis, undertaking comparative studies of the evolution of unicellular and multicellular organisms.

Leading Figures of Botany

Like most other Hungarian naturalists who came to prominence during the second half of the nineteenth century, Árpád Dégen (1866–1934) obtained his initial training at medical school, though botany was his chief preoccupation even then. He completed a pioneering survey of the flora of the High Tatra mountains and then of the Balkans, especially in Bosnia and Herzegovina. In addition to such basic research, however, he also took an active interest in the practical applications of botany, which led to his appointment in 1896 as head thc Seed Inspection Station in Budapest. In that post he became widely known in international agricultural research circles through his advocacy of the establishment of an International Seed Inspection Organisation. This came into being at Hamburg in 1905, adopting many of the screening procedures that Dégen had worked out in Budapest. He remained a passionate spare-time botanical collector and researcher, interested in all forms of plant life, from mosses to flowering species, and travelling around not just to nearby countries but as far afield as the Middle East, on whose flora he was an expert. His German-language survey of species collected along the 60-mile Velebit mountain range is still considered the most valuable of all available descriptive flora of the Balkans. The respect in which his work was held is marked by thc nearly 50 plants that have been named after him.

Sándor Jávorka (1883–1961) achieved similar renown as an exploratory botanist during the first half of the twentieth century with his studies of the flora of the Carpathian basin, Balkans and Alps. Some six of the sixty years of his working life were spent in the field, in the course of which he collected some 22,000 pages of herbarium samples, leading to his identification and description of 120 new species, while his name is preserved by 40 gymnosperm and angiosperm plants. Most of his published scientific work tackles the floristics and taxonomy of the plant life of the above-mentioned regions. His *magnum opus* was the 1,400-page *Flora Hungarica* (1924–25), the product of two decades of work, which provides a critical review of all flowering plants occurring naturally in Hungary. The morphological sketches for this include the early work of the outstanding botanist and illustrator, Vera Csapody, who subsequently also contributed the wonderful plates for the picture atlas *Iconographia florae Hungaricae*, which the two co-authored and which remains an indispensable work for the study of the flora not just of Hungary but the European continent as a whole.

The third of these major figures of Hungarian botany, Rezső Soó (1903–80) made his reputation in plant taxonomy and phytocoenosis (or plant ecology as it might be called today). Having obtained his Ph.D. from Budapest University, the young researcher started his scientific career at Berlin–Dahlem, then the world's leading centre for taxonomy, where with Keller he co-wrote what remains to this day an indispensable monograph on the orchids of Europe. On his return to Hungary, Soó published a monograph on the geo-botany of Kolozsvár (Cluj), one of the earliest works in this field anywhere. By the late 1920s, working in the Biology Research Institute at Tihany, on Lake Balaton, he embarked on the first systematic program of research into the ecology of Hungary's plant life, the principle product of which was his manual on the taxonomy and geo-botany of the Hungarian plant world.

In the Footsteps of Great Precursors

Biological research today in Hungary pursues three main fields of enquiry: molecular biology in a broad sense, ecology and neurobiology. The latter may be seen as continuing a tradition that began over a century ago by the likes of Lenhossék, Hőgyes and Szentágothai, whilst the other two fields are more recent, dating essentially from the 1960s and '70s.

Of the various lines of neurological investigation, it is the work of E. Szilveszter Vizi and Tamás Freund that has gained the widest international attention. However, the country also has groups looking into aspects of neurotransmission, learning and memory processes, behavioural studies, the typology of brain damage, and neuro-endocrinology, which supplement traditional methods with a wide range of molecular biological, genetic and neurophysiological approaches.

Molecular biology encompasses the whole spectrum of biophysics, biochemistry, enzymology, genetics, phytobiology and microbiology, with the Hungarian Academy of Sciences (MTA), Center for Biology at Szeged, established in the late 1960s, as its internationally recognised center. To pick out a few of the research lines being followed, one may mention, for instance, work into the fundamental pathways by which energy is transformed in living cells, that on biological membranes and on nucleic acids, and that on the molecular basis of physiological processes and enzyme function. The Center's molecular genetics research, specifically on the genetics of nitrogen bonding, in immunogenetics, and on cell and developmental genetics, is world renowned. Other programs of research into phytogenetics, photosynthesis and the mechanisms of plant stress are showing how basic molecular biological approaches can be harnessed to applied agrarian research. Leading experts in these areas include Dénes Dudits, Pál Venetianer, Péter Friedrich, Gyula Hadlaczky, István Raskó, Ervin Balázs, Zsuzsanna Bősze, and Ádám Kondorosy.

The pressing demands of environmental protection and nature conservation, with particular regard to preserving biodiversity, place great weight on ecological research. The recent death of Pál Juhász-Nagy has prematurely deprived the country of a pioneering figure in this field, but successful programs are running to study how dry-land plant habitats are organised, to identify variable scales in patterns of diversity, and into the country's hydro-biology.

Theoretical research is being directed towards studying the impact of habitat on the dynamic behaviour of populations and communities as well as into the explicit modelling of ecological problems. The theoretical problems of evolutionary biology are also widely studied, with Eörs Szathmáry a leading figure in this area. As part of the country's commitment to preserving bio-diversity, Hungarian scientists are active in international projects that are undertaking taxonomic and systematic studies of biological communities both within the Carpathian basin and on other continents. Finally, mention should be made of the research into animal behaviour that is being undertaken by the team around Vilmos Csányi, Hungary's leading ethologist.

5. Astronomy in Hungary: between Ógyalla and Budapest

(Mrs Domokos Varga)

The Beginnings

Accurately fixing the location and movement of the different celestial bodies went hand in hand with the progress that was made in telescope construction. Kepler's laws of planetary motion and Newton's law of gravity offered the means for calculating the exact size of our solar system, provided the mean earth–sun distance—the solar parallax—could be determined. As this could not be observed directly, because the sun is too bright, it had to be computed from the limited parallax of Venus during transits of the planet across the sun's disc, with international expeditions being mounted for this purpose to observe passages of Venus in 1761 and in 1769. The value for this parallax that was obtained by Miksa Hell (1720–92), director of Vienna's observatory, from his observations of the phenomenon on 3rd June, 1769, on the island of Vardø in the Arctic north of Lapland (then a Danish possession), is very close to the currently accepted one. Hell himself had studied astronomy at Hungary's only university in those days, at Nagyszombat (Trnava, Slovakia), where an observatory was already operating in 1756. The work there was devoted principally to a systematic study of the moons of Jupiter, though all the then-known planets and comets, solar and lunar eclipses, and many other events were regularly monitored. For instance, Nagyszombat looked at the polar lights that were visible there in 1761, and Ferenc Weiss published a theoretical explanation of the phenomenon in the French *Journal d' Étranger*.

In 1777, during Weiss's directorship, the observatory was relocated, along with the university, to a tower in Buda's Royal Palace on Castle Hill in Buda, recommencing operations in 1780. The site was then moved again to nearby Gellért Hill (Gellérthegyi

Csillagda), and the new observatory, equipped with the best instruments of the day (mostly manufactured in Munich), was inaugurated on 15th October, 1815, in the presence of the three sovereigns of the Holy Alliance. Few paintings or sketches of Pest-Buda of the beginning of the nineteenth century would have failed to include the attractive cluster of its buildings. By then the director was János Pasquich, already a well-known mathematician and astronomer, who had published a number of significant scientific books.

Landowner-Turned-Scientist: Miklós Konkoly-Thege

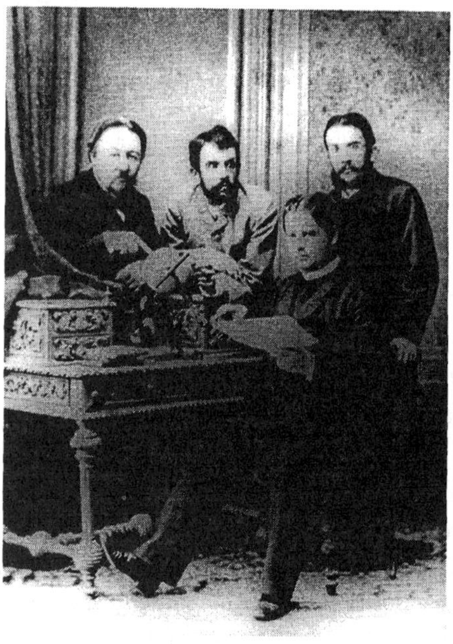

Miklós Konkoly-Thege,
Jenő Gothard, Sándor Gothard

With the Buda observatory left in ruins by the Austrian siege of 1849, astronomy in Hungary languished for two decades until, in the swell of patriotic and entrepreneurial fervour that was unleashed by the 1867 Compromise that Hungary reached with her Habsburg king, the trained physicist and astronomer Miklós Konkoly-Thege (1842–1916) in 1871 built a well-equipped private observatory on his own estate at Ógyalla (Hurbanova, Slovakia), just north of the Danube at Komárom.

By luck, Konkoly-Thege's involvement in astronomy coincided with the birth of three branches of astrophysics: spectroscopy, photometry and photography. Konkoly-Thege had studied solar radiation whilst a student of Ányos Jedlik at Pest University and subsequently, during a spell at Berlin University that was partly to further his knowledge of chemistry, gained some experience with spectroscopy, so his academic background predisposed him to join the development of the new disciplines from their very start. Although this required sophisticated adaptation of what was entirely novel equipment, Konkoly-Thege's special feel for technology enabled him to contribute massively to putting that ambition into working practice, designing and constructing several instruments himself, including a 'barrel-less' solar telescope (1877) and a heliograph (1884), as well as compiling three large manuals on the subject. At Ógyalla, in 1872 he commenced a long series of observations of sunspots and solar flares, which he continued until his death, co-ordinating the work and pooling data with Carl Vogel and Gustav Spörer, the Potsdam observatory's astronomers. From 1885 onwards, he sent regularly reports on these observations to the *Mitteilungen*... in Zurich, fully one quarter of whose data in 1894 came from this source.

Konkoly-Thege was also a keen planetary observer, with a particular interest in the surface features of Jupiter and the Mars, of which he made many valuable sketches, publishing them in the observatory's own yearbooks. His record of the drifting of Jupiter's 'Great Red Spot', which reappeared in 1878, enabled an estimate to be made of the planet's speed of rotation around its own axis. (A good number of his Jupiter drawings, incidentally, have remained unpublished to the present day, so it is quite possible they record other events that are still not public knowledge.) Konkoly-Thege was the first to observe spots on the surface of Venus, making it possible for him to draw conclusions on the atmospheric changes taking place on the planet. The appearance of the extra-galactic supernova S Andromedae in the sky, during August 1885, offered the first ever opportunity to study such an event by spectroscopy, and Konkoly-Thege's findings were duly published in the

Astronomische Nachrichten. As one of the first amateur photographers in the country, Konkoly-Thege also made some early experiments with astrophotography, preparing his own emulsion-coated glass plates until the dry-plate process became available.

In 1899 the entire Ógyalla facility was transferred to the state, thereupon becoming the Hungarian Royal Astrophysical Observatory. With Harvard's observatory having started photometry of the stars in the 1880s, employing photographic dry plates, the reorganised observatory immediately installed the necessary equipment so that it could undertake its main task of measuring the brightness of stars. A first catalogue of this work, published in 1916, contained the magnitude of 2,122 stars, though publication of a companion volume, including documentation on 129 variable stars (mostly long-period red stars), was delayed until 1925, by which time Ógyalla lay in the new successor state of Czechoslovakia and the observatory had been obliged to find a new home within Hungary's reduced borders.

Not content to rest on the laurels won at Ógyalla, Konkoly-Thege generously assisted others to set up observatories elsewhere in Hungary, employing his wide network of contacts to obtain instruments and help in the design of the facilities. A prime example was the observatory built by Bishop Haynald at the Jesuit-run gymnasium of Kalocsa in 1878, of which Gyula Fényi (1845–1927) became director in 1885. Fényi's photographically documented observations of solar flares, and their relation to sunspot activity, over the next thirty years, until his death, constitute the most substantial unbroken series of such data available, with two particularly spectacular flares that he observed in 1895 often still referred to in the modern literature.

On another private estate at Herény, just outside Szombathely in western Hungary, Jenő Gothard (1857–1909) and his brother, Sándor, in 1882 began regular observations using a 10.5-inch Browning reflector purchased from Konkoly-Thege. Co-ordinating its work, and occasionally undertaking joint projects with Ógyalla from the outset, the Herény observatory was already printing news about its work by 1884. Apart from being, like Konkoly-Thege, a highly skilled constructor of astronomical instruments, Jenő Gothard was also a master of astronomical photography and spectrography. His serial shots of the partial solar eclipse on 16 May 1882 became well known around the world, and his success in photographically recording the central stars of planetary nebulae was a considerable achievement for that time. He began to study the spectrum of ß-Lyrae in 1883, discovering the regular periodicity of the appearance and disappearance of the emission lines for hydrogen and helium that demonstrated that it was a binary star. From 1886 onwards he worked mostly on the spectral photography of clusters, comets and gaseous nebulae.

Several years later, in 1886, a new observatory was opened by Géza Podmaniczky and his wife, Berta Dégenfeld, on their estate at Kiskartal. Countess Berta's name had already become well-known among astronomers by then for spotting the appearance of an unusually bright but previously unknown star in the Andromeda galaxy during the night of 22 August 1885, whilst carrying out observations with a portable telescope in the castle grounds. By sheer chance, she was one of the first observers of what was to be called the S Andromedae supernova, the news of her discovery being communicated to the astronomical community by Konkoly-Thege, who himself undertook the spectroscopic examinations mentioned above, whilst Jenő Gothard documented it photographically.

After World War I

Konkoly-Thege had actually had to lobby the Hungarian state long and fiercely to induce it to take over his observatory; however, the wisdom of that transfer truly became apparent after the redrawing of national boundaries imposed by the Trianon Treaty, since Ógyalla no longer lay under Hungarian jurisdiction. Fortunately, the League of Nations, which was overseeing all property transfers out of the successor states, officially accepted that Konkoly-Thege's donation had been a private gift, and so granted permission for the Ógyalla facility to be removed to a new site on Svábhegy (Swabian Hill) in Buda. Most of the existing instruments went there, along with—even more crucially—the several decades of accumulated documentation.

Though the new observatory complex was not fully constructed until 1928, some observational work was already under way by 1923, with a scientific program practically the same as at Ógyalla, given that the equipment was from there; in other words, with photometry (particularly of variable stars) as the principal line, but now without regular solar and meteorite observations. Amongst other projects, György Kulin in 1934 initiated a series of observations on asteroids in 1934, discovering 22 of these small planets over the ten years to 1943, and developed new methods for computing asteroidal orbits. Károly Lassovszky undertook photometric studies of these tiny bodies, making 680 separate measurements of the brightness of Pallas.

László Detre (1906–74) became director of what, by then, was the Institute of Astronomy in 1943 and held this position until his death. Despite the ongoing war, its work continued until interrupted by the Red Army's encirclement of Budapest in early December 1944 but was resumed in July 1945, and by 1947 all its instruments were fully operational. Under the very changed circumstances that followed the Communist take-over of the country, the Institute passed under the supervision of the Hungarian Academy of Sciences in 1951. The International Astronomical Union (IAU), with the ratification of its 1955 general assembly in Dublin, signalled a wish to assist in developing the Institute as a world centre for the study of variable stars. What Detre had shrewdly realised was that the huge number of variable stars meant that, by choosing this as a topic, the potential workload could usefully offset his institution's lack of instrumentation. He and Júlia Balázs concentrated their activities mainly on period changes and the search for multiple periods, with the aim of gaining a better understanding of the inner structure of such stars and to develop pulsation theory further. At the IAU general assembly in 1961, Detre was asked to take on the editorship of the *Information Bulletin on Variable Stars*, which meant that Budapest became the first place to be informed of any news on the subject from around the world.

The Institute's standing was further increased when it opened a subsidiary station at Piszkéstető in the Mátra Hills, some 50 miles north-west of Budapest, in 1962. This was equipped with a 60/90/180 cm Schmidt telescope, which permitted detailed investigation of the Milky Way and thus enabled new research projects to be embarked on. A 50-cm Cassegrain refractor came into operation there in 1966 in a new dome, followed by a 1-metre Ritchey-Chrétien telescope in 1974.

An Uncertain Future

For all its distinguished track record over the past century or more, astronomy in Hungary has not enjoyed much favour since the 1989–90 change of regime. It is now more than a quarter of a century since the last big astronomical instrument was installed, and even operational costs have been held down tightly by successive governments. Because astronomy is not perceived as yielding quick returns (though the longer-term indirect benefits that have accrued in, for instance, satellite communications and defence intelligence are readily apparent), the private sector has shown no interest in funding basic research. The subject thus appears to have been trapped in an era of slow decay, and it is hard at present to see how it will be able to attract and nurture the outstanding scientists it needs to regain its former international standing.

6. Chemistry: from Gold Assaying to the Modern Chemical Industry (Ferenc Szabadváry)

Roots in Gold

Chemistry appeared in Hungary in the Middle Ages as an adjunct to the mining of precious metals, the country then being Europe's largest gold and second-largest silver producer, as well as having significant copper mines. Assaying (chemical analysis) of the metals, once extracted, was an essential safeguard against the adulteration and debasement of the coinage. A decree issued by King Sigismund in 1405 gives us reason to suppose that by that time they also used a 'wet' method for the purpose—"mit der Prob auf das Gold in Wasser," which in this case is almost certainly a reference to nitric acid (*aqua fortis*), which dissolves silver but not gold, rather than to water.

The foundation of a Royal School of Mines at Selmecbánya (Banská Btiavnica, Slovakia) in 1735, during the reign of Emperor Charles VI (Charles III as king of Hungary), was a milestone. Students could get a post in the industry only after successfully passing all six subjects taught at the school, one of which was assaying, in which the various tests prescribed in detail by ordinance were practised in the school's own laboratory. The science of chemistry in Hungary developed from this background when Queen Maria Theresa established a department of chemistry and metallurgy at Selmecbánya in 1763, and raised the school to college status in 1770.

Selmecbánya College soon won wide recognition, attracting students from all over Europe eager to acquire the skills it could teach them on the applications of assaying. Indeed, it was cited as a specific model by the committee that, on 28 September 1794, submitted to the French Revolutionary Convent a recommendation that an École Polytechnique be set up in Paris. The report on the

Convent's deliberations, published in the official *Gazette national ou Moniteur universal*, cites among the reasons for granting approval the following: "Hitherto, only the theory of physics and chemistry has been taught in France. The Selmecbánya School of Mines in Hungary is a spectacular example of how useful it is for students to have practical experience in the manifold procedures that form the basis of applied sciences. [In Hungary] laboratories were established where all necessary materials and tools were made available, so that all students might repeat experiments and witness for themselves the phenomena that occur during the fusion of certain materials."

The first chemistry professor at Selmecbánya was N. Jacquin, born in the then Austrian Netherlands, who was succeeded in turn by G. Scopoli, from South Tyrol, then Antal Ruprecht (1750–1810), born in Szomolnok, Hungary, and a past student at the school. Under the latter, the College reached the peak of its glory, with students thronging to it from around the world. Ruprecht himself engaged in various productive lines of research, which he published in the scientific journals of Germany and Austria, though his most notable successes were in being the first to melt platinum and also to suggest (in 1790) that the so-called 'alkaline earths' were actually metals—a view that was strongly attacked until Humphry Davy proved it correct by electrolysis in 1808. Ruprecht also had a hand in the discovery of tellurium—the sole chemical element to be discovered in Hungary (or to be more exact, in Transylvania)—by the Austrian Franz Joseph Müller (1742–1825), another Selmecbánya graduate and director general of the Transylvanian mine causing long lasting and heated debates.

For most of the eighteenth century, Hungary's only university, founded by Archbishop Péter Pázmány in 1635, was located at Nagyszombat (Trnava, Slovakia). When a medical faculty was added in 1770, during the reign of Maria Theresa, this set up a department of chemistry and botany. The Austrian Jakab Winterl, the first to be appointed to the new chair and occupying it until his death, played a key role in the development of both disciplines in Hungary, particularly after the university was relocated first to Buda (in 1777) and then across the river to Pest (in 1783). Winterl was internationally known above all for his magnum opus *Prolusiones ad chemiam saeculi decimi noni* (Preparing for the Chemistry of the Nineteenth-Century), published at Buda in 1800, which the renowned physicist Oersted, no less, translated into Danish.

Hungary's Reform Movement and the Language of Science

The imposition by Emperor Joseph II of German as the language of public administration throughout the Habsburg realms produced a major backlash in Hungary, prompting a vigorous campaign to use the native tongue not just in political life, literature and the arts but also in the natural sciences, for which Latin had been the traditional language of discourse. However, the lack of appropriate Hungarian vocabulary was a serious hindrance to its use in the sciences. As Ferenc Nyulas (1758–1808), one of Winterl's students and author of *Az Erdélyországi orvosi vizeknek bontásáról* (On the Analysis of the Medicinal Waters of Transylvania, 1800), the first book on chemistry to be written in the language, admitted in its Introduction, he had had a tougher job coining adequate Hungarian expressions than performing the chemical operations themselves, and he felt obliged to append a Latin–German–Hungarian glossary to the volume. Others too chipped in their own 'Hungaricisms', for the most part independently of one another, as a result of which the idiom of the chemistry textbooks written in the first half of the nineteenth century are harder to understand today than the foreign-language texts of the day. One of the neologists was János Schuster, Winterl's successor in the chair of chemistry, who was the first to 'christen' the chemical elements in Hungarian and was the first chemist to become a member (1831) of the embryonic Academy of Sciences.

Though the Hungarian Diet in 1844 ordered that the native language was to replace Latin for instruction in all subjects, German was widely re-imposed after the failed 1848–49 War of Independence, and it was not until 1860 that Hungarian regained its position in the country's educational system. With the boost to all spheres of life that was seen during the half-century following the 1867 Compromise, the science of chemistry forged ahead by leaps and bounds, with new schools and laboratories being opened under professors with Western-acquired (mainly German) modern learning.

The Legacy of Károly Than

During that transition period no small role was played by Károly Nendtvich (1811–92), who in 1845 was elected to the Academy as its second chemist member and the following year was appointed lecturer in chemistry at the brand-new Joseph Industrial School in Buda (named after the country's Habsburg palatine), which became a polytechnic in 1856 and thus formed the core of what, in 1872, became the Budapest Technical University.

The indisputable leading figure in chemistry of this era, though, was Károly Than (1834–1908). Having studied medicine and pharmacy at Vienna University, Than had gone on to learn chemistry under the great Wilhelm Bunsen in Heidelberg, obtaining his doctorate in Vienna (1859) and staying on there as a lecturer until he secured a chair at the University of Pest (1862). Over the next decade he fought successfully to get the pitifully antiquated departmental facilities enlarged and updated, moving into a new building with modern laboratories in 1872. Than's most significant scientific achievement was to define and introduce the concept of molar volume. The two volumes of his unfinished *Kisérleti chemia elemei* (Elements of Experimental Chemistry, 1898 and 1906)—perhaps precisely because, in line with his conviction that it was his duty to do everything in his power to popularise his subject in his own country, it was written in Hungarian—sadly failed to gain the wider recognition they deserved. On the other hand, virtually every chemistry professor who was active in Hungary during the first third of the twentieth century had been taught by him.

A prime example was Béla Lengyel (1903–90), who, having started his career as Than's assistant then worked for a while with Bunsen, accepted a new chair in chemistry that was set up at Pest when the department moved into its modern premises in 1872. Lengyel's early special interest was in spectroscopy, but he later became the country's first researcher to move into the emerging field of radiochemistry and pressed the case for the establishment of an institute of radiology along the lines of those in Paris and Vienna, though all he managed to secure for this, in 1916, was a single room.

The year of 1872 also saw the government deciding to re-establish a second university at Kolozsvár, where a university had originally been founded by Prince István Báthory in 1591 and operated intermittently until 1786, when Joseph II had closed it down. Though the convention in Central Europe, until then, had been to place the natural sciences administratively within a university's faculty of arts, Kolozsvár from the outset had a separate Faculty of Natural Sciences (and a department of chemistry within that)—an arrangement which persisted after the university was transferred to Szeged in 1920, and which was adopted at all other universities only after 1949. It was under its second professor, Rudolf Fabinyi (1879–1936), appointed in 1878, who had studied with Károly Than and had been a lecturer at the Technical University, that chemistry at Kolozsvár acquired a modern building that could vie with the facilities at Pest. At a time when organic chemistry had become the main area of chemical research all over Europe, Fabinyi was essentially the first in Hungary to venture into this field. His initial investigations were into the chemistry of the azarons, whilst in the course of later studies of salicylates he developed several new dyes. He and Gyula Farkas, professor of physics at Kolozsvár, also worked together on a carbon-based fuel cell, which has since gained importance in the space age due to its potential as a long-lived source of electric power, and in 1882 he founded *Vegytani Lapok* (Chemical Papers), the first Hungarian journal to be devoted to chemistry. Fabinyi published this at his own expense up till 1890, when he had to abandon it for financial reasons. When the Hungarian Association of Chemists was established in 1907, Fabinyi was chosen to be the first chairman, remaining in this position until his death.

The Polytechnic School in Buda meanwhile acquired a Department of Chemical Engineering in 1870, to which Vince Wartha (1844–1914) was appointed professor. Wartha was a versatile scientist, contributing to the huge water engineering projects of that era, to viticulture and to a wide range of other areas, but he is remembered today chiefly for his work during the early 1890s in developing the eosin red glaze that was employed so successfully by Vilmos Zsolnay (1828–1900) in the wares of the family ceramics factory that he directed from 1865 (the glaze has won the distinction of a Gold Medal at the Paris Exhibition of 1900). In that connection, Zsolnay himself had already produced a string of major innovations, most notably the 'porcelain faïence' process that brought him international fame with the Gold Medal and Ribbon of the Légion d'honneur that he was awarded at the Paris World Exhibition of 1878.

With the Polytechnic School's elevation to a Technical University in 1872 came what proved to be a temporary relocation to Pest until continued expansion necessitated yet another move to its present campus, on the Danube embankment in Buda, at the beginning of the twentieth century (1900–08). The Department of Chemistry occupied the first building to be completed on the site

(1904), followed in 1905 by a new Department of Electrochemistry (virtually the first such in Europe), and in 1908 by a Department of Agricultural Chemistry, with Elek 'Sigmond (1873–1938) as its first professor.

This phase of expansion at the Technical University was rounded off by the establishment of the country's first Department of Organic Chemistry, under the young Géza Zemplén (1883–1956). Zemplén had studied in Budapest before taking up a teaching post at Selmecbánya, then going on to work with Emil Fischer in Berlin, a world authority on sugar and protein chemistry, aniline dyestuffs and other areas of the burgeoning discipline. On his return home, Zemplén was not only responsible for schooling virtually all of the country's organic chemists but also, having seen at first hand, in Germany, the value of maintaining close contacts between academic research and industry, was a key figure in fostering similar links in Hungary. As testament to his importance, two reactions are still commonly linked with his name: Zemplén saponification and Zemplén's sugar degradation.

Outside the universities, by the end of the nineteenth century the Ministry of Agriculture had set up a network of 23 experimental stations undertaking applied chemical research within specifically targeted areas (e.g. dairy processes, alcohol manufacture, viniculture), with the National Chemical and Experimental Research Institute, founded in Budapest in 1887, playing a leading role. And as a sort of footnote to the competing priority claims of Jedlik and Siemens over the invention of the dynamo, a Hungarian secondary school teacher called Móric Preysz (1829–77) had successfully used the controlled application of heat to kill microbes in beverages (in his case, wine), and so arrest fermentation at the optimal stage, fully four years before Pasteur recommended the process to France's beer industry in 1865. Sadly, Preysz too had failed to publish in time, only doing so after reading of Pasteur's 'discovery'.

Virtually all the graduate chemists who had worked as assistants to the great figures of the early days (Than, Lengyel, Fabinyi) went on to become professors in their own right, having spent some years as lecturers or researchers at leading institutions abroad—chiefly in Germany, but also in France and the USA. (That was mirrored in a sudden spate of papers by Hungarian chemists that started to appear in the German and, to some extent, French specialist literature and, more modestly, in the Hungarian Academy of Sciences' own less widely circulated *Mathematische und naturwissenschaftliche Berichte aus Ungarn*, launched in 1882.) During the 48 years of Károly Than's professorship at Budapest University, for instance, a total of 25 lecturers were appointed, each spending, on average, 2–3 years in this position. Amongst them was the analytical chemist Lajos Winkler (1863–1939), who started to work for Than in 1886, becoming an unsalaried *Privatdozent* in 1908, and likewise physical chemist Gusztáv Buchböck (1869–1935). Winkler was a born researcher, who practically lived in his laboratory and demanded similar devotion from his colleagues, developing several new analytical methods and publishing widely, but he conservatively refused to accept much of the rapidly developing technology of modern analysis. Nevertheless, he—like Buchböck—counted on succeeding Than, when the latter died—a choice that was side-stepped by giving them both chairs but leaving them to nominally share responsibilities for an undivided curriculum.

In the wake of World War I, and the huge disruption that was caused by Hungary's territorial losses under the Treaty of Trianon, the once much-criticised centralisation of higher educational institutions in Budapest could be considered to have been fortunate in retrospect. Several provincial universities and colleges that were of major importance to chemistry, including the University of Kolozsvár, were marooned outside Hungary's new borders, though by good luck the School of Mines and Forestry at Selmecbánya had by then already been relocated (in March 1919) to the town of Sopron. The stream of Hungarian refugees who chose to leave the new successor states, however, compounded by the mass demobilisation of soldiers, fed a large post-war bulge in student numbers at the remaining universities. That alone meant that teaching had to be given priority. Scientific research at universities during the 1920s was further hampered by the fact that the imposition of war reparations on Hungary, as well as general economic recession, severely limited the state's ability to fund the patching up, let alone replacement, of rapidly deteriorating and outdated equipment at a time when enormous strides were being made in all scientific disciplines. Despite the difficulties, chemistry in Hungary during the inter-war period managed to chalk up highly creditable achievements that in many respects surpassed those around the turn of the century.

Hungary's Chemical Industry

Though large-scale manufacturing facilities began to operate during the latter half of the nineteenth century, the chemistry industry did not become significant in Hungary until after World War I. Thus, in 1856 a plant to produce coal gas for street lighting was established in Pest, with other provincial cities following suit later; in 1872 the Nobel Company founded a dynamite factory in Pozsony (Bratislava, Slovakia) and several sulphuric acid and artificial fertiliser plants elsewhere in the country; the Solvay company of Belgium installed the first soda factory (which entailed ammonia production); a Hungarian Rubber Factory Works (Ruggyantagyár), under American ownership, was inaugurated in 1889; whilst by the turn of the century an oil refinery had been set up in Fiume (Rijeka, Croatia) as a first step by the infant crude-oil industry.

As some half of Hungary's industrial plant was concentrated on the capital, the chemical industry did not suffer quite the scale of direct losses of other sectors under the Trianon Treaty. Production was nevertheless severely handicapped, because those factories too had been heavily dependant on raw material supplies from territories that had passed to the successor states. External controls on the economy and its access to many markets, as well as the World Slump that began in the late 1920s, further complicated the problems of recovery. During the late '30s, József Varga (1891–1956), as a professor of chemical engineering at Budapest Technical University, expended much effort in trying to mitigate the chemical industry's supply difficulties, attaining promising results with his experiments on the hydrogenation of lignite to produce benzene. As it turned out, however, Varga's process was never taken up industrially, because by then sufficient mineral oil reserves had been discovered in Hungary to make it unnecessary. In 1939 Varga was made State Secretary and shortly thereafter (until 1943) Minister for Industry.

The tribulations of industry at large did not preclude the emergence of new sectors, most notably that for pharmaceuticals. Gedeon Richter (1872–1944), a qualified pharmacist, in 1901 set up a laboratory in his own pharmacy to produce larger batches of many common medications that he and fellow pharmacists in those days normally formulated themselves as and when required. This quickly grew into a regular manufacturing enterprise, based in the Kőbánya district of Budapest since 1907, expanding continuously and even exporting its products to neighbouring countries. Though Richter himself was shot dead into the Danube in 1944—like many other Jewish inhabitants who were herded in the Budapest ghetto during the reign of Nazi terror unleashed by Hungarian Arrow-Cross militiamen when the country came under direct German occupation—the firm survived World War II and still flourishes today, under its founder's name and on its original site. Richter's enterprise indeed bred further success, when two of his chemists, Emil Wolf and György Kereszthy, left the firm in 1910 to set up their own pharmaceutical company, Alka: this was later changed to Chinoin, under which name it too continues to operate (under

French ownership). A third example of success in the same sector was the Alkaloida Chemical Works, established by János Kabay to exploit his own patented process for producing morphine out of poppy. These three firms formed the core of a pharmaceutical industry, which was to grow particularly important during the post-war Soviet era.

Hungarian Chemists of International Fame

As in other scientific fields, chemistry too had its share of emigrants who had already gained international reputations before they left the country, but there were more who only achieved fame after moving abroad.

Several of these figures have already been mentioned in earlier sections, most notably George von Hevesy. Michael Polanyi, on the other hand, originally qualified at Budapest's medical school but was primarily interested in physics and theoretical chemistry, initially absorption thermodynamics and later (in England) reaction kinetics. His part in the team at United Incandescent Co. that, during the early 1930s, developed processes for extracting krypton and xenon from air, which led to a factory being built for this purpose at Ajka, west of Veszprém, has likewise been covered earlier. He left for England on being appointed professor of physical chemistry at the University of Manchester (1933–48). Having gained international standing as a political thinker and philosopher during the 1940s, Polanyi then moved to the chair of social studies at the same university (1948–58), ending his career as a Senior Research Fellow at Merton College, Oxford (1959–61).

Many younger people of Jewish descent, of course, left Hungary during the 1920s and '30s to avoid the pernicious effects of the 'Numerus Clausus', which prevented them from entering university or the professions, a surprising number going on to have distinguished careers in their chosen new homes. Leo Szilard, John von Neumann, Eugene Wigner and Edward Teller are well known for the various contributions they made to the development of the atomic and nuclear bombs in the USA, but it is less well known that all four were originally chemists by training and had begun their studies in Hungary. Thus, Szilard initially studied architecture at Budapest's Technical University before switching to chemistry in Berlin; Neumann, on the other hand, graduated concurrently in chemistry at Zurich and in mathematics at Budapest, whilst Wigner, who received a Nobel Prize for physics in 1963, always considered himself to be a chemist, the subject in which he graduated at Budapest. Teller pursued his chemistry studies in Karlsruhe.

The upheaval of society that followed the 1949 Communist take-over in Hungary profoundly impinged on everything and everybody. Chemistry, like most other natural sciences, certainly did not suffer the disruptions in the acceptable content of the discipline that were seen in the humanities and social sciences; nevertheless, not a few leading figures in the academic community, including Gyula Gróh, István Náray-Szabó, Jenő Plank and Géza Doby, were quickly sent into early retirement. On the other hand, opportunities for chemical research opened up considerably; the pre-war problems over lack of equipment dramatically melted away, with most requests being met, at least initially (that idyllic situation did, however, deteriorate after a while); student enrolment at universities grew rapidly, so departments expanded and multiplied; institutions were given reasonable budgets to purchase foreign journals and textbooks and so could keep up with international developments, even if that was not quite so unproblematic as in academic libraries in the West. Proof that chemistry in Hungary was not backward was forthcoming with the big wave of emigration seen after the 1956 Revolution: qualified scientists generally did not experience much difficulty obtaining suitable posts in their new homes, and again a number went on to reach high-level positions and distinctions within their profession.

By and large, then, the conditions for chemistry within the academic world improved significantly under the Socialist régime. The main driving forces were, firstly, a restructuring of its institutional base, in tune with shifting ideological demands, and secondly, a pronounced shift to 'chemicalisation', as the servicing of the manufacturing sector was called in the jargon of the day. Against that, a growing swell of unofficial protest about the environmental impact of that policy emerged as the 1970s wore on, which became fairly vocal over the following decade.[14]

In 1949, the country's concentration on a few large universities was broken up somewhat with the foundation of the Veszprém Chemical University and two adjoining research facilities: the Hungarian Mineral Oil and Natural Gas Research Institute and the Heavy Chemical Industrial Research Institute. The Eötvös University of Budapest was also revamped to set up a separate Natural Sciences Faculty, with new departments of physics, chemistry, chemical engineering, etc., and parallel changes occurred at the Technical University as well as at the Miskolc University of Metallurgy and the other provincial universities in Szeged, Debrecen, and Pécs.

A restructuring of the Hungarian Academy of Sciences in 1951 resulted in chemistry finally winning its own section, with 16 full academicians and several honorary members who were domiciled outside the country (amongst others, Nobel laureates George von Hevesy and George Olah as well as highly respected figures such as László Zechmeister, Gábor Somorjai and Tamás Bárdos). One of the main features of the restructured Academy was that it set up and administered a growing network of research laboratories which were charged to undertake basic research alone (although their activities did in practice also meet some industrial applications) and specifically had no remit to teach. Thus, the Central Chemical Research Institute (KKKI) was established in 1952, the Isotope Research Institute in 1959, the Technical Chemical

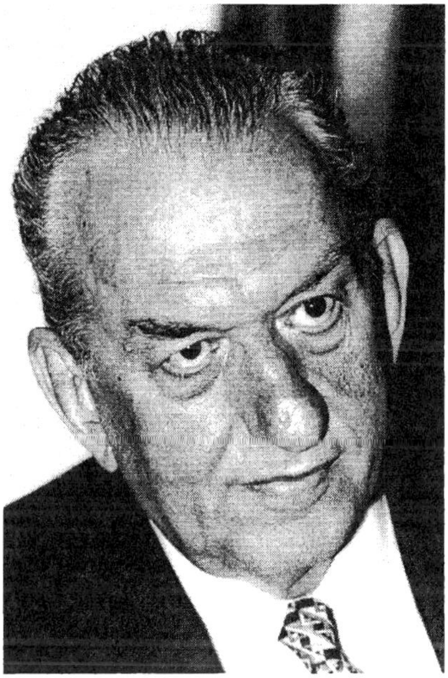

George Olah

Research Institute in 1960, and the Inorganic Chemical Research Laboratory in 1976. Besides these, industry established several of its own facilities (e.g. research institutes for pharmaceuticals and plastics) to support common interests in various branches.

The institutional set-up may have been transformed but the leading posts inevitably went to scientists who had been trained before the war, and they tended to stick to already familiar ground. Over the next two decades or so it was this generation that set the standards and the general tone against which emerging new areas, ideas and approaches were measured, often lending an inscrutable inner logic to what is anyway the complex process of scientific advancement.

At Eötvös University, Budapest, Aladár Buzágh (1895–1962) carried on his pre-war work on both the theory and practical applications of colloids, a branch of physical chemistry, specialising in the permanency of dispersed systems, mechanisms of dispersal and peptisation, the structure of gels, and adhesion; his former student and successor, Ervin Wolfram, pursued successful research into the wettability of surfaces and adsorption. Győző Bruckner (1900–80), a charismatic teacher of organic chemistry, built up an entire school in the area of peptide chemistry, whilst Sándor Müller (1903–66), likewise one of Zemplén's former students, was an outstanding theoretical organic chemist. Tibor Erdey-Grúz (1902–76), who was also Secretary-General to the Academy of Sciences for much of the 1950s and 60s, did research into electrochemical problems and, like Bruckner, wrote chemistry textbooks that were used by generations of chemistry students. In the field of analysis, Elemér Schulek (1893–1964) continued to pursue the more classical approaches of the pre-instrumentation era, especially for pharmaceuticals, as did László Erdey (1910–70) at the Technical University, though the latter did gradually adopt more modern techniques, employing potentiometers, polarography, spectrophotometry, etc.

At the Department of Chemistry in Pécs, László Zechmeister (1889–1972) had been undertaking outstandingly successful research into carotinoids and other vegetable dyes, as well as into the then new techniques of chromatographic analysis, before leaving the country in 1940, but both lines of work were continued there after the war by his student, László Cholnoky (1899–1967). At Szeged University, Árpád Kiss (1889–1968) was the Hungarian pioneer in the application of ultraviolet spectroscopy to chemistry, whilst Zoltán Szabó (1908–94) fulfilled that role in reaction kinetics, mainly studying reactions taking place in the gas phase though later also looking at the same processes in solid phase, and Gábor Fodor (1915) launched research in organic stereochemistry. Finally, at Debrecen, the organic chemist Rezső Bognár (1913–90) was a specialist in alkaloids, whilst Lajos Imre (1900–74) studied isotope tracing techniques. Within the KKKI, research programs were centred on organic reaction mechanisms, molecular structures, and chemical and molecular pharmacological processes in bipolymers. Radiation chemistry research commenced there in 1955, but subsequently transferred to the Isotope and Surface Chemical Institute of the Chemical Research Centre.

As far as theoretical chemistry is concerned, Hungarian researchers first became involved with quantum chemistry in the '50s, tackling the problems of describing molecular orbits, then moved on to molecular graphics during the '60s, whilst since the early '80s they have been studying molecular dynamics, protein crystallography and the modelling of proteins. Chemical information technology has also taken root, producing some excellent results in combinatorial chemistry.

Amongst more recent subdisciplines of chemistry, materials science has focused particularly on industrial ceramics and bioceramics, plastics, nanostructures, intelligent materials, solid-state chemistry and liquid crystals, whilst other research areas include co-ordination chemistry, metallo-organic chemistry, nuclear and radiation chemistry. Hungary is also now running intensive research projects on environmental chemistry, especially in connection with atmospheric studies, the reduction of harmful emissions, and the search for 'environmentally friendly' industrial processes.

7. Agriculture: the Scientist's Modernising Hand
(István Matskási)

From Serfdom to Modern Agriculture

The upsurge in agricultural production that occurred in Hungary during the second half of the nineteenth century was closely linked to the abolition, by the April Laws of the revolutionary parliament in 1848, of feudal relations that until then had tied most of the country's peasants to the lord of the manor, imposing on them frequently onerous dues and labour services. This emancipation from serfdom allowed peasants to acquire the land that they worked, leading to a gradual but profound shift in the structure of land ownership over the ensuing decades. At the same time, up to the turn of the century, the economy of Austro–Hungary (and the wider world) underwent a sustained boom, stimulating demand for agricultural produce, cereal crops in particular, and thus creating an expanded market for farmers. Though cereal production in Hungary had traditionally been extensive, and thus confined to the larger estates, ways of boosting yields with the aid of new technology became increasingly attractive—a process that was further accelerated by the marked drop in agricultural prices that was experienced during a prolonged slump in Europe around the turn of the century, which hit two of Hungary's principal staples—cereal and wool—particularly hard. Farmers, both large- and small-scale, strove to make up for lower prices with higher yields and in the livestock sector, in response to a still growing urban demand for meat and dairy products, with more intensive animal husbandry practices. Achieving such improved productivity called for better replenishment of the fertility of regularly cultivated land, the breeding of new, higher-yielding crop varieties, a shift in the pattern of livestock breeds that were raised, and minimisation of losses from epidemics and animal diseases. The greater complexity of production and rapidly growing demand in turn obliged farmers to gain ever more basic knowledge and new skills.

The huge advances that were made in the natural sciences, economics and engineering during the early part of the nineteenth century saw the agricultural sciences start to emerge as a distinct branch of knowledge around the middle of the century. By the latter half of the century, Hungary could already boast a number of outstanding agronomists who played a major part in modernising the country's agricultural base. Their work in soil science, plant breeding and animal health left an enduring imprint both on the way this field has evolved in Hungary right up to the end of the twentieth century and on the high esteem in which its achievements are held outside the country.

Two Soil Scientists of Renown

Elek 'Sigmond (1873–1938) was noted as the first occupant of the new chair of agricultural chemistry at Budapest's Technical University. The young chemist had first come to notice in 1899, when Sándor Cserháti, a teacher at the Mosonmagyaróvár Agricultural Academy, invited him to set up and head a chemistry laboratory for the school's experimental plant-breeding station. There 'Sigmond studied the nutrient requirements of maize as well as the characteristics of alkaline soils and the effects of irrigation. His first major achievement, reported in 1900, was to refine the celebrated French soil scientist Schlösing's method for determining the available phosphorus content of soils—a method that became standard around the world for the best part of 50 years. However, it was 'Sigmond's research into the chemical structure and physical characteristics of alkaline soils where he truly made his mark, with his book on the extensive domestic alkaline lands and their amelioration being widely published in translation (including a 1927 edition by the University of California

Press). An interest in soil classification and mapping from early in his career bore fruit in the classificatory scheme that he completed in 1932, the radically new theoretical underpinning for which was set out in *Általános talajtan* (General Soil Science, 1934). As the first systematic classification to encompass all of the world's soil types, 'Sigmond's 700-page treatise, which has been translated into many languages, has been an indispensable reference work for soil scientists ever since. Building on that, he also went on to work out the artificial fertiliser regimens that were indicated for land based on soil test results.

Intensive cultivation became crucially dependent on accurate knowledge of the local environment, its soils above all, and hence on detailed surveys and mapping of the geographical location of soil types, as well as developing a sound theoretical basis for effective soil conservation. Lajos Kreybig (1879–1956) was tireless and highly influential in both campaigning for and implementing this multifaceted and wide-ranging program in Hungary. Initially, though, in the wake of World War I and the country's ensuing economic difficulties, the climate was unreceptive to the pragmatic temperament he brought to these ambitions, and so, retiring to his family estate, Kreybig followed his own instincts in attempting to put his chemical insight to good use. He perceived clearly that it was little use the country having soils of intrinsically superior quality to those of its neighbours when incorrect conservation was resulting in substantially lower yields. This drove him to undertake agrochemical and soil biological research in an experimental laboratory that he set up on his own estate, carrying out comparative field trails of the efficacy of natural manure and artificial fertilisers in improving crop yields and soil quality.

Kreybig's crowning achievement, however, was to catalogue the characteristics of Hungary's soils, producing a preliminary map. He went on to make preparations for a national survey of soil samples, with the aim of publishing a 1:25,000 scale map of the findings. In 1934, he made a start on the job on his own initiative, having personally trained up the many technicians required to process the huge number of samples. From the outset, already in the 1930s, Kreybig paid great attention to catering for the various potential uses to which the maps might be put, marking features relating to the water management of an area as well as the chemical composition and physical properties of the soils. War interrupted the project, and it was finally completed only at the beginning of the 1950s. Many of the conclusions for soil conservation drawn in *Az agrotechnika tényezői és irányelvei* (Factors and Guidelines for Agricultural Engineering, 1953), his major synthesis of this work, are still valid today.

Improving Wheat Varieties

Cereal crops, especially winter wheat, have long been vital to Hungarian agriculture as a major export commodity. It was hardly surprising, then, that intensive efforts to raise wheat output should have got under way in the nineteenth century and particularly, following the slump in wheat prices at the end of that century, on the eve of World War I. Farming interests had become increasingly vocal in demanding that the government do more to support work aimed at improving wheat yields and quality. By timely coincidence, three Hungarian plant breeders who were to win international names for their improved wheat strains started their activities during this period.

Elemér Székács (1870–1938) originally sought, by careful selection and cross-breeding of strains, to develop an earlier-ripening, less rust-susceptible strain from amongst the varieties that were then widely used in large-scale production. The impressive results that he speedily achieved spurred him to extend the work, collecting new wheat types from over a hundred places and subjecting them to a battery of tests in a search for more disease-resistant types. By the 1910s, the higher yields and better quality wheat attainable with Székács's strains, by virtue of their rust-resistant properties, had led to their completely displacing the former varieties sown in the entire Great Plain region. As these strains were still being used in production even after World War II, this contribution to wheat improvement endured over more than half a century. Demonstrating his wide range of interests, Székács also undertook useful work on shortening the ripening time of maize and on improving barley, oat, grass and lucerne (alfalfa) varieties, and moreover takes the credit for creating and instigating the activities of a network of plant breeding stations around the country. He was also the prime mover behind a national exhibition of Hungarian plant breeders that became the model for subsequent agricultural shows where they could keep domestic and foreign producers informed about new varieties of crop seeds.

Better known abroad, however, is László Baross (1865–1938). Though that fame rests on his improved wheat strains, prior to this the young Baross had already produced excellent varieties of maize and sugar beet. At the beginning of the twentieth century, the bulk of Hungary's wheat lands were under an assortment of domestic and imported foreign strains but due to their poor characteristics and susceptibility to disease, yields were often low. Baross set out to selectively breed out these negative characteristics from the old strains, but his early results prior to World War I, whilst encouraging, were not conspicuously successful. Hungary was by then starting to lose her export markets for wheat because Europe was being flooded by the high-quality grain yielded by the Manitoba variety, and thus, in effect, from the wheat-producing provinces of Canada and the northern states of USA. Having obtained a small amount of the seed in 1915, Baross immediately realised that he could use this strain as a basis for eliminating the defects of Hungarian wheats. By crossing the Manitoba variety with his own earlier wheat strains, Baross succeeded in producing more frost-resistant winter and spring strains of what was marketed as the 'Bánkúti' variety, whose suitability for the local conditions, and the unsurpassed excellence of the flour that it gave, was borne out by over a quarter-century of subsequent use in Hungary (even as late as 1957, 57% of the land under wheat was still being sown with the Bánkúti strains). The summit of the recognition that Baross won for his work was the Gold Medal for 'Best Winter Wheat' that was awarded to the Bánkúti 1201 variety at the World Wheat Exhibition at Regina, Canada, in 1933.

Plant Breeders and Horticulturists

Rudolf Fleischmann (1879–1950) is considered to be the founder of modern plant breeding in Hungary, being responsible for introducing many new techniques into this area. The earliest phase of his activities began in 1906 with his appointment as steward to a large estate, where he worked on maize improvement, producing a strain that gave a vastly superior crop to the varieties being cultivated domestically at the time. He had also started to look at improved wheat and rye strains when his friend, Elemér Székács, persuaded him to head the Plant Breeding Station then being set up in Kompolt, between Gyöngyös and Eger in northern Hungary, where he continued to develop a wheat which was a cross between one of Székács's strains and an American variety, eventually coming up with the F-481 strain of winter wheat, which grew well on the country's alkaline soils and yielded a good bread-making flour and so enjoyed widespread popularity for some 20 years. The efforts to improve maize were also taken further, with Fleischmann experimenting with hybrid maize varieties already in the 1930s, even though the large-scale cultivation of such crops did not emerge until the 1950s, following the collectivisation of Hungary's agriculture under the Communists. In developing these varieties, he paid much attention in the first place to strains or hybrids which gave optimal yields under the widest possible range of soil and climatic conditions, though subsequently the nutritional content of the grain and disease resistance were increasingly the main targets. Many of the maize strains and hybrids that he introduced were still being cultivated decades later.

Though arable crops have always formed the dominant sector of Hungarian agriculture, horticulture has also become a major source of earnings. From the turn of the nineteenth into the twentieth century, in parallel with a surging demand brought by urbanisation and rapid population growth, intensive fruit-farming gained greatly in importance. Previously, domestic fruit production had been undertaken without bringing any particular know-how or expertise to the job, with few agronomists feeling inclined to tackle the task of addressing the sector's needs. One who conspicuously did was Máté Bereczki (1824–95), who, after military service during the 1848–49 War of Independence and two years as a fugitive from the Austrian reprisals that came in the wake of its failure, devoted the rest of his life to systematic studies of fruit species and their varietal characterisation. In the course of this, using his excellent connections with Hungarian and foreign fruit growers and breeders to gather the best that Europe had to offer, he put together what became the first national collection of fruit varieties from the specimen material that he amassed. By recording and publishing the massive body of knowledge he had acquired in the four volumes of *Gyümölcsészeti vázlatok* (Outlines of Fruit-farming), which included descriptions of some 800 different varieties, Bereczki laid the foundations for the discipline of pomology in his own country and offered international experts insight into the country's native varieties.

János Mathiász, Acclaimed Grape-Vine Ennobler

At the start of his career, János Mathiász (1838–1921) collected some 1,600 varieties of vinestocks within just a few years. Later he moved to Mád, in the Tokaj region of north-eastern Hungary, where most of his energies went into fighting the newly emergent threat of phylloxera. This vine pest had appeared in Hungary in 1875, and in just a few years destroyed more than 40% of the country's vineyards, wreaking particular devastation in the hill areas that were the source of its great traditional wines. After a decade of fruitless struggles, it became clear that new approaches would have to be adopted to achieve any enduring success. By then it was common knowledge that the pest did not spread amongst vinestocks planted on sandy soil, and so, encouraged by a vine-grower in Kecskemét, Mathiász bought a 30-hectare (75-acre) tract of land in the interfluvial region between the Danube and Tisza and became one of the pioneers of developing a sound, phylloxera-free basis for viniculture in sand, using his extensive collection of varieties to turn himself into a leading world expert in selecting suitable pairs to cross.

With the demand for dessert grapes rising towards the end of the century, Mathiász realised that varieties then being cultivated for that purpose in western and southern Europe could not be simply transplanted to Hungary. He therefore sought to create new varieties by taking their positive characteristics and crossing them with domestic stocks in order to fit them to local conditions, always with profitability as a prime consideration. His first significant dessert grape variety, which he named 'In Commemoration of Hungary's Millennium', met with instant international approval, after which Mathiász introduced a new variety practically every year. In the course of his working life, he created some 3,500 hybrids, providing detailed description of some 180 of them. So successful have they been that even today 67 of his varieties are still known, and twelve of them are found around the world, from

California to the Crimean peninsula. His 'Szőlőskertek királynője' (Queen of the Vineyards), the last variety that he introduced (1916), is now the most widely planted of all stocks in Hungary, as well as being widely cultivated in Italy, Romania, Yugoslavia, Bulgaria, France and the USA. Just as important as the surviving varieties, however, is the fact that Mathiász's varieties continue to be used as root stocks for new hybrids, both in Hungary and elsewhere.

Great Names in Animal Husbandry

The boost to the livestock industry that came with growing demand for animal products—initially wool, but later also meat and dairy products—in the latter part of the nineteenth century necessitated radically improved husbandry, triggering a remarkable upsurge both in the provision of modern veterinary training and in research to advance the knowledge base of the veterinary sciences.

An outstanding figure in both areas, Ferenc Hutÿra (1860–1934) devoted most of his attention during the early years of his teaching career at the Veterinary College in Budapest to reforming the standards that students seeking to qualify in the profession had to aspire to—modernising reforms which brought a major and, for the time, remarkably progressive overhaul of the content of the courses taught at Hungary's main veterinary school. Amongst several key textbooks that he wrote over his lifetime, the massive tome *Állatorvosi belgyógyászat* (Veterinary Internal Medicine, 1894–96), especially in its second, revised edition (1904), with his pupil József Marek as a co-author, had the signal virtue of offering new perspectives on the subject by providing a synthesis of the existing disputes on many topics, which made it an ideal basic manual and reference for veterinary training and practice in many parts of the world for over fifty years.

Hutÿra's research work covered a wide diversity of areas, but his most enduring achievements were in the field of infectious diseases.

Hungary's pig stock was decimated in the closing years of the nineteenth century when swine fever (hog cholera) was introduced from outside the country. Hutÿra was the first to demonstrate (in 1906) that the disease was caused by a viral infection and, working with his colleagues, went on to show that the best protection against it was bestowed by the hyperimmune serum won from pigs exposed to infection with an attenuated viral strain. They developed a cost-effective method for producing the serum, which led to its being marketed from 1909 onwards. For many years this was adopted as a prime defence against swine fever in Europe, until it was supplanted by a crystal violet vaccine (and that in turn by modern-day eradication programs).

József Marek (1868–1952) made his mark early on with his work on diagnostics, establishing a place for 'electrobioscopy' (the application of an electric current to diagnose morbid conditions) in veterinary practice around the turn of the century, but became more widely known as co-author of Hutÿra's above-mentioned book. Amongst his many original contributions to the discipline, he designed a number of superb instruments that are still used today for diagnostic purposes. These included a rhinolaryngoscope, which greatly facilitated the direct examination of respiratory problems in large animals, and, to assist his work on the differential diagnosis of gastro-intestinal diseases in horses, a stomach-tube (sometimes referred to as a Marek tube) for emptying the gastric contents. However, he is best remembered for his discovery of the avian tumour bearing his name, a viral disease whose outbreaks cause high mortality in poultry, most commonly amongst broilers between the ages of three and eight months. Ever since his first description of its symptoms and pathological lesions in 1907, it has been referred to as Marek's disease and its etiological agent as Marek's disease virus. He also devoted many years to searching for an effective measure to combat liver fluke, then a common and highly damaging parasitic infestation of ruminants, especially sheep and cattle. He eventually lighted on a highly effective substance known as filicin, extracted from the male shield fern, working out a method for effectively delivering this by mouth. The medication (which was marketed by the pharmaceutical company Chinoin under the name Distol) enabled deaths due to liver rot to be virtually eliminated between the two World Wars, bringing further recognition to the already well-known professor. Many of Marek's contemporaries, however, considered his greatest achievement to have been to elucidate the pathology of rickets in young animals.

Hutÿra's mantle as the dominant figure in veterinary teaching and research passed to Rezső Manninger (1890–1970), whose main fields were infectious diseases, bacteriology, immunology and epidemiology. Early on in his career, in 1918, he developed a vaccine against fowl cholera, and highlights of his subsequent work demonstrates that the viral agent of foot and mouth disease occurs in multiple strains and characterises the biological properties of numerous pathological strains of bacteria. Manninger also left an enduring legacy through his discovery and elucidation of the properties of a string of pathogenic viruses, whilst in the field of epidemiology he carried out significant work in investigating the spread of, and developing control measures for, fowl typhoid and paratyphoid, brucellosis, swine fever, rabies, and transmissible gastro-enteritis of pigs. He also contributed to the production and testing of vaccines against foot and mouth disease, tuberculosis, swine fever, swine erysipelas, anthrax and fowl cholera.

János Mócsy (1895–1976) likewise tackled an unusually wide range of research topics, including parasitic diseases, clinical toxicology, and a comprehensive study of equine infectious anemia before going on to explain the etiology of the paralysis seen in some animals after anti-rabies inoculations, to look at various aspects of viral biology, and to explore ways of controlling bacterial diseases in poultry stocks.

Agricultural Research Today

The basic directions that were adopted by agricultural research in Hungary during the nineteenth century have been pursued down to the present day. This largely stems from the fact that most of the country's land area is superbly endowed by nature for agricultural production. The overriding exigencies of maintaining food supplies have always required that agricultural production, and any research related to it, be permitted to carry on substantially regardless of momentary economic and political considerations. The change in régime at the start of the 1990s, however, was attended by a marked disruption, with the conjunction of dramatic shifts in the ownership of farmland, a drying-up of available capital, and falling output all working together to severely reduce the agricultural sector's demand for, and even its ability to take advantage of, research results. Meanwhile, drastic cuts in the state budget, though they primarily affected the funding of basic research, did not leave applied research unscathed. At the same time, an increasingly consumer-oriented food supply chain and environmental concerns have been setting a new agenda for which a strategy can only be elaborated by a concerted effort that involves the whole spectrum of research approaches, from projects in basic biology and chemistry, across the many branches of agricultural research through to economics and sociology. All civilised societies set improvements in the quality of life as a fundamental goal, and one of the cornerstones for that has to be the production of adequate supplies of wholesome food. Rational land usage and amelioration, alongside raised outputs of plant and animal products, are certainly prominent considerations here, but maintaining a balance between agricultural production and nature, in order to ensure a sustainable basis for future production, has to be equally significant.

The central goal of soil science research in Hungary is now to gain the understanding required to allow the controlled regulation of soil processes. This work encompasses accurately surveying the present status of the country's soils, continuously monitoring changes, learning more about the flows of materials and energy that are taking place, developing a framework for modern plant nutrition, and preventing soil contamination.

Research into plant cultivation and breeding is taking advantage of the latest tools that biotechnology offers, alongside the classical approaches of former times, with the aim of creating, characterising and improving new genetic sources. Biotechnological techniques utilising cell culture and selection of molecular markers have come into widespread application. In crop technology studies are being undertaken to ensure that production is sustainable as well as cost-effective. Ecological research has come to the fore, with the aim of showing the physiological, biochemical, genetic and agronomic impact on cultivated plants of adverse environmental effects that may ensue from any changes in the atmosphere or soil quality.

Biotechnology has been gaining an ever greater role in agricultural research since the 1980s. Basic and applied research in this area has been targeted principally at producing genetically modified varieties of cultivated plants and using reproduction biology to raise the fertility of livestock. Veterinary research is employing the findings of molecular biological investigations to pursue modern approaches to combating viral, bacterial, mycoplasmal and parasitic pathogens, as manifested by novel types of vaccines, immunological studies and the development of up-to-date diagnostic tools.

8. Engineering: a Repeatedly Reborn Phoenix (György Kovács)

The story of engineering in Hungary over the past century and a half built on a legacy from the immediately preceding era in several important respects. In the mid-nineteenth century, engineering was defined chiefly, even preponderantly, by the work of land surveying and regulation of the country's waterways. Large areas of Hungary's land surface were marshy, and with its abundance of rivers the country continually faced a threat of periodic inundation. Of all the initiatives made by Hungary's great modernizer, Count István Széchenyi, on resign his army commission, perhaps the single most important one was to drive through plans to regulate the most troublesome of the large river systems, that of the Tisza, to the point where the huge scheme was put in hand. The detailed work itself was planned by a number of distinguished engineers, amongst whom Pál Vásárhelyi (1795–1846) was a leading figure. Primarily an expert geodetic and hydrological surveyor, he had previously worked with Széchenyi on a scheme to make the Iron Gates on the lower Danube navigable, allowing uninterrupted shipping between Budapest and the Black Sea. The actual riverbed of the Tisza, due to its meanderings, was then twice the length of the valley itself, and Vásárhelyi's survey work allowed him to draw up a scheme to cut a shorter, deeper course by constructing a series of connecting canals and banked river defences. This not only substantially remedied the major problem of dealing with the run-off of floodwaters but also made the Tisza navigable. It is a testament to the soundness of Vásárhelyi's basic plan that not one breach of the defences occurred along its critical sections in the Great Plain area in the period between its final completion, in 1889, and 2000. His tragically premature death was largely brought on by the stress of fighting the obstructionist tactics put up by selfishly short-sighted feudal landlords of the region and the bureaucrats who served them.

Education was key to laying the groundwork. In that respect, Hungary has prided itself on keeping at the forefront of the discipline, at least within Europe and possibly even the world. The Royal School of Mines founded at Selmecbánya in 1735, supplemented by new chairs in mineralogy and chemistry in 1763 (forestry was added in 1846), was the first college of its kind anywhere in the world, which even at the close of the century was still regarded as a model for France's first polytechnic college. Amongst other distinctions, in 1786 it hosted the first international congress on mining and the extraction of precious metals on the initiative of Ignác Born (1742–91), its renowned professor of mineralogy. Born had made his name by devising an amalgamation process which utilised mercury to extract precious metals from ore. As well as being a man of learning and member of a number of prestigious scientific societies (including the Royal Society in London), Born did much to popularise science through his writings.

The Dawn of a New Era

It was through the foundation of an Institutum Geometrico-Hydrotechnicum by Emperor Joseph II at Pest in 1782 that Hungary first acquired a school of engineering, which had the right to confer degrees in the discipline. However, as the result of a reorganisation, four years after the Joseph Industrial School had been set up in Buda to remedy the country's chronic shortage of engineers (in reality the equivalent of the skilled mechanics turned out by today's technical middle schools), the Institute in 1840 was merged into the school, which effectively ended degree courses in engineering over the next twenty years, until József Eötvös, the great reform-minded minister of education obtained parliamentary approval in 1870 to establish a Technical University at Budapest. Teaching was already under way at the new university by the following academic year of 1871/72. It remains to this day one of the recognised world centres of excellence for the training it offers in engineering disciplines.

When wars rage, the Muses may be stilled but technology thrives. Hungary's 1848–49 War of Independence had the support not just of the general public but also many of the country's prominent scientists. The army's corps of generals, though not particularly large, included two qualified engineers, whilst the commander-in-chief, Artúr Görgey, was himself an excellent chemist. The majority of the army engineers (amongst others, the mathematician János Bolyai) who sided with the Revolution had studied at the Ludovika Military Academy in Budapest, which was established in 1808, but many had graduated in Vienna, so it was not uncommon for teacher and student to confront one another on the battlefield. Whereas Prince Ferenc Rákóczi II had no option but to turn to French military engineers for their assistance during the country's earlier liberation struggle, almost a century and a half earlier, by 1848 Hungary was able to call on the services of her own professionals. Over the intervening period, the science of constructing and equipping fortifications had advanced greatly. This was also when the country's first major armaments factory was built, to a plan conceived by István Széchenyi. In that connection, the figure of Áron Gábor (1810–49) became literally the stuff of legend and folk song during the revolution. An artillery officer and superbly skilled craftsman from Transylvania, with only rudimentary tools at his disposal, not even a boring machine, he contrived to furnish General Bem's army with seventy working cannons that played no small part in its initial victories against the Austrian army in 1849.

Ganz and Mechwart: Founders of an Industry

Industrialisation finally got under way in Hungary, albeit very slowly, only around the middle of the nineteenth century. In 1844, Abraham Ganz (1815–67), Swiss by birth but one of the towering figures in the country's industrial history, established a foundry that was to form the core of what became one of its largest industrial conglomerates. In 1853 Ganz introduced a new process—previously employed only in North America—for chill-casting railway wheels, and this became a major factor in the rapid construction of Hungary's railway network. Ganz's works manufactured over 100,000 wheels in 15 years, exporting them to 60 countries around the world. It also cast some of the elements used in building the Chain Bridge, another of István Széchenyi's pet projects, which still gracefully defines the Danube waterfront of Budapest. (The Scottish engineer Adam Clark who designed and constructed the bridge is remembered in the name given to the square lying at its Buda bridgehead.)

Born into a large family, Ganz was a foundry apprentice and then a carpenter in his native land before making his way to Pest at some point in the early 1840s, where he worked for a series of employers before setting up on his own account. Though ever an adapter by temperament, he did not shy away from purchasing patent rights where necessary. Running the company on strict patriarchal lines, he travelled extensively on its business, showing off its goods and making deals. He was a widely revered figure in his adopted home by the time that, faced with a grave but not hopeless crisis, he chose to take his own life.

On Ganz's death, András Mechwart (1834–1907), also Swiss-born but a qualified mechanical engineer who by then had been working eight years for the firm, was appointed general manager of the factory. Whilst maintaining his predecessor's style of management, Mechwart successfully combined superb business acumen with engineering insight in knowing when and where to take risks. Though he did not deflect the company from the path on which it was running, he was responsible for bringing in two further major innovations. In 1874, he purchased the patent rights on a flour roller-mill design but modified it substantially, most crucially by utilising in its construction steel made by the factory's own famed chill-cast process. So successful did this prove that by the end of the century the Ganz Works had built and sold some 30,000 mills—to this day, one of the greatest feats of Hungary's engineering industry. The second far-sighted step that Mechwart took was to set up an electrical engineering division. In so doing, he gave the company an early entry into what was then still a fledgling field of technology, setting it on a path of expansion into what would nowadays be called a 'global' enterprise. The Ganz Electrical Works, as it became, would export its products not only throughout Europe but also to Africa, India, and even America. Many of the markets that were gained then, at the turn of the nineteenth century, remain basic markets for the successor companies of today.

Missed Opportunities

A major factor in Hungary's technological advancement during the industrial boom of the latter half of the nineteenth century was undoubtedly the general excellence of the teaching staff that was then active in higher education. Even in that context, however, Ányos Jedlik stands out for his inventiveness in dreaming up a stream of ideas of the sort that nowadays keep whole branches of industry busy, yet his name remains largely unknown to posterity. He constructed a single-phase (i.e. direct-current) dynamo, working on the principle of self-induction, some six years before Siemens and Wheatstone, as well as electric motors and a high-voltage power source. Sadly, however, neither he nor those around him grasped their industrial significance; of all his achievements, perhaps the longest-lived were the extraordinarily fine-resolution optical gratings that he produced for astronomical use in the lens systems of refracting telescopes.

József Petzval (1807–91) followed a different career path, qualifying as an engineer at the University of Pest but then settling in Vienna, where he worked until the end of his life. Petzval's major achievements were in the field of optics, where he used very precise mathematical computations to develop a system of camera lenses (later named after him) which gave so little spherical and chromatic aberration that for a long time they were unsurpassed. This system was utilised by the Voigtländer company of Vienna in making their cameras, but Petzval gained no profit from this, having also failed to register a patent.

The Great Water Engineers

As was indicated in the opening paragraph of this section, surveying and water management have long had a special place in Hungary. Certainly down to the 1989–90 change of regime, river and canal engineering, flood protection, irrigation and water use have counted as the most successful and highly developed of all branches of Hungarian engineering, the area in which Hungarian experts have long been in highest demand from abroad, with the results of their valued work visible in Tunisia, Iraq, and many other countries.

Pál Vásárhelyi's skills have already been noted, but his close contemporary József Beszédes (1786–1852) was equally one of the pioneering water engineers. To his credit is a long list of sites within the old borders of Hungary where he undertook projects whose usefulness has endured to the present day. He was the first to build an industrial canal near Arad (Oradea, Romania); regulated the Sió–Sárvíz–Kapos river network; carried out flood protection and regulatory work on the Danube, reclaiming some 160,000 hectares (400,000 acres) of marshland in the process; was involved in the regulation of the lower Danube; and drew up the first plans for a Danube–Tisza canal. His versatility is illustrated by the fact that he even set up a railway rolling-stock repair workshop, which was later to form the core of the Hungarian State Iron, Steel and

Machine Works (MÁVAG), founded in 1868 (and thus of the huge Ganz–MÁVAG conglomerate of the Socialist era). Beszédes was the first engineer to become a member of the Hungarian Academy of Sciences.

The dominant figure in this field around the turn of the nineteenth into the twentieth century was Jenő Kvassay (1850–1919). The Institute of Agricultural Engineers that he proposed and established became the basis for uniform state regulation of the profession. He brought a holistic perspective, seeing waterways as an integral part of the soil and Nature, and always regarding social advancement as the real goal of his work. Those activities resulted in more than 3 million hectares (7.5 million acres) being brought up to the standards required for modern farming and the reclamation of close to a further 1 million hectares (2.5 million acres). No less important was his contribution to draft legislation, and in particular the first parliamentary act to govern water use in 1885).

The turn of the century was the great era of construction for Budapest's Danube bridges. The Margaret Bridge, completed in 1876, was designed by French engineers (as indeed was the city's other major technical feat of the time, the great hall of Western Railway Station, designed by no less than J. Eiffel). The original Elizabeth Bridge, however, which was opened to traffic in 1903 and for 20 years held the record as the suspension bridge with the widest single span in the world, was designed and built by Hungarians. (After its destruction during the siege of Budapest, towards the end of World War II, it was rebuilt in 1964 to a design by Pál Sávoly.)

The advance of river engineering in turn boosted the importance of hydraulic engineering—one of the fields in which Donát Bánki (1859–1922) rose to prominence. He had started his career in the foundry section at the Ganz Works before being appointed a professor at Budapest Technical University in 1899. A prolific inventor throughout his working life, he is most closely associated

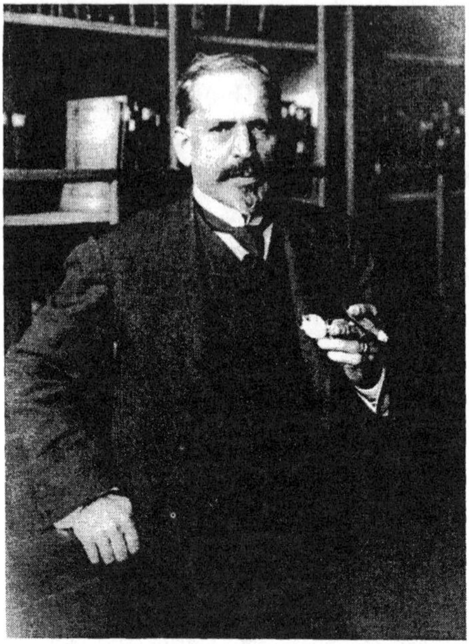

Donát Bánki

with two innovations in particular. The first was a simple and cheap radial-flow hydraulic turbine (1916), which permitted cost-effective power generation on rivers that provide only small or moderate heads of water (as on most of Hungary's rivers). In a Bánki turbine the water jet flows twice past the rotor blades, so the conversion of the water's potential energy into the mechanical work of the turbine is more complete. Many turbines of this type are still operated; indeed, the United Nations Organisation's aid agencies often advise developing countries to adopt this technology. Before that, Bánki did much work on improving the early internal combustion engine, being chiefly remembered, jointly with János Csonka (1852–1939), for being the first to describe a petrol carburettor in 1893, some six months before Rudolf Diesel in Germany. Five years later Bánki went on to invent a high-compression engine with a dual carburettor for vaporising fuel and water.

Electrotechnology: from AC Transmission to the Better Light Bulb

The electrical division at the Ganz works had the good fortune to have a large pool of highly talented engineers working for it, of whom four became widely known in their field. First of all, there was the famous team of Károly Zipernowsky (1853–1942), Ottó Titusz Bláthy (1860–1939) and Miksa Déri (1854–1938), whose sound insights on the superior merits of alternating current for public supply on electricity show that it can pay to go against the grain of accepted scientific opinion. During the early 1880s, most experts (including Thomas Edison) still considered that direct current was the key to the future of electricity supply. One of the main obstacles facing the Ganz team was that stepping an AC voltage up or down for different applications, until then, had been a tricky, highly inefficient business. The iron-core transformer that they designed elegantly solved the problem without incurring significant losses of

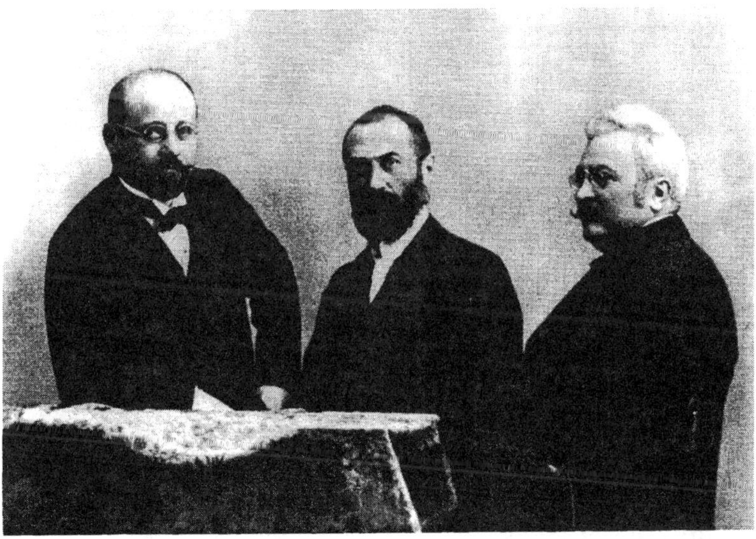

Miksa Déri, Ottó Titusz Bláthy, Károly Zipernowsky

power, thereby making it possible to plan and build electricity supply grids with relative ease, whilst Bláthy also invented a watt-hourmeter (1889)—based, in effect, on a small induction motor—to meet the practical need to measure the amount of electricity used by consumers.

With the initial problems of transmitting and distributing electricity overcome, the industry needed to drum up customers, and one of the most obvious early targets was street lighting in the larger towns. The first town anywhere in the world to install electric power for public lighting was Temesvár (Timişoara, Romania) in 1884. Although an Anglo–American firm implemented that project, by the time the town's citizens got the chance, four years later, to introduce electricity into their own homes, it was Ganz that undertook the work.

The adoption of AC for electricity distribution offered a fine opportunity to electrify the railways, even though DC has intrinsic advantages to AC for locomotive performance. In 1898, the Italian government decided to electrify a section of the Valtellina railway running beside Lake Como, and placed the order with the Ganz Works. Direction of the project was in the hands of Kálmán Kandó (1869–1931), the fourth of Ganz's top engineers, with his team developing, manufacturing and installing virtually all the major new items of equipment that this called for, from the high-voltage transmission lines and transformer stations to the electric locomotives and other rolling stock. So well did the three-phase 3,000-volt, 15 Hz scheme work when it went into operation in 1902 that the Italians ordered the electrification of a further 2,000 km of track using the system. Kandó stayed on in Italy to work for the Westinghouse factory at Vado Ligure, where he developed a single-phase 16,000-volt locomotive. The Ganz Works itself long manufactured (and, on a lesser scale, still does) a wide range of electric motors and generators.

The founding of the United Incandescent Lamp Co. in August 1896 proved to be a significant date in Hungary's industrial history, given the position it has enjoyed for many decades as the biggest earner of all the country's factories (in 1988–89, before its acquisition by GE, it had 5% of the world market for light bulbs). Its major success came after Ignác Pfeifer (1868–1941), the former professor of chemical engineering at the Technical University, was recruited to its staff in 1912 and proceeded setting up a world-class R & D laboratory, which became the basis for the firm's competitive edge in modern lamp design (e.g. the tungsten filament and krypton-filled bulb), as already sketched earlier. Much of the credit must also go to Lipót Aschner, who took over as the managing director in 1918, for his foresight in letting his brilliant researchers follow their heads as well as his gifts for spotting commercially significant innovations, organising production and selling the company's product. After World War II, four engineers working at this laboratory used their free time to experiment with a thermionic valve design that retained its performance with half the waste heat losses—a trick that other international companies only mastered five years later. That sort of ingenuity helped the company to secure regular supplies of import items vital to its production (e.g. the electron-releasing material with which the cathodes of vacuum tubes are coated even during the depths of the Cold War).

The telecommunications industry in Hungary was very quick off the mark. Budapest had a telephone network as early as 1881, and by 1893 (almost three decades before the first radio broadcasts) users could even subscribe to a unique telephone news service—the idea of Tivadar Puskás (1844–93) and his brother Ferenc (1848–84)—which offered news bulletins and even performances of live music broadcast from the central exchange. The suggestion that telephone subscribers be linked by a central switchboard, incidentally, had originally come from Tivadar Puskás when he was working with Edison. Even though all the early telephone equipment in Hungary (as elsewhere in Europe) came from America, local companies manufacturing the various items of equipment, and even entire systems, sprouted like mushrooms, some becoming targets for major international companies as they tried to establish a footing in the country. The line taken by the government, particularly during the period when Gábor Baross proved to be a far-sighted and energetic minister, first at Public Works (1886–89) then at Commerce (1889–92), was highly supportive of that development. Baross recognised the importance of telecommunications, and gave a massive boost to the industry in its early days by placing substantial public orders for its equipment. It was he who granted the state postal service a monopoly in telephony and in 1891 set up its substantial research institute, which continues to serve the industry to this day. (It was here that Georg von Békésy carried out the work on the physics of human hearing for which he received the Nobel Prize.) By 1908 Budapest could communicate with sea-going Hungarian ships by wireless telegraphy; in 1914 the country opened its own first telegraph transmitter station; and by 1928 the first automatic telephone exchange was in operation.

Success in the Machine Industry

Given the country's long tradition in mining and metallurgy, it is perhaps not surprising that the machine industry in Hungary took off at much the same time, and with similar entrepreneurial verve and speed, as the electrical and telecommunications sectors. One measure of the rapidity of growth is that by the early 1900s as much steel was produced and consumed as raw iron. Two of the pivotal firms, the Ganz Works and MÁVAG, have already been mentioned. In the seven years after the 1867 Compromise, capital investment in the Hungarian machine industry was five times what it had been over the previous seven years. Substantial inroads were made in export markets, which stood the industry in good stead during the

Europe-wide economic crisis of 1873 and indeed, later on, in helping it pull rapidly out of the World Slump of 1929–34. Ever since then the machine sector as a whole has been the major contributor to the national revenues. More or less from the start, it has concentrated on three main areas; the manufacture of motor vehicles, agricultural machinery and engines, but it is possible here only to flag some major milestones of this development.

The first steam locomotive to be made in Hungary was built in 1873; the 1,000th came off the production line in 1896. Automobile manufacturing started in 1906 (led by Bánki and Csonka), whilst by 1913 aircraft were just starting to be manufactured. There were also several significant shipyards, some even building vessels for customers abroad. The initially fragmented agricultural machinery sector underwent the usual string of mergers to end up with a few larger firms, the Kühne Works in Mosonmagyaróvár prominent amongst them. Ede Kühne, who came to Hungary from Berlin in 1862, constantly strove to keep his factory and its products at the state of the art, picking up medals and markets at a long series of major international exhibitions.

László Láng (1837–1914), born in Pozsony (Bratislava, Slovakia), set up a workshop in Budapest in 1868 that within five years he managed to build up into a major enterprise, manufacturing steam boilers, steam engines and steam turbines (the latter were capable of developing up to 10,000 h.p. after 1910). The factory still exists under the ownership of Sweden's ABB group.

Joining the Ganz works as an engineer, and later becoming its director, György Jendrassik (1898–1954) was one of the towering industrialists who was swept into exile in Britain in 1947 (where he died fairly young). At the beginning of his career he produced internal combustion engines, including diesel motors, for track inspection railcars and ships. The extraordinary demand these enjoyed helped significantly in buoying the company through the Great Depression, but perhaps Jendrassik's crowning engineering achievement was the design he presented in 1938 for the world's first low-output gas turbine with an independent combustion chamber.

An earlier section has already referred to the tragic fate that befell one of the country's big industrialists, Gedeon Richter (1872–1944) after the German occupation in 1944, and this also overtook Leó Goldberger (1878–1945; he was amongst the very first Jews to be arrested and deported to an extermination camp, dying there shortly before it was liberated). Yet it was Goldberger's achievement to have turned the Hungarian textile industry from a net importer on to a net exporter, principally through his success with the then-innovative concept of creating a vertically integrated operation and by employing the best experts and latest technology that money could buy. The firm's products were well known in Europe, America, the Near and Far East, whilst Japanese companies, taken to law for duplicating its copyright patterns, had to pay out substantial fines in compensation. His factory in Buda was almost the only textile works in Hungary not just to survive the Great Depression but actually to expand and pay its shareholders dividends.

Technical Flair Cut off from the Wider World

The 1950s were a period when, for most Hungarians, contacts outside the country were infrequent and even then difficult, if not impossible, but nevertheless it marked a second major phase of growth for the country's industrial base. A whole string of R & D institutions and new factories were established; industrial output increased many times over; and Hungary became a net exporter of industrial goods, albeit in the highly artificial trading terms of the Comecon system of trade within the Soviet bloc. The new scientific schools and manufacturing traditions that evolved were robust enough to weather the hard jolts experienced after the 1989–90 change of régime and so lay the foundations for the present promising trends. It has become abundantly clear that foreign capital is not attracted to Hungary primarily for its low-wage labour face but rather for the creativity and professionalism that are found there (those who have other ideas tend to depart fairly soon). Naturally, that culture has continued, as before, to throw up particularly outstanding figures throughout the post-war era. To mention just a few examples: Ottó Benedikt, with his pioneering work on magnetic forces; Miklós Uzsoky, designer of the first Hungarian television transmitter; György Almásy, a superlative engineer who instigated the development and manufacture of microwave equipment in Hungary; as well as a string of first-rate professors teaching at Budapest's Technical University, including József Gruber, Géza Á. Pattantyús, Ádám Muttnyánszky and László Verebély.

Notes

1. A legal antecedent to the Technical University, the Institutum Geometrico-Hydrotechnicum, was established at Pest in 1782 as the world's first university-level engineer training institution, twelve years before the Paris École Polytechnique. This became an independent Technical University in 1871 (again, the first in the world), which was still the third largest institution of its kind in Europe at the end of the nineteenth century.
2. To commemorate their father's work, Kőnig's two sons set up what was the most prestigious mathematics prize in Hungary up till 1944.
3. The method for measuring surfaces produced by Geőcze in 1906, along with his subsequent investigations, laid the foundations for the modern theory of surface measurement.
4. The concept of the topological group links all three main branches of mathematics (algebra, arithmetics and geometry) through its interpretations of algebraical structure (group) and topological space (i.e. a set endowed with geometrical attributes), in which the basic arithmetic concepts are also defined.
5. The theorem states that all spaces of Lebesgue integral functions are capable of forming full limiting values within a certain interval; the limiting function of all convergent series of functions also belongs to the space.
6. Any two polygons of identical area can be cut up into congruent pairs of polygons.
7. How can a large number of points be placed within a square in such a way that the shortest distance between them is also the longest?
8. A whole library shelf could be filled with books on Erdős, but two in particular are accessible to general readers: László Babai's "Paul Erdős: In and Out of Hungary," in D. Miklós, V.T. Sós, T. Szőnyi, eds., *Combinatorics: Paul Erdős Is Eighty* Vol. II. (Budapest, 1996), pp. 7–95 and Paul Hoffman's *The Man Who Loved Only Numbers* (London, 1998).
9. This is an edited version of the chapter appearing in *The Voice of the Martians* (Budapest, 1997).
10. When Zipernowsky travelled to the USA to promulgate its advantages, the American electrical industry exerted pressure to ensure

that the country's first execution by electric chair was performed with an AC current in order to underline just how dangerous it was!

11. As professor of physics at Heidelberg, Lenard received the 1905 Nobel Prize for his demonstration that the electrons of cathode rays pass through atoms, and thus much of the space within atoms is empty. He also later showed that in the photoelectric effect electrons are released by the action of light, which Einstein subsequently interpreted by introducing the concept of photons.
12. Norman Macrae, *John von Neumann* (New York, 1992).
13. Quoted from Péter Hanák, "Az alapítások kora" (The Gründerzeit), in Péter Hanák, and Zsuzsa Szentgyörgyi, eds., *Gazdaság és társadalom a dualizmus korában* (Economy and Society in the Era of the Austro–Hungarian Monarchy), *Magyar Tudomány* 10–11 (1989), p. 806.
14. This final section was compiled by Gábor Palló, drawing particularly on information presented in two lectures delivered at a meeting organised to celebrate the fiftieth anniversary of the Department of Chemical Sciences of the Hungarian Academy of Sciences (MTA Headquarters, Budapest, 9 May 2001): Sándor Görög, *Kémiai Tudományok Osztálya: múlt, jelen, jövő* (The Department of Chemical Sciences: past, present and future), and Endre Kőrös, *Honnan és merre? A hazai kémiakutatás fél évszázada: emlékezés a múltra* (Where has it come from and where is it going? A half century of chemical research in Hungary: remembering the past).

Ferenc Vámossy

ARCHITECTURE AND MODERNISATION DURING THE PAST CENTURY AND A HALF

Hungary's Early Period of Modernisation: 1820–1867

In Hungary, the latter half of the eighteenth century brought a spurt of economic growth to what was still a largely feudal country, by then firmly lodged within the Habsburg Empire, a growing prosperity of culture, industry and trade that was accompanied by an advance of the urban middle classes already seen elsewhere in the Empire. The Enlightenment era (1780–1820) generated a demand for a classicising modernisation, a faster pace of urbanisation and urban dwellings, alongside the provincial world of the country mansion. The ensuing three decades of the 'Reform era', which set itself the goals of overhauling Hungary's feudal society, preserving and renewing the Hungarian language and national culture, and István Széchenyi's program for developing the industry, trade and transport, culminated in the struggle for liberty and restoration of the country's historic sovereignty and unity in the 1848–49 revolution and ultimately failed war of independence.

The wholesale cultural transformation that went ahead in Hungary during the nineteenth century was also manifest in architecture and its regulation. Archduke Joseph of Habsburg (1776–1847), Emperor Francis II's younger brother and initially vice-regent then (from 1796) palatine of Hungary, in 1808 set up a Municipal Beautification Commission for Pest in order to ensure that the first comprehensive planning outline for the town, drawn up by Austrian-born János Hild (1766–1811) in 1805, should be implemented within a framework of building permits. In 1839 the

Committee laid down the first regulations prescribing permitted building heights, the alignment of street façades, and spacing of window axes—a set of rules that would give the Classicist-style Pest then under construction a unified townscape and, as Archduke Joseph noted in one of his reports, make it the most splendid city in the Austrian empire. The most distinguished city builders of that period were Mihály Pollack (1773–1855), who designed many mansions, residential blocks and the *National Museum* (1837–46), and János Hild's son József (1789–1867), who designed numerous public buildings and almost a thousand dwellings. One of the jewels of this ever more metropolitan Classicist Pest, besides the Museum, was the unified development of the Danube bank by Hild, which, together with the Chain Bridge (*Lánchíd*) of Scottish engineer William Thierney Clark (1783–1852), István Széchenyi's choice of designer, was the outstanding achievement of the first half of the century.

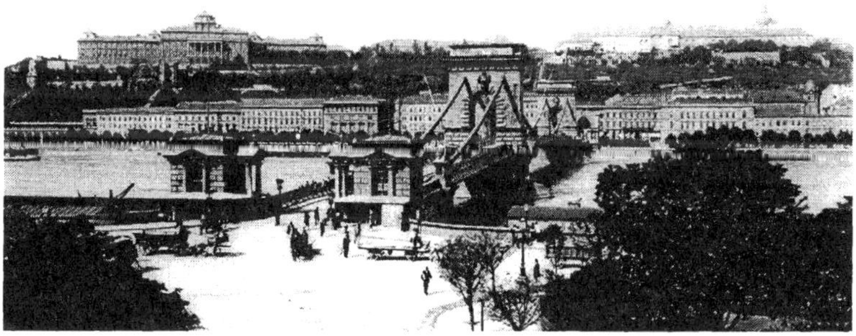

Chain Bridge

The restrained, pure forms of this Hungarian Classicism were to become distinctive marks of urbanisation across the country as the building of county halls, country houses, multi-storey town dwellings and a few big church complexes laid a kernel for future development. The *cathedrals* of *Esztergom* (1822–56) and *Eger* (1830–37) were the work of József Hild, whilst the Calvinist *Great*

Church of Debrecen (1805–21) was the creation of architect and military engineer Mihály Péchy (1755–1819). This characteristic *couleur locale*, which became widespread in domestic urban and rural architecture alike, was a diluted form of late Baroque—stuccoed brick buildings, with relatively little carved stone, serene and harmonious, rather than cold and ceremonial—which represented a tempering of the Reform era's pure, innovative spirit by scarcer available resources, both of finances and building materials, as compared to western Europe.

The Romantic architecture of the 1850s and '60s assumed a similar character. The historical Buda, on the Danube bank facing Pest, began to flourish similarly, with the construction of a goodly number of churches, dwellings and schools bearing the stylistic traits of a round-arched neo-Romanesque or a Gothic revivalist idiom. The most representative work of this short-lived period of National Romanticism was the *Vigadó* (Assembly Rooms, 1860–65), designed by Frigyes Feszl (1821–84), on the Pest bank of the Danube, built to replace Mihály Pollack's Redoute Building, which had been destroyed in 1849 by Austrian artillery firing from Buda Castle, reconstruction of which had grown into a national cause over the intervening decade. The ruins on the site were cleared away in 1860, and the new building born out of this desire to create a national architectural style was completed in 1865. With its imposing façade, with its arcade and elegant superstructures, its well-balanced proportions, its Moorish flavour and decor, his *Vigadó* is the most impressive example of Romantic architecture. A prolific architect who came from a German-speaking patrician family in Pest and was to turn against Classicism and later historicism as well, Feszl sought, in a manner not devoid of all originality, to express his Hungarianness in intellectual terms. Over the ensuing decades, its orientalising elements would place a stamp on the ornamental language of Hungarian synagogues. Attempts to devise a national architectural style based on vernacular forms were to be pursued as a program for more than half a century and were present in Hungarian architecture up to the end of the 1920s.

The search for expressions of national character, which had its origins in the eighteenth century, was followed only belatedly by Eastern Europe and Hungary, but it had turned into a nastily inflated flaunting of nationalistic sentiments by the time of the 1896 celebrations marking the millennium of the Hungarian state. The first attempt to devise 'Hungarian-style' capitals was made by Johann Schauff (1757–1827), a Czech-born drawing teacher at Pozsony (Bratislava, Slovakia), in a book he published as early as 1790. This naïve concept went unnoticed at the time, but the emotional drive for national forms became more cogent in the aftermath of the failed war of independence.

Capitalism was still at a rudimentary stage, whilst neither the press nor informed criticism yet had a role of any significance, and general assessment of architectural achievements was a long way off. It is characteristic of how far Hungarian culture had come, however, that all three above-mentioned architects, though coming from German-speaking families, accepted a Hungarian milieu and culture, and pitched their works to serve its nascent, distinctive self-assertiveness, at once national and middle class. The middle classes of Pest and Old Buda and some of Hungary's other towns were still German, but the Hungarian language had been renewed, and the receptivity of the culture prompted a switch in identity or cultural loyalty, a form of assimilation through identification with Hungarian culture which in due time was to feed through to the vast majority of the Jewish middle classes.

Architecture and Modernisation under the Austro–Hungarian Monarchy: 1867–1918

The Seventies and Eighties—The Founding Period

After eventually reaching the 1867 Compromise with Austria after the failed 1848–49 War of Independence, Hungary entered a period of vigorous boom that was sustained until 1914. The impetus and scope of that growth were epitomised by Budapest's birth as a new metropolis. The founding fever of the 1870s and '80s, the expansion of industry, trade and transport, and the urbanisation that was predicated on these, generated a pace of development similar to that of Chicago, making Budapest a European counterpart to that triumph of modernisation in America. The wave of change of this *Gründerzeit* marked its beginning with the unification of the hitherto separate municipalities of Pest, Buda and Old Buda into a single capital in 1873, underpinned by a general economic boom throughout the eighties, with the rapid build-up of the country's railway network amongst many other things.

A Metropolitan Board of Public Works, set up with prudent foresight in 1870 (and staying in existence until 1949), already the next year announced an international competition for a comprehensive city plan, with the creation of a system of ring roads and radial avenues on the Paris model, which was to define the direction of Budapest's development. Work on the key thoroughfare, first called *Sugár út* (Radial Avenue) later, and again now Andrássy út, began in 1871 and was essentially completed by the time the finishing touches were being put on the *Royal Opera House* (1875–84), the 'Palace on Radial Avenue'. Construction of the *Nagykörút* (Great Boulevard) started in 1872 but only really forged ahead from 1884 and continued into the 1890s.

Architecturally, this was the period when the Romanticism that had characterised the repressive 'Bach era' was supplanted by

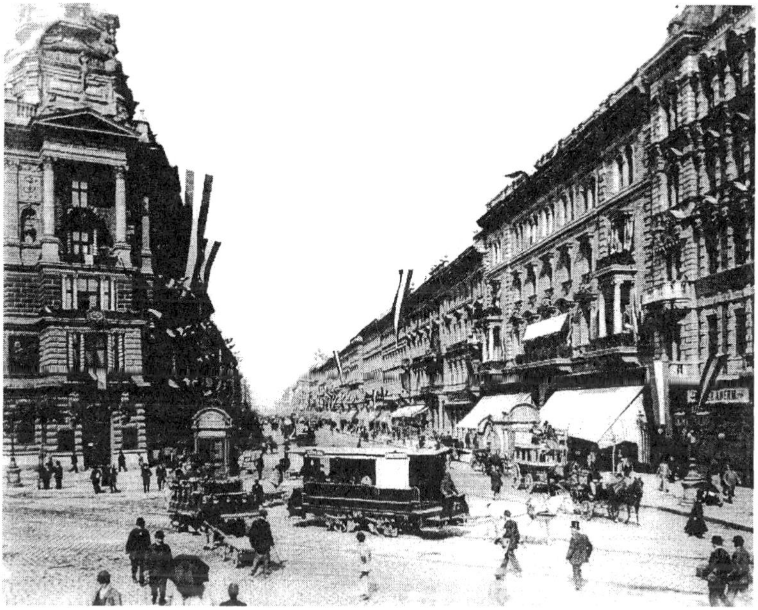

Radial Avenue

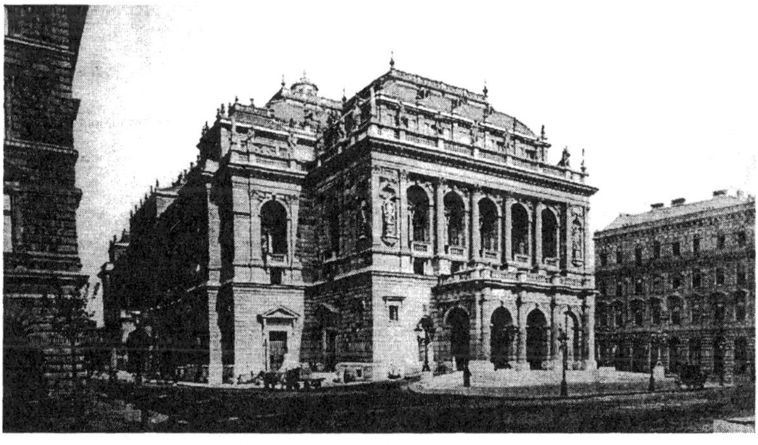

The Opera

Historicism. The Opera House by Miklós Ybl (1814–91), supreme master of the style, is the finest example of strict Hungarian Historicism. His death on 22nd January 1891 symbolically marked the end of an age the creative horizons of which had been defined by masters born in the early nineteenth century who had links to the world of guild craftsmen and a profound respect for historical traditions and ways of life. As Virgil Bierbauer (1893–1956) put it in 1937, in the first major synopsis of the history of architecture in Hungary, Frigyes Feszl was "the belatedly born last great son of Romanticism,"[1] and that Romanticism was indeed a spent force by the mid-1860s. Romanticist-inspired entries for an 1861 competition for new headquarters for the *Hungarian Academy of Sciences* on its present site, beside the Pest end of the Chain Bridge, including a plan from Feszl himself, were passed over. Instead, the competition was won by Berlin-based Friedrich Stüler (1800–65), who thus got to construct his neo-Renaissance building (1862–65), which inaugurated a new stylistic era even as Feszl's *Vigadó* was still rising. Since Stüler was unwilling to oversee the construction work itself, the Building Committee assigned the task to Miklós Ybl and the young Antal Szkalnitzky (1836–76).

Ybl was by then a fully fledged architect, with a string of work behind him. As a young man he had studied at Vienna's Technical University then worked for four years with Mihály Pollack, who had an enduring impact on him. His first commission, given by the Károlyi family, was the *Parish Church at Fót* (1845–55), which is still in Romanticist style, as was the *Unger House* on the Kiskörút (Little Boulevard), near the National Museum. His first masterpiece, the *Main Custom's House* (1870–74) on the Danube embankment, at what is now the Pest foot of the Liberty Bridge, still bears discernible affinities to the Academy building's concept, which the British architectural historian Nikolaus Pevsner dismissed as provincial during his visit to Budapest in the 1960s. Built on the opposite Danube embankment, at the foot of Buda's Castle Hill,

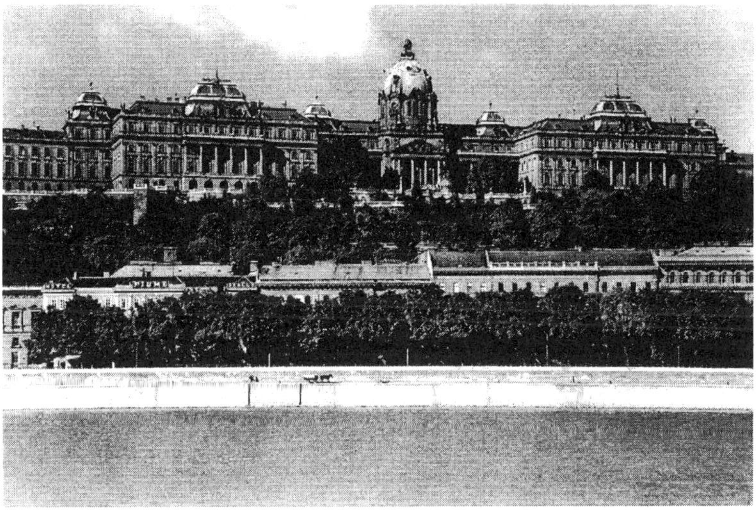

The Royal Palace on Castle Hill and the Danube embankment

Ybl's picturesquely dynamic *Várkert* (Castle Garden) *Bazaar* (1875–78) whose two pavilions and linking flight of steps, originally intended for commercial purposes but soon given over to sculptors' studios, invokes the colourful majolica world of the Italian Renaissance. Ybl's most important work, nonetheless, is the *Opera House*, whose calm and compact masses, its neo-Renaissance forms, the elegant treatment of its interior spaces, staircase, the main auditorium and the second-storey smoking room are typical of the mature art of a master who was by then in his sixties. In 1867 Ybl took over from József Hild the construction of the *Basilica of St. Stephen* in Pest's Lipótváros (fifth) district, but this protracted project was only finished by József Kauser in 1891–1905. In 1883 he produced a plan for the remodelling of the *Royal Palace* on Castle Hill, with the addition of a west wing facing Buda and a main courtyard. Ybl's concept for this grandiose, state-financed project was to be completed over 1891–98 by Alajos Hauszmann. (The palace was almost totally destroyed in the 1945 siege of Buda;

restoration work dragged out for over two decades until it reopened in 1972 to house its present art gallery, museum and library functions.) The valedictory address by the president of the Society of Hungarian Engineers and Architects at Ybl's funeral, sentimental though it may sound to modern tastes, aptly sums up his life's work and the importance of the era in which his masterpieces were built: "You were the one who raised architecture to an art in our country."[2] Since 1953, the highest distinction in Hungarian architecture, the Miklós Ybl Prize, has taken his name.

1871 marks the end of the period when architects could only gain professional qualifications and training outside Hungary. By 1890, Budapest's new Technical University, in those days named for the former palatine, the Archduke Joseph, was offering architectural training, which by the turn of the century was at the standard of the similar institutions at Munich and Zurich.

Architecture at the Turn of the Century

The 1890s opened a three decades long fin-de-siècle period which was to place a decisive stamp on the entire twentieth century and was thus creatively the most significant spell in the past two hundred years. This was Hungary's great age of modernisation, differing even qualitatively from the *Gründerzeit* in that it brought a genuine breakthrough after four centuries of relative decline and historic setbacks.

Preparations for the millennium celebrations (which included continental Europe's first underground railway, constructed by the 'cut and cover' technique, and the introduction of tramway travel), its growing attractions as a magnet for migrants, and an accumulation of various other effects further increased the tempo of construction in Budapest. Development of the *Danube embankments* (which, along with the Castle, have been listed as World Heritage sites since 1990) proceeded unbroken from the time work on the

Vigadó began up till the First World War. Hand-in-hand with this went the construction of three more bridges over the river—the Margaret Bridge (1872–76), Franz Joseph (Liberty) Bridge (1894–96), Elizabeth Bridge (1897–1903)—along with the squares at their respective ends and extension of the ring-road system to Buda. On the Pest side there was the setting out of many of its major squares: *Kossuth tér* (1895–1904), on the eastern front of the Parliament building; *Szabadság tér* (Liberty Square, 1896–1908), built to cover over the site of the 'Neugebäude' barracks, demolished on account of their unhappy associations with the repression and executions that had followed the 1849–49 revolution; *Hősök tere* (Heroes' Square, 1896–1908), the culmination of the Radial Avenue; and *Ferenciek tere* (Franciscans' Square, 1909–13). On the Buda side, besides completion of the Ybl–Hauszmann plans for the Royal Palace (1890–1905), there was the building of the *Technical University* campus (1903–09) and of the *Gellért Hotel* (1909–18), as a first stage in the city's southwards expansion. These may not vie in sheer size with the Eiffel Tower and its surrounds or the re-engineering of Manhattan in New York, nevertheless the broad scale of architectural unity achieved in Budapest, not

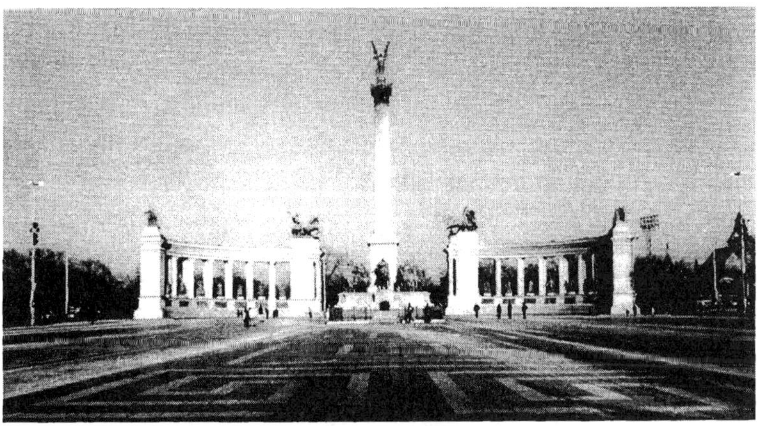

Heroes' Square

least in view of its historicist European character, is comparable in many respects. This was as much a tribute to the activities of the Metropolitan Board of Public Works as to any acceptance by the *Zeitgeist*—above all in Budapest, now capital of a country of almost twenty million souls—of the goals of modernity as its natural element. The reconstruction of Szeged after the disastrous 1878 over-flooding of the River Tisza, the rapid development and remoulding of Kecskemét, Debrecen, Pécs, Szabadka (Subotica, Croatia), Nagyvárad (Oradea, Romania), Marosvásárhely (Tîrgu Mureş, Romania), Arad and Temesvár (Timişoara, Romania), and the urbanisation and rebuilding of many other towns around the turn of the century were all manifestations of a country-wide boom.

With the economic boom in the Monarchy and Hungary alike, modernisation of building materials, structures and technologies proceeded apace. Alongside bricks, wood, lime and sand, came cast iron and glass as the new structural materials of the nineteenth century, then steel, concrete and, at the turn of the century, ferroconcrete, opening up new spatial forms and broad-spanning structures. The new technology of an iron skeleton with glass canopy was at the core of one railway station, shopping arcade, market-hall, exhibition pavilion and factory workshop after another, though many of these structures still preserved the character and craft values of traditional building techniques. It was not uncommon for craftsmen of different nationalities from the Monarchy's multiethnic population to work together on major construction projects. In a still predominantly agricultural country, almost one hundred thousand people worked as masons, joiners, and carpenters in the construction industry.

Having basked in the glow of being sought out by hordes of domestic and foreign visitors during the 1896 Millennium, the Magyar populace's awakened awareness of their state's thousand-year existence soon began to disfigure the ethnic and social cohesion that had begun to emerge during the *Gründerzeit.* By the turn of the century the internal tensions and social conflicts that, at the end of the First World War, were to lead to the break-up of the Monarchy, the disintegration of a socially and economically advanced but ethnically mixed state formation that encompassed the greater part of Central Europe, were already present. Expression was given to the clash of conservative views and cultural diversity by the odd tenor of the *Millennial Exhibition* itself. Here the main building was an assemblage of replicas, designed by Ignác Alpár (1855–1928)—originally built (1895–96) as a temporary structure but by popular demand rebuilt in more permanent form (1902–07) to fulfil its present function as an Agricultural Museum—which illustrates the country's architectural history, ranging from the portico of the thirteenth-century Romanesque Benedictine abbey church of Ják though the mid-fifteenth century Vajdahunyad (Hunedoara) Castle to motifs from the Late Gothic Zápolya Chapel at Csütörtökhely and the gatehouse at Segesvár (Sighişoara, Romania), which are of similar vintage. The intention of this montage, an outcome of the conservative view of history that had developed by the 1890s, was already that of proclaiming Magyar cultural supremacy.

The economic growth of the first decade of the century was reflected by an architecture of European quality. The most important representatives of this significant period in Hungary's architectural history included, on the one hand, conservative-thinking masters like Imre Steindl (1839–1902), Frigyes Schulek (1841–1919) and Alajos Hauszmann (1847–1926) who identified deeply with European Historicist trends and, on the other, innovative spirits such as Ödön Lechner (1845–1914) or Béla Lajta (1873–1920), the pioneer of reinforced concrete, as well as István Medgyaszay (1877 1959) and Károly Kós (1883–1977).

Continuous change, with a shifting and loosening of formal idiom, is discernible in top-class works even of such prominent proponents of Historicism as Hauszmann, Schulek or Albert

Schickedanz (1846–1915). The 1890s taste for clusters of cupolas and turrets and vigorous shaping of volumes was still feeding on Classicist and Baroque mannerisms, as in the case of Hauszmann's *Supreme Court* (1893–96) on Kossuth Square, next to the Parliament building, or his *New York Palace* on the Great Boulevard (1891–95), and a little later, in a brief revivalist period, the Hellenistic style of Schickedanz's *Museum of Fine Arts* (1900–06) on Heroes' Square, or Schulek's *Fishermen's Bastion* (1900–03), decorating the Danube side of Castle Hill, whose meld of neo-Romanesque and National Romanticist styles was a logical complement to the approach he had adopted in his major restoration work on the *Matthias Church* (1874–96).

The late-Historicist phase also threw up a *fin-de-siècle* German neo-Baroque in Ignác Alpár's *Stock Exchange Building* (1899–1905) and *Austro–Hungarian* (now National) *Bank* (1903–05), both in Szabadság tér, and later still the bizarre eclecticism of the overpoweringly proportioned *Anker Palace in József nádor tér* (1907–10), now the Ministry of Finance, with its interweaving of Mesopotamian and Hellenistic motifs, was an indication that stylistic miscegenation could be taken no further.

Ignác Alpár exemplifies the career of European *fin-de-siècle* entrepreneurs who piled one success on another, with a talent, aplomb, stamina and verve that marked him to accomplish great things. He started as a stone mason, became Hauszmann's draughtsman and later, after graduating from Schinkel's Bauakademie in Berlin, a colleague and teaching assistant. Prompt competition success led to winning a string of commissions, enabling him to maintain the biggest architectural office in Budapest, and handle the biggest design projects, such as the simultaneous construction of the above-mentioned Stock Exchange Building and Austro–Hungarian Bank. In the 1899 competition for the latter, his office won both first and second prize out of 33 entries, beating off Vienna's great avant-garde master Otto Wagner, the Fellner–Helmer duo, who designed around forty theatres across the Monarchy's territory, as well as Ödön Lechner. Up to 1916 he constructed 124 buildings. A combative, dynamic personality, Alpár lived and worked at an American tempo, finding time away from his numerous projects to travel extensively abroad and play a leading role in public life, holding the presidency of the Master Builders Corporation for the first decade and a half after its foundation in 1910. His buildings were distinguished by their well-organised ground-plans, his office was exemplary and prolific in its output. He got to see America in his old age; returning from New York in 1928, he caught a chill on the train and died in a Swiss sanatorium. In terms purely of his career and wealth, he was Hungary's most successful architect, a real self-made man.

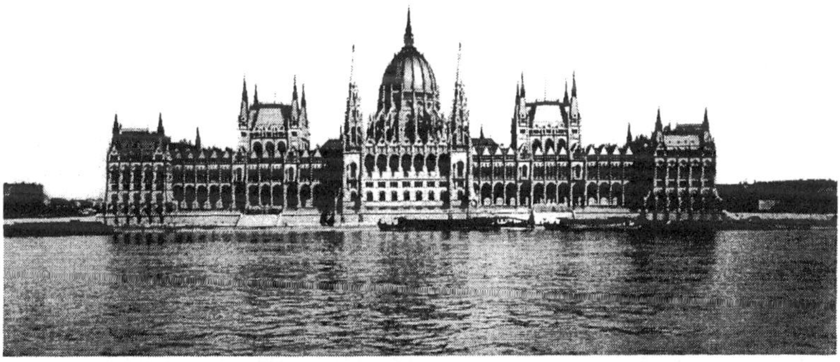

The Parliament

The conservative, eclectically historicising approach runs from Imre Steindl's use of Late Gothic forms to dress the Baroque ground-plan and spatial organisation of the *Houses of Parliament* (on which inordinate care and money were lavished during its construction between 1885 and 1902), through the synagogues of Lipót Baumhorn (1860–1932), the finest being that at Szeged (1902–04), with their medley of Romantic, Byzantine and Moorish elements, to the *Votive Church* at Szeged (1914–30), designed by Frigyes

Schulek and finished by Ernő Foerk (1868–1934), whose use of Romanesque elements in the brick and stonework switches over to another, more Romanticist concept. As illustrious professors at Budapest Technical University, Steindl and Hauszmann also placed their stamp on the architecture of the period through their students.

Amongst those who started to keep a less fettered eye on what was going on in Paris and Europe more widely were Flóris Korb (1860–1930) and Kálmán Giergl (1863–1954), who worked with Hauszmann on the Supreme Court and New York Palace buildings. Their work is informed by a metropolitan elegance, a toning down of historic reminiscences, a looser treatment of historic styles. The twin blocks of the *Klotild Palaces* (1899–1901), named after the Archduke Joseph's daughter-in-law and standing between Ferenciek tere and Váci utca, at the Pest end of the Elizabeth Bridge, and the *'Royal' Apartment Block* (1900–01), on the south side of the square, are an integral part of one of Budapest's key

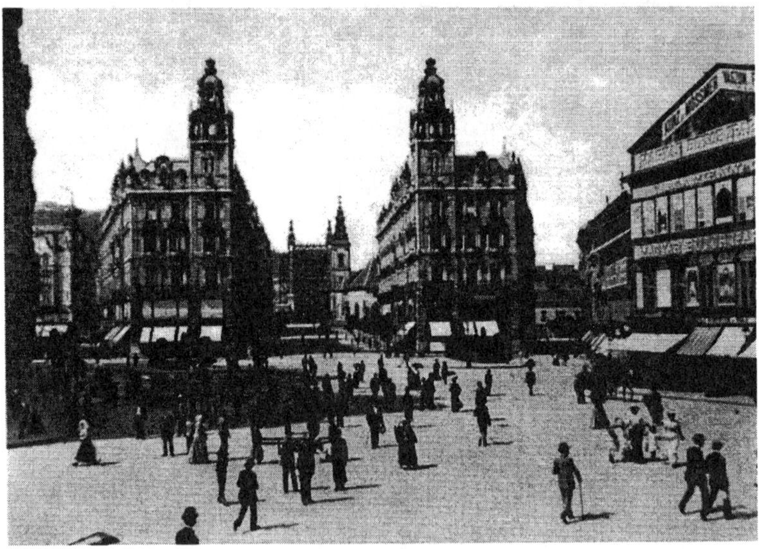

The two Klotild Palaces on Apponyi Square

architectural assemblies. The north side of the square is occupied by the building now generally known as the *Párisi Udvar* (Parisian Arcade, 1909–13), originally commissioned by the Inner City Savings Bank from Henrik Schmahl (1846–1913), who turned himself into an architect after starting work under Miklós Ybl as an assistant foreman on the Main Customs House then as site manager on the Opera House. Completed by the *Király Bazaar*, designed by Géza Kármán (1871–1939) and Gyula Ullmann (1872–1926), the most interesting tenement block in Károlyi Mihály utca, this assemblage of buildings, passages, arcades, coffee-houses and shops generated a big-city ambience and 'feel' that make the way of life that could be indulged here at the turn of the century so attractive, even looking back from the present day.

Despite a rather bleak, ponderous external architecture, the interior of Korb and Giergl's *Academy of Music* (1904–07), with its reinforced concrete structures and its elegant Art Nouveau in-fillings, maintains the high standard of their earlier work in Ferenciek tere. A break with the late-Eclecticist approach is also manifest in the interiors and in details of the attendant artwork, such as the two figures of 'Genius' by Géza Maróti (Rintel) that stand atop the main façade, the honest concrete forms, the frescos by Aladár Körösfői-Kriesch (1863–1920), and the structural nature and restraint of its modelling. Flóris Korb had started work on the plans for the *Medical School of Debrecen* before the First World War, though the *Main Building of the Kossuth University* there was built only after the war. Its most impressive feature is the galleried central hall with its glass vault that continued to keep alive values in the thinking about the handling of space that had emerged at the turn of the century.

A similar stance to that of Korb and Giergl is evident in the sweeping forms adopted by Zsigmond Quittner (1857–1918) in the *Gresham Palace* (1905–07), built for an English insurance company, though this embodies the creative views not so much of Quittner himself, who had practised for a long time in England and previously taken a neo-Renaissance and later a Baroque historicizing approach, as of his younger colleagues, László Vágó (1875–1933) and, especially, József Vágó (1877–1947), who was to acquire an international reputation after leaving Hungary in the 1920s. Standing virtually spot at the centre of Budapest and the country and providing the backdrop to the Chain Bridge on the Pest side, this building would become the most important element in a restructuring of what is now Roosevelt tér that had begun back in the Classicist period, its passageway intimating a big-city atmosphere that posterity was unable to live up to.

The above-mentioned works denoted that a fundamental departure from the conservative approach was already in progress: the roots of their modernity lay in a metropolitan middle-class identity. Inner evolutionary processes that exhibited, aside from French Art Nouveau influences, a coherent, multi-componential development of structures—a distinctively liberal democratic line of evolution—was now pervading the vocabulary of architecture, for this assimilation had already taken place in functions, structures and ordering of spaces.

The advance of modernity, with its transformation of everyday life, the typically swift change that engulfed the whole of Europe, and its growing inherent tensions, was attended by a confusion in traditional forms of identity, greater permissiveness in old systems of norms, clashes between urban lifestyles and the traditional values of the feudal world. With that same advance, the Historicism that had hitherto constituted the ideology of democratic transformation lost its footing, and this induced a change of tack on the part of more sensitive proponents of the conservative view. Hauszmann himself supplies evidence of this modest rejuvenation in his design for the *Main Building of Budapest Technical University* (1903–09), with its shift in scale, its handling of large masses, its placing of corridors and lecture halls, the coherence of the brickwork, and the proportioning of surfaces, though for all the changes in detail, the hallmarks of Historicist conventions—orders, cornices, accustomed proportioning of surfaces—still persist.

The unleashing of a hitherto unrecognised imaginative wealth led original architects like Ödön Lechner into a much more deliberate search for new directions. Individual modelling—with new materials and structures, freed from the rules of the Historicist canon—for Lechner was to represent a chance of progressing as he matured creatively. His journeyman years and a joint study trip to England that he made with the ceramist Vilmos Zsolnay in 1890, the heyday of the Victorian age, shook his faith in the articles of Historicism. He became aware of the opportunities and legitimacy of a more personal approach to modelling, selective and more imaginative in the wider sense even as it sustained and renewed the world of styles. The *Town Hall at Kecskemét* (1893–96), which he designed together with Gyula Pártos, the masterpieces of his *Museum and School of Applied Arts* (1893–96), the *Institute of Geology* (1896–99) and the *Postal Savings Bank* (1899–1901), all in Budapest, mark Lechner's exploration of possibilities—by then a conscious search for a specifically Hungarian language of forms that raises him to the ranks of the great European innovators. The loose handling of Baroque formal masses and free patterning of openings on the Museum, and the similar principle that inform the façade of the Institute of Geology; the huge wall planes, the controlled pattern of openings, the florid vernacular motifs on the crenellations along the roof line on the façades of the Savings Bank; the innovative steel and glass structure covering the main exhibition hall and the inter-linking of spaces in the Museum—all these things, coupled with a highly imaginative use on roofs and façades of pyrogranite tiles developed by the Zsolnay ceramics factory, are the ingredients of Lechner's Secessionist language, which in its novelty and individualism turns against conservative authority and late-Historicist beliefs in the universal.

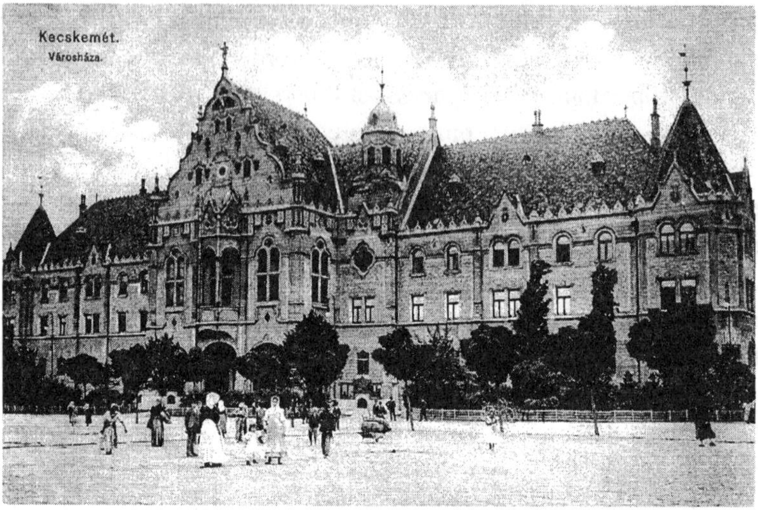

Town Hall, Kecskemét

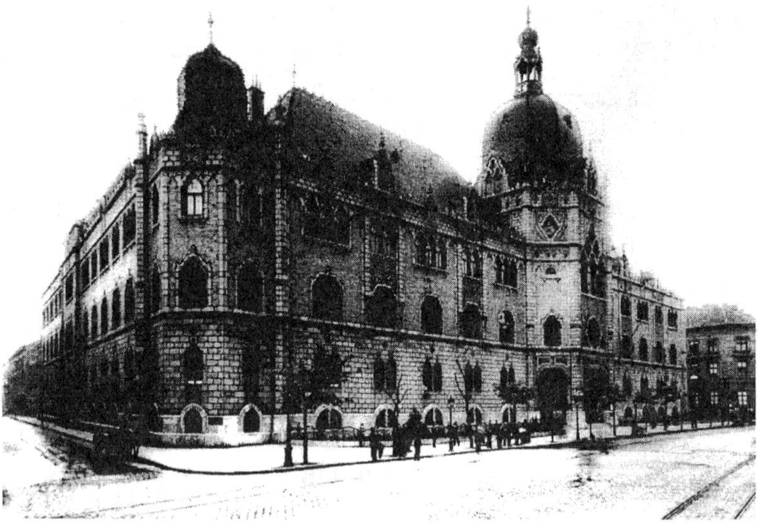

The Museum of Applied Arts

Lechner's captivating artistry of forms is built, among many other things, on oriental motifs. In their playful use of ceramics, his works are allied to the language of Antoni Gaudí. But the Lechnerian stance did not break completely with Historicism; it was merely imbued with the possibilities of modelling in a way that was different, more original and more self-sufficient than anyone else. During the last decade of his life the work dried up; he obtained no major commissions and thus shared the fate of Louis Henry Sullivan, at much the same time and in similar circumstances.

Lechner's disciples were mostly of a younger generation and included Sándor Baumgarten (1864–1928), producer of the standard designs from which a thousand of Hungary's elementary schools were built, Albert Körössy (1869–1955), Marcell Komor (1868–1944), Dezső Jakab (1864–1932), Béla Lajta in his early works, and to some extent also József Vágó and Artúr Sebestyén, who was one of the designers of the Gellért Bath in Budapest. Most typical of the endeavours to emulate his values in original works of their own are a clutch of buildings by Komor and Jakab: the *Town Hall* (1907–09) and *Palace of Culture* (1911–13) *at Tîrgu Mureş*, decorated with frescos by Aladár Körösfői-Kriesch, mosaics by the painter Sándor Nagy (1868–1950), and stained glass windows by designer Miksa Róth (1865–1944), or their *Town Hall at Subotica* (1900–10) and the *Bath at nearby Palics* (1910–12). One indication of how complex Lechner's pursuit of a Hungarian identity was is the fact that a goodly proportion of these disciples came from the assimilating Jewish middle class and were drawn by the oriental aspect of Lechner's thinking.

Ede Thoroczkai Wigand (1870–1945), a Transylvanian Saxon by background, stands apart from these figures but is a good example of the striving to articulate a Hungarian national identity. A designer turned architect who had worked with Steindl on the interior of the Houses of Parliament, he was among the first (1902) to take an interest in the peasant's way of life and during World War I published a series wonderfully illustrated hand-written books to commemorate the period he spent in Tîrgu Mureş between 1907 and 1914. Likewise representative of another camp of Lechnerites were Géza Márkus (1872–1912), designer of the *Cifrapalota* (Ornamental Palace) in the centre of *Kecskemét* (1903), and Valér Mende (1886–1918), who produced the plans for the same city's *Calvinist High School* and *Law Academy* (1910–13), now known as the New College, though the sparer concept that he adopted in the *Lutheran Courtyard* (1910–11) there shows certain affinities to the aims of Finnish architecture at that time.

Like Mende, the previously mentioned architect and ornamental sculptor Géza Maróti (1875–1941) was also close to the Finnish standpoint. A friend of Eliel Saarinen and Akseli Gallén-Kallela, after World War I Maróti worked on important buildings in Mexico City (the interior and exterior decorations of the Palacio de Belles Artes) then in the United States: from 1921 to 1923 on Saarinen's Cranbrook School then up to 1929 on a long string of works of Detroit-based Albert Kahn (Fisher Building, Foreman Bank Building, Hudson Motor Co. Building, Times Building, Livingstone Memorial). He connects with the search for a national idiom through his interior for the *Hungarian Pavilion at the Milan World Exhibition* (1906) and his design for *Hungary's permanent Pavilion at the Venice Bienniale* (1908–09). The *Hungarian Pavilion at the Turin International Exposition of Industrial Art* (1911) was the work of Emil Tőry (1863–1928) and Móric Pogány (1878–1942), who ran a joint practice from 1908, and Dénes Györgyi (1886–1961). What these official pavilions register is that ideological considerations had by then come to the fore in the search for a national architectural language. Tőry and Pogány's master work was the *Adria Insurance Company Building* (1912–18) in Budapest's Deák tér (after 1945 this housed the city's police headquarters and since 2001 has functioned as a hotel), its puritanical elevations and clear development of masses making it a forerunner of modern architecture.

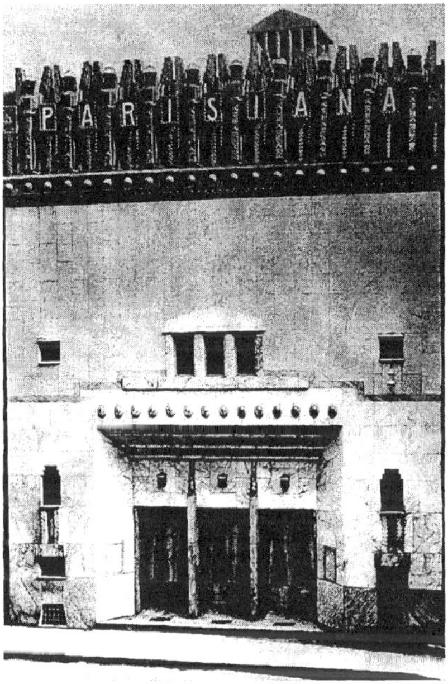

Parisiana Night-club

Béla Lajta was one of the early Lechner adherents who went on to distinguish himself as one of the most important architects of the early twentieth century. He set up his practice along lines that he had observed during a study trip to England, using his connections to gain major commissions. The high-pitched roofs and other features of his early buildings—the *Jewish Institute for the Blind* on Mexikói út (1905–08), the *Parisiana Night-club* (1908) and the *Jewish Charity Home at No. 57 Amerikai út* (1909–11), both in the fourteenth district—showed the influences both of Lechner and of his own English experiences, but the *Office Block* at *No. 5 Martinelli* (now Szervita) *tér* (1911–12) in the Inner City, the ground floor of which houses the *Rózsavölgyi Music Shop*, the interior of which was the work of Lajos Kozma (1884–1948), and the *Vas utca Commercial School* (1909–12), just off Rákóczi út, likewise with Kozma's decorative motifs, display an approach more akin to contemporary work in Chicago and Germany, making them important models of pre-modern design trends in Hungary. A further example of this pre-modern approach is the *Czech–Hungarian Industrial Bank in Nádor utca* (1912–13), the joint work of Béla Málnai (1878–1941), another Lechner disciple in his early days, and Gyula Haász (1877–1945).

The turn towards folk arts (Bartók and Kodály undertook their first collecting trips around 1904), the first volumes of poetry by

Endre Ady (1906–07), and the launching of the periodical *Nyugat* (1908) were among the major new signposts that influenced a fair number of these creative minds and they had their counterpart within architecture in a wide-ranging revival of vernacular approaches. The first major manifestations of this folk-rooted 'National Romanticism' were the works of István Medgyaszay (1877–1959), one of its most original exponents and also, even in international terms, an early pioneer of the use of reinforced concrete. During a two-year tour of Hungary to collect examples of folk style, he was struck by the structural purity of vernacular architecture. His horizons were widened by subsequent spells in Vienna (where he attended Otto Wagner's masterclass at the Academy of Fine Arts) and Paris, recognising the potentialities of ferroconcrete as a vehicle for a national architecture. His *Theatre for Veszprém* (1907–09), the plans for whose side elevation date from when he was still working in François Hennebique's Paris office, his modelling of the *Sopron Theatre* (1909), and the *Parish Church at Rárosmulyad* (Mul'a, Slovakia; 1908–10), with the stressed-skin construction of its cupola, are acclaimed pioneering works of reinforced concrete architecture, on a par with those of the Perret brothers. Medgyaszay travelled widely, researching in museums and going as far as India in a search for archetypes for his work. He fantasised about the Magyar race's origins in the East and an ancient Turanian art, coming to the conclusion that the ultimate, purified Hungarian taste is characterised by sobriety, oppositions and serenity. After World War I, however, with the advances of structural modernisation and the puritan aesthetic of the Modern Movement, Medgyaszay's treatment of forms and his mythopoeic notions became contentious. Though he won many design competitions between the wars, he mostly built residential units and lesser public buildings, the most prestigious of which was the *Baár Madas Reformed Boarding College for Girls* in Buda (1929). After 1945, towards the end of his life, he became impoverished and burned most of his archives.

Now recognized as one of the major international figures of this era, Aladár Árkay (1868–1932) produced one masterpiece kindred in approach to Medgyaszay's conception, the *Calvinist Church in Városligeti fasor* (1913), in Pest's seventh district. After early works in an Eclecticist vein, he was already incorporating Hungarian vernacular elements in the *Babockay Villa* (long the Yugoslavian Embassy), built in 1905 at the corner of Andrássy Avenue on Heroes' Square. Between 1903 and 1913 he built close to forty residences in a Buda villa community for judges and lawyers, in a style that has affinities to the weird Cubism being pursued by some Czech architects at that time. Between the wars, by now influenced by the Roman School, he became a pioneer in Hungary of the modernist application of ferroconcrete to church architecture.

The most consistent advocates of the true spirit of the revival of folk architecture were the *Fiatalok* ('Young Ones') group, the core of whom were students at Budapest Technical University, all born around 1885, who banded together at the time musicians and writers were likewise turning to folk culture. With Károly Kós as their leader, they included Béla Jánszky (1884–1945), Dezső Zrumeczky (1884–1917), Lajos Kozma, Dénes Györgyi and several more, such as Valér Mende and, albeit indirectly through Mende, József Bábolnay (1886–1983), who emigrated to the United States in 1924 and worked there as an architect and furniture and set designer. By mounting exhibitions of their works and carrying out surveys of folk architecture, the students proclaimed the perpetuation of that architecture as their moral and aesthetic principle, but they also paid close attention to new developments in contemporary European architecture, particularly in England and Finland, and the English Arts and Crafts movement. The buildings they designed around 1910, including Kós and Jánszky's *Parish Church at Zebegény* (1908–09), Kós and Zrumeczky's *Deer House* and *Aviary* in Budapest's *Zoological Garden* (1909–12), and Zrumeczky's

Elementary School in Áldás utca (1911–12), on Buda's Rózsadomb (Rose Hill), are mature masterworks of this period, a fresh and singularly vital branch in Hungary's architectural culture that was to be abruptly snapped off by the First World War. In a search for a way out of the problems of big-city slum areas, the *Wekerle Housing Estate* (1912–13) was put up in the working-class Kispest (nineteenth) district of Budapest, shaping it on the pattern of England's garden cities, with Károly Kós as one of the architects who contributed to planning the central area. A number of barrack-block estates were built in the capital during the war, such as the Mária Valéria Estate on the Üllői Road in Budapest, to serve as military hospitals for war casualties and invalids.

The glory of *fin-de-siècle* Budapest was set as a petrified past during the war and the revolutions and misery that followed it. Zrumeczky, amongst the '*Fiatalok*', was killed in action. Kós followed his conscience: despite being appointed professor at the School of Industrial Art in 1919, and with the capital offering innumerable possibilities, he chose to move to Transylvania and endure the vicissitudes brought by the fall of empire and the consequent border revisions. For him, an architect, as a representative of culture, should not let his head be turned by the desire for success but should share the fate of his nation.

The half-century of civil democratic evolution under the Austro–Hungarian Monarchy pervaded the whole of Central Europe. The outcome of economic permeability and co-operation of the huge swathe of territory from Slovenia and Croatia across to the Bohemian Basin and southern Poland was a socio-economic advancement, which set the regional components of the territory on a more or less comparable level, but its most conspicuous consequence was the success in building up Budapest—a success measurable not just in financial terms or cultural and artistic values, in the architectural and aesthetic quality of the townscape, but also in the progress made in developing the urban environment and the factors that determine quality of life.

In certain respects, admittedly, Budapest's *fin-de-siècle* architecture may pale in comparison with Vienna's late-Historicist and Secessionist splendour, since the likes of Theofil Hansen or Gottfried Semper were historicizing architects of European standing, whilst Otto Wagner and Joseph Olbrich were pioneers of the Modern Movement. Undeniably, the development of the Ringstrasse and at least some of the grandiose public buildings that were constructed in the Imperial city, the epicentre of the Monarchy, are more imposing than their counterparts in Budapest. Nevertheless it is not incongruous to pair Lechner with Wagner as an architect of equivalent value, and Hungarian architecture, both in its early adoption of reinforced concrete and the formal innovations of its pre-modern architecture and folk-inspired National Romanticism, is an integral part of the early phase of Modern Movement.

Hungarian Architecture Between the Wars: 1920–1945

The cataclysm of 1918–19 split the first half of Hungary's twentieth century, the period from the Millennium up to 1945, into two essentially antithetical periods. Historic Hungary, with its advanced stage of development, was lost. Though it had at long last regained political independence and sovereignty, what remained of the country, along with the newly emergent East European countries that had realised their own dreams of nation-statehood and independence at its expense, was plunged into a long crisis. Hungary's socio-economic system may not have changed, but it became unviable for a protracted period after World War I. Not only did the exceptionally favourable economic role and position that it had attained in Central Europe during the decades preceding the war end, but grave political tensions poisoned relations with the surrounding successor states.

The fate of Hungarian architecture and architects sadly mirrored that of a country now reduced to one third of its former area, burdened with economical problems, and set back in its development.

The best creative talents either left the country, as in the case of the subsequently renowned László Moholy-Nagy (1895–1946) or Marcel Breuer, or else scraped along on minor commissions, as in the case of Farkas Molnár (1897–1945) or Lajos Kozma. Due to the public roles they had played during the Hungarian Soviet Republic, Kozma, Béla Málnai and Rezső Hikisch (1872–1934) were held in disfavour and for a long time could only obtain small jobs, while József Vágó and Móric Pogány were forced to emigrate. Moving to Italy, Vágó entered and won the competition for the design of the League of Nations Palace in Geneva in 1926–27. During the thirties it became possible for him to return home, and in 1936 he even exhibited his plans for developing Budapest: despite 'ostracism', he had retained his idealism and continued to dream about the 'city of the future'.

By the latter half of the 1920s, the *Fiatalok* generation were getting into their forties and assuming an ever-growing role, with few of the older generation still active. Amongst the latter, and representing more conservative approaches, were a number of the professors at Budapest's Technical University and the likes of Dezső Hültl (1870–1945), who took on some of the bigger government and communal projects, Gyula Sándy (1868–1953), Róbert Kertész K. (1876–1951) and Gyula Sváb (1879–1938), who together directed the country's school-building program, Jenő Kismarty Lechner (1878–1962), whose designs resuscitated Hungary's domestic version of a battlemented late-Renaissance style, and likewise such earlier exponents of National Romanticism as Árkay, Medgyaszay and Thoroczkay Wigand. Other older masters who had established successful practices before the war, working with somewhat more innovative approaches, included the likes of József Boros (1875–1952), Alfréd Hajós (1878–1955), Hungary's first Olympic champion (winner of gold medals for 100 and 1,200 meters freestyle swimming at the 1896 Athens Olympiad) and in his professional life the architect of the *National Indoor Competitive Swimming*

Pool on Margaret Island (1930), and Béla Rerrich (1881–1932). Gáspár Fábián (1885–1953), officially blessed as the most successful architect of this period on account of his ecclesiastical commissions, was likewise in the conservative mould of Hültl and Sándy; his many buildings, including churches, and his architectural writings constitute a body of work similar (albeit more modest) in many aspects to that of Ignác Alpár.

Hungary had dropped back severely not just in economic and material terms but intellectually too in relation to modernisation. The social crisis of the twenties pulled Hungary's culture back in a conservative direction: this was a 'neo-Baroque' society, displaying an ideologically hidebound conservatism that had essentially lost its liberalism. One of the main driving forces of a new consolidation was the cultural policy of Count Kunó Klebelsberg, who between 1922 and 1931 was Minister for Education and Religious Affairs in the government of István Bethlen. Through his program of university and school construction, his wringing of state support for culture, and his advocacy of the practice of public subscription to causes, Klebelsberg was instrumental in helping the country find its feet again intellectually and in promoting social renewal through adult education.

Hungary was obliged to start once more from the bottom, as a result of which the typical modernising processes were reiterated all over again in the course of this quarter century. On the other hand, the conservative architecture which characterised the latter half of the twenties sprang not so much from any specific intentions on Klebelsberg's part as genuinely from the frame of mind of the established architects who gained the commissions, which was a function both of their age and the backwardness of the *Zeitgeist* in the broad sense. During the inter-war period, the only architects who had graduated abroad were the older ones, and initially there was little chance for any architect to gain experience outside the country. The Great Depression of the late twenties brought renewed unemployment within the profession, so fuelling further generational rivalry, with recovery coming only from 1934 onwards.

Architectural training at Budapest Technical University was likewise conservative in outlook and only altered slowly compared with the reform plans of the immediate post-war period. Its most notable teachers of design were Gyula Wälder (1884–1944), who took a Baroque revivalist line, and Iván Kotsis (1899–1980), who encouraged a more liberated approach. One sign that things were looking up was that the 12th International Architectural Congress in September 1930, organised by the Society of Hungarian Engineers and Architects, was associated with an international design exhibition held in the *Műcsarnok* (Exhibition Hall) in Heroes' Square. The secretary-general of the Congress was Béla Rerrich, whose *magnum opus*, Szeged's *Dóm tér* (Cathedral Square), had been finished by then, which was recognised at the event by Rerrich's election to membership of the Royal Institute of British Architects.

The 'Wälder industrial baroque' style that evolved in the twenties, with the typical prefabricated cast-stone details on the many of Wälder's buildings that were put up in Miskolc, Gyöngyös, Szeged, Eger and Szombathely, is best exemplified by the *Cistercian Order High School on Villányi út* (1927–29), on the western side of Gellért Hill in Buda. Almost a decade later, a complete change of approach, with the clinker brickwork cladding of its façade is displayed by another of Wälder's major works, the much-discussed Madách tér complex of the massive *Metropolitan Board of Public Works Headquarters and apartment block* (1935–38), planned as the gateway to a never-implemented radial avenue. He eventually became the professor of architectural history then, in the early fifties, the first rector of a newly separated Technical University for the Building Industry and Transport and first president of the Union of Hungarian Architects, founded in 1952. Wälder's most important collaborators were György Kardos (1902–53), who ran his office but passed on at the peak of his carrier, László Lauber (1902–53) and István Nyíri (1902–55).

The most significant city complex to be built in the inter-war period is Béla Rerrich's design for *Dóm tér in Szeged* (1927–30), undertaken around the start of the Great Depression and conceived as a brick structure and even what stone and clinker brick were utilised are only a cladding. The somewhat Romantic effect of the details likewise bears the traces of Northern models, but this in no way diminishes its value. The Klebelsberg-inspired project bound Szeged's new *University Buildings and Bishop's Palace*, set round Schulek's pre-war Romanesque cathedral, into a huge unifying courtyard, arcaded on three sides. The archway street entrances confer an intimacy on what rates as one of the most splendid twentieth-century urban squares in Central Europe, which since 1934 has been the site for the musical and other performances of Szeged's annual Open Air Festival.

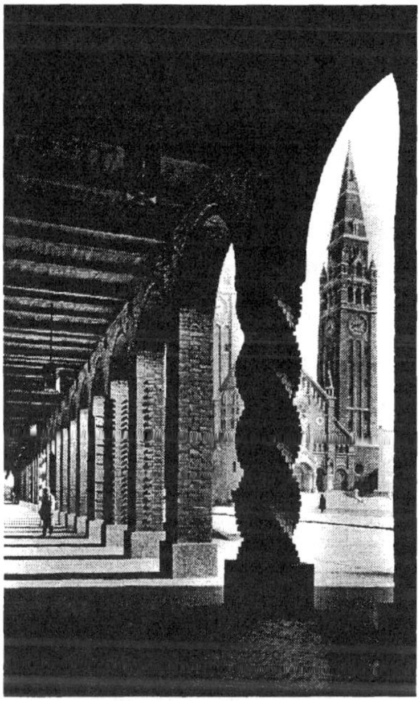

Votive Church, Szeged

The architects who fostered the use of brick, another original master of which was József Borsos, chief architect of Debrecen—notably the *Nagyerdő Crematory* (1931) on the northern outskirts of the city—were relating to a craft tradition that is the connecting link between the rich formal language of European Historicism and modern architecture's puritanical handling of form and materials.

The pre-war search by the *Fiatalok* for a national style underwent a change in approach, well illustrated by Dénes Györgyi's *Electric Works Co. Apartment Block on Honvéd utca* (1926–31) and the *Transformer house* (1931) behind it, the joint work of himself and Ernő Román: a rigorously articulated structure, this differed from the previous two examples in following the north German brick architectural style. Among the older generation, Jenő Kismarty Lechner's office produced a stream of competition-winning work that still preserved his earlier design illusions, including a city-centre plan for Belgrade (1922) and designs for the *Chicago Herald Tribune Tower* (1922; this jointly with Imre Forbáth and László Warga), an *Ethnographic Museum in Budapest* (1923; jointly with Warga) and a *Town Hall* for *Mezőkövesd* (1927–28), though none of these was actually implemented. By the time of the *Heroes' Monument* (1932) for Debrecen's public cemetery, designed with Pál Szontágh, his work was bold but fully resolved, its elements essentially modern in spirit. Kismarty Lechner also taught at Budapest's Academy of Applied Arts, while from 1929 László Warga (1878–1952), through his town-planning work, was the first professor of urban design at the Technical University.

István Medgyaszay's work in that period was the product of a singularly imaginative handling of forms: the complex of dramatically modelled buildings, organized around a huge courtyard, of the *Baár Madas Reformed Boarding College for Girls* (1928–29), nestling in the hills of Buda, and the Eastern-influenced, pagoda-like roof of the *Mátraháza Sport Hotel* (1927), set in the woods of the Mátra hills. The severely disciplined, stone-clad façade of the *TÉBE Bank and Apartment House* (1932–42) in Budapest's inner-city Hild tér preserves only traces of the earlier search for a national idiom, but the sanitising, puritanical spirit of the age was no longer (or not yet) receptive to its strange, even bizarre originality.

Debate about National Romanticism was rekindled with the foundation of an Ödön Lechner Society in 1928 by a group of architects and other intellectuals (Béla Bartók and Zoltán Kodály were among its members), and the publication of a history of the movement by Béla Jánszky, the proceeds from which were to fund the erection of a statue to Lechner, but the only upshot of this were the statue itself in front of the Museum for Applied Arts in Budapest and an annual Lechner Prize.

By then, the need for a new approach that took full account of technological and functional advances had become glaringly obvious. The first great period of the Bauhaus under Walter Gropius at Weimar had already passed. In addition to Moholy-Nagy and Breuer, who both taught there and, from 1928, by now in Dessau, edited the Bauhaus journal, this school had drawn in a fair number of other Hungarians into its activities, amongst them Ernő Kállai (1891–1954). Farkas Molnár spent several years at the Bauhaus (1921–25) before graduating from Budapest Technical University. Several other artists and architects who went on to establish names for themselves outside Hungary, including György Kepes (1906–2001) in the USA and Alfréd Forbát (1897–1972) in Sweden, were likewise Bauhaus students.

1928 was also the year in which modern architecture made its breakthrough in western and northern Europe, and the year in which the International Congress of Modern Architecture—CIAM—was founded, with Molnár and his circle associating themselves in 1929. The sense across Europe that lifestyles were modernising at a breakneck pace inevitably fed through to Hungary too. There, the turning point, after which it became impossible to think and build in the old manner, can be put at around 1930.

Hungary's real problems at the threshold of the thirties were conflicts in social conditions that could only be addressed through a sustained process of modernisation. The economic slump, however, only intensified the conflicts, the most acute of which were a still unsolved housing shortage and the growing obsolescence of the housing stock that did exist. Two distinctive versions of how modernisation represented a way out of these dilemmas, based on opposing ideologies, found expression in Hungarian architecture with the emergence, at much the same time, of the Roman School and the CIAM group. A growing religious revivalist movement triggered a spate of modern church-building. The CIAM group, by contrast, set itself the program of emulating the rationalist architectural aims of the Modern Movement in Europe in order to achieve a complete revitalisation of house-building and town-planning and a radical transformation of social conditions.

Following Pál Teleki and István Bethlen's failed attempts to win political allies in the English-speaking world, the country began to swing towards Mussolini's Italy—a process whose start was symbolised by the opening of the *Academia d'Ungheria* in Rome (1927). Amongst the first students to benefit from scholarships there were Bertalan Árkay and, later to become distinguished professors, Alfréd Bardon (1904–86), László Gerő and György Korompay (Krompecher); somewhat later on, between 1935–36, were the Olgyay brothers, Aladár and Viktor. Yet others, like Gyula Rimanóczy (1903–58) and Iván Kotsis, were to ally with its aspirations, which embraced the visual arts and a sort of renewal of esthetic intentions, a modernisation of compliance with tradition.

This change in approach is readily discernible in the differences between Aladár Árkay's first (1919) plan for a *Small Church* in Buda's *Városmajor Park*, the Transylvanian-style elements of which bear an obvious kinship with the idiom of the group around Károly Kós, and the actually implemented ferroconcrete structure of the *Parish Church of the Holy Heart of Jesus* (1932–33) and its later *bell tower* (1937) that he and his son, Bertalan Árkay (1901–71), jointly redesigned for the site and the latter finished. With the world slump under way, welcome opportunities for work were offered by commissions for the Catholic *Church at Balatonboglár*

(1932–1933), won by Iván Kotsis, the *Franciscan Church of St. Anthony of Padua on Pasaréti tér, Buda* (1931–34) by Gyula Rimanóczy, and the *Pauline Church and Monastery at Pécs* (1937) by Károly Weichinger (1893–1982), with its Italianate campanile, round-arched openings and rubble-masonry façade. That series is closed by Farkas Molnár's bold *Holy Land Church* (1942–49) in Buda's Hűvösvölgy district, still unfinished to the present day.

In its late phase, the Roman School (nowadays often referred to as 'the other modernism') encompassed aspirations that went wider than undertaking projects within a purely ecclesiastical domain. Its approach pervades Rimanóczy's Italianately classicist details on one of his later works, the *Administration Building for the Post Office and Telephone Company, Dob utca* (1939–40), in Budapest's seventh district, now the Ministry of Transport, Communications and Waterways, whose partners on that project, Lajos Hidasi (1902–70) and Imre Papp, were also responsible for the *Árpád High School in Old Buda* (1940), whilst Hidasi and Tibor Kiss (1899–1972) designed the Italianate *High School and Dormitory annex to the Benedictine Abbey of Pannonhalma* (1940–47) to blend in with the original thirteenth-century Italian architecture of this historic hill-top complex.

The Hungarian section of CIAM, calling itself the CIRPAC group (for Comité International pour la Réalization des Problèmes d'Architecture Contemporain), was a major force between 1929–38 in several respects. On the one hand, it represented the modernist mainstream in Western Europe; on the other hand, in Farkas Molnár, one of its leaders, it had one of the towering figures in the Modern Movement. Molnár and József Fischer (1901–55) were this loosely bound CIRPAC group's delegates to CIAM, which kept them in direct contact with a considerable number of exiled Hungarian architects, who also contributed to CIAM's work. The CIRPAC group's social concerns were reflected in their plans for affordable family houses, exhibitions and articles written for the

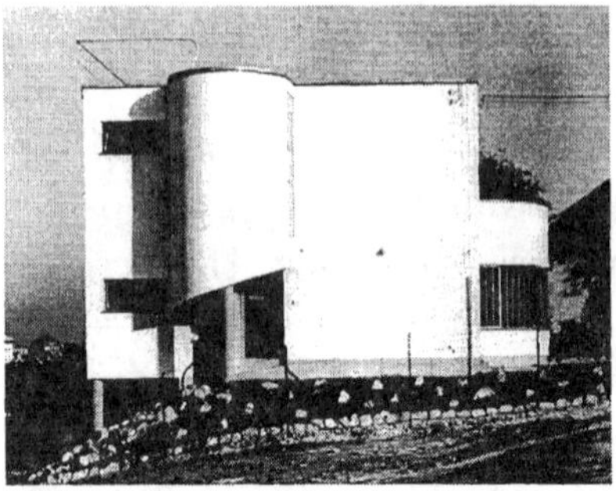

Villa on Lejtő Street

progressive periodical *tér és forma* (Space and Form) that Virgil Bierbauer edited from 1928. Two of the exhibitions they mounted at Budapest's Tamás Gallery, in March and September 1932—the depths of the depression—drove home the necessity for social architecture with such force that Molnár and six of his colleagues were taken to court for 'anti-state agitation'. As a result, they received almost no major commissions, for the most part only jobs designing small private houses. Molnár's design for the *Dálnoki–Kovács Villa, Lejtő utca* (1932) in Buda's twelfth district, won the first prize at the 1933 Milan Trienniale and was widely publicised. His *Villa on Darányi* (now Hankóczy) *utca* (1933) and the *Condominium Block on Lotz Károly utca* (1933), both in Buda's Pasarét district, were used as examples by Gropius in his first Bauhaus book (1939). József Fischer's *Villa on Bajza utca* (1936) and Molnár's *Condominium Block on Pasaréti út* (1937) are amongst the enduring masterpieces of Hungarian architecture.

Despite the international recognition that was won with these small commissions, the political pressure applied on members of the CIRPAC group grew to the point that even this modest livelihood was threatened. Molnár dissolved the group, and now made himself an enemy in the eyes of many of his colleagues, partly on account of a single article that he published at this time but also because they felt that Molnár's acceptance of a commission from the Franciscan Order, and a study trip that he made to the Holy Land, represented a betrayal of their modernist principles. His unfinished *Holy Land Church*, with its elliptical shell dome, raised on twenty columns, which were elements of the collar interfered by two rows of square openings, would have composed an interesting inner space. A reinforced concrete surface of comparable strength to this collar block, realised in 1940, was only achieved by Le Corbusier in 1947 on the external walls for the staircase–elevator modules of his Unité d'Habitation at Marseilles. The war cut short further building work on the church and its architect's life as well. József Fischer, the co-leader with Molnár of the CIRPAC group and a convinced Social Democrat, having gained an appointment as a government commissioner for reconstruction in the immediate aftermath of the war, was declared an 'enemy' by the Communist régime when it assumed sole power in 1949 and debarred from further work as an architect. Eventually leaving the country, he resided in New York between 1966 and 1975 before returning home. Typical Hungarian careers—but, for all their merits, they can hardly be called models of unclouded success.

More fortunate than fellow-Hungarian CIAM members who stayed at home was Marcel Breuer (1902–1981). He studied at the Bauhaus from 1920 until 1925 as a furniture designer, and was then put in charge of the joinery and cabinet workshop until the school's closure in 1932. Except for a few months spent in Hungary in 1934, then in London in 1935, he thereafter settled in the USA. The tubular steel chairs that he designed in his Bauhaus days, his subsequent teaching activity at Harvard University (1937–46), the villas that he designed in his partnership with Walter Gropius during that period, and then his independent practice in his New York studio after 1946 all won Breuer a place as one of the most renowned creators in the Modern Movement. Much of his architectural work was achieved with partners, including his most notable works, the *UNESCO Headquarters in Paris* (1953–58, with P. L. Nervi and Zehrfuss), the basic idea of which was also successfully applied to the *IBM Research Center in La Gaude* (1960–61), and the *Whitney Museum of American Art, New York* (1963–66).

Though not himself associated with CIAM, Pál Ligeti (1885–1941) separately partnered with both József Fischer and Molnár on villa designs for the *Housing Estate in Napraforgó utca* (1931–32), in Buda's Pasarét district, the 21 dwellings of which (one a duplex apartment house) were built by the firm of Fejér and Dános, with a subsidy from the Metropolitan Board of Public Works, as a showcase for progressive architects of the era in order to popularise their ideas about how the new architecture could be applied to the construction of affordable middle-class homes. Amongst other notable figures who became involved in the project, besides the above, were László Vágó, Gyula Wälder, Virgil Bierbauer, Róbert Kertész K., Alfréd Hajós, Gedeon Gerlóczy (1895–1975) and Károly Weichinger.

Lajos Kozma, whilst remaining outside CIAM, nevertheless became an advocate of the new architecture and designed three houses for the Napraforgó utca estate. Kozma's villas for the Rózsadomb district of Buda and his apartment blocks, such as the *'Átrium' Apartment Building and Cinema* (1936) on the Margit körút in Buda, or the *Apartment House in the Régiposta utca* (1937) in Pest's inner city area, along with his own weekend cottage on Lupa Island (1935) are the best examples of the smart but not conspicuously luxurious dwellings that the country's upper middle-class were starting to look for. In his progression from a Lechner-inspired Art Nouveau through his own Baroque revivalism, Kozma evolved into one of the leading figures of modern architecture in Hungary with the care he bestowed on modelling the spaces and furnishings of his public buildings and private dwellings; his swivel armchair and a good few of his other pieces vie with the furniture of Mies van der Rohe or Breuer. His book *Das neue Haus* (1941), published in Zurich, is one of the pioneering works in modern house design. He survived the war to become director of the Academy of Applied Arts in Budapest for three years after 1945 and pass on his painstaking insights to his interior design students.

Other significant examples of the modernist approach to house design were Nyíri and Lauber's five-storey *Apartment House at No. 10 Kékgolyó utca* (1935), to the west of Castle Hill, and Rimanóczy's *Szakáts Villa in Pasaréti út* (1934). Also outstanding are the designs for the apartments in *'Georgia House' at No. 4 Rákóczi út* (1935), part of the complex diagonally opposite the Astoria Hotel in Pest, designed by the partnership of Béla Barát (1885–1945) and Ede Novák (1888–1951), whose work had attracted considerable interest at the *Architecture d'Aujourd'hui* exhibition in 1933, and on the far side of the same block, at the intersection of Dohány utca with Károly körút, the *Broadway Cinema* (1938), by Ferenc Domány (1899–1939). This formed part of a still developing ensemble with the nineteenth-century Moorish-style Central Synagogue across the way in Dohány utca, built some eighty years before by the Viennese Ludwig Förster (1797–1863), with Frigyes Feszl's assistance on the interior design, which in 1931 had acquired the adjoining *Arcaded Courtyard and Temple of Heroes* by László Vágó and Ferenc Faragó. Fitting neatly into this setting at the Astoria intersection, with the recessed mass of its façade acting, so to say, as a symbolic portico, is the *Hungarian Academy of Sciences Apartment Block at No. 1 Rákóczi út* (1939–40), built by the conservative Hültl four years after Georgia House, whose modern architecture thus had a determining influence on its design.

The most interesting of all the apartment buildings from this period nonetheless are the twin blocks and arched staircase of the *Apartment House for the Manfred Weiss Company on Zárda* (now Rómer Flóris) *utca* (1937), at the corner of Margit körút in Buda, built by the firm of Béla Hofstätter and Ferenc Domány with their then partner Dezső Cserba. Also of a superb standard is the *Apartment House on Párizsi utca* (1942–44), in Pest's inner-city area, designed by Gedeon Gerlóczy, a block of magnificently laid

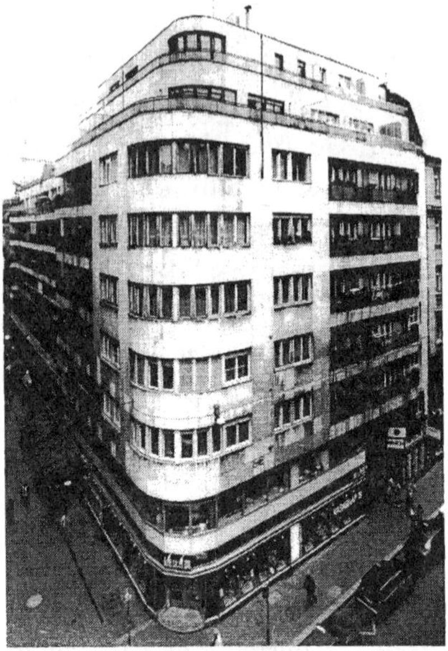

Apartment house
on Párizsi Street

out modern flats which incorporates an elegant passage and row of shops, with generous and modern interior structure. One measure of Gerlóczy's feel for quality is that it was he who saved the works of Tivadar Csontváry Kosztka from destruction after the painter died in 1919, thus preserving for posterity the truly unique mythico-symbolic world of a then barely recognised artist. After 1950 Gerlóczy, like many others of his generation, was relegated to the shadows, only to start to regain a measure of professional acclaim from the late sixties onwards.

By the latter half of the thirties the generation born around the turn of the century had begun to establish names for themselves and were to go on to create some of the outstanding works of the ensuing decade. Amongst those representing of the mainstreams of European architecture, with their own individual twists, were Gyula Rimanóczy, with his affinities to the Roman School; Károly Dávid (1903–73) and János Wanner (1906–87), who both started their careers in Le Corbusier's atelier (in 1940 Dávid won the competition for the main terminal building of the new *Ferihegy Airport*, though he got to build it only after the war); István Janáky (1901–66), who was influenced both by the Lechner tradition and by Le Corbusier; László Lauber and István Nyíri, who drew equally from Le Corbusier and the Bauhaus. They, together with Tibor Brestyánszky, Nándor Körmendy and Jenő Padányi Gulyás (1900–82), formed a group who called themselves the Little

Robbers' Society, a social circle, the name coming from the Buda restaurant where they met. They expressly banded together to fight the unfair monopolisation of public commissions by what they called the 'Big Robber' clique of officially favoured figures, such as Hültl, Sándy and Wälder, and helping one another to win a string of subsequent major open competitions with their modern designs. Besides Gedeon Gerlóczy and Károly Weichinger, the Olgyay twins, Aladár (1910–63) and Viktor (1910–70), were amongst the distinguished architects of this period. They designed the *Hungarian pavilions for the New York World Exhibition* (1939) and the *Izmir International Fair* (1943), as well as a pavilion at the 1941 Budapest International Fair and, in the same year, the *Stühmer Chocolate Factory* in Budapest. In 1947 they emigrated to the USA, obtaining teaching posts at Princeton University, where they became recognised experts in climate design.

Around then there were also some significant accomplishments in fields allied to architecture. István Menyhárd (1902–61), and Jenő Padányi Gulyás worked together on the design for the *Hamzsabégi út Bus Depot* (1939–40) in Buda's Kelenföld (eleventh) district, the 72-meter span of whose 6-cm-thick shell vault was at that time the broadest such structure in the world. The most extensive, and also architecturally the most notable public building projects were those carried out in the health care area, with the health insurance companies that were the main channels for statutory benefits funding the construction of a spate of specialist clinics, hospitals and sanatoria. The biggest of these projects was the OTI's *Magdolna Trauma Hospital and Central Clinic on Fiumei út* (1939), in Pest's Józsefváros (eighth) district, a nine-storey steel-frame block designed by Gedeon Gerlóczy and Nándor Körmendy. Another important building was Gyula Rimanóczy's *Mester utca Polyclinic* (1938), in the Ferencváros (ninth) district, whose bold handling of brick was to provide a prototype for the architecture of the late forties.

Comparable in scale was the *Building for the Central Corporation of Banking Companies at No. 5–6 Szabadság tér* (1938–40), at the corner with Október 6. utca, in Pest's inner city. The design by Nyíri and Lauber which won the 1937 competition for the project came under much critical fire. Taking their lead from the horizontal division of the overdone late-Historicist style of Alpár's 1901 Hungarian National Bank Building across the street, the two composed the façades of their own edifice as smoothly and severely as possible, in line with its intended office use, so that its austere frontage stands out from any vantage point in the square. It eschews symmetry, and even the entrance is not from the square as might be expected: instead, on the square frontage there is a glass-brick wall running behind columns at ground level. Whereas Rimanóczy's above-mentioned Dob utca administration building for the Post Office and Telephone Company has an Italianately classicist showiness, Nyíri and Lauber's building is more an embodiment of Bauhaus principles or some species of *Neue Sachlichkeit* spareness. Similarly ascetic, but more malleable, are János Wanner's designs for the *Condominium Building at No. 61 Szilágyi Erzsébet fasor* (1936), in Buda's Pasarét district, and for the *Headquarters for the Chamber of Architects and Engineers in Szalay utca* (1940), as well as several of the buildings on which the Olgyay brothers worked, including the *Villa on Városmajor utca* (1941), again near the Pasarét district. More relaxed, despite avowing similar principles, are the hotels and holiday homes built around then; for instance, Lauber, Nyíri and Andor Nendtvich's *Mecsek Hotel, Pécs* (1935–37), or Károly Puskás and György Uray's *Grand Hotel at Galyatető* (1936).

Several of István Janáky's projects are also amongst the typical public buildings of the period. He won the competition for the big open-air swimming-pool complex of the *Palatinus Strand on Margaret Island, Budapest* (1937; with György Masirevich), whilst his clinker-brick clad *Office for the Ministry of Supply and Price*

Control on Fő utca, Buda (1941–42; with Jenő Szendrői), clearly shows an influence of Le Corbusier. By this point, Hungarian architecture in the domains of public building and town planning had again caught up, both technically and conceptually, with the advanced countries of contemporary Europe—a state of affairs that was to be knocked back, for the second time in a quarter of a century, by a war that was forced on the country and the ensuing defeat.

In the inter-war period, alongside construction work governed by the Metropolitan Board of Public Works, a number of other major urban development schemes forged ahead during the thirties, which transformed the Buda side in particular through new bridge construction, demolition of much of the seedy Tabán quarter, nestling between Castle Hill, Naphegy (Sun Hill) and Gellért Hill, and the rapid building-up of the adjacent Lágymányos and Kelenföld areas (Buda's eleventh district) to its south.

For complex reasons, and ever more oppressively in its final years, this quarter of century lacked the liberal creative atmosphere which had made life tolerable for most people during the last third of the nineteenth century. The steady drain of talent alone represented a serious loss, and that is to say nothing of the architects who stayed in the territories ceded to the successor states after 1920.

Architecture and Modernisation after 1945

The Trianon Peace Treaty signed after World War I on 4th June, 1920, gave Hungary its independence but also cause for tensions with the successor states. The treaty had cut off nearly one-third of the ethnically Magyar population in territories that were ceded to foreign states.

The at least partial regularisation and then recovery that took place over the two decades between the wars were curtailed by the devastation and total social and political collapse of World War II. Defeat and a redistribution of the Great Power spheres of influence in Central Europe consigned Hungary to the Soviet zone, leading within a few short years to a total transformation of its social order, the termination of its capitalist economy, and the establishment of a régime of state socialism. For the next forty-five years its fate was bound to a Great Power bloc whose states, for the most part, were far less socially advanced than Hungary had been in 1938, and this conditioned the forms of, and opportunities for, any further socio-economic development. The short period of post-war reconstruction (1945–48) still offered some scope for a market economy and broadly democratised the right-wing conservatism of the pre-war régime, but the dogmatism of the Communist dictatorship that came after the turning-point of 1949 reimposed a rigidly conservative social system of its own. The largely artificial, inflated and sketchy modernisation that a misguided central planning process imposed on the economy (and architecture too) was followed by a renewed severe slump after the collapse of the USSR's zone of political power in 1989. The distinctive periods of socio-political crisis that the 'Sovietized' régime itself underwent—in 1956 and 1968, for example—allow the history of the era (and its architecture) to be further subdivided.

The changes in lifestyle, outlook, attitudes, fashion and taste over the sixty-some years since 1938 have been so profound as to make a ranking by value virtually impossible, since any scale of values repeatedly proved self-contradictory or, at given phases, acutely divisive amongst different social groups. Within architecture, the Modern Movement surged ahead at first until, after the temporary intervention of a period of socialist realism, it became generally accepted, only to be ultimately discredited—a process that went hand in hand with the changes in values over the past sixty years that, over and above any internal contradictions, as a matter of course set one generation against the next.

Despite several decades of determined efforts, renewed after the war, the developmental imbalances between the individual regions within Hungary continued to grow, and this has given rise to social tension—perceptible in relations between the capital and the provinces—that can only be resolved by carefully thought-out regional support. Even 45 years of subjugation did not completely stamp out the basic human social instincts, cultivated over a thousand years of contact with Europe and particularly the preceding 250 years of modernization, that have been the preconditions for Hungary's existence and continued survival. The 1956 revolution forced the Soviet camp itself to reassess its stance towards that process.

Architecture in the Period of Crisis

Design work took off again only slowly in the war-ravaged country, with the first works to emerge all being conceived in the spirit of the Modern Movement, many of them expressly seeking to serve social goals of the new regime. A group of architects including Lajos Kozma, Máté Major, Lajos Gádoros (1910–91), Pál Németh (1913–95) and Jenő Szendrői (1913–2000), with István Janáky and József Schall as associates, formally banded together for this purpose into a New Architecture Circle on 28th November, 1946, and they picked up most of relatively few commissions that were to be had. Ideological guidance for architects was not slow in coming: József Fischer, the government commissioner for reconstruction had wanted to have all damaged pitched roofs cleared away. All too soon, however, dogmatism became an obstacle to further progress in architecture; after 1948 even architects such as Kozma were considered bourgeois 'class enemies'. With architects generally in a precarious and even dire situation, the centralisation of design and planning work into government offices during 1948–49 was all the speedier. The New Architecture Circle too was eventually forced to disband when the Communist régime assumed full control of power on 12th September 1949.

The USSR-style party state was now in a position to exercise ideological coercion; by the time it had entered its second phase, from 1951 onwards, only design and building in the archaising spirit of Stalinist–Zhdanovite 'socialist realism' was permitted. New university complexes were planned, new 'socialist cities' arose as part of the country's drive to heavy industry, and even the *Népstadion* (People's Stadium) went up in Budapest, but the Cold War was already putting industrial priority on military needs, the centrally planned economy ensuring that resources were diverted away from the consumer to meet those goals. The large-scale design office made its appearance, which was not necessarily to the detriment of standards of architectural practice, for although the offices were prompted by the Soviet example, they were actually implemented along the lines of big American practices and so represented a progressive division of labour to handle the increased flow of work.

Several masters from the pre-war era, such as Rimanóczy, Nyíri and Lauber, as well as staff from the Dávid and Kiss office that had been established during the war to work on the designs for Ferihegy Airport and the above-mentioned members of the New Architecture's Circle became the chief designers in the big state offices. Amongst the younger architects, some—András Ivánka (1911–2000), Kamill Kismarty Lechner (1914), György Szrogh (1915–99), Árpád Szabó (1916–91), Jenő Juhász (1918–99), or Ferenc Kiss (1918–93)—had graduated before the war, but the first intake into the offices also included others who had only qualified during the war, amongst them Tibor Farkas (1922) and the 'Danes'—Zoltán Farkasdy (1923–89), György Jánossy (1923–98) and Lajos Zalaváry (1923)—who had formed part of a group of sixty students who managed to dodge military service by being evacuated to Denmark, whence they had returned home in 1946. Having subsequently built themselves up with post-war graduates, the staffs of these design offices were again decimated by the wave of emigration that followed the 1956 revolution.

By establishing a Union of Hungarian Architects in 1951, with 218 architects as founding members, the dictatorship seemed to be in the curious position of trying to present itself as accepting a genuinely professional scale of values even as was driving up tensions in other areas. As a result, the scale of values that prevailed in the period between 1951 and 1956 was more or less the one that had developed between the wars, with the outstanding masters of that era gaining leading roles, for instance, IPARTERV, the state project office for industrial buildings, set up by Jenő Szendrői, employed Lauber, Gerlóczy, Schall and Juhász, while KÖZTI, the state project office for public buildings, was the workplace of Janáky, Rimanóczy, Ivánka, Szrogh and the aforementioned three 'Danes'.

Even the works of that depressing decade or so from 1945 to 1956 include some outstanding pieces of architecture—amongst others, a Ministry of Defence complex, several clinics and a few new university and industrial buildings. The first wing of the *Ministry of Defence Building*, part of the response to the onset of the Cold War, was designed by Jenő Szendrői to fit in with József Hild's Classicist Valero Silk Factory at 26–30 Honvéd utca, with László Lauber responsible for the main building (1949–53), the most splendid and important of all Budapest office blocks, though publication of it was prohibited for half a century. Lauber and Szendrői were partners on the *Horticultural and Food Industry University on Ménesi út* (1949–50), in Buda's Lágymányos district, while István Nyíri, Lauber's partner before the war, designed the most controversial building of that period, the professedly 'modernist' *Coach Station in Erzsébet* (then Engels) *tér* (1949), and the *Head Office for ÁMTI* (1950–52), the design office for subterranean structures, near Feszl's Vigadó on the Danube embankment.

This initial period saw the construction of a trade union headquarters that came to typify the whole era, the *Office Building and Assembly Hall for MÉMOSZ*, the Hungarian Construction Workers' Trade Union (1947–49) at the edge of the City Park on György Dózsa út, the main route for the many festive parades of this era.

For many years the site of Communist Party congresses, it was formally a private commission though in fact the product of a collective design by Imre Perényi, Lajos Gádoros, Gábor Preisich and the young György Szrogh.

During the period of 'socialist realism' several university buildings of enduring value, the elements of subsequent larger ensembles, were built. Gyula Rimanóczy's redbrick *Building R for Budapest Technical University* (1951–55), continuing Hauszmann's main building complex on the Lágymányos Danube embankment southwards, sought models in the Adam brothers' well-mannered Classicism of the later eighteenth-century England rather than Soviet archaising. István Janáky's plan for the *Main Building for Miskolc Technical University of Heavy Industries* (1952), on a campus in the Tapolca Valley, being conceived as a huge castle in a manner deriving from a modern Nordic Classicist influence, came under heavy fire, but it proved a clearly formulated design when it was finally constructed in 1960. Janáky's partners at this period, the 'Danes' Jánossy, Farkasdy and Zalaváry, were to turn out to be the most important designers of the next decade. Jánossy himself was to plan a set of six uniform student residential blocks for the Miskolc campus and his *Water Tower for the University of Agricultural Sciences, Gödöllő* (1952–54), designed together with István Czebe, is likewise more a synthesis of the modern and the traditional in the spirit of contemporary Danish architecture than of any socialist realist postulates; its simple brick cylinder, with the water-storage tank at its top surrounded by a glasshouse, still retains its value in today's architectural environment. Zoltán Farkasdy was the designer of the *High School for Adult Education* (1952–53) in Buda's Zugliget (twelfth) district, now the University of Applied Arts, was prompted by similar considerations in drawing on the simple, clean forms of Hungary's own Reform-era Classicism to bring a country-house feel into this valley's rich mix of ancient woods and Classicist villas.

Farkasdy roundly rejected the socialist realist approach to design in an address that he made to the general assembly of the Union of Hungarian Architects in May 1954. However, he and everyone else had to wait until the night of 23rd October, 1956, for the eight-meter Stalin statue which then stood across the way from the MÉMOSZ Headquarters, to be toppled and the rigors of socialist realism swept away. At no point during the long years of post-revolutionary consolidation or the heyday of the ensuing Kádár era did the régime ever again dare resort to the levels of physical and intellectual duress that were applied during the years of dictatorship in the fifties.

Modernisation within the Constraints of a Planned Economy: 1957–1989

The three decades of modernisation that ensued from 1957 to 1989 brought progress in many respects for Hungarian society at large, and architecture with it, but the inherent contradictions that were to promote the collapse of this social order were present and continued to multiply. During this soft dictatorship of 'goulash socialism', the big design offices became even bigger, excessively so, and the 'value' of the individual architect diminished inversely as the size of the office and its work burden increased. A professional design ethic was rendered impossible by the serial assembly of standard designs, short-sighted technological constraints, obligatorily low technical and cost standards for implementation, cramped apartment dimensions, the spread of prefabricated housing during the sixties, that reduced the creative process to a production line.

Still, the management bodies within the Union of Hungarian Architects had a certain role in maintaining the scale of values within the profession, and peer review, albeit only through the grapevine, was able to offer a certain objectivity. Struggles between technocratic and organic architecture in the 'tulip debate' of

1974–75, and its administrative conclusion, the growing internal contradictions and eventual disintegration of the institutional system—these all served only as further proof that the arrangements were rotten to the core and incapable of reform. The wounds inflicted on the disciples of organic architecture would only be mended a decade and a half later when they alone represented Hungary at the Fifth International Architectural Bienniale at Venice in 1991.

In assessing the achievements of this era consideration has to be given to the evolution that went on within the big design offices and to the institutional in-fighting between different interest groups. This yields several natural landmarks by which the period can be segmented into various phases, such as the sudden emergence of, and international recognition accorded to, architecturally notable work in the early sixties, followed by general ossification and the 'underground' struggles of organic architecture in the seventies, then the stagnation of the eighties.

Successes and Debates: the 1960s and '70s

In the early sixties, with the depressing isolation of the fifties behind it, Hungarian architecture turned afresh towards the developed world, becoming active internationally in the work of the UIA. Those who were by then in mid-career quickly latched on to what had been going on outside the country, whilst in 1964–65, through Ernő Goldfinger (1902–87), seventeen of the country's best young architects were given the opportunity to work for a year in England, giving a chance both for new architectural values that diverged from those of the 1940s to be implanted in Hungary and, moreover, for them to be publicised beyond the Iron Curtain. Renovation of the holiday resorts around Lake Balaton and replacement of the housing stock on Castle Hill that had been destroyed during the siege of Buda at the end of World War II were the most typical areas for this regeneration of a modernist approach which both retained what was good from the older era and gave creative scope to a new generation. Amongst the specific projects in Buda are the *Apartment House at No. 32 Úri utca* (1957–59) by Zoltán Farkasdy and the *House in Bástya Sétány* (1957–63) by György Jánossy; several smaller public buildings in the regions, such as the Institute of Astronomy's new *Observatory on Piszkéstető* (1960–62), in the Mátra Hills north-west of Budapest, by György Szrogh, Lajos Zalaváry's formal ideas for the *Swimming Bath at Jászberény* (1958–63), the *Mortuary in Szeged* (1958–60) by Béla Borvendég (1931), or Miklós Hofer's (1936) *TV Tower at Miskolc-Avas* (1964–66); a number of designs by Zoltán Gulyás (1930–2000), especially the *Office Building for the Chemolimpex Trading Co. at No. 4 Bécsi út* (1962–64), on the corner of Deák Ferenc utca in Pest's inner city area, an edifice that cites Alvar Aalto's work; and in the field of industrial construction, the *Factory for BHG Telecommunications Co. on Fehérvári út* (1963–64) by Lajos Arnóth (1929) and Jenő Szendrői.

Preparation of the most promising young architects was helped by the two-year 'Master School' courses in theory and practice that the Union of Hungarian Architects organised in the offices of established masters. The first cycle of this postgraduate training started in 1953 under István Janáky and Gyula Rimanóczy, and the next two cycles were held under Janáky, Jenő Szendrői and Pál Granasztói. The program then had to be suspended for a decade due to the 'overcritical' creative outlook it was perceived to be nurturing until Jenő Szendrői finally managed to resuscitate it in 1970 under the name of *Fiatal Építészek Köre* (FÉK), or Young Architect's Circle (though the ironic implications of the acronym, which means 'brake' in Hungarian, were not lost on the authorities). Since then about four hundred architects have participated in the program, which since 1990 has had its original name restored, a good many of the former students going on to become masters in their own right in an arrangement that for a long time was the sole avenue for professional postgraduate architectural training in Hungary. In 1990 the Peripatetic School of the Károly Kós Association was set up to serve a similar purpose.

Beside the sixties' achievements in regional planning and industrial design and some architectural successes during the earlier phases of constructional industrialisation, such as Tibor Tenke's (1924–84) cast concrete *Point-Blocks for the Experimental Housing Estate in Budafok* (1963–65) or the rubble-walled cast concrete of Zoltán Farkasdy's three 18-storey *High-Rise Apartment Blocks in Kelenföld* (1967–69), and a few terrace houses in the Buda hills, the most important result was the redeveloped town centre of Salgótarján, including the *Karancs Hotel* (1960–64), by György Jánossy and his partner József Hrecska, the *House of Culture* (1962–65) by György Szrogh, the nine-storey city-centre *Apartment Blocks* (1963–65) by Géza Magyar, and the *Pécskő Department Store* (1966–68) by József Finta (1935). This complex, which bears intellectual affinities to modern Japanese architecture of the day, set standards for other Socialist countries for a good decade and a half thereafter.

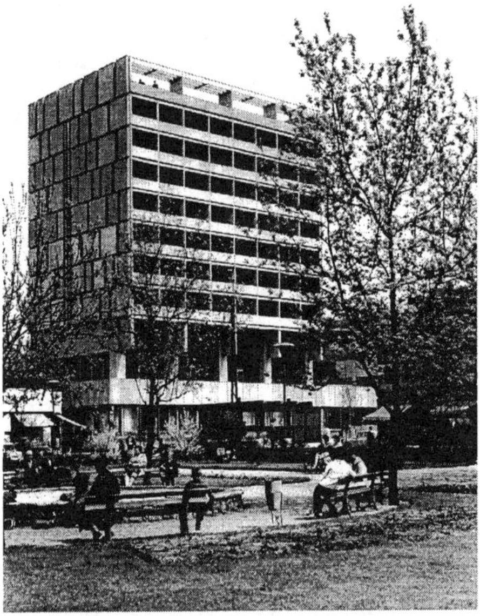

Town centre with Hotel Karancs, Salgótarján

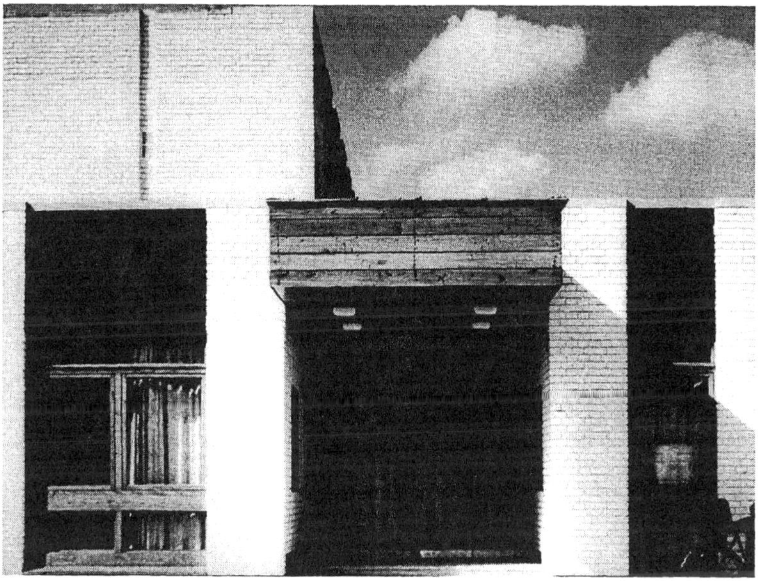

Cultural Centre, Orgovány

Salgótarján marks the start of the urbanisation of Hungary's previously neglected provincial towns, above all its county administrative capitals, with redevelopment and renovation of their centres and construction of new residential districts, for the most part using prefabricated panel systems (see the picture above). An outstanding exception to the rule from the late sixties is *Szekszárd Town Centre* (1968–72), planned by Károly Jurcsik (1928) and Levente Varga, its brick architecture clearly drawing lessons from British architecture of the day and comparable in standard to Rerrich's Dóm tér in Szeged.

During the first half of the seventies Jurcsik was still taking a particularly exciting approach to designing with standard prefabricated systems. The LAKÓTERV office that he worked for, developed a whole family of plans for nursery schools, utilising a 'panel skeleton system' of small modules that allowed interesting variants to be assembled by permutating the standard modules for individual functional units to cater for variable class sizes. These plans were adopted to construct small but smart-looking public buildings on dozens of housing estates; however, when the firm was reorganised to meet new government building guidelines, that put paid to further use of the fruits of close to a decade of work on a design methodology that had been internationally recognised as innovative and promising. His career at LAKÓTERV broken by this insensitive bureaucracy, the highly talented Jurcsik chose to leave architecture and move to Győr.

In 1968, winds of reform began to blow through Hungary's economy and society to reinvigorate them from the sense of stagnation that had meanwhile been developing. A signal year internationally too, in view of the events in Czechoslovakia, it ushered in a brief spurt of recovery. It was also around then that the two most distinguished representatives of the younger generation shot to prominence at more or less the same time: József Finta, with his design for the former *Duna–Intercontinental* (now Marriott) *Hotel* (1966–69) on the Danube embankment in Pest, near the recently renovated Elizabeth Bridge (1964), and Imre Makovecz, with the early version of his organic architecture displayed by the *Wine Bar at Balatonszepezd* (1966), the *Gulyás Tavern in Kőbánya* (1967), near the HUNGEXPO site in Pest's tenth district, and the *Csákányos Tavern in Tatabánya* (1968), which not much later were to be followed by masterpieces like the interior refurbishment of the *Mortuary at Farkasrét Cemetery* (1975) in Buda, and the *Sárospatak Cultural Centre*, built in 1974–77, to great international acclaim.

By then another trend in organic architecture had emerged, that of Group Pécs, an atelier formed by György Csete (1937), an important early work of whose was *Spring House in Orfű* (1974), a resort on the north-western outskirts of Pécs. Csete had founded this group of young architects within the Pécs regional design office in 1973, and amongst the first manifestations of their distinctive curvilinear formal conceptions and motifs, which came to be known as the 'tulip' style, were a set of proposals for enhancing dwellings on an existing housing estate for workers at the Paks Nuclear Power Station. The official attacks that they came under in the ensuing 'tulip' debate, however, resulted in disbandment of the group and an end to an innovative struggle against late-modernism that has since become more widely familiar in the developed world, too, under the flag of postmodernism. The monotony and architectural barrenness of the country's housing estates were already considered depressingly anomalous by 1975—a year that marked the peak of the construction boom (with 100,000 dwellings being built). In late 1973 the developed world had been plunged into the crisis induced by the cuts in oil supplies by the Arab petroleum-exporting countries, but whereas in the wider world this was a spur to a radical rethinking within architecture, in Hungary short periods of boom were to alternate with ever-longer periods of bust for the next fifteen years.

The start of the seventies, when the official program was still to encourage the use of light structures—György Kévés' (1935) terraced *Nursery for the Hungarian Radio and Television* (1971–74) was widely publicised abroad—saw the erection of the country's first high-tech building, Csaba Virág's (1933) *Electric Load Distribution Centre Buda Castle* (1971–73), then, in the middle of the decade, Miklós Hofer's complex for *Győr Technical College for Transport and Telecommunications* (1969–77; now the István Széchenyi High School), which drew on British experience with campus construction in the sixties. Two young architects in the IPARTERV office, Antal Lázár (1941) and Péter Reimholz (1942), both students of Zoltán Gulyás, showed a new design approach in the capital with their *Domus Furniture Store* (1970–74) at the corner of Lehel út and Dózsa út.

The mid-seventies saw the completion of a new *town-centre development in Kecskemét*, embracing the famous but by then dilapidated Secessionist town hall built by Lechner and Pártos and the seven churches in its environs, which József Kerényi (1939) renovated with a vernacular approach close to that of the organic school. Making a start with the Romantic, Moorish-style former Synagogue (1972–73), this was completed in 1974 with his redesign of the former Franciscan monastery to house the Zoltán Kodály Institute of Musical Education, with a new pedestrian passage opened up alongside it. For this and subsequent work in the southern part of the Great Hungarian Plain, including his *Chapel of Our Lady of the Hungarians at Cegléd* (1980–84), Kerényi received the Sir Robert Matthew Diploma at the 1985 UIA Cairo Congress. The rehabilitation of the centre came one year before the UNO announced its goals for the 'Year of European Architectural Heritage' program in 1975, which goes to show that Hungarian architecture in the mid-seventies had a good sense of which way international trends were moving in more than a few areas.

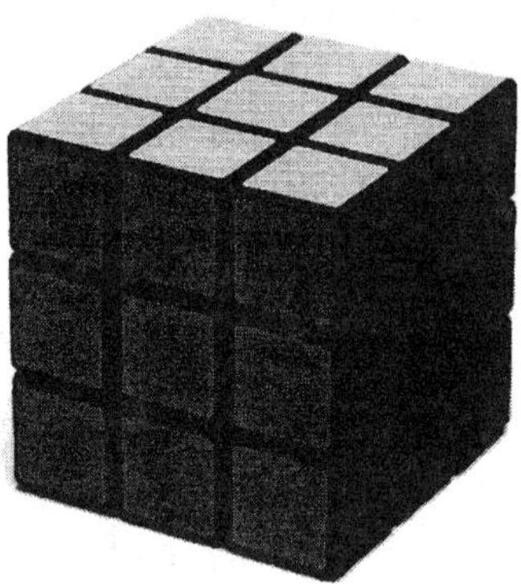

The Rubik Cube

The greatest success achieved by a Hungarian architect during the seventies was not with a building, however, but with the cube invented by Ernő Rubik (1944). Rubik had begun his studies in the sculpture department at a specialist middle school for the arts before qualifying as an architect then working as an interior designer under György Szrogh at the Academy of Applied Arts.

In the generally more liberal climate of those days, there were no inhibitions on a designer with an interest in constructivist trends to study and play with the artistic aspects of basic spatial geometry, which allowed Rubik to combine his accumulated design experience with a native inventive gift (his father had been a well-known glider designer) in the experimental studio work which resulted in the 'Rubik cube'. The world-wide success that followed its introduction in 1977 gave a substantial boost to Hungarian design at the time.

The 1980s and the Evolution of Organic Architecture

By the time János Gerle and György Szegő, two young architecture critics, mounted their exhibition of 'Architectural Trends in Hungary 1968–1981' at the Óbuda Gallery in early 1982 it was already obvious that the approach of the Modern Movement had become discredited, especially in the eyes of the young, over the preceding decade and a half. The change in attitude was typified by a renewed concern for fitting into the urban setting, for linkage and accommodation to historic contexts, a rethinking of architectural design methods, a systematic approach to analysing and classifying functional requirements, and a recognition of the significance of form and intellectual values that architecture can mediate, all of which was coupled to a growing high-tech emphasis. This was happening at the very time the post-modern stance was beginning to feed through, with its ambiguities, proliferation of forms, legitimation of historic citations, playfulness, and ironic distancing coming to the forefront in the creative processes of many architects.

The appeal of high technology was particularly characteristic of Csaba Virág's work during the eighties, as in the *Hungarian News Agency (MTI) Centre in Naphegy tér* (designed 1983, built 1987–90) and the *Konzum Shopping Centre in Tatabánya* (1981–84 and 1984–88), but the *Buda Centre Office Building on Hegyalja út*, at the foot of Naphegy (completed in 1990, with Árpád Marillai as partner) already shows deconstructionist features. Similarly high tech in approach are Ferenc Bán's (1940) *Cultural Centre for Nyíregyháza* (begun in 1969, finished in 1981) and, adjacent to it, his *Headquarters for the Central Council of the Hungarian Trades Union* (SZOT) (1982–84), with its double cube—one an enclosed glass curtain, the other a perforated assembly room. Taking a different view by exploiting the adaptive possibilities of high technology is the approach of Ferenc Callmeyer (1928) in his dynamic *Building for the American School* (1986–89) on Buda's Szabadság-hegy (Liberty Hill), for which Callmeyer was elected an honorary member of the Union of American Architects. The work of István Ferencz in converting a condemned mill into a *Rehearsal Hall for the Miskolc Symphony Orchestra* (1980–84), by contrast, is a good example of the sort of retrieval of values that was inherently opposed to the earlier destruction that had attended modernisation.

By the start of the eighties, through his work in Vienna, Finta had fully absorbed international high-tech planning methodology and its concomitant materials and functional organisation. At the same time, the formal hallmarks of post-modernism appear on many of his buildings: the trend-setting triple-curved glass façade of the *Taverna Hotel in Váci utca* (1982–85), the shopping precinct in Budapest's inner city, or the ironic post-modern motif of the stone-clad arcades set in front of the high-tech glass façade of the *International Trade Center* (1982–84) opposite it; the similar gestures on the façade of the disrupted mass of the *Representative Office Block in Erzsébet tér* (1988), and in the same square on the dynamic block of the *Grand Hotel Corvinus Kempinsky* (1988–92; planned jointly with Antal Puhl).

The best of Hungary's post-modern works are typified by a reformulation and reinterpretation of values much more than any radical eclecticism or individualism. That holds just as true for Ferenc Bán's buildings in a post-modern vein—for example, his

Cultural Centre for Mátészalka (1984–88) or the *Council Offices for Tiszavasvár* (1989)—as it does for Péter Molnár (1925–2000) and János Kóris's 1987 restoration of the storage wing of the historic (late eighteenth-century) former Zichy palace in Old Buda whilst transforming it into the *Vasarely Museum*, which demonstrates how linkage to the past and post-modern (here very much the exception) can be fused into an organic whole. The further elaboration of Ferenc Bán's philosophy evidenced in his winning entry in the competition for the National Theatre design for Budapest, judged in the spring of 1997, was strikingly elegant—but the plan was never realised.

István Janáky Jr.'s (1938) *Butterfly House* plan, which in February 1990—just before the formal change of regime—won the design competition for the Hungarian pavilion to be built at Seville's EXPO '92, expressed the desire for change with a peculiarly potent and poetic metaphor, as if to announce that the time had come for the country to break out of its chrysalis-like confinement. By the time for EXPO '92 itself, however, another sort of metaphor had been born: the work of organic architect Imre Makovecz, which not long before had been stifled by all possible means, had become a focus of international attention at the 1991 International Architectural Biennale at Venice, and so at Seville in 1992 it was his pavilion design that represented a Hungary now in search of its identity.

Having gained recognition with the award of the Ybl Prize in 1969, Makovecz was to face two decades of struggles. The essentially illegal summer camps that he organised for architecture students (1981–90) initiated a widening band of his disciples into ancient ways of building and a human world of genuine community. By the early eighties his works had become internationally known, with the 1985 UIA Cairo Congress listing the Cultural House for Sárospatak as one of the ten most important buildings

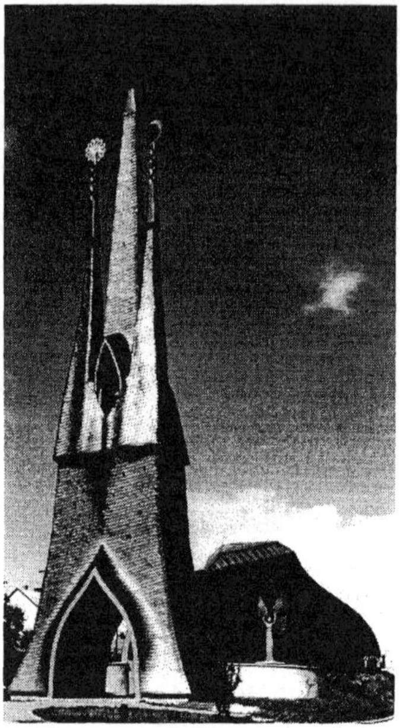
Catholic Parish Church, Paks

realised in 1983. The *Catholic Church at Paks* (1987–91) was widely publicised in architectural periodicals across the world. By around 1989, the Makona private office that he established was providing him with security for his creative freedom.

Makovecz's architectural philosophy connected folk culture to high culture, integrating rural elements into the world of high technology. The communal sensitivity of the organic ideal that he propounded gave it a peculiar appeal to the young generation but, by the same token, provoked antipathy on the part of the authorities. His life and work inevitably became a model for the aspiration for social renewal of an intellectual movement that, being founded on individuality and on architecture's general values and rejecting technocracy's one-sided domination, could consider itself a legitimate heir of the opportunity, hitherto denied, for a change in régime. Through its appropriation by those seeking change, then, the organic approach came to represent a form of continuity during the period when the régime did in fact change.

Makovecz incorporated his own thoughts on this in the means of spatial expression that he adopted for the Hungarian pavilion at Seville. Through the captivating design of the building, with its citation of tradition, the tree-of-life motif and the particularly graceful lines of the structure shaped by it glued wooden struts, the ideal of an organic space garnered wide-ranging approval. The dynamism of the work as a whole, with the fairy-tale atmosphere of its seven towers, sought to reflect, by historical argument, by evocation of a past world, the sense of belonging to European civilisation and of returning to it after an absence of five hundred years.

Amongst architects of the Makovecz school, the work of Dezső Ekler (1953) stands out for its sensitivity and intellectual fecundity. His *Dance Barn in Nagykálló* (1986–89), in the north-east of Hungary, is typical of his Kaposvár period, while the *Lecture Building for the Pannon University of Agricultural Science* in Kaposvár (1988) ranks him as one of the pioneers of an architectural approach that was to be adopted after the change.

During the nineties some 150 young architects professed adherence to the organic approach, participating in group exhibitions or in the collective—notably, László Vincze, Ferenc Salamon, Lőrinc Csernyus, Attila Turi, Tibor Heil and others, including Makovecz himself—that from the eighties onwards has regenerated the centre of Csenger, a small market town lying close to the Romanian border at Szatmár that had long been left in a state of suspended animation, with a string of creatively interesting buildings in the organic style.

Hungarian Architecture After the Change of Regime: 1989–1999

Despite the difficulties that followed 1989, with old contradictions giving way to equally serious new contradictions as Hungary abruptly lost its markets in the other former Socialist states of East Europe, precipitating a severe economic slump, compounded by the anomalies that accompanied the profound structural changes of the privatisation process, the tensions and threats to social cohesion that arose from the seismic shifts in patterns of employment and the impoverishment of broad strata of the populace within the space of a few years, a new scale of values was nourished and did make headway in the overall process of regeneration.

In the early nineties such construction work as went ahead was still mainly to designs that had arisen within the big state offices, or were in that spirit. One of the first significant buildings to be finished was Lajos Zalaváry's design for the *East–West Trade Centre* (1991) at the Astoria intersection on Pest's Rákóczi út, which has the virtue of being carefully conceived to echo the masses of Hültl's apartment block on the opposite corner, whilst the doubled articulation of its façade and the covering of the interior atrium have been widely copied.

Sándor Dévényi (1948), who created his own version of organic architecture over the last twenty years, stands out amongst architects practising outside the capital (early works include the *Wedding-Party House* (1983–84) and *House Struck by Lightning* (1978–84), both in Pécs). His work combines post-modern and deconstructivist-type formal elements with the vigour of ancient Roman architecture. The curious modelling of the *Roman Courtyard* (1991), fashioned between two streets in Pécs, cites local relics of the antique world, while the street front is permeated by a deconstructivist approach that intriguingly blends forms with elemental force. In the *Munkácsy Courtyard* (1992), by emphasising continuity with the past through his retention of the remains of the medieval walls and reconstructing part of its timber-beamed roof, he succeeds in creating a distinctively individual architectural world.

József Finta, the most successful architect of the decade, is at odds with the organic movements yet the defining stamps that his large-scale commissions make on their urban surroundings are still driven by a similar desire to create a modern Hungarian regional architecture. The Finta Studio has continued to draw on its accumulated international experience with high tech in realising big new projects in Budapest, such as the *International Bank Center in*

Szabadság tér (1991–95), the new *Central Police Headquarters in Deák tér* (1994–97), or the *Informatics Building for the Technical University and Eötvös Loránd University* (1995–97), on a Buda embankment site south of the Petőfi Bridge which was originally designated for an EXPO '96 that never took place and now forms part of a new University campus. Their most recent work is the block-long *Westend City Center* (1998–2000), a huge complex of offices, commercial units, a multiplex cinema and a Hilton Hotel, with a main entrance building and travellers' interchange that knits the vicinity of Nyugati (Western) Railway Station into a well-integrated scheme. Finta and his leading partners on this project, Antal Fekete and Alfréd Peschka Jr., headed a team of around 18 other architects and interior designers—a return of the big-office approach required for projects on this scale to ensure that international professional standards are maintained.

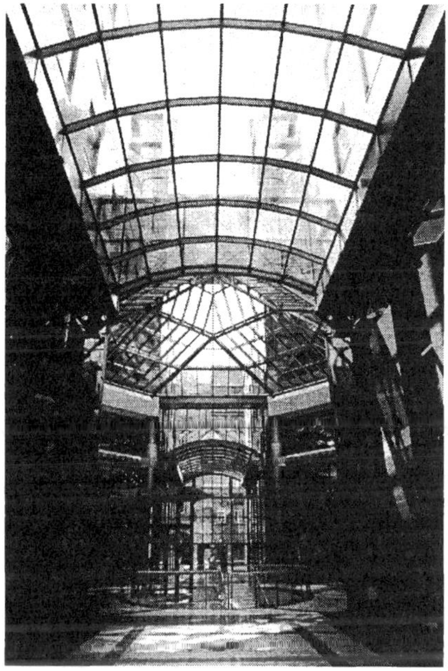

Bank Centre Offices

György Vadász (1933) left the state planning office in 1985. In his planning of a new housing and shopping precinct for *Zalaegerszeg Town Centre* (1991) he employed a system of playful and colourful stylistic elements to transform the old town atmosphere in a kind of anti-high-tech spirit. The *Hungarian Pavilion for Hannover EXPO 2000*, on which Vadász worked with Péter Basa

(1964) and Gergely Fernezelyi (1971), was one of the crowd-pulling highlights of the exhibition with its interesting shape and the witty matter in which it presented the various themes that were on display.

Even before the change of régime, Péter Reimholz had risen to prominence amongst the architects then in mid-career with his designs for a new *High School and Cultural Centre for Balassagyarmat* (designed 1983–86, built 1990–92), which are the most typical Hungarian responses to the post-modern approach. One of his most recent successes is the *Hapimag Apartment House* (1996–2000) in the residential quarter of Buda Castle, built to replace a destroyed corner building, which makes masterly use of carefully preserved historical remains to create a noble architecture fully worthy of its setting.

As one generation handed over to the next during the nineties, a fair number of new masters began to emerge from amongst the younger Hungarian architects and join the above-mentioned representatives of the middle generation. Peter Wagner (1951) had already made a name for himself with his meticulous designs for several bank buildings when he got together with Tamás Kőnig (1951) in the IPARTERV office to plan the restoration of Béla Lajta's 1920s building (formerly the Parisiana Night Club) for the *János Arany Theatre* on Paulay Ede utca in Pest, just behind Andrássy út on the side opposite the Opera House. The care with which this was remodelled to set it back in its original context was rewarded with a Europa Nostra Prize. Ferenc Ságoly's (1948) *Housing Complex on Számadó utca* (1991–93; with Richárd Hőnich), on Buda's Gellért Hill, is the most interesting residential project of the new era to date. The severely formulated architecture of his latest work, the glass and brick *Headquarters for the Graphisoft Information Park* (1997–98), in Old Buda, is the best local example of a neo-modern approach, eschewing all Romantic elements to yield a distinctive intellectual clarity and economy.

Gábor Turányi (1948), too, has created a series of interesting buildings over the last twenty years. His *School in Visegrád Wood* (1996), with its broken mass accommodating to the hillside and valley, perpetuates the best traditions of the Hungarian modern approach. His latest work, the *Porcelain Museum on Kossuth utca, Herend* (1998–99), the commercial and tourist centre for the Herend Porcelain Factory, is a strikingly mature building that enhances the little town with a Nordic touch.

The finest constructions of the brief decade that has elapsed since the change of régime were in the domain of sacral architecture, most of them ecclesiastical commissions. Imre Makovecz and József Finta both built new churches, at the same time, at Százhalombatta, on the western bank of the Danube just south of Budapest, whilst György Csete's *Museum* and ecumenical *'Church of the Forest' for the National Memorial Park at Ópusztaszer* (1992–96), the site about 30 km north of Szeged where Árpád and the other Magyar tribal chieftains are supposed to have met to distribute their newly conquered territory in the year 896, and his petaloid *Chapel of Reconciliation at Beremend* (1997), near Hungary's border with Serbia, are more cultic memorials. In addition many younger architects have contributed work of their own, including Ferenc Török (1936), with his *Church of the Hungarian Saints* (1994–95, partner Mihály Balázs), built on the aforementioned site of the cancelled EXPO 96 and Mihály Balázs, with his equally well-considered *Greek Catholic (Uniate) Church at Kazincbarcika* (1992–94), north of Miskolc. Tamás Nagy (1951) runs through the gamut of possibilities for modelling spaces and textural treatment offered by brick architecture in his *Lutheran Church at Dunaújváros* (1992–96), the elliptic ground-plan (in the tradition of old Transylvanian churches) and composition of the brickwork generating a sense of tranquillity and intimacy.

The commissioners and executants of architecture, those who give the orders to build and those who do the building, the cultures of patronage and of work, the system of social institutions and the regulatory role of culture—all play a fundamental role in, and share responsibility for, bringing anything of true value into existence. After its second millennium, there is a need for the architectural culture of Hungary to create an intellectual climate that creatively encompasses both the continuity and handing on of traditions and innovative approaches to meaningful, effectual ways of living. Architects and society must together establish the conditions for a Hungarian architecture that evolves along its own chosen path and is thereby able to contribute valuable colours and flavours of its own to the culture of a United Europe.

Notes

1. Virgil Bierbauer, *A magyar építészet története* (The History of Hungarian Architecture) (Budapest, 1937), p. 254.
2. Gáspár Fábián, *Nagy magyar építőművészek* (Great Hungarian Master-Builders) (Budapest, 1936), p. 9.

Mihály Szegedy-Maszák

GREAT LITERATURE IN AN INACCESSIBLE LANGUAGE

The Hungarian Enlightenment (1780–1790)

Although the origins of Hungarian literature can be traced back to the oral culture of the period before 1000 A. D., and the poetry written by Janus Pannonius (1434–72) and Bálint Balassi (1554–94) belong to the peaks of Renaissance literature, the Enlightenment was the first period to produce a wide range of genres and institutions. The eighteenth century ushered in a profound turn not only in Hungary's economic and social development but also in its culture. The reason for this was quite simple: towards the end of the seventeenth century the country, which had been divided into three parts for 150 years, won back its unity.

If one wished to encapsulate in a single word the main trend in the eighteenth century, it would have to be 'secularisation'. Faith was gradually yielding its central position to knowledge. Although the Enlightenment had certain indigenous roots, and the beginnings of the Hungarian Enlightenment cannot be ascribed to any single writer, the turning point can be considered to have come around 1765, when Joseph II became co-ruler, with his mother, Maria Theresa, of a Habsburg empire that included the kingdom of Hungary and György Bessenyei (1747?–1811) was admitted as a member of the Viennese Royal Bodyguard. The Edict of Tolerance (1781), issued by Joseph in the year after he became emperor, permitted Protestant as well as Greek Orthodox believers to hold public office and widened access to education for much of the population. Secularisation was further encouraged between 1781 and

1784, when the monarch dissolved all monastic orders not engaged in medical or instructional activities and redistributed their property and funds. In 1785 Protestants were given permission to establish schools wherever they wished. By 1788 it became compulsory to set up a school in every settlement that had at least 100 children living within an hour's walk.

The changes introduced in elementary and secondary school education created a demand for changes in related fields. Bessenyei outlined a plan for the modernisation of intellectual life. In *Magyarság* (The Hungarians, 1778) he paved the way for the language reform later promoted by Kazinczy, while in *Egy magyar társaság iránt való jámbor szándék* (A Humble Proposal for the Establishment of a Hungarian Learned Society, written in 1781 but published only in 1790) he made suggestions for the organisation of science and scholarship. His *Holmi* (A Miscellany, 1779), a collection of works in several genres, shows a profound understanding of the different trends within the French and English Enlightenment. In aesthetic terms, however, his output is less remarkable. His lyrics and plays are inferior to his essays, and his most important novel, *Tarimenes utazása* (The Travels of Tarimenes, 1804), written at the end of his life, was not published until the first half of the twentieth century.

Romanticism and Biedermeier (1790–1849)

After Joseph II's death, the Viennese court—prompted in part by a fear of the influence of the French Revolution—turned away from the ideas of the Enlightenment. In 1794 there was a round-up of intellectuals with suspected Jacobinist sympathies. Among those arrested was the translator, prose writer and poet Ferenc Kazinczy (1759–1831), who, after his release in 1801, organised a large-scale language reform. The two major lyric poets of the turn of the century, Mihály Csokonai Vitéz (1773–1805) and Dániel Berzsenyi

(1776–1836), made the transition from the then prevailing Baroque and Classicist styles to a subjective interpretation of sensibility. Csokonai's comic epic *Dorottya* (Dorothy, 1804) is full of grotesque humour and stylistic innovation. Combining late Baroque stylisation with elements of the Classicists' Horatian ideal of *ut pictura poesis* (poetry is like a painting), the tradition of popular culture, and a subjectivity influenced by Jean-Jacques Rousseau's cult of sensibility, Csokonai created an individual style in his two collections of lyrics, *Lilla* and *Ódák* (Odes), both published in 1804.

Whereas Csokonai was of humble origin, Berzsenyi belonged to the middle noble gentry. At his best, he combined the use of Greek and Latin verse forms with originality of diction and metaphor. The second version of his ode *A magyarokhoz* (To the Hungarians, completed in 1810) lamented the decay of the nation, highlighting the need for political reforms. Although a landowner, he became alienated from the traditional life of his class. His finest poems, *Levéltöredék barátnémhoz* (Unfinished Letter to My Lady, composed 1804–08), *A közelítő tél* (The Approaching Winter, 1804–08), and *Életfilozófia* (A Philosophy of Life, 1811), are elegiac expressions of a sense of decay. In his later years, however, he made an effort to participate in social life and composed poems that were either didactic or served as self-commentary. He blamed Bonaparte for deserting the cause of liberty in the epigram *Napoleonhoz* (To Napoleon, 1814), whereas in *Dukai Takács Judithoz* (To Judit Dukai Takács, 1815), an epistle addressed to an interesting woman poet, he wrote of *ut musica poesis* and the use of symbols as the characteristic features of his own work.

In 1817 a largely unfavourable review of the second edition of Berzsenyi's poems appeared in *Tudományos Gyűjtemény* (Scholarly Anthology), a periodical with 600–800 subscribers that played a formative role in Hungary's intellectual life between 1817 and 1841. The review's author, Ferenc Kölcsey (1790–1838), proved to be the first major figure in the history of Hungarian criticism. When Berzsenyi's reply was rejected by the editors, he started working on *Poétai harmonistika* (A Treatise on Poetic Harmony, 1832), a long essay on poetics that can be read as a Romantic manifesto.

Like Berzsenyi, Kölcsey was a poet whose work represented a transition from Classicism to Romanticism. He was a fine lyric poet when at the top of his form, but most of his verse has no great originality; as an essayist, however, he was superior to most of his contemporaries. History became his main interest, and one of his poems about the Hungarian past, *Himnusz* (Hymn, 1823), was adopted as the text of the Hungarian national anthem after Ferenc Erkel set it to music in 1844. Preoccupied with the idea of predestination, Kölcsey was inclined to pessimism. His best poem, *Vanitatum vanitas* (1823), gives a nihilistic view of history. A vision of the death of the Hungarian nation is presented in *Zrínyi dala* (The Song of Zrínyi, 1830) and *Zrínyi második éneke* (Zrínyi's Second Song, 1838). Inspired by Herder and Friedrich von Schlegel, in *Nemzeti hagyományok* (National Traditions, 1826), an essay that served as a starting-point for later poets, he compared the organic culture of Greece with the inorganic, imitative culture of Rome, urging Hungarians to draw inspiration from folklore in developing their own organic culture.

Several of Kölcsey's essays discussed the dramatic genres that were starting to appear as professional acting took its first, tentative steps in the country. The first Hungarian theatrical company was one assembled by László Kelemen in 1790. The first theatre in Pest opened in 1837, sixteen years after an establishment in Kolozsvár (Cluj, Romania), six years after one in Nagyszombat (Trnava, Slovakia), and three years after one in Temesvár (Timişoara, Romania). The swelling numbers of travelling companies and, later, of repertory theatres, as well as the rich variety of plays, are indices of Hungary's expanding bourgeoisie. In 1815 the editor of the periodical *Erdélyi Museum* (Transylvanian Museum) organised a competition calling for an historical drama to mark the opening of the theatre in Kolozsvár. József Katona (1791–1830), actor, translator, and adapter of foreign plays for Hungarian companies, submitted his five-act historical tragedy *Bánk bán* (1813–14). Katona's play was not even listed among the commended works, but he rewrote the text and published it in 1819. Although it was not performed until the 1830s, later generations came to regard it as the most important Hungarian tragedy of the first half of the nineteenth century, a work presenting a conflict between foreigners and natives, as well as between central power and feudal anarchy. Each of the characters speaks a highly idiosyncratic language, a wide range of mental states is represented, and tension is often heightened by clashes between individual styles. Such are the features that make *Bánk bán* a work of lasting value.

In the first half of the nineteenth century literature was evolving more rapidly than most other components of culture, a fact that could be ascribed to the intimate connection between politics and the Hungarian language. Along with the economic boom triggered by the Napoleonic wars, the opening of theatres in the larger cities, and the spectacular rise of the periodical press, the progress that had been made in the field of education also contributed to the increasing weight of the middle class, which grew in size as both nobles and more prosperous peasants joined its ranks. Social changes were facilitated, in part, by a reform movement whose advocates played an important role in a series of parliamentary sessions. The National Casino, a gentleman's club on the English pattern established in 1827 by Count István Széchenyi (1791–1860), set an example in promoting social intercourse, while his book *Hitel* (Credit, 1830) offered a program for both economic and intellectual advance.

The son of Ferenc Széchényi, a prominent representative of the Hungarian Enlightenment and founder of the National Library in 1803, István Széchenyi (who adopted this slightly variant spelling of the old family name) travelled extensively in Western Europe and the Near East, having distinguished himself as a young officer in the Napoleonic Wars. In 1825 he offered one year's income from his estates towards establishing an Academy of Sciences in Pest. Over the next 35 years he exerted a decisive influence on Hungary's economic, political, social, and intellectual renewal. The thesis promulgated in *Credit*, that the criterion of social good is the happiness of the greatest number of people, and its program of seeking a market economy, capitalist enterprise, and an end to the laws of entail on property would suggest an acceptance of Bentham's ideas, but Széchenyi's insistence on the intellectual independence of the individual represents a departure from the principles of utilitarianism. He explored the ontological implications of his Romantic liberalism in *Világ* (Light, 1831), and summarised his political agenda in *Stadium* (The Stage We Are At, 1833). In *Hunnia*, written between 1831 and 1835 but not published until 1858, he outlined a relativist definition of culture, insisting that the survival of even the smallest, most primitive nation is of utmost importance for mankind.

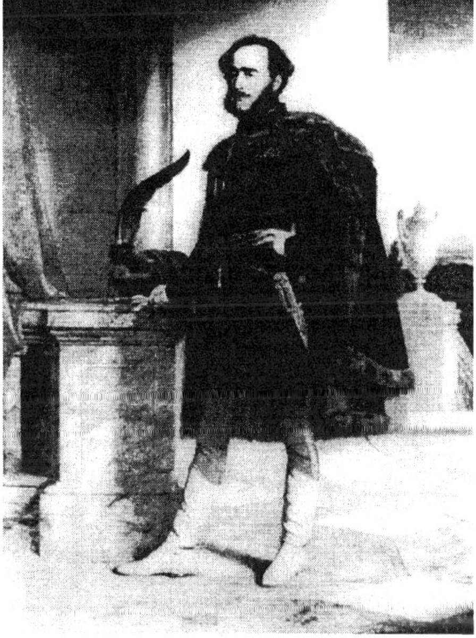
István Széchenyi

Széchenyi was the most important Hungarian essayist of the first half of the nineteenth century. The books and articles he wrote in Hungarian show him to be a brilliant stylist. His diary, written mainly in German and published in six volumes in 1931–39, is a *journal intime*, one of the great Romantic confessions. The other major genre of the Hungarian Romantics was the visionary lyric, as represented by Mihály Vörösmarty (1800–55).

Vörösmarty was unquestionably the finest poet of Hungarian Romanticism. Although the public expected him to compose epic and dramatic works, his greatest poems are apocalyptic lyrics in which he proved a highly original creator of metaphors and syntax. *Zalán futása* (The Flight of Zalán, 1825), the work that brought him fame, is a long verse epos about the Magyars who settled in the Danube Basin in the ninth century. It became an important source of national consciousness. Historically less significant but artistically more successful are the much shorter *Tündérvölgy* (Valley of Fairies, 1826) and the fragmentary *A Délsziget* (A Southern Isle, 1826), which testify to the poet's interest in the fantastic and the supernatural. *Magyarvár* (Fortress of the Magyars, 1830), another narrative fragment, is an attempt to re-create Hungarian pagan traditions. Like most Romantics, Vörösmarty was attracted to the Middle Ages, but as a liberal he viewed feudalism as self-destructive. *A két szomszédvár* (Two Neighbouring Castles, 1831) is a powerful interpretation of the family feuds that dominated medieval Hungary. Among his plays, *Salamon* (Solomon, 1827), *A bujdosók* (The Fugitives, 1830), and *Czillei és a Hunyadiak* (Czillei and the Hunyadis, 1844) are dramatisations of incidents from Hungarian history, inspired by Shakespeare, whose *Julius Caesar* and *King Lear* he translated into Hungarian. His most original play, *Csongor és Tünde* (Csongor and Tünde, 1828–31), is a dramatic fairy tale that sets out to interpret the history of mankind. The Night's monologue, marking the climax of this allegorical poem, is a frightening vision of the darkness and nothingness surrounding existence. *A Rom* (The Ruin, 1830) is a verse parable about a destructive God, and *Gondolatok a könyvtárban* (Thoughts in a Library, 1844) is a long meditation on the moral implications of knowledge. *Az emberek* (On Mankind, 1846) was inspired by the failure of the Polish peasant rising of that year to liberate Galicia from Austrian rule, whereas *Előszó* (Prologue, 1850) was composed under the influence of the tragic end of Hungary's own revolution. Although Vörösmarty was an active supporter of the anti-Austrian revolution that broke out on 15 March 1848 and was crushed the following year by the army of the Russian tsar that had intervened on the Habsburg emperor's side, his pessimism was not the consequence of any specific historical event. Long before that, *Szózat* (Appeal, 1836) already contains a prophecy of the possible death of the Hungarian nation. Among Vörösmarty's major poems *A vén cigány* (The Old Gypsy, 1854) is the only one to offer, in the final stanzas, a glimmer of hope.

The works of Sándor Petőfi (1823–49), the poet who died in action against that Russian army on the last day of July 1849, represent an epitome of the alliance between political liberalism and literature. Liberals wanted to make the country aware of the conditions of peasant life, and they found support from the poet whose ambition was to close the gap between oral and written, popular and high culture. Petőfi's descriptive poems, from *Az alföld* (The Lowlands, 1844) to *A puszta, télen* (The Puszta in Winter, 1848) and *Kis-Kunság* (Little Cumania, 1848), represented a radical transformation in the self-image of the Hungarians.

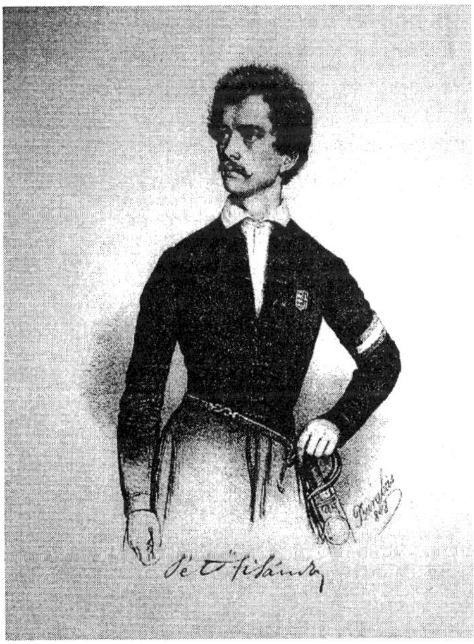
Sándor Petőfi

In these works the peasantry of the Hungarian plains is presented as the most significant element of the nation. The longer epic poem *János vitéz* (John the Valiant, 1844), inspired by fairy tales, suggests that national culture should be based on folklore, and it implicitly calls for the abolition of serfdom. Petőfi was a republican, imbued by the spirit of the French Revolution of 1789. *Egy gondolat bánt engemet* (One Thought Keeps Going Round My Head, 1846) is a prophecy of the triumph of liberty in the entire world, *Az ítélet* (The Judgement, 1847) predicts a radical social change, and *A tizenkilencedik század költői* (The Poets of the Nineteenth Century, 1847) contains a vision of universal equality. In 1848 he sought election as deputy for his native lowlands, but was rejected by the people. His bitter disappointment is reflected in the verse tale *Az apostol* (The Apostle, 1848). The poems he composed during the revolution were a running commentary on the events.

The style of some of Petőfi's poetry may be interpreted with the help of the concept of the Biedermeier. An influence of painting and lithography cannot be excluded, and the poet himself admitted that his early poem *Vándorélet* (A Vagrant's Life, 1844) was inspired by a painting of Miklós Barabás, the leading artist of the Hungarian Biedermeier. Petőfi's *oeuvre*, however contains a much wider range than descriptive lyrics, political verse, and genre or conversation pieces. *Tündérálom* (The Fairy Dream, 1846), a longer elegiac lyric imbued with a search for the self, shows him as a worthy follower of Vörösmarty. In the sequence *Felhők* (Clouds, 1846) he emerges as a most original master of the fragmentary Romantic epigram.

Absolutism and the Rise of the Novel (1849–1867)

The crushing of the civil revolution of 1848–49 was undoubtedly a watershed in Hungarian history. The losses were enormous. Many were forced into internal or external exile; the activities of institutions were drastically restricted. Still, it was at this time that some writers who had started their careers in the preceding Reform Era produced their finest works. In a two-volume work entitled *Der Einfluß der herrschenden Ideen des 19. Jahrhunderts auf den Staat* (1851–54), Baron József Eötvös (1813–71), whose liberal breadth had already been well attested by earlier essays *Szegénység Irlandban* (Poverty in Ireland, 1837–40) and *A zsidók emancipációja* (The Emancipation of the Jews, 1840) as well as by his fictional works—*A karthausi* (The Carthusian, 1839–41), a Romantic, confessional-style novel, *A falu jegyzője* (The Village Notary, 1845), a satire on the corruption of local authorities, and *Magyarország 1514-ben* (Hungary in 1514, 1847), a historical novel about a peasant uprising—presented a sophisticated analysis of the ways in which the ideals of equality, liberty, and nationality mutually contradicted each other.

Like Eötvös, Baron Zsigmond Kemény (1814–75) distinguished himself as a publicist. In 1842–45 his articles appeared in *Erdélyi Híradó* (Transylvanian Herald). When the liberals were defeated in his native Transylvania in 1849, he started publishing in *Pesti Hírlap* (Pest News) and moved to the Hungarian capital. Kemény was unquestionably the most serious Hungarian novelist of his century. *Gyulai Pál* (Pál Gyulai, 1847) is an attempt to realise Friedrich Schlegel's ideal of the novel as a synthetic genre, a dialogue of very different discourses. *Alhikmet, a vén törpe* (Al-Hikmet the Old Dwarf, 1853), is a 'dream vision' inspired by Orientalism and E.T.A. Hoffmann's tales, and his romance *Ködképek a kedély láthatárán* (Hazy Visions on the Horizons of the Mind, 1853) displays consistent Romantic irony and employs a *Doppelgänger* motif. The structure of this work is complex. Since the chronology of the events is very different from the order in which they are narrated, and the different parts of the story are told by different narrators, the reader is forced into interpretation. *Özvegy és leánya* (A Widow and Her Daughter, 1855–57) focuses on the psychology of a self-destructive elderly woman, *A rajongók* (The Fanatics, 1868–69) is a large-scale presentation of irrationalism in history, and *Zord idő* (Bleak Times, 1862), a novel about the Ottoman conquest of central Hungary, suggests that the goal of the oppressor is to convince the oppressed that he has been raised to a higher level by the occupation.

Kemény was obsessed with the dilemma of free will and determinism. János Arany (1817–82), the finest Hungarian poet of the period between Romanticism and the twentieth century, had a similar preoccupation. Viewing himself as a late-comer, he made irony and elegy the most important elements of his works. Although he was an excellent story-teller, even his narrative poems are dominated by lyricism and self-irony. Of the three verse tales he wrote about the legendary character Miklós Toldi—*Toldi* (1846), *Toldi estéje* (Toldi's Eve, 1847–48) and *Toldi szerelme* (Toldi's Love, 1879)—*Toldi estéje*, centred on the tension between the claims of progress and nationhood, is the most significant. The mock-heroic *A nagyidai cigányok* (The Gypsies of Nagyida, 1852) is a grotesque portrayal of the 1848–49 Revolution. *Buda halála* (The Death of Buda, 1863) is the only finished part of a trilogy he planned to write about the Huns, a poem focusing on the self-destructive nature of power. Fate and freedom of will are at the heart of Arany's ballads, which reconcile a desire to combine story-telling with dramatic tension and lyric tone. The peasant woman in *Ágnes asszony* (Mistress Agnes, 1853) believes that she has no choice but to help her husband; the heroine of *Zách Klára* (Klára Zách, 1855), a medieval lady-in-waiting, is raped by the younger brother of the queen. Both the pseudo-historical ballad *Szondi két apródja* (Szondi's Two Pageboys, 1856), and *A walesi bárdok* (The Bards of Wales, 1857), a free adaptation of Thomas Gray's Pindaric ode *The Bard*, have political undertones; they concern singers who refuse to praise their oppressors. His last ballads, such as *Tengerihántás* (Corn-husking, 1877) or *Vörös Rébék* (Red Rébék, 1877), in which polyphonic arrangement and fragmentation play a major role, manifest even greater structural complexity. Tormented by a sense of decay, Arany had doubts about any teleology. Life is viewed from the perspective of death, both in the short lyric *A lejtőn* (On the Slope, 1852–57) and in the self-reflective meditation *Mindvégig* (To the Very End, 1877), and existence is compared to endless circularity in the dramatic monologue *Az örök zsidó* (The Eternal Jew, 1860).

While the structural complexities of Kemény's prose fiction and the wordplay characteristic of Arany's verse are closely tied to the history and language of the Hungarians, *Az ember tragédiája* (The Tragedy of Man, 1859–60), a lyrical drama by Imre Madách (1823–64), contains hardly any reference to the history of the nation and has been translated into more languages than any other masterpiece of Hungarian literature. The opening scene portrays the Creation and the Fall: Adam is eager to know the fate of his race, and Lucifer gives him a chance to have a vision of the future through a long dream sequence of eleven episodes. Egyptian tyranny, Greek democracy, nihilistic Rome, the Christian Middle Ages, the scholarly Renaissance, the French Revolution, free-market capitalism (a London scene inspired by Hogarth's engravings), and centralised socialism are presented as failures. The deliberately unresolved ending of the drama has something in common with Nietzsche's reflections on the disadvantages of history for life.

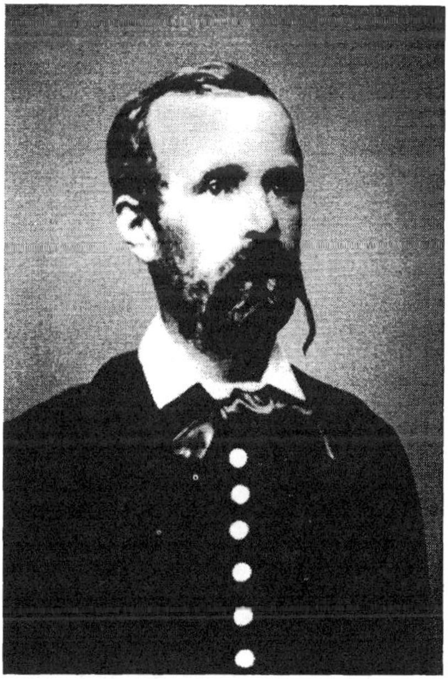
Imre Madách

This vast dialogue between Adam's Romantic liberalism and Lucifer's Positivistic determinism has attracted the attention of writers as different as Maxim Gorky, Virginia Woolf, and James Joyce. It is usually regarded as the last work of Hungarian Romanticism, although the Romantic legacy continued to attract the attention of the public until the end of the century, thanks to the activity of Mór Jókai (1825–1904), who was one of the most prolific Hungarian writers. This writer's popularity was enormous, not only in his lifetime but also in the decades following his death, although such critics as Pál Gyulai (1826–1909) and Jenő Péterfy (1850–99) held strong reservations about the aesthetic qualities of his works. Among his shorter works the best are the visionary short stories published in *Forradalmi és csataképek 1848 és 1849-ből* (Hungarian Sketches in Peace and War, 1850) and *Egy bujdosó naplója* (Diary of a Fugitive, 1851), inspired by the tragic end of the Hungarian revolution. Most of his adventure stories abound in clichés. *Egy magyar nábob* (A Hungarian Nabob, 1853–54) has a loose structure, but some of the anecdotes are well written and even show psychological insight. His best novel, *Az arany ember* (translated as Timar's Two Worlds and The Man With the Golden Touch, 1872), expresses nostalgia for a complete withdrawal from the commercialism of contemporary capitalism. In this work genre painting is in harmony with visionary description and there is an attempt to portray a split personality.

The Golden Age of the Dual Monarchy (1867–1890)

In the period of absolutism that followed the defeat of the 1848 revolution people lived under harsh conditions, yet great intellectual achievements were made. In the summer of 1866 war broke out between Austria and Prussia for the hegemony over German unity. Austria's defeat forced Vienna to sign a treaty with Hungary's political leaders. Austria may have regarded this *Ausgleich*, or

Compromise, as far too indulgent toward Hungary, but for the Hungarians it still represented an abridgement of their historic independence. This contradiction may explain the inconsistencies in the attitude of the writers of the next decades. Expressing contempt for the provincialism of those who represented passive resistance during the years of political oppression, János Vajda (1827–97), through two long pamphlets published in 1862, *Önbírálat* (Self-criticism) and *Polgárosodás* (Embourgeoisement), helped lay the ground for the Compromise, yet after 1867 he became one of its most violent critics. This apparent self-contradiction alienated him from many of his contemporaries, but his nonconformity made him influential in later decades. His Romantic cult of the subjectivist sublime was often at odds with his interest in biological determinism, but the historical significance of his lyrics cannot be questioned. His early cycles of poems, inspired by an unhappy love affair, were patchy; after 1870, however, he succeeded in introducing a highly personal form of love poetry and topics from urban life, thus marking a transition from Romanticism to twentieth-century poetry. *A vaáli erdőben* (In the Forest of Vaal, 1875) is an expression of irresistible nostalgia for the idyllic, and in *Húsz év múlva* (Twenty Years Later, 1876) unrequited love is turned into a symbol of the impossibility of existential fulfilment.

If we had to identify the salient features of Hungarian culture in the decades after the Compromise, we could say that they were the rise of capitalism and the advance of science and technology rather than an age of literary innovation. Vajda was a much more uneven poet than his great predecessors. The same is true of the verse of Gyula Reviczky (1855–89). He, too, tried to come to terms with conflicting views of the world, being drawn simultaneously to Schopenhauer and the positivists. A personal reason also explains his uncertainties: he had grown up believing he was of noble birth, but on his father's death it turned out that he was illegitimate. The autobiographical hero of his novella *Apai örökség* (Paternal

Heritage, 1884) has striking similarities with the 'superfluous man' in the works of Lermontov, Turgenev and other nineteenth-century Russian novelists. Accepting exclusion, he rejected the belated representatives of populism and called for greater openness to foreign literature and became one of the early translators of Baudelaire. Tone was his central interest, as most clearly shown in the poem *Számlálgatom...* (I Keep Counting..., 1889), composed shortly before his untimely death.

The rise of Hungarian capitalism led to radical changes in the political and cultural life of the country. In 1875 Kálmán Tisza became prime minister and the leader of a Liberal Party that united the two main opposing forces. As a result, the multiparty system lost much of its significance. The ruling party enjoyed such a large majority in Parliament that it was able to run the country uninterruptedly until 1890. Some writers, in their disillusion, went into opposition. János Asbóth (1845–1911) proclaimed his alternative vision of a new conservatism in his essays, and elaborated a psychology of disenchantment in his remarkable novel *Álmok álmodója* (Dreamer of Dreams, 1876), and László Arany (1833–98), son of János Arany and author of *Délibábok hőse* (The Hero of Mirages, 1873), a verse novel also influenced by Russian portrayals of superfluous men, underlined the decline of intellectual and moral standards in a series of essays on shallow liberalism.

The rapid growth of the Hungarian economy was greeted with optimism by some nationalists. The conflict between these and their opponents became obvious when Zsolt Beöthy (1848–1922) published his six-hundred-page *A tragikum* (On Tragedy, 1885). A spokesman of continuity with the national traditions of the nobility, Beöthy claimed that the tragic hero commits a transgression for which he must find expiation; hence tragedy, in the final analysis, is predicated upon the restoration of a universal moral order. Jenő Péterfy (1856–99), in his studies *A tragikumról* (On the Tragic, 1885), *A tragédiáról* (On Tragedy, 1887), and *A költői igazságszolgáltatás a tragédiában* (Poetic Justice in Tragedy, 1891), refuted this moralistic view and defined the tragic as an aesthetic quality in its own right. Jenő Rákosi (1842–1929) was also sharply critical of Beöthy's didactic approach to art. His two-hundred-page *A tragikum* (The Tragic, 1886) offers an ontological interpretation: in his view, the tragic is an inherent quality of existence. Retrospectively, this debate may be regarded as a starting-point of the opposition between aesthetic conservatism and modernism.

Innovation and Crisis at the Turn of the Century (1890–1919)

If the Age of Absolutism witnessed a tension between some great writers and public taste, the 1870s and 1880s saw the rise of mass culture. Theatres were opened, translations proliferated, and the education system was developing at a steady pace. Between 1881 and 1892 the number of periodicals published in Hungary rose from 504 to 734. Because of the rise of the press, the representative genre of the final decades of the nineteenth century was the short story. The Transylvanian tales of István Petelei (1852–1910), who came from a well-to-do Armenian merchant family, suggest a fatalism through the perfect fusion of the characters' interior monologues with a symbolic representation of nature. Regionalism is no less characteristic of the works of Kálmán Mikszáth (1847–1910), born in Upper Hungary, in a village of mixed Hungarian and Slovak population. Already in the four novellas of *Tót atyafiak* (Slovak Kinsmen, 1881) his powerful mode of presentation follows oral patterns and creates a distinctive ambience, but compression and ballad-like elisions became a feature of his art only in the shorter stories of *A jó palócok* (The Good People of Palocz, 1882).

Although a government representative until his death, Mikszáth was fully aware of the ethnic tensions that became apparent at the end of the nineteenth century. In *Club és folyosó* (Club and Corridor, 1887), a collection of political sketches, he provided a graphic description of the great powers' interests in manipulating inter-ethnic relations. Mikszáth's art was inspired by popular anecdotes. Theodore Roosevelt was said to have admired *Szent Péter esernyője* (Saint Peter's Umbrella, 1895), a short novel based on a popular superstition, and visited its author during his European trip in 1910. Mikszáth's other works reveal the ambiguities of the rise of capitalism. *A gavalérok* (Gentlemen, 1897) presents members of the impoverished gentry class who live in a dream-world. Count Pongrác, the hero of *Beszterce ostroma* (The Siege of Beszterce, 1895), represents the anachronistic illusions of the aristocracy, and yet he is above the petty utilitarianism of the more advanced world that surrounds him. Similarly, Pál Görgey in *A fekete város* (The Black City, 1910) is voluntarism incarnate, but his violent emotions and tormented soul make him a fascinating personality, in contrast to the impersonal puritanism and lack of imagination characteristic of the bourgeoisie of Lőcse (Levoça, Slovakia), the city that opposes his voluntarism.

The intimate world of Mikszáth's best tales was matched only in *Az én falum* (My Village, 1898) by Géza Gárdonyi (1863–1922), though this combination of psychological insight and personal confession is almost unique in the latter author's *oeuvre*. Among his other works, *Az egri csillagok* (Stars of Eger, 1901), a historical novel for children about the Ottoman occupation of Hungary, still enjoys well-deserved popularity. The novella *Az öreg tekintetes* (The Old Squire, 1905), which chronicles the increasing isolation of an old man forced to move from his provincial home to Budapest, is a moving portrayal of urban alienation.

Gárdonyi wrote far too much, and that could also be said of Sándor Bródy (1863–1924). Bródy mined a number of genres, but his best works were the short stories in which he identified with the elderly, outcast Rembrandt. These works were written at the end of his life, when, after the fall of the 1919 Commune, he was forced to live in exile. His earlier works—plays and narrative prose—are mainly of historical interest, marked by a strong influence of Western naturalism.

There was only one unquestionable great talent attracted to this artistic movement, Zsigmond Móricz (1879–1942), whose first novel, *Sárarany* (Dirt and Gold, 1910) is an attempt to focus on the role of instincts in the life of the peasantry. The tension between the environment and the individual in *Az Isten háta mögött* (Godforsaken, 1911), a short novel about the young wife of a school teacher, 'a provincial Madame Bovary', concealed an opposition to determinism, making it possible for the author to distance himself from naturalism.

The reaction against biologistic determinism was initiated by those who denied positivism in the spirit of a neo-Romanticism comparable to what is known as Secession or Art Nouveau. A search for the first signs of this trend would take us back in time, perhaps as far as the monthly illustrated *Magyar Szalon* (1884–1918), to which Vajda, Reviczky, and Mikszáth contributed. Certainly it would take us at least to *A Hét* (The Week, 1890–1924), edited by József Kiss (1843–1921), a poet of Jewish descent who was slow to free himself from the influence of János Arany's ballads. It was in this journal that the aristocratic Minka Czóbel (1855–1947), an eclectic follower of the French decadents, published her free-verse poems. The intellectual opposition included those committed to Naturalism as well as the representatives of the Hungarian Secession. Conservative taste was represented mainly by *Új Idők* (New Times, 1896–1949), an illustrated weekly edited by Ferenc Herczeg (1863–1945).

It is instructive to contrast Herczeg's career with that of the most innovative prose writer of the Hungarian Secession, Gyula Krúdy (1878–1933). Herczeg came from a bourgeois family of German origin in the south of historical Hungary (now the Voivodina region of Serbia) and attracted early notice. His first volume of short stories, *Mutamur* (1892), showed him to be a caustic critic of the gentry who, well aware of his own limitations, had an excellent sense of form and a flair for pithy evocation of atmosphere. Krúdy, by contrast, was an illegitimate child but he inherited much of his father's pride in his noble birth. Most of his early volumes were hastily written and derivative. In later years, Herczeg often lost his self-discipline, trying his hand at the grand historical novel, for example, and his works were marred by the shallowness of his portrayal of character. Krúdy, on the other hand, renewed the short story, especially in the collections from *Szindbád ifjúsága* (Sindbad's Youth, 1911) onwards. *Napraforgó* (Sunflower, 1918) shows him to have been one of the great Hungarian novelists of the twentieth century. By fragmenting both story and character, he managed to instil new life into the genre. Of the Secessionist prose writers, only Krúdy succeeded in creating longer works of enduring value; others were at best minor masters by comparison. The fantastic tales of Viktor Cholnoky (1868–1912), for instance, are permeated by a Secessionist decadence and mysticism. Géza Csáth (1887–1919) might be another example. His talent was diffuse. As a music critic he was quick to appreciate the value of new initiatives, including those of Béla Bartók, whilst *Anyagyilkosság* (Matricide, 1908) testifies to his interest in psychoanalysis.

In contrast to prose fiction, drama was dominated by the values of a successful entertainment industry rather than artistic innovation. The Vígszínház (Gaiety Theatre) opened in 1896, the Magyar Theatre and Kisfaludy Theatre in 1897, and the Király Theatre in 1903—all in Budapest. In was in these theatres that the plays of Herczeg and the Jewish-born Ferenc Molnár (1878–1952) were staged. Herczeg's pseudo-historical *Bizánc* (Byzantium, 1904) testifies to its author's awareness of the inner conflicts of the Dual Monarchy, whereas Molnár's *Az ördög* (The Devil, 1907) or *Liliom* (1909)—the latter a 'folk play' that became a successful musical in Rogers and Hammerstein's adaptation (*Caroussel*, 1945)—belong to popular culture, in the same way as his *A Pál utcai fiúk* (The Paul Street Boys, 1906), a novel later given mass appeal by several film directors.

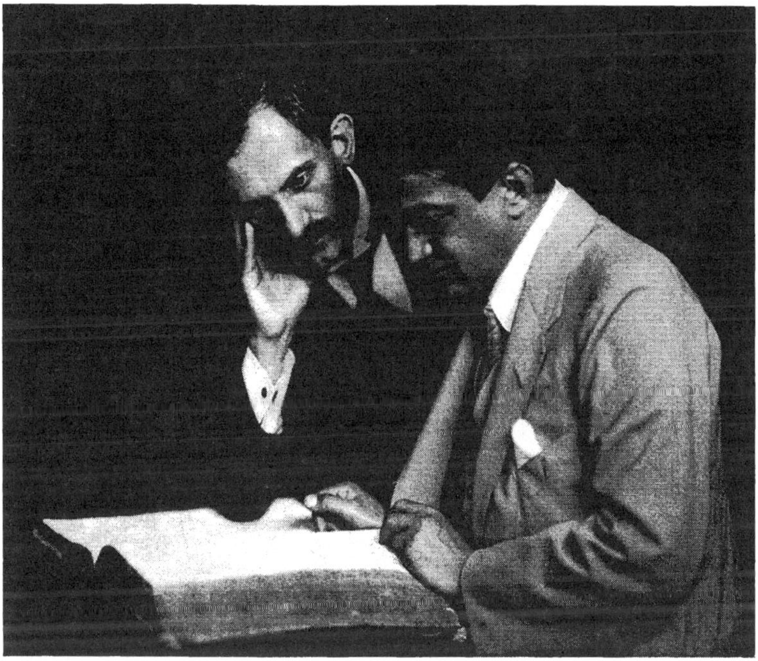

Mihály Babits and Endre Ady

Although the international impact of Molnár's outstanding stagecraft is undeniable, it would be futile to seek in Hungarian play-writing anything comparable to the profound change that had come over lyric poetry with the title poem of *A homályból* (From the Dark, 1895), a volume by Jenő Komjáthy (1858–95) reflecting his re-formulation of the Romantic cult of the self under Nietzsche's influence. No less profound was the German philosopher-poet's influence on Endre Ady (1877–1919), Mihály Babits (1883–1941), and Dezső Kosztolányi (1885–1936), three major poets attracted to Symbolism, Secession, psychoanalysis, and new trends in philosophy. Ady's third collection, *Új versek* (New Poems, 1906), followed by the even more significant *Vér és arany* (Blood and Gold, 1907), represented a new start in Hungarian literature, which he himself explored in a further seven volumes.

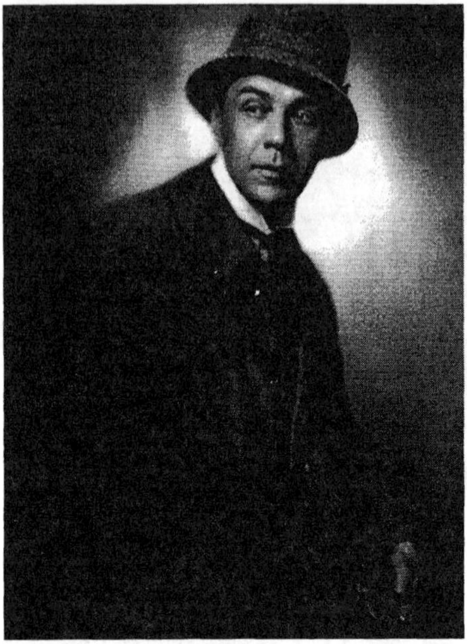

Dezső Kosztolányi

The highly stylised, Secessionist language of the early love poems, the prophetic pathos of political visions, the intentional obscurity of such pieces as *A fekete zongora* (The Black Piano, 1907), highly complex poems like *Az eltévedt lovas* (The Lost Horseman, 1914), written in a Symbolist style inspired by Baudelaire and Verlaine, and the tragic connotations of his devotional lyrics represent a wide range that make Ady the most influential Hungarian poet of the years before World War I. Not for nothing does *Így írtok ti* (Here's How You Write, 1912), a highly successful collection of parodies by Frigyes Karinthy (1887–1938), begin with texts caricaturing the idiosyncrasies of Ady's style.

Ady's formidable personality forced the other significant poets of his generation to seek their own, distinctive voices. Babits, who also wrote outstanding essays on literature and philosophy, forged a more objective lyric, starting with his very first verse collection, *Levelek Iris koszorújából* (Leaves from Iris's Wreath, 1909). His interest in William James's idea of stream of consciousness, Bergson's interpretation of the flux of time (*durée*) and Nietzsche's idea of eternal recurrence helped him create a poetry with ontological implications. While Babits turned to English poets for inspiration, Kosztolányi's chief source was Rilke. His second verse collection, *A szegény kisgyermek panaszai* (Laments of a Poor Little

Child, 1910), presents a young boy's vision of a world of plenitude, in contrast to the alienation and materialism of the adult world.

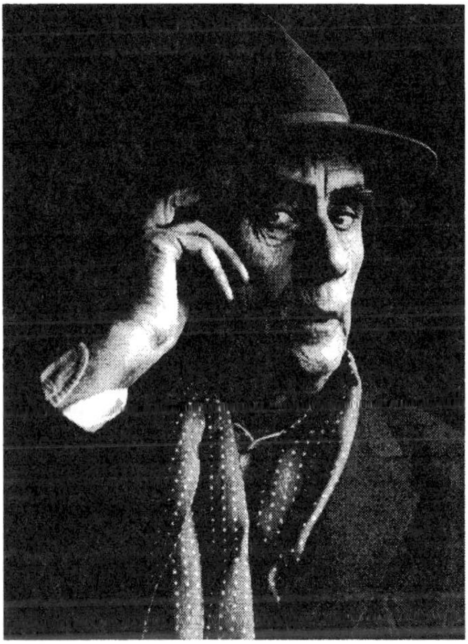

Lajos Kassák

Hungarian Modernism became institutionalised with the publication of four journals. *Huszadik Század* (Twentieth Century, 1900–19) was edited by the political scientist Oszkár Jászi (1875–1957). *Nyugat* (West, 1908–41) became associated with Ady, Babits, and Kosztolányi, *A Szellem* (Mind, 1911) was the organ of a group of urban intellectuals (mainly of Jewish parentage) that included the poet, playwright, essayist, and film critic Béla Balázs (1884–1949), the philosophers Lajos Fülep (1885–1970) and György Lukács (1885–1971), the sociologists Arnold Hauser (1892–1978) and Karl Mannheim (1893–1947), and the art historians Frederic Antal (1887–1954) and Charles de Tolnay (1899–1981). *A Tett* (Action, 1915–16), edited by the writer and visual artist Lajos Kassák (1887–1967), was the first journal of the Hungarian avant-garde movement.

Whereas Ady, Krúdy, Babits, and Kosztolányi were deeply rooted in the traditions of the lesser nobility and the provinces, Kassák identified himself with the working class. *Mesteremberek* (Craftsmen, 1915) is a conscious departure from the lyric ideal of *Nyugat* in that the poetic self speaks on behalf of an international community. The free verse Kassák used was radically different from the language of Ady and his contemporaries, with the exception of Milán Füst (1888–1967), whose first volume of poems, *Változtatnod nem lehet* (Nothing Canst Thou Change, 1912), was inspired by the language of the Old Testament.

Kassák labelled his avant-garde movement Activism, indicating that he wanted to implement an artistic revolution simultaneously and in conjunction with a social revolution of the working class, striving to create an Expressionist art with socialist commitment. He was openly opposed to World War I, which resulted in *A Tett* being banned for its pacifist articles in early October 1916. Unfazed, the next month he launched a new journal, *Ma* (Today), which played a significant role in preparing the ground for the civil revolution of October 1918. When a Communist Republic was proclaimed on March 21st, 1919, however, it was not only non-socialist writers but also Kassák who found themselves at odds with its totalitarian régime. At a national party meeting, the Communist party leader Béla Kun called the trend represented by *Ma* a product of 'bourgeois decadence', a statement that was printed in the June 14th issue of *Vörös Újság* (Red News). Kassák replied in the July 1st issue of *Ma*, expressing respect for Kun's political experience but questioning his competence in artistic matters and rejecting the view that art should serve any particular class. József Pogány, a deputy people's commissar, revoked the journal's permit.

Neo-conservatism and Its Liberal Opposition (1919–1930)

In the turmoil of the years following World War I Hungarian culture suffered severe losses. After the collapse of the Commune on 1 August 1919, Balázs, Lukács, and other leading participants left Hungary. The most influential early work of Lukács's, *Die Theorie des Romans* (1920), and his first, and possibly most original, contribution to Marxist theory, *Geschichte und Klassenbewußtsein* (1923) came out in Berlin, while Vienna became home to Kassák while he wrote *A ló meghal a madarak kirepülnek* (The Horse Dies the Birds Fly Out, 1922), a poem which fuses visions of the Russian revolution with memories of his 1909 travels in Western Europe.

The Treaty of Trianon, signed on 4 July 1920, not only cut off three million Hungarians from their native land, together with two thirds of the country's territory, but also rendered much of the Hungarian cultural heritage inaccessible. The reorganisation of the country's intellectual life became possible only upon the appointment of Count István Bethlen as prime minister on April 14th, 1921, and the entrance of Count Kunó Klebelsberg as minister of religious and educational affairs in 1922. Klebelsberg unveiled an ambitious cultural and educational program based on a conservative ideology outlined by the historian Gyula Szekfű in *Három nemzedék* (Three Generations, 1920), a book suggesting that Liberalism should be blamed for the political dead end of 1919. The remarkable influence of this work can be detected even in Móricz's most important historical novel, *Tündérkert* (Fairy Garden, 1922), inspired by the post-1849 revolutionary novels of Zsigmond Kemény. The replacement of visionary Gábor Báthory by pragmatic Gábor Bethlen as leader of Transylvania is here portrayed as inevitable. An explicit criticism of revolutionary utopia was also formulated by Dezső Szabó (1879–1945) in *Az elsodort falu* (The Village That Was Swept Away, 1919), a novel excoriating the urban population for its political influence.

In 1923 a new periodical was started in the spirit of the neo-conservatism of Klebelsberg and Szekfű. *Napkelet* (East) was intended to serve as a counterweight to *Nyugat*. Its editor, Cécile Tormay (1876–1937), was the author of Secessionist short stories and *A régi ház* (The Old House, 1914), a historical novel about the assimilation of Budapest's German bourgeoisie which was translated into numerous languages. The essays published in this journal, overseen by the outstanding literary historian János Horváth (1878–1961), provided trenchant support for Klebelsberg's cultural policies. Indeed, so pervasive was the Neo-conservatism of the political establishment in the 1920s that the representatives of prewar Liberalism were often put on the defensive. In his *Tanulmány Adyról* (A Study of Ady, 1920) Babits was constrained to refer to his friend (who had died a year before) as a follower of János Arany, and several authors turned to the genre of the historical novel in their search for an explanation for the collapse of the historical kingdom of Hungary. Even Krúdy tried to experiment with this genre, highlighting the roles of Habsburgs in the sixteenth century, in the trilogy *Három király* (Three Kings, 1927–30). Of his vast output from this period, what now stands out as having more enduring value had its roots deep in the experimentation of the previous decade. With their technique of providing two conflicting versions of a story, *Utolsó szivar az Arabs szürkénél* (A Last Cigar by the Arab Grey) and *A hírlapíró és a halál* (The Journalist and Death), both from 1928, are remarkable achievements. Krúdy's late novel *Boldogult úrfikoromban* (My Late, Lamented Times as a Young Rake, 1930) has an extremely slow rhythm: action is not just delayed but denied.

The other major novelist of the decade was Kosztolányi. Through the characters of Nero, Seneca and Britannicus, the protagonists of *A véres költő* (Darker Muses: The Poet Nero, 1922), a work whose German translation appeared with a prefatory essay by no less a figure than Thomas Mann, Kosztolányi attempted to answer the question of how far man's fate could be influenced by the irrational, a yearning for the supernatural, and stoicism. *Pacsirta* (Skylark, 1924), a short novel that appeared in Richard Aczel's first-rate English translation in 1993, is about a spinster living in a provincial town called Sárszeg ('muddy place') in 1899, an epitome of the post-historical state, of a loss of purpose that the whole world must face eventually. Values are no less ambiguous in *Aranysárkány* (The Golden Kite, 1925), a second novel about Sárszeg that hints at the irrelevance of enlightened rationalism in the twentieth century. The message is similar in *Édes Anna* (Anna

Édes, 1926), in which a peasant girl who seems to be the perfect maid murders her employers. The relation between the oppressor and the oppressed is portrayed as damaging to both sides.

The thinking of both Babits and Kosztolányi had been shaped by nineteenth-century Liberalism, which also provided a cultural-historical background to the Hungarian exponents of Diltheyan *Geistesgeschichte*, the concept of a broad cultural conspectus (*Zeitgeist*) in a given age that, for many areas of the humanities under German influence, came to form the most potent intellectual reaction to the prevailing positivism of previous decades. In Hungary, this movement also owed much to Klebelsberg's efforts to decentralise and his aim of providing the universities in the provinces with a distinctive character. Thus it was that the University of Pécs became a stronghold of *Geistesgeschichte*, captained by Tivadar Thienemann (1890–1985), head of the German Department, under whose editorship the journal *Minerva*, launched in 1922, became a workshop for a new generation of essayists.

Bourgeois and Peasant Culture (1930–1944)

The ultimately global recessionary effects of the Wall Street Crash, triggered by panic selling of securities on the New York Stock Exchange on October 24th, 1929, soon made themselves felt in Hungary. On August 18th, 1931, Bethlen's government resigned; Klebelsberg's plans to establish a network of cultural institutions thus remained only partially fulfilled. On October 1st, 1932, Gyula Gömbös became prime minister, and on January 30th, 1933, Hitler came to power in Germany. Cultural policy in Hungary changed, with a further widening of the gap between the political establishment and its Liberal opposition.

It would be wrong to deny strong elements of continuity, however. In a long essay entitled *Lenni, vagy nem lenni* (To Be or Not to Be, 1930), Kosztolányi identified with Széchenyi's Romantic

Liberalism, while in *A magyar nyelv helye a földgolyón* (The Place of the Hungarian Language on the Globe, 1930), an open letter addressed to the historical and general linguist Antoine Meillet, he outlined a theory of language-based cultural relativity. *Esti Kornél* (Kornél Esti, 1933), published in French translation in 1999, and a later section called *Esti Kornél kalandjai* (The Adventures of Kornél Esti) in the short-story collection *Tengerszem* (Mountain Tarn, 1936) display a passionate fascination with language games and a deliberate fragmentation of narrative continuity. *Ének a semmiről* (A Song About Nothing, 1933), the last poem in Kosztolányi's last verse collection, *Számadás* (Reckoning, 1935), comes close to giving an existentialist interpretation of nothingness. Of his shorter poems, the three-line *Októberi táj* (October Landscape), composed some time between 1926 and 1935, was not appreciated until decades later, while *Őszi reggeli* (Autumn Breakfast, 1929) and *Vörös hervadás* (Crimson Withering, 1931) inspired Attila József (1905–37), the most important Hungarian poet of the 1930s.

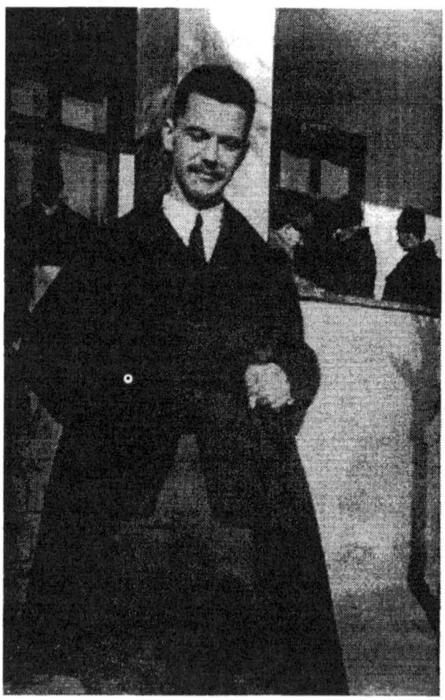
Attila József

József, whose poetry was admired by his friend Arthur Koestler (1905–83), a prose writer who left Hungary and wrote first in German and later in English, and is widely translated into Western languages, travelled an incredibly long artistic road in a matter of a few years. In the highly cryptic cycle *Medáliák* (Medallions, 1927–28) he merged, in a singularly original manner,

Expressionist style with montage. József's aspiration was to reconcile the ideas of Marx and Freud. On joining the underground Communist Party, he wrote such explicitly political verse as *Munkások* (Workers, 1931), yet his vision of the Hungarian cultural legacy was based on a misleading dichotomy: in *Ady emlékezete* (The Memory of Ady, 1930) he praised Ady for his revolutionary attitude, but in *Egy költőre* (On a Poet, 1930) dismissed Babits for his scepticism. Later on, however, he distanced himself from the Communists and developed a style of great complexity. While the influence of Croce can be detected in his aesthetic writings, some of his poems, such as *Reménytelenül* (Without Hope, 1933), reveal an Existentialist orientation. In *Eszmélet* (Consciousness, 1934), probably his greatest poem, he struck a powerfully intellectual note. Of the poems composed just before his suicide, the self-addressing *Karóval jöttél...* (You Came with a Stake..., 1937) and *Tudod, hogy nincs bocsánat* (You Know There Is No Forgiveness, 1937), or *Talán eltűnök hirtelen...* (I May Suddenly Disappear..., 1937) testify to the poet's desire to understand nothingness and his ability to write with great economy. His ode *A Dunánál* (By the Danube, 1936) appeared in a special issue of *Szép Szó* (Fine Words). This literary magazine, initially edited by Pál Ignotus and Attila József, later by Ferenc Fejtő, offered its columns to writers of various persuasions in an issue entitled *Mai magyarok régi magyarokról* (Today's Hungarians on Hungarians of the Past).

The very fact that Babits was among the contributors to the latter issue suggests that his position had strengthened since the early twenties. In 1927 the aesthete Ferenc Baumgarten bequeathed his estate to a literary foundation and asked Babits to act as judge of an annual prize. Two years later, following the suicide of Ernő Osvát, Babits had the chance to become editor of *Nyugat*. While these two events placed additional burdens on him, they also provided him with a greater say in the country's cultural life. He rose to the challenge, reiterating his opposition to violence and his condemnation of irrationalism. In his collection of poems *Versenyt az esztendőkkel* (Racing the Years, 1934) he unequivocally subordinated the individual to tradition, and he wrote the synthesis *Az európai irodalom története* (A History of European Literature, 1934–46)—later published in German, in 1949, as *Geschichte der europäischen Literatur*—with the obvious intention of educating his public to focus on a supranational heritage. *Ősz és tavasz között* (Between Autumn and Spring, 1937), his late poem on death, invokes various tonalities from the past, *Balázsolás* (Supplication to St. Blaise, 1937) advocates the Christian acceptance of death, and *Jónás könyve* (The Book of Jonah, 1939) is a remarkably sustained exercise in the classicising mode, a poetic rewriting of the biblical text.

Neo-classicism certainly had links with the neo-conservative spirit of the 1920s, but the former also had an important international dimension. Only a few years after his return to Hungary in 1926, Kassák himself published poems in traditional verse forms. The avant-garde seemed to have run its course, and there arose a demand for non-fiction. A series of autobiographical works were written at this time, starting with *Fekete kolostor* (Black Monastery, 1931) by Aladár Kuncz (1886–1931), a chronicle of the five years the Transylvanian author spent in a prison camp because he happened to be in France at the outbreak of World War I. This moving work, detailing the sufferings a great admirer of French culture had to endure due to the chauvinism of the nation he loved so much, was followed by Kassák's eight-volume *Egy ember élete* (One Man's Life, 1927–35), then *Egy polgár vallomásai* (Confessions of a Citizen, 1934–35) by Sándor Márai (1900–89), scion of a family of German background in the Upper Hungarian city of Kassa (Košice, Slovakia), and the three-part *Erdélyi történet* (The Writing on the Wall—A Transylvanian Story, 1934–40) by Count Miklós Bánffy (1873–1950). Although each gives an account of a different stratum of society, these works have one thing in common: they all view their respective communities from the outside. Kassák observes the working-class environment he was born into with grotesque humour; Márai sees a conflict between the bourgeois and the anarchic 'upholders of civilisation'; and Bánffy portrays the aristocracy as a class doomed to fall.

The move away from the avant-garde, which characterised the 1930s, can even be felt in the poetry of Lőrinc Szabó (1900–57). The poet of the volume *Különbéke* (Separate Peace, 1936) was fascinated by the reworking of borrowed material. Unlike his former master Babits, however, Szabó was stimulated not only by occidental models, as can be seen in his poem *Dsuang Dszi álma* (The Dream of Chuang Tzu, 1936). In fact, in the 1930s Lőrinc Szabó became interested in the ideology of a movement that is usually called Populism.

The origins of this 'back to the roots' agrarian movement were political rather than aesthetic. It was a response to social tensions. Among representatives of the older generation, Móricz was one of the first to call the attention of his readers to the miseries of the life of the peasantry. His superb short story *Barbárok* (Barbarians, 1931) was written with typically balladistic elisions; the longer *A boldog ember* (A Happy Man, 1935), based on interviews with a poverty-stricken peasant who is content with his lot, constitutes a transition to non-fiction. Of those Transylvanian writers who began their careers in Romania after World War I, it was Áron Tamási (1897–1966) whose outlook most resembled that of the Populists. He made his mark with short stories suggestive of the world of peasant oral culture, and with his novel *Ábel a rengetegben* (Abel Alone, 1932), the first part of a grotesquely humorous trilogy about a young peasant boy. Dezső Szabó's intellectual creed loomed large for the Populist writers, especially after his series of pamphlets called *'Ludas Mátyás' füzetek* ('Ludas Mátyás' Booklets), launched in 1934, offered a comprehensive critique of political and cultural life in Hungary. Another author who came to support the movement was the essayist, novelist and playwright László Németh

(1801–75). At the start of his career, in the late 1920s, Németh was an independent critic who held reservations about both the spokesmen of urban modernisation and those who upheld the values of traditional rural life. He found himself alone with his views, as he states in the confessional *Ember és szerep* (Man and His Role, 1934). Taking the idea of national character from Széchenyi, about whom he published a book-length essay in 1942 and a play in 1946, he reached the conclusion that Hungarians could develop their culture only through organic evolution. Regarding the development of the Hungarian bourgeoisie as inorganic, he decided to join the Populists. The starting-point of his long pamphlet *Kisebbségben* (In the Minority, 1939) is that the intelligentsia cannot do its job unless it proceeds by degrees from the masses. This idea provoked Babits and the Liberals, who attacked Németh for considering the assimilation of foreigners—both Germans and Jews—as superficial.

Practitioners of the other arts, historians, and ethnographers may also have inspired the Populist movement. The genre of social reportage revealed a convergence between literature and ethnographic description. Lajos Nagy (1883–1954) gave the impetus to this initiative. His tautly composed *Kiskunhalom* (1934), portraying an imaginary village in Central Hungary, was followed by Péter Veres' (1897–1970) documentary book, *Az Alföld parasztsága* (Peasants of the Alföld Plain, 1936). *Puszták népe* (People of the Puszta) by Gyula Illyés (1902–83) should be mentioned as the most original contribution to this genre. This confessional piece, combining a description of the poverty of farm labourers with a nostalgic evocation of the author's own childhood, has seen several editions in G. F. Cushing's excellent translation since 1967. Other examples of this type of non-fiction include *Csángók* (The Csángó People, 1941), a scholarly account of the life of the Magyar-speaking community of Moldavia by László Mikecs (1917–44), and *Baranyai utazás* (A Journey to Baranya County, 1941) by the novelist János Kodolányi (1899–1969). Meanwhile, György Sárközi

(1899–1945), a Jewish-born poet who moved from a neo-conservative position toward that of the Populists, in 1937 launched a series of sociographic studies under the banner *Magyarország felfedezése* (The Discovery of Hungary).

These were the years when a division between 'Populists' and 'Urbanists' began to crystallise—most visibly in the running clashes between *Kelet Népe* (People of the East), a journal established in 1935 and edited by Móricz from 1939, with *A Toll* (The Pen), a weekly headed by journalist Béla Zsolt from 1919, and *Szép Szó*. One reason for the debate was that the Populists were not at all united politically, embracing as they did antagonistic factions of crypto-Communists and right-wing radicals holding anti-Semitic views. The German threat, the outbreak of World War II, and the anti-Jewish laws of 1938, 1939, and 1941, of course, had a profound effect on Hungarian culture. Although political pressures intensified, some tried to preserve the autonomy of literature. Márai embraced the ideal of the multi-ethnic state of King (St.) Stephen, the monarch who turned Hungary into a Christian kingdom in *Röpirat a nemzetnevelés ügyében* (Pamphlet on the Issue of Nation-Building, 1942).

It is little short of miraculous that works of art were still being created in these bleak times. This was the period when Milán Füst composed *A feleségem története* (The Story of My Wife, 1942), a novel that appeared in English translation, some thirty years after French and German editions, in 1987. Among the novels that Márai wrote during the war, *Szindbád hazamegy* (Sindbad Goes Home, 1940), a moving testimony to the art of Krúdy, may be the best-composed in the original, although two others, *Vendégjáték Bolzanoban* (Guest Appearance in Bolzano, 1940) and *A gyertyák csonkig égnek* (Embers, 1942), proved to be surprise commercial successes in the late 1990s, the latter becoming a best-seller in Italy (*Le braci*) and Germany (*Die Glut*). The poet Sándor Weöres (1913–89), who started his career as a child prodigy of fourteen, possibly reached the peak of his art with the volume *Meduza* (Medusa, 1944). The striking imagery of one poem in particular, the twenty-stanza *Dalok Naconxypanból* (Songs from Naconxypan), may be viewed as some compensation for a previous general absence of Surrealist influences in Hungarian poetry. Miklós Radnóti (1908–44), who by contrast was moving away from Expressionism and Surrealism towards a more Neo-classical style, wrote his most significant poems whilst on a forced march from a German labour camp in Yugoslavia towards Germany, ending with him being shot in western Hungary. His last volume, *Tajtékos ég* (Foamy Sky, 1946), thus appeared posthumously—a dismal monument to a period that had been so rich in intellectual and artistic endeavour.

The Rise and Fall of Bourgeois Culture (1945–1948)

The German occupation of Hungary on 19 March 1944 not only put an end to the country's independence; it also paralysed its intellectual life. The losses sustained turned out to be severe. It was not only those of Jewish descent who met death in extermination camps or left the country; many from other backgrounds went into exile or were killed as the Russian army occupied the country in late 1944 and early 1945. Continuity was broken also by the arrival of Communist apparatchiks from Moscow. Béla Balázs, György Lukács, and the critic József Révai (1898–1959) belonged to a generation that had been active in the movement since the early twentieth century but, unlike younger colleagues, had lived in the Soviet Union before and during World War II. Largely ignorant of the conditions in Hungary, they made an immediate attempt to impose Soviet ideology on the country's cultural life.

Despite the intimidating back-up of Soviet troops, the Muscovite Communists met with considerable resistance. The surviving representatives of the literature of the inter-war period, with few resources at their disposal, tried to restore continuity by starting new periodicals. In 1946, a group of the younger generation of poets, foremost amongst them Weöres, János Pilinszky (1921–81) and Ágnes Nemes Nagy (1922–91), decided to start the monthly *Újhold* (New Moon), which, by adopting the title of Radnóti's 1935 collection, made open reference to the tragedy of the Holocaust. Posthumous publication of Radnóti's last collection specifically served as a point of departure for Pilinszky, the most important poet to emerge in the immediate post-war years. *Trapéz és korlát* (Trapeze and Parallel Bars, 1947) represented a new start: identifying himself with the fate of the Holocaust victims, the young Catholic poet focused on a cosmic homelessness to create a language of great complexity in short and cryptic pieces. His influence was felt even in the work of Weöres, who included a series of one-line poems in his *A fogak tornáca* (The Colonnade of Teeth, 1947).

Révai had already been working on the ideology of a popular front back in the 1930s. His chief objective on returning to Hungary became to strike a deal with compliant middle-class writers and others claiming to represent the interests of the peasantry. There can be little doubt that the reasons for favouring these Populists were mainly political; whatever the case, though, a campaign was launched against bourgeois values.

One of the most important books published in Hungary in 1945 was the *Napló* (Diary) that Márai kept during 1943–44. "The Hungarian middle class is so rotten that it still does not want and has no face to courage reality," comments an 1943 entry on those who believed in a German victory. After March 19th, 1944, when Germany directly occupied the country, Márai lived in internal exile, and when the persecution of Jews started, he identified himself with the victims. Yet Márai's diary was characterised by Lukács, in a lecture delivered in December 1945, as 'vulgar bourgeois individualism'. The text of this long lecture was placed in a wider context when Lukács' *Irodalom és demokrácia* (Literature and Democracy) was printed in April 1947. To promote Realism, the author asked for the introduction of tighter controls and outlined a program with the aims of "destroying the reactionary thinking of imperialism."

After the Hungarian Communist Party had coerced the Social Democratic Party to merge with it on June 12th, 1948, many publications were banned. One of these was Márai's *A Garrenek műve* (The Work of the Garrens), the fifth part of a *roman fleuve*, which amongst other things presents the figure of a visionary 'Leader' addressing a public demonstration. Although this part was based on Márai's article about Hitler's 1933 speech at the Berlin Sport Palace, it is possible in retrospect to assume that the Communist authorities saw it as a general criticism of all totalitarian systems.

Stalinism (1948–1956)

The period between 1948, the year when Márai was forced to leave Hungary, and 1956 represent the darkest period in Hungarian literature. Most of the country's important writers were silenced, including the avant-garde poet Kassák, the Roman Catholic Pilinszky, the Populist László Németh, and István Bibó (1911–79), who advocated a 'consensual revolution' in his long essays *A magyar demokrácia válságáról* (On the Crisis of Hungarian Democracy, 1945), *A kelet-európai kisállamok nyomorúsága* (The Distress of the East European Small States, 1946), *Eltorzult magyar alkat, zsákutcás magyar történelem* (Distorted Hungarian Character and the Dead-End of Hungarian History, 1948), and *Zsidókérdés Magyarországon 1944 után* (The Jewish Question in Hungary After 1944, 1948). Those attacked by the Communists in the post-war years were forced to such resorts as writing children's books or translating from Russian to make a living. The authorities' cultural policies demanded that all artists should offer a socialist realist picture of the changing world around them, yet this vaunted socialist realism was, at best, little more than a figment of sterile academicism. Révai accused Tibor Déry (1894–1977) of giving a distorted picture of the workers' movement of the 1930s in the unfinished novel cycle *Felelet* (Reply, 1951–52). In *Die Zerstörung der Vernunft* (1955) Lukács branded as forerunners of Fascism the very German writers and thinkers whom he had so enthusiastically admired in his youth; in the essay *Probleme des Realismus* (1955) he went on to reject the modernism of Kafka, Musil, Gottfried Benn, and the avant-garde virtually *en masse*. The important works written in this period remained unpublished, including *Egy mondat a zsarnokságról* (One Sentence on Tyranny), a poem scathingly critical of totalitarianism that Gyula Illyés composed in 1950 but did not publish until October 1956.

Goulash Communism (1957–1989)

The suppressed revolution of 1956 proved a watershed in culture as well as politics. Bibó and Déry were imprisoned; Illyés fell silent for several years; many younger writers left the country. Literary activity recovered only slowly from the resultant paralysis. With a subsequent gradual softening of official cultural policy towards the intelligentsia, a continuity of sorts was restored with the heritage of pre-1948 Hungary.

Weöres was increasingly drawn to the imitation of earlier modes of writing, prefiguring with his *Merülő Saturnus* (Saturn Submerging, 1968) the eclecticism of the post-modern works of more recent decades. *Psyché* (1972), presenting the works of an imaginary poetess of the early nineteenth century, marks the high point of these 'imitations'. Pilinszky's masterpiece, *Apokrif* (Apocrypha), invoking the figure of the Prodigal Son in an apocalyptic vision, considerably antedates *Psyché*. It was written between 1950 and 1954, but was published only in *Harmadnapon* (On the Third Day, 1959), a collection that lends itself to both a tragic and a Christian interpretation. Pilinszky subsequently composed epigrammatic pieces, the fragmentary overheard monologues of the volumes *Szálkák* (Splinters, 1972), *Végkifejlet* (Dénouement, 1974), and *Kráter* (Crater, 1976).

While Pilinszky drew upon the poetry of Attila József, Ferenc Juhász (1928) and László Nagy (1925–78) defined their identity in relation to the Populist writers. The former was inspired by Bartók's *Cantata profana* to produce *A szarvassá változó fiú kiáltozása a titkok kapujából* (The Boy Changed into a Stag Clamours at the Gate of Secrets, 1955), a re-writing of a folk ballad that was hailed by W. H. Auden on the appearance of its English translation in 1970. Fragments of oral culture are also incorporated into Nagy's long dramatic monologue *Menyegző* (Wedding, 1964).

Verse, so vital a part of Hungary's culture in the 1960s, lost much of its influence in more recent decades. Dezső Tandori (1938), the most gifted poet of the post-1956 generation, started as a worthy follower of Pilinszky, with *Töredék Hamletnek* (A Fragment for Hamlet, 1968), while in *Egy talált tárgy megtisztítása* (Cleansing of a Found Object, 1973) he explored the possibilities of anagrammatic self-reflectiveness. In the sonnets of *A mennyezet és a padló* (The Ceiling and the Floor, 1976) he began a phase marked by a proliferation of quotations which has since stimulated him to astonishing poetic fertility.

In the twenty years after 1956 the parable came to be the preeminent prose genre. Outstanding examples are *Iskola a határon* (School at the Frontier, 1959), a highly innovative novel by Géza Ottlik (1912–90) suggesting the self-preserving power of internal autonomy; *A látogató* (The Case Worker, 1969), a work by György Konrád (1933) that combined a radical critique of Communism with a style somewhat reminiscent of the *nouveau roman*; and *Sorstalanság* (Fatelessness, 1975) by Imre Kertész (1929), an unusually ironic novel about the Holocaust. Along with these, the fiction of Miklós Mészöly (1921–2001) anticipated the endeavours of the coming generation. While his early short story *Jelentés öt egérről* (A Report on Five Mice, 1958) was a pure allegory of defencelessness, his later works are more implicit: the protagonist of *Saulus* (1968) persecutes his own consuming despair through other people, and the multi-layered structure of *Film* (1976) almost masks its vision of history as an endless succession of violent acts.

In the 1970s a new generation of prose authors emerged. Written between 1959 and 1972 (but published only in 1977), *Egy családregény vége* (The End of a Family Story) by Péter Nádas (1942) was at attempt to test the validity of tradition through the fate of the Central European Jewry. His second large-scale novel, *Emlékiratok könyve* (A Book of Memories, 1986), focuses on a divided Berlin and Europe, portraying Central Europe as a world that awaits redemption through its invocation of a more general applicability of the historical tragedies of the recent past and its simultaneous questioning of the traditional dichotomy of life and art. Péter Esterházy (1950) similarly attempts to blend autobiography and fiction. The first part of his *Termelési regény* (A Novel of Production, 1979) is a parody of the officially prescribed genre of socialist realism, while the second part, relating the author's personal history, is written in the form of notes to the first part—a structure that breaks up the usual linearity of the reading process. *Bevezetés a szépirodalomba* (An Introduction to Literature, 1986) offers an assemblage of seemingly incongruous and incommensurable modes of writing which include a highly intertextual section about Kosztolányi and his cousin, Géza Csáth; a collection of jokes about the 1950s; a short narrative, abounding in quotations from Pascal, that seems to be about a rape but can be read as an allegory of the Soviet occupation of Hungary; and a sublime, even religious testimony to the memory of the author's mother.

Post-communism (1990–)

There is a general feeling that Hungarian literature was more exciting in the 1980s than it has been since the political changes that occurred around 1990. The shift from Communism to Post-communism may have led to a decline in the prestige of literature, and the need for reading between the lines, so important in the period of goulash Communism, has lost its justification. For some time even Esterházy seemed to have no word for the new conditions; the articles he published were topical, rather than of enduring value. Recently, however, he published another major work. The two-part, 712-page *Harmonia caelestis* (2000) can be read in two different ways; on the one hand, as a book that undermines the distinction between fact and fiction, on the other, as a chronicle of the Esterházy family which encapsulates the whole of Hungarian history.

A similarly retrospective orientation characterises the other memorable work published in the last decade, *Sinistra körzet* (Sinistra District, 1992), by Ádám Bodor (1936). This short novel is a frightening allegory of the political nightmare of Ceauşescu's state with none of the structural tricks and metafictional or self-referential devices characteristic of Esterházy's prose. The Transylvanian-born author's prose is extremely economical, free of literary allusions, and shows barely a trace of anything but the most bleakly ironic humour, dealing with a form of existence in which there is no individual freedom and human fate is directed by invisible forces. The hero is visiting the region—a sort of penal colony inhabited, except for a few bureaucrats, solely by people who have been condemned to captivity there—with the purpose of finding his adopted son, who, it emerges, has followed the example of many others by committing an astonishingly ugly suicide, prompting the hero to escape.

Thus the most interesting literature coming from Hungary in recent years is closely related to the sufferings of those who spent their formative years during the Communist era in this part of the world. Bodor may have left Romania and Esterházy may have survived internal deportation with his parents as a child, but the former is haunted by memories of the activities of the Securitate and the latter cannot forget the persecution of the family after World War II and in 1956. The conclusion seems inescapable that the best works written in a Post-communist world have been inspired by memories of totalitarianism.

Gergely Bikácsy

HUNGARIAN CINEMA IN EUROPE

Hungarian Silent Films: Success and Depression (1908–1924)

Exploring the history of the silent film in Hungary puts one in the mind of an archaeologist, except that archaeologists at least have tangible, visible relics to deal with. That is rarely the case for scholars of early film, since the copies of most silent films have been lost, so they themselves no longer exist and all that can be investigated is the trace they have left in the press of the day or subsequent memoirs.

Budapest audiences got their first chance to view a motion picture in April 1896, less than four months after the Lumière brothers' first public exhibition of the cinématographe in Paris (though the first permanent cinema was only to be opened ten years later). Film historians consider the first Hungarian-made film to have been one entitled *Tánc* (Dance, 1901). Lasting no more than a few minutes, this was shot on the roof terrace of Budapest's imposing Uránia Theatre and documented the history of dance and dancers at the Royal Opera House, featuring prominent Hungarian artists of the day. Regular film production, though, did not start in Hungary until as late as 1908 and for years consisted of little more than 're-makes' of mock newsreels or comic sketches, which were projected during the interludes of live stage shows. Many of these were written by Frigyes Karinthy, then a fledging writer and humorist, who had a fine eye for the technological and intellectual innovations of his time. He himself parodied these motion-picture interludes in exquisite short prose pieces that are still enjoyable today and, indeed, preserve a record of those lost reels—a case of writing being more durable than film stock.

The first Hungarian feature film of any length to be shot (some 900 meters, with a playing time of 20 minutes) was by choreographer-turned-film producer Ödön Uher (1897–1953). Almost two dozen film production companies were formed between 1912 and 1916, though most soon went out of business and only four survived. One was a studio set up in Kolozsvár (Cluj), in Transylvania, and headed by National Theatre director Jenő Janovics (1872–1946). In co-production with the French film company Pathé, and with a French director, they shot a film entitled *Sárga csikó* (Yellow Colt, 1913), an adaptation of a then well-known *népszínmű*, or 'folk play' (this distinctive genre of mawkishly romantic rural melodra ma, with set-pieces of dancing and singing, enjoyed incredible popularity in Hungary during the nineteenth and early twentieth centuries, but the texts now read more like Karinthy-style parodies and have long been unstageable). *Yellow Colt* was distributed by Pathé with considerable success all over Europe.

Film-making developed apace up to the end of the First World War: by 1913 there were 114 movie houses in Budapest. Far from knocking it into a slump, the war gave a further boost to both production and distribution, to the point that one can speak about a genuine Hungarian film industry. Films produced by hostile countries—France and, later, the USA and Italy—were banned, prompting a heavy demand for Hungarian films. One of the biggest wartime hits was *János vitéz* (John the Valiant, 1916), adapted from Sándor Petőfi's verse romance and directed by Jenő Illés (1877–1951), who had returned home a few years before from a spell of working in Germany—it was, in fact, this old hand's hundredth film.

By the middle of the decade, two figures had emerged as leading directors in the blossoming industry: Mihály Kertész (1888–1962), and Sándor Korda (1893–1956). Mihály Kertész, who achieved world fame in Hollywood under the name Michael Curtiz, started his career in Hungary as an actor but soon became a key force in domestic film production. Director of around fifty films, he left for Austria in the spring of 1919 and, after a few years working in Europe, eventually settled in America. Amongst his early successes were two melodramatic crime thrillers, *Ma és holnap* (Today and Tomorrow, 1912), starring Arthur Somlay, and *Rablélek* (Captive Soul, 1913), one of the first full-length Hungarian feature films. The ability to keep an audience on the edge of its seats and move it to tears witnessed by these two films was something Kertész had picked up abroad, having already in 1912 made a study trip to the Danish studio of Nordisk, then a leading film company, where he had also done a bit of directing on one of their sets. Although he was not in the habit of resorting to classics or well-known literary works, it was not only for movies set in the present day that Kertész was acclaimed. Starring Mari Jászai (1850–1926), one of the leading actresses of the time, in the role of Queen Gertrude, his adaptation of *Bánk bán* (1914), the tragedy that József Katona had written almost a century before, and the first major drama to be written in Hungarian, proved a resounding success, albeit only in its native land.

Mihály Kertész soon went on to become the leading director of a new film production company, ‘Phoenix’. Strange as it may seem, during the period when American films were proscribed, a growing number of films were being produced in wartime Hungary—102 in 1918 alone—and five new studios were built within just two years. Apart from a few directors with a background of experience and success in the theatre world, most of the Hungarian films cranked out in this period of four or five years were made by Kertész and Korda. Sadly, however, almost all of the output from these two directors was subsequently lost.

The short-lived Hungarian Soviet Republic nationalised film production in April 1919, just a couple of weeks after it came to power—the first state to do so, preceding even the USSR. A total of 31 films were shot during the Commune, but the only one now extant is *Tegnap* (Yesterday, 1919). The films themselves do not seem to have reflected the rapid political and ideological changes of that period. Bolshevism or not, the few months for which the Commune endured were just a continuation of the previously established trends of pre-Bolshevik times: apart from Korda, the directors of these films were all seasoned theatrical hands. As for literary adaptations, Hungarian films continued to mine successfully the works of the best-known foreign authors, such as Émile Zola, and, amongst domestic writers the novels of Sándor Bródy, the most distinguished of the realist-naturalistic school of writers around the turn of the century. The central authority did nevertheless map out a plan for several years ahead—never implemented, of course—in which the 'main line' was to have been the adaptation of modern classics—Zola again, but also Gorky, Shaw and Ibsen—with no suggestion of radical breaches or rejecting the legacy of the recent past.

Mention should be made, if only briefly, of the luminaries of Hungarian silent films. Sári Fedák (1879–1955) was already a celebrated operetta diva, but the screen made its own stars from the likes of Kamilla Hollay (1899–1967), Ica Lenkeffy (1896–1955), Ila Lóth (1899–1975)—'the Hungarian Mary Pickford'—and Lucy Doraine (her stage-name; 1898–1989), as well as from such leading men as Oszkár Beregi (1876–1965), Béla Lugosi (1882–1962), Michael Varkoni (1891–1976), Paul Lucas (1894–1971), and Peter Lorre (1904–64)—most of whose names were soon to become internationally known as, over the next few years, they transferred their careers abroad, generally landing up in Hollywood.

As the fact that 31 films were made during the Commune's 130 days suggests, most Hungarians in the film business carried on actively and even enthusiastically. They managed to put out a specialist journal, *Vörös Film* (Red Cinema), and 19 editions of a socialist newsreel, *Vörös Riport Film* (Red Report Film). There was little time for any alarming intimations of censorship to raise their head. The fall of the Soviet Republic, on the other hand, saw a shoal of film directors, actors and producers leave the country in 1919 and 1920, in the wake of Hungary's temporary occupation by the Romanian army and the reprisals taken in the 'White Terror' that followed as the authoritarian Horthy régime took over government. The greatest loss, after that of Mihály Kertész, was undoubtedly Sándor Korda, but mention should also be made of Lajos Biró (1880–1948), a talented young short-story writer and dramatist, who later worked as a screenwriter on Korda's films in England.

Mihály Kertész left Hungary in April 1919, almost as soon as the Soviet Republic was declared, leaving behind unfinished his adaptation of Ferenc Molnár's *Liliom* (the play on which Richard Rogers and Oscar Hammerstein's musical *Carousel* is based) and not even waiting for the premiere of his *Jön az öcsém!* (My Brother's Coming Home!, 1919), a recently finished anti-war short film. Whilst he was in Vienna, he was engaged to direct several big-budget costume spectaculars. The most important of these was *Sodom and Gomorrah* (1922), featuring many Hungarian émigrés, such as Lucy Doraine and Michael Varkoni, which, despite the ambitions of its producers, fell some way short of W.G. Griffith's *Intolerance*, the film that had served as its model for both the screenplay and the directing. Even so, this Austro–German production was the biggest blockbuster to be shot in Europe up to the middle of the decade. It is intriguing to consider how it was possible for such lavish film productions to be mounted in cash-strapped Austria, in the grips of destitution and recession, when much the same conditions all but put the Hungarian film industry out of business. Another of those big-budget epics, *Samson and Delilah* (1922), was produced by Korda, likewise an émigré by then, and directed by Kertész. The huge success of that earned Kertész an invitation by Warner Brothers to come to Hollywood, where he started to direct films as Michael Curtiz. Thus the man who was arguably the most prominent figure in early

Hungarian film-making was before long to make himself part of American film history, and his abilities to be known around the world, as the creator of such swashbuckling adventure films as *Captain Blood* (1935), *Robin Hood* (1938), *The Sea Hawk* (1940), and *Casablanca* (1942).

Although he excelled in several other stock genres of the American cinema (film noir, thrillers and crime stories) besides the adventure film, it is the psychological melodrama *Casablanca* with which Curtiz is most indelibly associated in film history. To jump ahead a little in order to assess that enduring monument, it would be over-simplistic merely to label *Casablanca* as the best known kitsch melodrama in a hundred years of film, as there is more to it than just kitsch or clichéd melodrama. It is no coincidence that both Woody Allen, who has spliced cuts from *Casablanca* into several of his own films, and Umberto Eco should make a point of emphasising the film's influence on mass culture. If there is any truth in the notion that love is basically nothing more than pure kitsch, then *Casablanca* is the most durable of all love films—a monument to unhappy love set against the unhappy historical background of the Second World War. The stoical lead figure played by Humphrey Bogart projects an air of both naïveté and cynicism, and it is this, over and above the brilliantly scripted dialogues and convoluted yet clear plot line, that abides as the film's real substance.

To cut back to the Hungarian silent film, Sir Alexander Korda, as he was to become, was born in Hungary and was one of the most prominent producers in the silent film era. He started his career there as a journalist and, having had access to the Pathé company whilst working as a correspondent in Paris, one of the country's earliest significant film critics.

From 1908 onwards, Budapest had its specialist magazines devoted to film, such as *Mozgófénykép Híradó* (Cinematography News), even if this was little more than a disguised promotional vehicle for the film production company that published it. With

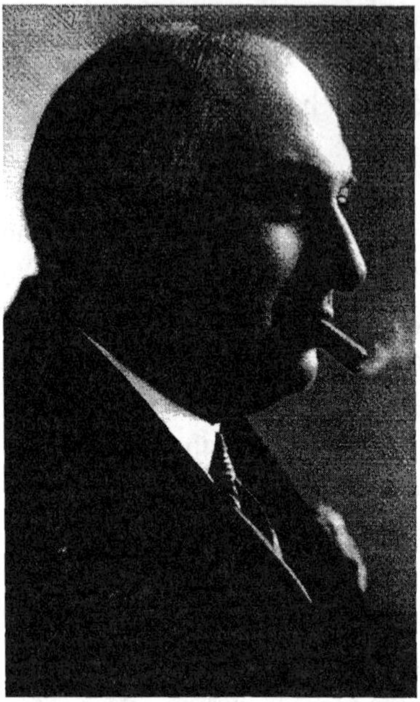

Sir Alexander Korda

young writers like the aforementioned dramatist Lajos Biró also contributing notable articles to it, Sándor Korda came to notice in the columns of the *Híradó* during its first few years with his reports on the films that Nordisk was bringing out. By 1912 Korda had his own regular film review column in a daily newspaper then went on to set up his own weekly magazine, which often featured articles by prominent writers of the day. For three years he also edited *Mozihét* (Cinema Week, 1915–22), possibly the most important and certainly longest-lived of these specialist journals. It was during this period that *Nyugat* (West), launched in 1908 and from its start the most pivotal Hungarian literary magazine of the first half of the twentieth century, began to give film serious consideration.

Korda first stood behind the camera at the end of 1914 then carried on his directorial work with the Kolozsvár-based film company for a while before purchasing its facilities in Budapest to set up his own 'Corvin Film Studio', which, largely through his own activity as a director, became an important company in its own right. As at other Hungarian studios, the bulk of the films he produced were adaptations of literary works—amongst others, *Gólyakalifa* (The Stork Caliph/The Nightmare, 1917), based on a psychological novel of that title by *Nyugat* writer and poet Mihály Babits. With Korda directing, a screenplay by Frigyes Karinthy, and Oszkár Beregi, who later emigrated to the United States, taking the double role of the main protagonist, the film had a very warm critical reception.

This was just one of many works by well-known Hungarian writers of the day to be adapted for the screen during this period, with films on 'contemporary' subjects distinctly more common than costume dramas, though film versions of several literary classics also count as significant milestones. Amongst these was *Mire megvénülünk* (When We Get Old, 1916), based on one of the books of Mór Jókai, the leading nineteenth-century Hungarian novelist, who in translation was also a best-seller in England and the USA around the turn of the century (this novel, originally published in 1865, appeared in England under the title *Debts of Honor* in 1900). Directed by Ödön Uher, this was Hungary's first 'full-length' feature film, around 90 minutes long (*c.* 3,000 meters). *Aranyember* (Man of Gold), the most famous of all Jókai's novels (published in England in 1888 under the title of *Timar's Two Worlds*), was adapted for the screen by Sándor Korda in 1918. Lasting almost three hours, this was the longest Hungarian feature film to date and is often referred to as the first Hungarian 'epic'. The biggest box-office success of the period was a film comedy, *Harrison and Barrison* (1917), also by Korda, which gained wide distribution in the United States.

Having been a leading figure in filmmaking during the Soviet Republic, Korda was one of many forced to emigrate when the régime fell in August 1919. After a spell in Vienna, he set up his own company in Germany in 1923. Whilst the sprawling *Samson and Delilah*, directed by Kertész, from the Vienna years has already been mentioned, several of the films he produced in Germany, including *Madame Dubarry* (1926), are arguably of greater artistic significance. From 1926 Korda began filming in Hollywood then finally took up residence in England, forming London Film Productions in 1932. Over the next twenty years (with a break in the USA during 1940–43), he applied his formidable skills as a film maker, producer and director—with the assistance of his brothers Zoltán (1895–1961) and Vince (1896–1979) Korda and, in no small measure, the scripts provided by Lajos Biró—to raising the British film industry to technical and artistic heights that rivalled those of Hollywood.

Korda has a rightful place in every encyclopaedia and history of cinema not just for the huge part he played in the blossoming of British film production but also for establishing new genres with films such as *The Private Life of Henry VIII* (1933), *The Private Life of Don Juan* (1934), and *The Ideal Husband* (1948). Major directors and famous actors and actresses went to work in his studio. He was a master in the clever blending of spectacular and thrilling action with fastidious, romantic yet at times nigh-realistic attention to historical and psychological detail. *The Thief of Baghdad* (1940) and *The Jungle Book* (1941) are two illustrative examples. Although the former was directed by Michael Powell, the latter by Zoltán Korda, in both style and atmosphere both films bear the hallmarks of Alexander Korda's influence in being half-way between children's tales and romantic action films with adult appeal; leavened with humour and even the occasional touch of irony, they create an air of legend not only solely through colours and inventive trick shots but also through their narrative and pictorial technique, an approach that blends innocent astonishment with zippy excitement. When they could be shown again in Hungary in 1956–57, after some years on the banned list, they helped audiences escape for a short while from a bleak everyday world under dictatorship.

Film production in Hungary carried on after 1919–20, albeit at a slower pace and, with so many of its creative talents having left the country, to lower artistic standards. Some value is placed on a few works by Márton Garas (1885–1930), who had already counted as a top director before the war, but in dealing with this period it is usual to speak of 'crisis' even though nearly 90 films were completed during the five years from autumn 1919 to the end of 1924.

The best-known and most successful director of that period was Béla Balogh (1885–1945), who, although he directed several films during the Soviet Republic, only became a prolific director after his more talented colleagues had left the scene. His better-known works from the early 1920s include *Elnémult harangok* (The Bells Are Silenced, 1922), *A megfagyott gyermek* (The Frozen Child, 1921), *Fehér galambok fekete városban* (White Doves in a Black City, 1923), and *A Pál utcai fiúk* (The Paul Street Boys, 1924)—the latter, based on a Ferenc Molnár novel, being of specific interest for scenes shot on location in the streets of Budapest. Balogh had difficulty finding a distributor for his films and in the end, during the Great Depression, emigrated. In the same way, Alfréd Deésy (1877–1961), a successful director who had not blotted his copybook during the Soviet Republic, preferred to work in Vienna. Deésy returned to Hungary in 1922 to direct an eagerly anticipated film about the poet Sándor Petőfi, to mark the centenary of his birth, but despite starring popular stage artists such as Tivadar Uray (1895–1962) and Gizi Bajor (1882–1951)—it was a terrible flop. Its hollow pathos did not go down well with the cinema's petty-bourgeois public.

From its flourishing position just a few years before, Hungary's film industry after 1923 entered into a massive funding crisis that put most companies out of business. A mere 14 feature films were released between 1925 and 1930. By the end of the silent film era in 1929, regular film production in Hungary had virtually ceased to exist. Amongst the performers who left Hungary to find success in Europe and Hollywood were the actresses Vilma Bánky (1903–1991), one of Rudolf Valentino's screen partners, and Lia von Putty (1897–1931), a big star of German films of the 1920s.

Pál Fejős (1898–1963), a significant figure in cinema history and ethnographic film documentary in particular, deserves rather more attention than the relatively minor role that he was accorded

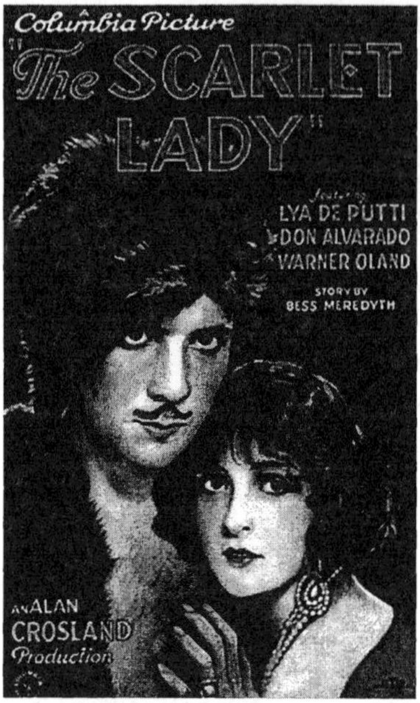

Lia Putty with Don Alvarado
in *The Scarlet Lady*

in his native land might suggest. Even as a young and talented director, with no political reasons forcing him to emigrate after the fall of the Commune, Fejős found it hard to establish himself in recession-stricken Hungary. He was obliged to stop shooting a half-finished *Egri csillagok* (The Stars of Eger) and leave the country in 1923. After gaining experience on some valuable American films, he returned briefly to Hungary at the beginning of the talking film era. In line with his original profession, he had embarked on medical research at the Rockefeller Institute in New York, but the opportunity then arose in 1927 to make *The Last Moment* (1928), which portrays the visions that flash through the head of a drowning man in the last few seconds before he dies. This attracted considerable attention (Chaplin himself was amongst those praising it), prompting Universal Film Studios to engage Fejős to direct *Lonesome* (1928), a masterpiece of historic significance whose tone foreshadowed the neo-realist and *nouvelle vague* trends of the fifties.

After several more American films, the by now internationally well-known Fejős went on to France and made French versions of several Hollywood productions. In the early 1930s, he was taken on by Osso, a Franco–Hungarian co-production venture that had made a number of films in Hungary, to shoot two of their films. Fejős thus got a chance to direct again in Hungary. Of the two productions, *Tavaszi zápor* (Spring Shower, 1932), which was distributed internationally under the title *Marie, une légende hongroise*, stood out from the run of Hungarian films at the time in using the devices of folk legend to depict the tragedy of a 'fallen' young village girl. It was precisely this avowedly symbolic, lyrical film's overdoing of the symbolism that proved its weakness, for the fairy-tale aspect is intercut with a harsher delineation of village life without the two levels binding together, the ersatz folklore sequences being smuggled into the film for the benefit of French audiences. Whilst not, perhaps, as masterly as the films that Fejős had made in Austria, France or America, it was still considered a model of pictorial form in Hungarian cinema, with its innovatively bold stylisation, calculated use of montage, changes of pace, and triumphantly spectacular imagery, even if its 'lyricism' dated it rather rapidly. On the other hand, it was certainly enhanced by the acting contribution of Annabella, its French star. The critical reaction to *Marie* in Hungary was acerbically provincial, with the conservative press perceiving it as a slur on the country's rural life.

Fejős's other Hungarian film, *Ítél a Balaton* (The Waters Decide, 1932), likewise received few appreciative reviews and was not a hit with the public either. He again left the country, this time for good. The most durable creation of his next artistic period was shot in Vienna, using a Hungarian screenwriter's script. *Sonnenstrahl* (1933), a European counterpart of *Lonesome*, is a bittersweet comedy about an unemployed pair of Austrian lovers that gives a memorably refreshing portrait of the vicissitudes of everyday Viennese life. Fejős was one of the pioneers of early sound films, employing artistic techniques that most other directors only truly began to latch on to later on. He shot several more films in Denmark then, in the mid-1930s, gave up making feature films altogether to join various archaeological and anthropological expeditions to Latin America and the Far East. The scientific documentaries that he made of these trips again made him a pioneer of a distinctive new film genre as well as an acknowledged ethnographer.

Low point though it may have been for domestic film production, the period after 1920 witnessed the emergence of a significant body of film criticism. Iván Hevesy (1893–1966) wrote a regular film column for *Nyugat*, providing film reviews and theoretical articles that vied with the best in Europe at the time. The same magazine also gave space for the occasional essay on major films of the day by the likes of Artur Bárdos (on V.I. Pudovkin's *Chinghis Khan's Successor*), poet-artist Lajos Kassák (on Walter Ruttmann's *Berlin—Symphony of a City*) and art critic Ernő Kállai (on Eisenstein's *Battleship Potemkin*). Hevesy also edited an independent film periodical, but with its uncompromisingly high standards it ran to just two issues in the autumn of 1925. Mention should also be made of a short film made by the émigré avant-garde artist László Moholy-Nagy and the curious 'visual screenplay' that Kassák published in his periodical *Ma* (Today).

The 1920s were also a time when several books on the aesthetics of the motion picture made an appearance. Especially noteworthy is *A film útja* (The Path of the Film, 1927), a slim booklet by the Marxist Lajos Gró, and more particularly Iván Hevesy's *A filmjáték esztétikája és dramaturgiája* (Aesthetics and Dramaturgy of Cinematography, 1925). Though Hevesy is still largely unknown outside Hungary, most of the conclusions that he reached in this prescient work remain valid and in many respects have stood the test of time better than the work Béla Balázs (1884–1949) published in Berlin, *Der sichtbare Mensch, oder die Kultur des Films* (1924), which was soon being hailed world-wide. Oddly enough, though he is Hungary's best-known film aesthetician, Balázs showed little interest in films prior to his emigration in 1919. He had been regarded as an intriguing, highly controversial poet and dramatist when he was living in Budapest, providing the 'mystery play' that formed the libretto for Béla Bartók's *Duke Bluebeard's Castle*. The philosopher György Lukács wrote an important review on what he saw as the significance of his friend Balázs's poetic *oeuvre*.

Besides his literary works, Balázs also published an interesting collection of essays on art theory—*Halálesztétika* (The Aesthetics of Death, 1908)—but it was only after leaving Hungary that he discovered cinema for himself whilst writing more or less regular one-off film reviews for Austrian and German publications. All the films to which he makes reference in *The Visible Man* are ones that he saw after 1919, either in Vienna or in Berlin.

Balázs does not propound any systematic aesthetics of film in this book; its approach is that of a fervent poet, the text charged with emotion, passion and lyricism. Though this is not an academic work, he parades his thinking on film aesthetics throughout, suggesting, for example, that close-ups and montage are the two most important characteristics of the film: "The close-up is the film's way of emphasising details. It directs the viewer's attention towards important, essential matters—not just depicting but evaluating. Two films with the same content, with identical long shots and the same actors, may reflect completely different views of the world if the close-ups are different,"[1] he wrote, making it clear in the process that he was looking at the visual world of a film from the perspective of the film's world view and content. He legitimately points out that the film is the only art form that is root-and-branch a product of capitalism: this determines not only its production system but its very nature, its artistic essence. Balázs's views on film were to exert a great impact internationally on the aesthetics of the genre over subsequent years. In *Das Geist des Films* (1930), he somewhat modified his earlier views on the overwhelming importance of montage in light of his experiences of sound films, but it was precisely his advocacy of the montage that made him the pioneer that he undoubtedly was, and an aesthete with the most profound understanding of the visual language of silent film. As the Italian film aesthete Guido Aristarco has commented: "Eisenstein and Pudovkin deserve credit for the view which espouses and proclaims that film is the art of montage, but it is also due, in very large measure, to Balázs. We are unable to establish who has priority for the theory of montage, Eisenstein or Balázs. Balázs's book certainly preceded those by the Russians, but the truth is also that they arise from one and the same root, were conceived in one and the same intellectual and artistic climate."[2]

Balázs was also active in Germany as a screenwriter and a film director. He worked with Reinhardt and Piscator, wrote a screenplay for Sándor Korda and several for German films, including Pabst's *Die Dreigroschenoper* (1931) and also co-directed *Das blaue Licht* (The Blue Light, 1932) with Leni Riefenstahl, then at the beginning of her career. Following Hitler's rise to power, Balázs emigrated to Moscow, where he taught at the film school. He engaged in an interesting theoretical debate with Eisenstein on the essence of montage, but in 1945 both of them were accused of formalism.

After the war, Balázs returned to Hungary. With his undeniable authority and Marxist worldview, he could count on gaining an exemplary role. He edited a film periodical. He played a major role, both as screenwriter and artistic adviser, in the production of Géza Radványi's (1907–86) historic masterpiece *Valahol Európában* (Somewhere in Europe, 1947). He was a founder of the School of Film Art in Budapest and also a part-time teacher at the College in Prague yet, despite his Marxist stance and Communist views, he soon found himself the target of attacks from Stalinist critics. Party ideologists vilified his literary and his theoretical work alike and blocked him from teaching. For all that he had been a recipient of the state's Kossuth Prize, Balázs died in 1949, an embittered, rejected man. A Budapest studio for young film artists has borne his name since 1959, but it was not until decades after his death that the full texts of his two crucially important books on film aesthetics were published in Hungarian.

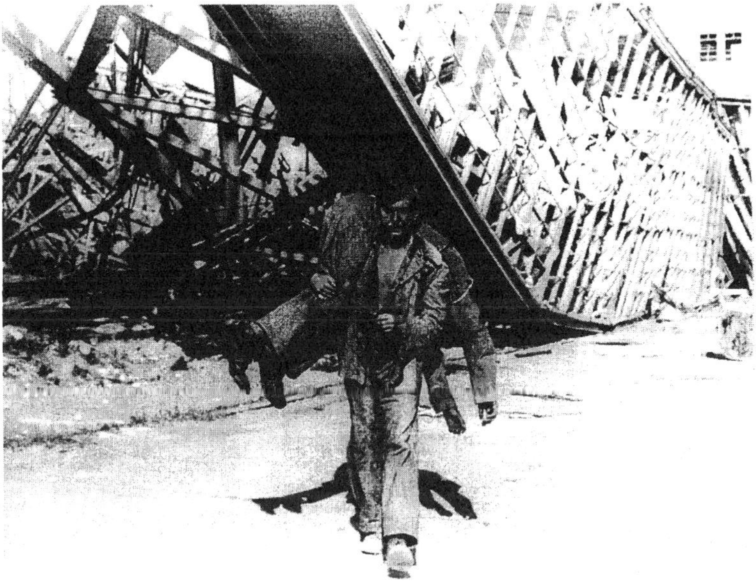

Scene from *Somewhere in Europe*

'Passengers in a Wonder-Car': Mass Entertainment and Artistic Endeavours, 1930–1945

In Hungary at the start of the 1930s there was no sign or hint that its near-defunct domestic film industry was at the threshold of a decade and a half of triumphant revival. The very first Hungarian-made sound film, *A kék bálvány* (The Blue Idol, 1931), was, admittedly, a failure, but the second, *Hyppolit, a lakáj* (Hippolyt, the Lackey, 1931), was a somewhat unexpectedly huge hit. The screenplay, though based largely on situation comedy and verbal humour, also gives scope for ironic portrayal: the main character is a parvenu businessman who is unable to discard his old habits and so gets into a series of comic scrapes, being lorded over in his own house by a butler who was once in aristocratic service. The film's director, the young István Székely (1899–1979), was launched on a highly successful career, which enabled him to secure regular work for many years thereafter.

Hippolyt, the Lackey's "unexpected and sweeping success had a lasting effect on the genre, subject-matter and acting-style of Hungarian cinema. In effect, the Hungarian film industry between the two wars attempted to pull off another success like that of *Hippolyt*,"[3] is the apt comment of István Nemeskürty (from whom the title of this chapter has been borrowed). He also adds that Károly Nóti, *Hippolyt*'s screenwriter, was also in the script-writing team on half of the Hungarian films of the 1930s, and that István Székely was the busiest director of that decade. The true glory of *Hippolyt*, however, is the quality of the acting. While the film itself offers little more than stock wisecracking and stage farce, the actors playing the two main characters draw on deeper levels of humour. The great stage actor Gyula Csortos (1883–1945) certainly turns in an exquisite performance as a butler mimicking an aristocratic mien, but the real star is Gyula Kabos (1887–1941).

For Kabos too *Hippolyt* was the lucky break that made his career. With his impersonation of an archetypal little man who strikes it rich, a nonentity unable to shake off his gaucheness but who slyly revolts against domination by his snobbish wife and haughty butler, refusing to be shamed any longer in his new milieu, he formed the mould for the sort of loveable mediocrity that now came to be a defining type in Hungarian film comedies, lending a more lifelike dimension to otherwise vacuous, artificial films that drew on the lower forms of slapstick humour or shallow imitations of the stereotyped conventions of *fin-de-siècle* French farce. The lower middle-class townspeople who made up the bulk of cinema audiences had little interest in the inauthentic world of landowners and high-ranking military officers as portrayed by film. For Hungarian films presented a completely anachronistic, non-existing feudal world whose main figures were mainly country squires carousing to gypsy music and grand ladies; where a peasant would only appear on the screen if it was cap in hand before his landlord or the bailiff, much as in folk-plays; whilst chauffeurs, ticket-inspectors, postmen and tradesmen were mere supporting roles, and factory workers were nowhere to be seen. It was as if Hungary were comprised entirely of countryside and, at most, the royal palace and old city on Buda's Castle Hill; even a metropolitan setting was extremely rare. One such exception was *Lila akác* (Wisteria, 1934), based on the 1919 novel of that title by Ernő Szép, an over-candied István Székely confection, though engaging enough for all that, with a handful of authentic outside street locations, a wan dash of irony and an unmawkish tone.

In most of his films Kabos played a bungling yet astute and sober small functionary who has the audience laughing with him; less often he was a chief accountant or, once only, a company manager. His milieu is that of secretaries and accountants or, amongst his social betters, at most engineers, who come across on the screen less as engineers than cashiered army officers, down-at-heel aristocrats and, above all, as *actors*—false to the very core, though maybe a touch more true-to-life when Antal Páger (1899–1986) and possibly Pál Jávor (1902–59) began to take such roles, but still light years away from any sort of realism as it would be understood today. In this bogus film world, Gyula Kabos livened up a mundane reality that was to be endured with a whistle or an alarmed giggle. By carrying audiences with him, he managed to pass off a string of shoddy films that, craftily made and undeniably witty in their repartee as they often were, were all in much the same pattern.

The second film that, in tandem with *Hippolyt, the Lackey*, decisively shaped the output of Hungary's film industry during the 1930s was first shown to the public at Christmas in 1934. *Meseautó* (Wonder-Car), directed by Béla Gaál (1893–1944), confirmed that a story about social climbing which spliced humour with tear-jerking kitsch in a petty-bourgeois setting whilst preserving a modicum of everyday reality was the real formula for box-office success, and finally dislodged the open apologies for a world of feudal civilities and hussar officers from production schedules. *Wonder-Car* is about a company manager who, in the end, marries his secretary and takes her off on their honeymoon—a Hungarian lower middle-class dream that did not take itself too seriously, perhaps, asking only that its audience wallow in the music and lyrics of the film's constantly recurrent hit tune. The film itself, in Nemeskürty's words, "mixed the proven formula of petty-bourgeois comedy with the sentimentality of a Hollywood success story and thereby gave the semblance of a moving, serious drama."[4] In assessing its merits, it should not be ignored that this was one of the first to feature shots of the streets of Budapest by day and night—in itself a distinct plus for an industry that relied on studio sets.

Founded as they were on comic acting, verbal wit and sometimes cabaret-style humour, tried-and-tested comedic clichés, Hungarian sound films depended for box-office success on well-known stars who were popular with public. *Wonder-Car* had Gyula Kabos again as the trump-card, here as a kind-hearted chief accountant who is getting on in years. (The girl, a naïve but thoroughly 'modern' miss, was played by Zita Perczel (1917–66), who took up residence in Italy after 1945, though she did play again in Hungarian films towards the end of her life.) To cite Nemeskürty: "in a four-year period between 1934 and 1939, 75 comedies in the *Wonder-Car* mould were shot in Hungary's film studios, and very many of those 75 copied that winner down to the casting and the storyline."[5] Kabos was a must for any hit film, whilst the love interest was from now on supplied by Imre Ráday (1905–83) and Klári Tolnay (1914–98). During the latter half of the thirties Pál Jávor took the romantic lead in 14 light comedies, while Antal Páger got to play a growing number of increasingly bigger roles. "Broadly speaking, a joke is cracked at least every five minutes: if you don't have jokes, you don't have a film," the ever-perceptive Nemeskürty comments. "What we get is not films but an hour and a half's worth of scenes, a loosely connected string of cabaret sketches that are stitched together merely by the fate of a pair of lovers who pop up every now and then, rather than by any resort to psychology or plot."[6]

Even Kabos's incredible popularity did not spare him from anti-Semitic attacks (not as yet life-threatening), and these intensified after the release of István Székely's *Lovagias ügy* (An Affair of Honour, 1937). Székely was clearly the most astute (but never tasteless) craftsman of what was becoming the 'mainstream' of the time, a near-standard Hungarian film genre. The occasional scene of modest but praiseworthy artistic merit would still be discernible in his films, an example being the First World War setting of *Café Moszkva* (Only One Night, 1935). After his earlier string of surefire hits, then, the furore that burst over *An Affair of Honour* was all the more unexpected: a section of the press accused it of denigrating 'the Hungarian upper classes'. Despite his having directed no less than 13 successful films in 1936–37 alone, that put an end to his filming career in Hungary, and he left the country in 1938. Gyula Kabos, the quintessential star of Székely's films, had already received an invitation from the United States in 1936 to undertake a theatrical tour there and then a film shoot in Hollywood, but instead of making the trip he had made a further 26 films in Budapest, including *An Affair of Honour*. István Nemeskürty evokes Kabos's acting style in the following terms: "The way he raises his hand in alarm to shield his face, because he still expects another blow even when apologising to the well-born stripling who has already slapped him once; the way he tries in vain to give orders to his daughter and strong-minded wife; the way he is totally at home in his professional life, in book-keeping, and that is evident in every move he makes; the way he attempts to fight the duel and seeks out a fencing master; the way he is moved when the apologies are tendered on behalf of his office…—all this is more than just a carefully worked-out character portrayal, it is a calculated critique of an age, an age which drives in the essential precariousness of existence for any citizen not of upper-crust or gentry descent... If all the Kabos sequences in his 39 extant films were to be taken out and shown spliced together, that would give us a more accurate sociography of the Hungarian common man in the thirties than any scholarly assessment."[7]

Up to 1935, an average of roughly 25 films were completed every year in Hungary. A Ministry of the Interior decree then prescribed that Hungarian films had to make up at least 10 (and later 15) percent of the films on the bill of every cinema. That gave domestic film production a substantial boost. Fledgling directors made their debuts, amongst them László Vajda, who was later to make an international name for himself directing in Spain. Vajda strove above all to emulate the formalistic meticulousness of French films, as did Artur Bárdos (1882–1974), similarly European in outlook, who, in collaboration with Viktor Gertler (1901–69), shot *Én voltam* (I Did It, 1936) an uncommonly challenging psychological crime thriller. Although, with Székely now gone, the majority of films were entrusted to old hands (Béla Balogh, Béla Gaál), a few young directors did emerge, including Márton Keleti (1905–73), the most successful maestro of the next two decades.

For Hungary's film industry, World War II was virtually a repeat of the boom that had occurred during the First World War, and for much the same reasons (American films were no longer distributed from the end of 1942), with the additional factor of Hungary's renewed expansion when it re-annexed most of the territory that had been lost under the Treaty of Trianon. In 1939 28 full-length feature films were made, but by 1943 that number was up at 53. To mark the fact that it was officially proclaimed the 200th Hungarian sound film, *Pista Dankó* (1940)—directed by László Kalmár (1900–80) and with Pál Jávor and Margit Lukács (1914–2002) in the main roles—was accorded a ceremonial first-night screening in the presence of the Regent, Miklós Horthy himself. With its feudalistically conservative perspective, the film has a plot of distressingly clichéd banality which is in grotesque contrast to a handful of more 'modern' formalistic devices, which are not integrated properly into the film but stick out as contrived, at times ridiculously so.

Kalmár is, to this day, indelibly associated with a single film that was a *succès de scandale* at the time its was released: *Halálos tavasz* (Deadly Spring, 1938), an adaptation of Lajos Zilahy's 1922 novel, in which a weak and irresolute young man, a landowner-cum-official (Pál Jávor), unable to get over a passionate love affair, takes his own life. Through scenes that now strike one as risible but were boldly erotic for that era, and its striving for psychological portrayal, as well as the highly charged presence of Katalin Karády in the female lead role, the film stirred up a great deal of interest. Right-wing papers and societies protested vigorously, demanding that the 'decadent', immoral film be banned. By the outbreak of war, Kalmár, its controversial director, was in great demand, turning his indubitable filmmaking flair to projects of utterly discordant styles and viewpoints.

The most important development of the late 1930s was undoubtedly what is best called the 'Karády phenomenon'. Katalin Karády (1912–90), though a singer of some talent, could hardly be said to be a subtle portrayer of character as an actress; she nevertheless became a key figure in the Hungarian film industry, her personality stamping its mark on the entire output of that era. She happened to come on the scene at just the time when the industry was seeking, albeit tentatively, to move away from its reliance on screenplays built on verbal slickness and traditional themes and on obsolete modes of visual expression. French films were starting to exert a growing influence, with René Clair's comedies having a perceptible impact on several films. Of particular note were the elegantly clean lines of several of the pre-1945 films that Géza Radványi made during a spell when he temporarily moved back to Hungary, for instance, *Európa nem válaszol* (Europe Doesn't

Answer, 1940), with its presentiment of looming world war. *Deadly Spring* was a precursor of a series of Hungarian copycat films of the early '40s that sought, with more or less skill and taste, to reproduce what was considered to be a European flavour. Karády created a new female type: not the naïve, little-girl-lost of the preceding era but an independent, self-willed woman who would get the better of men rather than give way to them, shaping her destiny by standing on her own two feet rather than being the marrying kind. (Her main rival, embodying the alternative but also hugely popular, was Zita Szeleczky (1915–99), one of whose films, *Sziámi macska* (Siamese Cat, 1943), adumbrated that other possible female persona of wartime films in its very title.) Karády was shrouded in mystery. The strong and dangerous sexuality that she emanated stamped and added some weight to flimsy or far-fetched stories. She mostly played modern-day characters but also the mystery women in several pseudo-historical costume melodramas, such as László Kalmár's *Halálos csók* (Deadly Kiss, 1942) or the Venetian chanteuse in *Sziriusz* (Sirius, 1942), a witty adaptation of a fantastic-ironical historical novel by Ferenc Herczeg. It is particularly noteworthy that Lajos Zilahy, one of the most successful Hungarian writers at the time, founded a film production company and commissioned a successful stage director to shoot an adaptation of his 1928 novel *Valamit visz a víz* (Something Drifting in the Water, 1943). With Karády in the role of the mysterious woman and an above-average script, this stands out as one of the higher-quality works of that period.

Hungary's further shift to the right and harsh new anti-Jewish legislation forced many artists, screenwriters and directors out of jobs. Gyula Kabos left Hungary in 1939; a few authors of trite bestsellers managed to get work as ghost writers (for example, producing screenplays for trashy Italian films); Pál Jávor went to Italy to shoot films; and even Karády was harassed as a liberal and Jewish 'hireling'.

On the positive side, the film industry began to show a growing social interest that it had barely displayed previously. *Földindulás* (Landslide, 1939) was the screen adaptation of a play of that year about peasant life by János Kodolányi. In *A harmincadik* (The Thirtieth, 1942), set in the world of miners and the poor, Antal Páger, in the leading role of an elementary schoolmaster, gets to cuff the despotic chief engineer in a film that mixed populist right-wing demagoguery with a commendable requirement for realism. The figure of an intellectual son of ordinary folk who battles against the oppression of the 'nouveaux riches', as played by Antal Páger in *Dr. Kovács István* (1941), radiated an even more unequivocal right-wing desire to 'do something about it', with references that were only too clear, if never overtly anti-Semitic, in sentiment. The film was directed by Viktor Bánky (1899–1967) who, with so many other directors excluded from working by the anti-Jewish laws, now became one of the busiest in the industry. He picked Páger as his lead in virtually all his films, including *Bors István* (1939), a piece of populist demagoguery in 'folk-play' style, and *Őrségváltás* (Changing Of The Guard, 1942), with its crudely anti-Semitic, near-Nazi propaganda.

One figure who stood out in this sea of trash and ambiguous artistic endeavours, was István Szőts (1912–98), the first real *auteur* of Hungarian cinema as we would now understand that term. He had been born in Transylvania, as was József Nyírő, whose short story was the basis for *Emberek a havason* (Men in the Alps, 1942), the first full-length film that Szőts wrote and directed, which was quite unprecedented unless one considers his own short films as forerunners. *Men in the Alps* is conspicuous not only amidst the multitude of run-of-the-mill, more or less well-crafted mass-entertainment products of the '30s and the bigoted, populist drivel of a somewhat later date but also comfortably outstrips the few timid efforts that had been made to emulate a more European approach. It is simply the finest movie that was made in Hungary

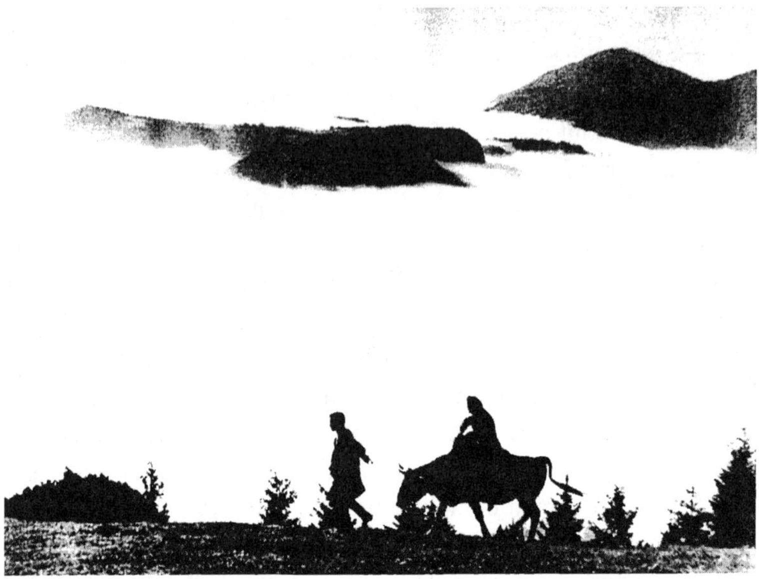

Scene from *Men in the Alps*

prior to 1945. Blending the world of Hungarian folklore and folk traditions with the new formal devices of the modern cinema, it tells a balladistic story in ballad-like manner, the jump-cut elisions of the editing style allowing brutal realism to be filtered through the unfettered, expansive visual lyricism of a mythical-poetic idiom. *Men in the Alps* picked up a major award at the Venice Film Festival in 1942 and won praise from Italy's more discerning critics (including Carlo Lizzani, the later neo-realist film director).

Somewhere in Europe: 1945–1956

Due to the devastation and adverse conditions that prevailed in the immediate aftermath of the war, little filmmaking was undertaken in Hungary up to the Communist take-over in 1948. Production and distribution facilities were shared out amongst the ruling coalition parties. During the period running from summer 1945 to the full nationalisation of the industry in the spring of 1948, four privately financed and five party-sponsored films were produced. Only two of those are of any importance, but both have a place amongst the enduring achievements in the history of Hungarian and European cinema.

The team that worked on *Somewhere in Europe* included many individuals who were shortly to emerge as the country's leading creative talents. Apart from director Géza Radványi and screen-writer Béla Balázs, who have already been mentioned, Félix Máriássy (1912–75) was assistant director. Catching the mood of the times, it is an affecting story about the friendship that a vagrant gang of orphaned children strike up with a lonely, unsociable musician (Arthur Somlay, 1883–1951), portrayed by Radványi with the best expressive devices of contemporary French film but also, courtesy of Balázs, drawing on earlier German and Soviet expressionist traditions. The montage technique blends romantic, lyric and comic elements with great virtuosity, and the impact is only heightened by the emotional power of the anti-war message, the children's sufferings, and the tragic death of one of them. The film was extraordinarily well-received both in Hungary and internationally.

After 1945, István Szőts was hampered from working first by the total paralysis of film production and the country's economic plight then—despite the fact that Béla Balázs, back in Hungary from his long exile in Moscow, did not just support him but became a close collaborator—by the obstacles placed in his way, primarily for ideological reasons, by the cultural policies of the ascendant Communist party. It blocked the release of his second feature film, *Ének a búzamezőkről* (Song of the Wheat-fields, 1948), adapted from a 1927 novel by Ferenc Móra, on the grounds that it was 'detrimental' on account of its 'clerical' outlook and 'narodnik' (i.e. peasant) ideology. Peasant mysticism and trust in miracles do indeed play a big role in the film (which opens with a procession of people praying for rain), but this and similar sequences, both separately and taken together, are plainly more indicative of the superstitious, forbidding, backward-looking and tragic nature of religious mysticism. An extraordinarily intense creation, it combines realism of a kind never seen before in a Hungarian film with the director's already familiar balladistic voice and visual language. (The main character, an unbalanced individual who returns home from a prisoner-of-war camp in the Soviet Union, was played by János Görbe (1912–68), who also took the leading role in *Men in the Alps*.) Again, what is so striking about it is how singular Szőts's work is, perhaps the sole precedent one might cite—apart from *Men in the Alps*, of course—being the lyrical documentarism of Austrian cameraman Gustav Höllering's *Hortobágy* (1936), with its scenario by the outstanding writer Zsigmond Móricz.

The masterpiece was not shown in public until decades later. Now subjected to sustained criticism and his work impeded, when his friend and collaborator, Béla Balázs, who had likewise come under Stalinist attack, died in 1949, the non-Marxist Szőts, branded an 'ideological enemy', found his planned projects running into one dead end after another and was not allowed to take any significant part in the renewal of Hungary's film industry. All he could direct in that period were a few short films: *Kövek, várak, emberek* (Stones, Castles, People, 1954) and *Melyiket a kilenc közül?* (Which of the Nine?, 1957). Having no faith in getting backing for one of his later film projects, which was to deal with Transylvania during the 1848 War of Independence, he chose not to return to Hungary after the Venice Film Festival in the autumn of 1957. It is unfortunate indeed that his plans fared no better in the West (at one point he was competing against Rossellini for a large-scale documentary project about India). He eventually settled in Austria, becoming a director of respected popular science and nature films. Following the 1989–90 change in Hungary's régime, he made frequent visits back to the country, but nothing came of any of his last grand film project, the screenplay alone appearing in print; his life's work was already well and truly completed some time before his death in 1998. Slim as it is, in relation to what his talent would have been capable of, given the chance, that *oeuvre* nevertheless had a major impact on several artists of a younger generation of filmmakers, Miklós Jancsó, Ferenc Kósa and István Gaál above all, so that he was still a 'presence' in Hungarian cinema. His life and work have still not been accorded the credit they genuinely deserve, even if only in retrospect, either inside Hungary or abroad. His success is, therefore, merely symbolic: a triumph of ethically irreproachable conduct on the part of a highly talented creative artist.

The nationalisation of the entire Hungarian film industry in March 1948 produced a more favourable state of affairs financially, organisationally and, or so it seemed for a short while, culturally as well. Géza Radványi and others left the country, it is true, but there was a big influx of young talents into films, though few new directors—aside from Félix Máriássy and László Ranódy (1919–83)—were actually given a chance before 1953–54. The latter was co-director, with Kálmán Nádasdy (1904–80), of *Ludas Matyi* (Mattie, the Goose-Boy, 1949), the first full-length Hungarian colour film, with an up-an-coming Imre Soós (1930–57), himself of peasant origin, in the title role. It is highly symbolic that this talented young actor, whom the Communist régime's cultural tsars tried to mould into a model figure of the 'new era' should have taken his own life in 1957.

The first and, for a very long time, greatest success of Hungary's nationalised film industry was notched up by Frigyes Bán (1902–69), who was by then a fairly experienced older hand. *Talpalatnyi föld* (A Patch of Earth, 1948) drew on motifs from a 1942–43 trilogy by peasant writer Pál Szabó that depicts the world of impoverished peasants during the 1930s in the context of a romantically tense story of a young couple's fight with those around them. With scenes now of heightened drama, now of a documentary flavour, it evokes a sense of reality previously experienced only with Szőts's films, though its devices and visual language are more traditional than there.

Still, no film released over the next few years came near to replicating that success until Zoltán Fábri's (1917–94) *Körhinta* (Merry-Go-Round, 1955)—which resembled it in many respects—came out. Not long after it had been completed, the dead hand of censorship and political coercion put a stranglehold on creativity. Films that were barely disguised propaganda dressed up as fiction, typically dealing with rooting out imperialist spies and internal enemies, did not go down at all well with the public, though light entertainment films that merely gave a socialist gloss to thirties clichés and banked on old-stagers like Kálmán Latabár (1902–70) were hugely successful. Foremost amongst them were *Dalolva szép az élet* (Sing and Enjoy Life, 1950), a 'factory comedy' that used a portrait of working men's choruses as a thinly veiled metaphor of

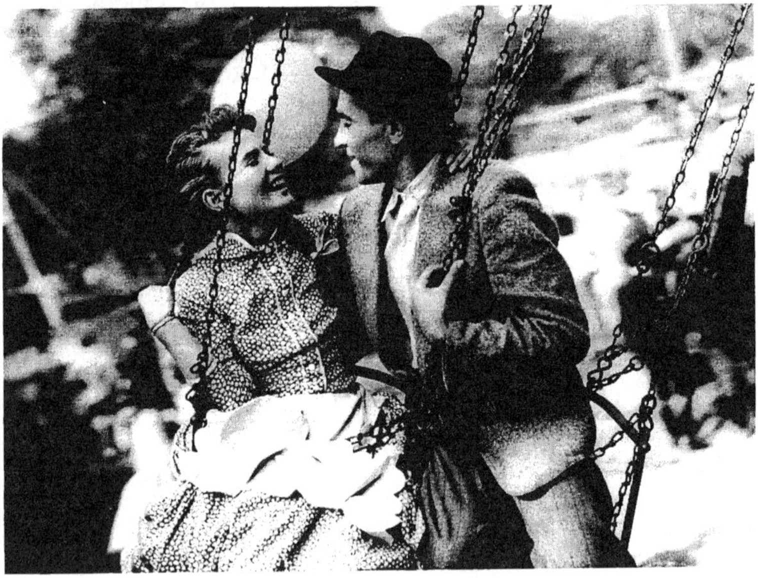

Mari Törőcsik and Imre Soós in *Merry-Go-Round*

Stakhanovite competition in the workplace, and *Civil a pályán* (A Rank Amateur, 1951)—both entrusted to the sure hands of Márton Keleti, a much-employed pro during this era—or *State Department Store* (Állami Áruház, 1952) by Viktor Gertler. These films are fun to watch even today, in part because they are illuminating documents of the period, in part because they are masterpieces of the sort of professional 'polish' that can give a varnish of credibility even to falsehoods. In artistic terms, however, the mendacity, the varnishing, the propagandistic manner in which subjects are treated as stark choices between black and white, the contrivedness of the screenplays, and the general obsoleteness of directing styles disqualify films of the period from serious consideration up till 1953–54. From then on, with the political thaw that ensued after Stalin's death, some liveliness returned to cultural life and chances opened for more realistic portrayals of people and reality. The first signs of that, after enforced work on a number of unpalatable projects, were the confirmation that Félix Máriássy gave of his genuine talent with a nuanced and fastidious adaptation of Zsigmond Móricz's 1932 novel, *Rokonok* (Relatives, 1954), and particularly *Életjel* (Fourteen Lives in Danger, 1954) by Zoltán Fábri, about the events following a mine-shaft collapse, along with *Simon Menyhért születése* (The Birth of Menyhért Simon, 1954), based by theatre director Zoltán Várkonyi (1912–79) on a long short story by Tibor Déry. *Liliomfi* (1954), a comedy by Károly Makk (1925), featuring Iván Darvas (1920) and Mariann Krencsey (1931) in the main roles, brought a new voice and colours with its light, tasteful humour and self-assured control of narrative pace and acting.

In freeing itself from the constraints of schematism and political coercion, the new Hungarian cinema strove to profit from the influence and stance of Italian neo-realism. By the time of *Budapesti tavasz* (Springtime in Budapest, 1955), his 'hymn of liberation', with Miklós Gábor (1919–98) taking the leading role, Félix Máriássy had overcome the stiff, cardboard-cut-out approach, gaining praise for his meticulous camera angles and direction of the cast with this attempt to depict a real human drama. His next film, *Egy pikoló világos* (A Half of Light, 1955), in which Éva Ruttkai (1927–86) endeared herself with the compelling zest she put into her role, was better received by the public and also stands up better artistically for the modern viewer. Screened to packed houses for months after its release at the Christmas of 1955, it was the first chance that Budapest's inhabitants had been given to see their own everyday lives and concerns, their own streets and squares, portrayed in an identifiable and unvarnished manner on the screen. (The film was awarded the Grand Prix at the Karlovy Vary Film Festival in 1956.) Similarly neo-realist in inspiration but altogether more dour, with its endeavour to lay open abuses of political authority, is Károly Makk's *9-es kórterem* (Ward 9, 1955).

The pinnacles of artistic achievement and international success in this brief but important period between 1953 and 1956 were two films directed by Zoltán Fábri. *Merry-Go-Round* places the issue of farm collectivisation within the setting of a dramatic love story. The sheer power and authenticity of the romance permit the viewer to forget the contrived point of departure and the laboured plot development as a strong human drama comes to life, and even the political content takes on a convincing colouring. The extraordinary tension, along with the true-to-life portrait of village life, that is generated by the powerful acting of Mari Törőcsik (1935) and Imre Soós (1930–57), makes this a truly memorable film, with Fábri proving to be a master of traditional montage, the tension in most of the crucial scenes (e.g. the merry-go-round scene and the wedding dance) being built up by careful calibration of the pace at which shots are intercut. Even though it took no prizes, *Merry-Go-Round* was the big discovery of the Cannes Film Festival (Truffaut wrote a laudatory review). Zoltán Fábri's other film, *Hannibál tanár úr* (Mr. Hannibal, Teacher, 1956), won the Grand Prix at Karlovy Vary the following year (in Hungary itself the film was screened for only a few days during the week running up to the outbreak of the 1956 revolution). Set in the 1920s, *Hannibal* is about the campaign of vilification to which an insignificant secondary school teacher is subjected for publishing an essay on the revolts with which the death of the Carthaginian general was associated; of course, this is merely a pretext to stoke up right-wing mob hysteria. Fábri limns his caricature figures with the devices of tragicomedy, biting irony and vitriolic sarcasm, for his film is clearly not just about the twenties but the dreadful pressures that every dictatorship exerts on the individual: anyone who saw the film at the time would have been in no doubt that its message applied equally to the Stalinist era and the present.

Hungary was now poised on the brink of revolution. *Hannibal* only just made it to cinema screens (but the outbreak of the revolution meant that it vanished unnoticed), whereas Zoltán Várkonyi's hard-hitting *Keserű igazság* (Bitter Truth, 1956) was still in post production, and since it was banned after the revolution was put down, did not even make it that far. History intervened to cut Hungarian film short just when it had entered a highly promising phase of recovery—an upswing that was only to be resumed in the early sixties.

Film Art in Hungary 1960–2000

The Hungarian film industry flourished in the 1910s and again during the '30s, though in neither case can one speak of art, merely of a successful branch of mass entertainment; István Szőts was a solitary figure years ahead of his own time. A few isolated experiments and achievements excepted, it was only from 1953–54 that creating work of artistic merit seemed to become a possibility, but that closed down when the 1956 revolution was suppressed. The artistic revival that cinema in Europe underwent in the fifties was naturally something that Hungarian directors, too, were eager to join in, albeit a little belatedly. That delay was mostly a consequence of the rigor of political dictatorship and the country's provincial isolation: the revival was only able to commence in Hungary around 1962–63.

As in other areas of cultural life, though, the post-revolutionary period did also usher in some beneficial changes, even as the political reprisals were taking place. In 1959, the Hungarian Institute of Film and Film Archives was set up on an independent footing, controlling a nation-wide network of film clubs and putting out both popularising and scholarly publications and a high-quality specialist periodical. Over this period an average of 15–20 domestic feature films were completed annually, which was an advance on the preceding era, but 1959–62 also happened to represent an alarming low point for modern Hungarian cinema in terms of artistic quality. By now its slickly crafted and high-mindedly conservative style, the habitual resort to French poetic realism or Italian neorealism for its conceptual approach, were beginning to look stale and dated.

Conspicuous amidst the general run of either insipid or careful but mediocre films was the *oeuvre* of Zoltán Fábri. When he resumed his directing career in 1957, after the international acclaim that had attended *Merry-Go-Round* and *Mr Hannibal, Teacher*, discussed above, the political climate in Hungary was initially even more unfavourable than before. Following a comedy of little substance, *Édes Anna* (Anna Édes, 1958), based on Dezső Kosztolányi's 1926 novel of the same title, represented a return to his top form, combining a portrait of the 1920s with penetrating psychological analysis and Fábri's expressionist-inspired formal devices, which swing between the stiltedly contrived and the boldly original. *Két félidő a pokolban* (Two Halves In Hell, 1961) is about a football match that a team drummed up from a company of (mostly Jewish) forced-labour servicemen plays against a German team, and its tragic ending. With a meld of grotesque and dramatic elements, the film attempted a searching examination of the moral dilemmas and humane choices implicit in that situation.

The one other work that stood out for its sure handling of means of expression and its evocative power was *Angyalok földje* (Land of Angels, 1962), put together by director György Révész (1927) from motifs drawn from Lajos Kassák's 1929 novel, which is set in the Angyalföld district of Budapest. Treatments of 'current issues', by contrast, would drift into prevarication, if only because this was the area on which the régime's cultural watchdogs kept the closest eye, thereby pushing directors into a possibly even more corrosive self-censorship. However, a sense of impending shift in the political winds—as signalled by the amnesties given to those who had been imprisoned as part of the reprisals taken in the wake of the revolution, with a smaller first wave of releases in 1960 and a larger one in 1963—began to be felt. Not for the first time in the twentieth century, this shift gave scope for a more truthful representation of things as they were and less fettered artistic experimentation. For some time to come, Zoltán Fábri continued to be the most important film auteur, although the talents of both Károly Makk and Miklós Jancsó (1921) now began to emerge almost in tandem.

Makk's *Megszállottak* (The Fanatics, 1961) and *Oldás és kötés* (Cantata, 1963) by Jancsó (who has already attracted attention with his expressive documentaries) were the two landmark works of a new era or, rather, approach in Hungarian cinema. Neither drew on Fábri and Máriássy, the successes of the brief spell immediately prior to 1956; the neo-realism of that period was now supplanted by the *nouvelle vague* of Truffaut and Resnais, and the new Italian style of Antonioni and Fellini, as evidenced by a demand for more profound portrayal of the contradictions inherent in human relationships.

Károly Makk was still in his early twenties when he entered the film industry, working as an assistant on *Somewhere in Europe* to Géza Radványi, whom he has considered his master ever since. He had already proved himself to be a promising master in a variety of film styles during the brief cultural thaw that preceded 1956, with his period-costume comedy *Liliomfi* providing an interesting example of an easy-going modern approach to classic stage-plays. After the post-1957 nadir, *The Fanatics* received widespread coverage in the press, stirring up a big controversy. The screenplay was based on a newspaper report about innovations in hydraulic engineering and their opponents. Makk completely changed the focus of the subject by creating a real human and social drama, instead of the usual didactic 'film about production'—a portrayal of the fervent friendship of the two main characters, a hydraulic engineer and a village leader who stands by him. The 'fanatics' put themselves body and soul into their work, whilst the bureaucrats and their superiors plot against them. The psychology of the protagonists, the portrait of their inner growth as they tussle with pessimism and despair, became the real content of the film; it is no coincidence that Makk and his actors were accused of 'existentialist' influences, 'alienation', and employing self-serving devices. Makk certainly made deliberate use of some of Antonioni's procedural idioms but was no slavish imitator: no other Hungarian since Bán, in *A Patch of Earth*, had captured the hard, coarse, clinging reality of the Hungarian land so well. In his next film, *Elveszett paradicsom* (Paradise Lost, 1962), Makk attempted a deeper psychological portrait of alienation, further overhauling and refining his technique and idiom.

Jancsó's *Cantata*, adapted from a short story by József Lengyel, a veteran Communist who never lost his naïve faith, despite 18 years of imprisonment in the penal camps of Stalin's Russia, likewise portrays an individual in crisis who has lost all hope. The protagonist, a physician, as played by Zoltán Latinovits (1931–76), is a close kin to the disillusioned doctor he performed in Makk's *Paradise Lost*. Both characters are first-generation intellectuals who, in their native villages, try to confront the reality of their pasts and futures and their own selves. For Hungarian cinema this was a completely new tone and subject matter, which of course called for a new mode of expression, so it is hardly surprising that both these directors were strongly influenced by the latest European trends, even though the Hungarian public at large was still screened from them.

The two-part *Párbeszéd* (Dialogue, 1963) by János Herskó (1926), with a storyline that spanned the period from 1945 to the present day, was the first attempt to give a responsible and honest portrait of the personal and public dilemmas of its main characters: Communists who, out of naïveté, had given active support to the Stalinist régime and then came into conflict with themselves in 1956. However, the weakest points of this otherwise not unworthy film, with its demonstrable sensitivity to the predicaments it investigates, were precisely the perfunctory, highly piecemeal sequences dealing with 1956.

An important role in the renewal of Hungarian cinema fell to the Béla Balázs Film Studio (BBS), which although founded in 1959 did not make its presence strongly felt until 1960–61. A shoal of young talents who had been attending the Budapest School of Film Art, where classes were taught by the likes of Zoltán Fábri and Félix Máriássy (and later on Károly Makk and János Herskó), were now emerging to start their careers. It was the BBS that gave them the first chances to prove themselves on documentaries and short films: Sándor Sára (1933) with *Cigányok* (Gypsies, 1962), István Szabó with *Te* (You, 1961) and *Koncert* (Concert, 1962), Zoltán Huszárik with *Elégia* (Elegy, 1965).

A third important creative contribution to the renewal came from a director who was already working at the BBS. *Sodrásban* (Current, 1963), by István Gaál (1933), a story of a holiday by the River Tisza that takes a tragic turn, consciously makes use of the artistic legacy of István Szőts. The landscape, the river, the village—these are not just backdrops but living, breathing characters in a film, which pullulates with the dramas of friendships, loves and betrayals that are part and parcel of growing up. Gaál, who had graduated from the Rome Film School, was thoroughly conversant with the Italian cinema's very latest formal devices and approaches to characterisation, and that leaves a clear stamp on his film. *Current*, which won considerable international acclaim, as well as prizes at several festivals, was a harbinger of the different way of viewing the world that was adopted by younger directors than the Makk–Jancsó–Herskó generation.

After his attention-grabbing short films, István Szabó (1938) likewise garnered much critical kudos both at home and abroad with his feature-length *Álmodozások kora* (The Age of Day-Dreaming, 1964) and *Apa* (Father, 1966). Like other directors of his generation, such as János Rózsa (1937), Ferenc Kardos (1937–99), and Zsolt Kézdi Kovács (1935), Szabó was greatly influenced by the highly individual, 'camera-as-pen' approach of the French *nouvelle vague*, with its predilection for incorporating autobiographical elements to establish personal credibility. *The Age of Daydreaming* is a loosely linked sequence of episodes about young intellectuals trying to find their places in the world. "Wake up, please," says a voice at the very end of the film. Szabó's second film, *Father*, is the story of a 'belief': pervaded by the memory of a beloved father-figure who is aggrandised and embellished into myth and legend, but out of whose shadow it is necessary to step to become an independent, grown-up individual. In these films Szabó succeeded in weaving a spell of individuality that had not hitherto been experienced in the Hungarian cinema. Contemporary newsreels and documentary inserts, the evocation of memories of 1944 and 1956, shards of everyday life and of the past—these are all fused together into artistic truth.

Szerelmesfilm (Love Film, 1970) was a continuation and, as it were, conspectus of this preoccupation. Its male protagonist is only able to meet again with his childhood love, who left Hungary in 1956, in Paris. This encounter evokes, in a sometimes seemingly illogical order, but one that obeys an inner, subjective logic of its own, the tormented memories of two decades of post-war Hungary as reflected in the personal fates of two young people. This marked an end to the first phase of Szabó's career; he was now to turn to other means and approaches.

For twenty years from 1945, Ferenc Kósa (1937) engaged in balladistic renderings of the epic story of the Hungarian village. His best film, *Tízezer nap* (Ten Thousand Suns, 1965), on which Sándor Sára was his cameraman, similarly bears traces of Szőts's handling of reality, as well as the principles behind Bartók's music, though Kósa's tone is more agitated and hesitant, less purely naïve, than Szőts's. A tragedy- and pain-laden ballad of the 'socialist transformation' of the Hungarian village, its lyricism not always free of sentimental overtones, and not shrinking from heavily embroidered, even at times exquisitely overindulged images and localities that intensify out of powerful landscapes and frescoes into almost live rock paintings—the film's merits were recognised by a prize for the best producer at the 1967 Cannes Festival.

Sándor Sára was already an acknowledged cinematographer when he directed the first of his own works, *Feldobott kő* (The Upthrown Stone, 1967), a portrait of the manipulation of well-meaning and honest people during the period of the Stalinist cult of personality, showing examples of the abuse of authority that were still prevalent at the time the film was being made.

In *Így jöttem* (My Way Home, 1964) Miklós Jancsó, the most significant formal innovator in Hungarian cinema, and its definitive auteur of the sixties and seventies, portrayed the Red Army's liberation and occupation of Hungary through the eyes of a naïve young man in a quite different manner from any previous approach. His next films, *Szegénylegények* (Round-Up, 1965), which attracted huge international attention and acclaim, followed in

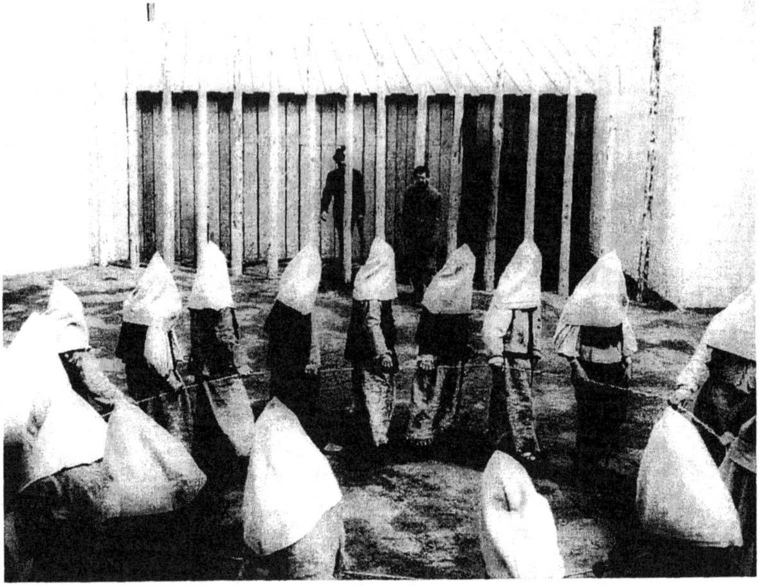

Scene from *Round-Up*

fairly quick succession by *Csillagosok, katonák* (The Red and the White, 1967), with its Russian civil war frame, and *Csend és kiáltás* (Silence and Cry, 1967), with its small cast and its setting in the period of right-wing retribution after the 1919 Hungarian Soviet Republic was suppressed, turned Jancsó into the best-known and most admired of all East European art film directors. A seminal master of the choreographed crowd scene, his films are characterised by dance, the 'body language' of his characters, wide vistas, the endless horizons of the landscape, and a symbolism reminiscent of Andrzej Wajda, that other great creative figure in East European cinema.

With Jancsó, unnamed authority confronts often equally unnamed and defenceless but resisting protagonists. His is an irrational, illogical world in which, as we pick up from various signals, constant intimidation rules. For many, particularly for East

Europeans, the references, the 'messages' for the present to be drawn from the films, over and above their own story or the concrete historical situation they depicted, were all too plain. Symbol-laden, allusive 'parable' became one of the dominant approaches in Hungarian cinema, and that intensified still further after the 1968 Soviet occupation of Czechoslovakia. With the open-ended, multi-layered shooting scripts that he wrote in collaboration with Gyula Hernádi (1926) as his platform, Jancsó elaborated an incomparably powerful and totally individual expressive vehicle. His hallmark is the long, intricate take: this is when a film is born, when the shooting is in progress, with the complicated camera manoeuvres 'painting' on the screen.

At the same time as Jancsó's works were taking a new and responsible view of history with their disillusioning, anti-romantic perspective, others were producing high-budget romantic costume films. Zoltán Várkonyi became a master of this genre, directing technically well-made picture-book spectaculars, with unmistakable 'goodies' and 'baddies', adapted from some of the great nineteenth-century romantic historical novels of Hungarian literature: *A kőszívű ember fiai* (Men and Banners, 1964), *Egy magyar nábob* (The Last Nabobs, 1966) and *Kárpáthy Zoltán* (1966), *Egri csillagok* (The Lost Talisman, 1968). These were films of a type that cinema-goers had long been deprived of.

The rapidly growing critical and festival success of Hungarian cinema in Europe was accorded to films that were addressed to a narrower public. For a decade and a half from the mid-sixties, Hungarian art films—basking in unwonted acceptance and praise, both at home and, at least amongst critics and fellow professionals, also abroad (although in Hungary too the support came from a rather small segment of the intelligentsia)—started to be seen by the country's by now increasingly flexible cultural policy-makers, perhaps somewhat perversely, as jewels, for they chose to regard the foreign success of these critical films as validation of their own authority. Annual production of films in Hungary rose to around 20–25, not including the BBS's productions. The annual 'Hungarian Film Week' at Pécs (and later Budapest) became a festival not just of culture but also of official cultural policy. With each major new project, even the screenplay was frequently a subject of widespread debate in the press and industry circles, not just in the case of Jancsó's films but others too, such as *Ítélet* (Judgement, 1970), co-written by Sándor Csoóri with Kósa and Sára, which was nominally about a peasant uprising in the Middle Ages but probed current issues in the process.

Fresh social issues likewise stood at the heart of the journalistic approach that András Kovács (1925) brought to his polemical films. In the documentary *Nehéz emberek* (Difficult People, 1964), which aroused considerable attention, scientists and inventors spoke to camera about how officials, often at the highest level, were blocking their discoveries and innovations. His best feature film, *Hideg napok* (Cold Days, 1966), based on a 1964 novel of that title by Tibor Cseres, was conspicuous for its sharp self-examination of a Jancsó-style historical 'borderline area', that of false illusions and where the line is drawn between national and individual responsibility. Beautifully shot by cameraman Ferenc Szécsényi (1922), with superb acting and masterly build-up of dramatic tension, the film examines the moral aspects of a notorious 1942 massacre perpetrated by the Hungarian gendarmerie in Novi Sad. Apart from the furore that always attended Jancsó's works, the biggest controversy in the press during this period was sparked by the polemical squib that Kovács planted in *Falak* (Walls, 1968). The film examined the prospects for economic and political reforms at the very time (in early 1968) when the Hungarian government had nerved itself to embark on the liberalised 'New Economic Mechanism'. More than a few critics questioned the value of Kovács' work in filmic terms, given their 'radio play' character, but his supporters and disciples saw him as the master of the 'interrogative film' (the term 'active film' was also fashionable at that time as an epithet for a whole trend in Hungarian cinema).

Zoltán Fábri was another who was palpably seeking a different film language. *Húsz óra* (Twenty Hours, 1965), which won the Grand Prix at that year's Moscow Film Festival, follows the lives of its one-time cotter protagonists with tense, realistic simplicity and at the same time within a complex yet precisely worked out temporal structure. The character who confronts the Communist regime in the autumn of 1956 is not an 'enemy' but a landless pauper. Fábri subsequently changed style again, amalgamating (or rather just mixing) a diversity of grotesque and traditional dramatic devices with visionary cut-ins to produce *Utószezon* (Off-Season, 1968), which was a flop at home but highly lauded, for instance, by Italian critics. An interesting offspring of his treatment of grotesque subjects is his 1969 filmed version of István Örkény's tragicomedy *Tóték* (The Tóth Family), a hit play of two years earlier that was staged in a number of European countries. He successfully struck a balance in *Ötödik pecsét* (The Fifth Seal, 1975), the tautest of Fábri's late-period films, a painfully forthright parable about maintaining human integrity in 1944—and eras of terror in general.

One of the early films directed by Péter Bacsó (1928), an experienced playwright and screenwriter who had worked with Fábri, was *Nyár a hegyen* (Summer on the Hill, 1967). In this the ex-commander of a Communist penal camp meets and is drawn into an argument with a former inmate, now a scholar. As seen through the eyes of two young intellectuals who are witnesses to their dispute, he may have lost his illusions but not his faith.

For a short time following the occupation of Czechoslovakia, in the autumn of 1968, it seemed that the adverse foreign political climate might have no impact on either Hungary's economic reforms or its cultural policies. However, it was only after much procrastination that shooting was allowed to go ahead on one of the most outstanding of all the creations of modern Hungarian cinema, *Szerelem* (Love, 1970). Directed by Károly Makk, this reshapes two short stories by the internationally well-known writer Tibor Déry into a motion picture with the feel of a reverie, with the main roles being played by Mari Törőcsik, Iván Darvas and Lili Darvas (1902–74), who returned to her native country specifically for the purpose. Makk's creative partner on this masterpiece of cinema history was his cameraman János Tóth (1930), a true artist at capturing images of the memory's traces of material objects, human faces, and exquisite interiors. The man is serving a sentence as a political prisoner and only appears in the final twenty minutes of the film, which hinges on the relationship that is developed between his sick mother and his wife, who sustains her by feeding her a constant stream of colourful and compassionate fibs. This does not yield a 'political' film but something else, something more than that: a real peak of Hungarian and European art.

A similarly enduring value was achieved by Zoltán Huszárik (1931–81) with *Szindbád* (Sindbad, 1971)—a peerless and unprecedented masterpiece in Hungarian cinema. One of the most celebrated characters in Hungarian prose literature, the protagonist of several cycles of consummate short stories by Gyula Krúdy—immortalised in the film by the acting genius of Zoltán Latinovits (who tragically died a few short years later)—Sindbad muses nostalgically on the tranquil 'music of transience' of a past life that, whilst bearing witness to much human unhappiness and pain, was somehow still liveable and more splendid than the present day. Almost a decade later, in the hugely ambitious *Csontváry* (1980), Huszárik used Tivadar Csontváry Kosztka, the greatest Hungarian painter of the twentieth century, as the central figure of a film which, by sliding between two temporal planes, explores the area where art and stylisation collide with crude reality.

István Gaál, one of the younger directors who made his mark as the 'new wave' was getting under way in Hungary in the early sixties, brought out *Magasiskola* (The Falcons, 1970), based on a 1956 novella by Miklós Mészöly, a strongly symbolic film about the power of constriction, compulsion and being in command. Sándor Simó (1934–2001) introduced an intriguingly new voice, the engineers and technocrats of his films being reminiscent of the youngsters in Szabó's earlier films, only not so enthusiastic, more sober and disillusioned, with a visual treatment, itself not entirely unironical, in the same vein.

The huge and protracted debate that raged in the press and on television following the release of Jancsó's *Fényes szelek* (The Confrontation, 1969), a historical parable set in 1947–48, whilst it was highly manipulated, did provide a channel for a *relatively* free airing of views, though the director was obliged to issue repeated declarations emphasising his 'allegiance' to socialism. Yet the occupation of Czechoslovakia also circumscribed public life; any illusions that had been cherished in the effervescent sixties were punctured. Some of the young who had recently passed through film school, left the country, whilst the defection to Sweden of János Herskó, an established director, manager of one of the state-run MAFILM's four feature-film studios and one of the film school's teaching staff, had a dampening effect on the entire film industry. Herskó's youthfully saucy *Szevasz, Vera* (Hello, Vera, 1967) had been a hit, but the intricately woven *N.N. a halál angyala* (N.N.—Angel of Death, 1970) had a hostile reception. Herskó's departure was the overture to an altogether more discouraging decade.

With the restoration of a highly conservative, neo-Stalinist line in both the economic and cultural sphere that took place in 1972–73, reforms were brought to a halt. Miklós Jancsó was one who could no longer find a place for himself. After *Égi bárány* (Agnus Dei, 1972), a thought-provoking portrait of the hysteria and mythico-superstitious rituals—the near-magic, so to say—bound up with the 1919 White Terror, which of course also held more general lessons, Jancsó took himself off to Italy for the best part of six years, directing several possibly less cogent films, one with Monica Vitti in the main role.

Whilst Hungarian directors, even the Jancsós and Fábris, had long experience of being given an official go-ahead on an agreed screenplay only to have it rejected before shooting began (and Kósa, for instance, was at the receiving end of a constant stream of paternalistic 'friendly' advice on *Ten Thousand Suns*, from conception on through the shooting and post-production work), Péter Bacsó's *A Tanú* (The Witness, 1969) was the only film to be completely finished and then not make it from can to screen. Strongly satirical, this is a worm's eye view—that of a simple lock-keeper who is coached to give a false eyewitness report in the trial of a local functionary—of Hungary's own Stalinist-style show trials, and in particular the trial and execution in 1949 of leading Communist party figure and Minister of the Interior László Rajk after his fall from grace (this was even mentioned in the history books in schools by then). Now virtually any film with a critical edge faced this threat, which not uncommonly led a radical reworking and even post-production shooting of extra material. The authorities were most reluctant, for instance, to allow public screening of *Fotográfia* (Photography, 1973), a fascinating document by Pál Zolnay (1928–95) on the irreconcilability of reality and any varnishing of truth, until a second prize picked up at the Moscow Film Festival changed their minds. It was likewise not a good time for Ferenc Kardos to be shooting *Petőfi '73*, with its many improvisations and use of non-professional actors, with the censor unable to refrain from meddling even in this peerless dramatic essay and polemic. Some films were already under the threat of a ban at the time they were made and then subjected to a barrage of criticism on their release, as in the case of Sándor Sára's *Holnap lesz fácán*

(Pheasant Tomorrow, 1974). The discerning periodical *Filmkultúra* was at the receiving end of a blast from the Party's 'Ideological Working Co-operative', leading to the removal of its editor-in-chief, the film aesthete Yvette Bíró (1930), who had worked as a writer on several of Jancsó's films (not long after that she was forced to emigrate and now lives in New York, where she is a university professor).

Some worthwhile work was still being allowed through, even in this climate. Though long obstructed, Károly Makk's *Macskajáték* (Catsplay, 1974), an adaptation of Örkény's 1966 play, did not disappoint with its rich texture, historical irony and humanity. In the area of comedy, since 1957 the reliably proficient Márton Keleti had kept up a regular flow of acceptable works, including *Hattyúdal* (Swan Song or Villa Negra, 1963) and the box-office hit of the decade, *A tizedes meg a többiek* (The Corporal and the Rest, 1965). This ironically relates the adventures of a resourceful army deserter in 1944 and introduced the tricky genre of war comedy as a new shade on the palette of Hungarian cinema. That aside, the general run of comedy films, for a good while to come were little more than crass botch jobs. It was against this background that the work of Lívia Gyarmathy (1912) and Géza Böszörményi (1924) in the early 1970s came across as freshly innovative. Their comedies, *Ismeri a szandi-mandit* (You Know Sunday-Monday, 1969) and *Madárkák* (Birdies, 1971) had nothing to do with the vacuous polish of hit movies of the thirties, with their cabaret-based traditions, and all the more with the inherent grotesqueries of reality.

At the point when the seventies were passing through a nadir of almost unrelieved tedium and conservatism, Márta Mészáros (1934), who had made her debut in feature films during the late sixties with *Eltávozott nap* (The Girl, 1968), the story of a strong-willed working-class girl's search for her father, achieved her greatest international triumph with *Örökbefogadás* (Adoption, 1975). *Kilenc hónap* (Nine Months, 1976) and *Ők ketten* (Women, 1977) were hard-hitting, at times abrasive, yet not unlyrical examinations of women's fates and opportunities that were widely greeted abroad; indeed, throughout the seventies Mészáros had a far higher standing outside Hungary, especially in France, and it was only with her autobiographical trilogy of the period immediately before the 1989–90 change in régime (as will be discussed below) that she won a fair measure of general recognition in her own country.

While László Ranódy (1919–83), one of the better directors of the older generation, produced little gems during the 1960s with his traditional handling of *Pacsirta* (Skylark, 1964) and *Aranysárkány* (Golden Kite, 1966), his adaptations of two of Dezső Kosztolányi's classic novels from the mid-'20s, which he later extended to *Árvácska* (Little Orphan Girl, 1975), based on one of Zsigmond Móricz's last works from the early '40s, Péter Bacsó was experimenting with a journalistic approach to the problems that were being faced by the country's technical experts and also with satires on current topics, such as *Zongora a levegőben* (A Piano in Mid-Air, 1976) and *Ereszd el a szakállamat!* (Let Go of My Beard!, 1975). He carried on along that track through the ensuing decade, with *Banánhéj-keringő* (Banana Skin Waltz, 1986), whilst *Titania, Titania* (1988) was particularly noteworthy as a historico-political slapstick farce set in a fictional state readily identifiable as Ceauşescu's Romania.

An important development in the 1970s was the emergence of a highly gifted documentary school among a 'second generation' now at the Béla Balázs Film Studio. Gyula Gazdag (1947) made his debut with *Sípoló macskakő* (Whistling Cobblestone, 1969), an intriguingly ironic feature with a redolence of '68. Some of Gazdag's medium-length films—notably *Válogatás* (Selection, 1970), which cast a satirical (i.e. natural) light on the Young Communist League organisation, and *Határozat* (Resolution, 1972), co-directed with Judit Ember (1935), which documents election rigging by a party organisation in the provinces—never received permits to be shown in public, any more than did

Bástyasétány 74 (74 Bastion Promenade, 1974), a quirky absurdist political satire that Gazdag dressed up as an operetta.

Ferenc Kósa's *Nincs idő* (No Time, 1974) was likewise a valuable exemplar of the 'parable', posing the question "Is it possible to conceive of a benevolent jailer?". *Küldetés* (Mission, 1975), a portrait of András Balczó, the Olympic champion modern pentathlete, shows that much more than just sport is at stake in a thought-provoking document of lonely heroism, perseverance, and fidelity to oneself.

There is space for only the briefest of overviews of the major, internationally acclaimed achievements of the Hungarian animated film, which would richly deserve a chapter of their own. Gyula Macskássy (1912–71), who had already produced the finest prewar cartoons, was the director of *A két bors ökröcske* (Two Little Magic Oxen, 1954), which drew on Hungarian folklore traditions, and also *Párbaj* (Duel, 1960), which gained even greater accolades abroad. Among younger directors, Attila Dargay (1927), József Nepp (1934), with his *Gustav* series, and the duo of György Kovásznai (1934–83) and Sándor Reisenbüchler (1934), with *A nap és hold elrablása, 1812* (Stealing the Sun and the Moon, 1812, 1968), created enduring works of art. The biggest animation venture of all, the feature-length *János vitéz* (John the Valiant, 1975), was directed by Marcell Jankovics (1941), who was also responsible for the folklore-inspired *Fehérlófia* (White Mare's Son, 1978), likewise a full-length masterpiece. The output from the Pannonia Film Studio during this period was highly appreciated both in Hungary and around the world. A new wave of more modern trends in animation only started to emerge in the early 1980s, but it is noteworthy that Hungarian animated cartoons for decades managed to maintain a reputation as being amongst the best in Europe.

Hungarian cinema had 'lost touch with the public,' was the line taken by party officials. By the late seventies, although box-office attendances at home, with few exceptions, were sharply down, Hungarian films were still gaining a good critical reception abroad. To pick a typical year, 8 out of the 21 feature films released in 1978 were awarded prizes at international festivals—an impressively high proportion. The most noteworthy of these were Márta Mészáros's *Women*, Fábri's *Magyarok* (Hungarians, 1977), Gaál's *Legato* (1977), Pál Schiffer's *Cséplő Gyuri* (Gyuri, 1978), and Imre Gyöngyössy and Barna Kabay's *Két elhatározás* (A Quite Ordinary Life, 1977). The latter pair's subsequent *Jób lázadása* (The Revolt of Job, 1982), a story about an elderly Jewish married couple who want to adopt a peasant child in their village in 1944, when Hungary's Jews were being persecuted, was an even more resounding international success a few years later.

Perhaps the most interesting development of that decade was the strengthening and enrichment of the documentary approach coming from the Béla Balázs Film Studio. From the release of *Gyuri* and *A Quite Ordinary Life* right up to the early nineties, documentaries gained hugely in importance to become virtually the foremost vehicle of expression in Hungarian cinema. All the films made by the Gulyás brothers, Gyula (1944) and János (1946), were of key importance for scrutinising the reality that was passed over in silence by television news bulletins and official documents, then after the transition they would plug on just as vigorously with their interrogations of survivors of tragic political events of the recent past.

One of the major fruits of this revitalised genre from the beginning of the period in question was *Egyszerű történet* (A Simple Story, 1975) by Judit Elek (1937), who was later to apply her training in the documentary approach to shooting *Vizsgálat Martinovics Ignác szászvári apát és társai ügyében* (The Investigation of the Case Martinovics, 1981), a chamber drama dealing with the uncovering of an anti-Habsburg plot by Hungarian Jacobins in the 1790s, at the end of the Enlightenment era. Critical opinion abroad soon took note of this new Hungarian wave, with French and Italian critics labelling the blend of fiction and documentary 'the Budapest school'. *Jutalomutazás* (Holiday in Britain, 1974) by István Dárday (1940) and Györgyi Szalai (1940)—a satirical look at the bureaucratic shenanigans that go on when a party of schoolchildren is selected for a foreign trip—was the first resounding success of this genre as feature film. The same pair also directed the psychological fiction-documentary *Filmregény—Három nővér* (A Film Novel—The Three Sisters, 1975). Others followed in their path, amongst them the first films of Béla Tarr (1955), now an internationally acclaimed director, whilst somewhat later Pál Erdőss (1947) began producing his important documentary fictions, including *Adj király, katonát!* (The Princess, 1984), winner of a Golden Camera Prize in Cannes. Even much later on, Dárday was to stick doggedly, albeit with varying success, to his explorations of the possibilities of the genre.

Another career that took off in parallel with these, after showing strong talent in the documentarist approach, was that of György Szomjas (1940), who, not unironically, started off directing the first Hungarian 'Easterns', before switching to search for a wider viewership with portraits of young misfits and grotesque social anomalies.

Jancsó's first project on returning from Italy, was the two-part (pseudo-)historical fresco *Magyar rapszódia* (Hungarian Rhapsody, 1978), which for all the typical bravura and innovativeness of its formal devices, had only a slight reference to the present day, as if Jancsó were having trouble casting off a literal engagement with pre-war Hungarian nationalism, although he had quite clearly intended that it should be more than that and should relate to the period when the film was being made.

Among the 'old' auteurs, András Kovács provided a valuable and honest portrait of an extreme borderline situation during the Rákosi dictatorship in *A ménesgazda* (The Stud Farm, 1978). The real discovery of the late 1970s was *Vera Angi* (1978), by Pál Gábor (1932–87), a story set in a Party school of the early fifties, in the course of which the male protagonist, a teacher there, falls victim to the young female protagonist, a pupil and his lover, who is gradually transformed into an amoral liar. No allusive or symbolic parable, the film embeds the pressurised atmosphere of that era in a much more direct, less abstract, traditional love story and psychological drama.

The greatest of all successes in Hungarian cinema was the American Academy of Motion Pictures 'Oscar', and countless other festival prizes, awarded to István Szabó's *Mephisto* (1981). Running up to this, Szabó, whose career up to *Love Film* was outlined above, had made a number of films that studied Budapest life, its past and present, with an amalgam of dreams, visions and symbols, before turning to a quiet but tensely smouldering psychological chamber-piece with *Bizalom* (Confidence, 1979).

Mephisto is an enquiry into the individual's responsibility towards himself. With its story of a talented actor who is unable to

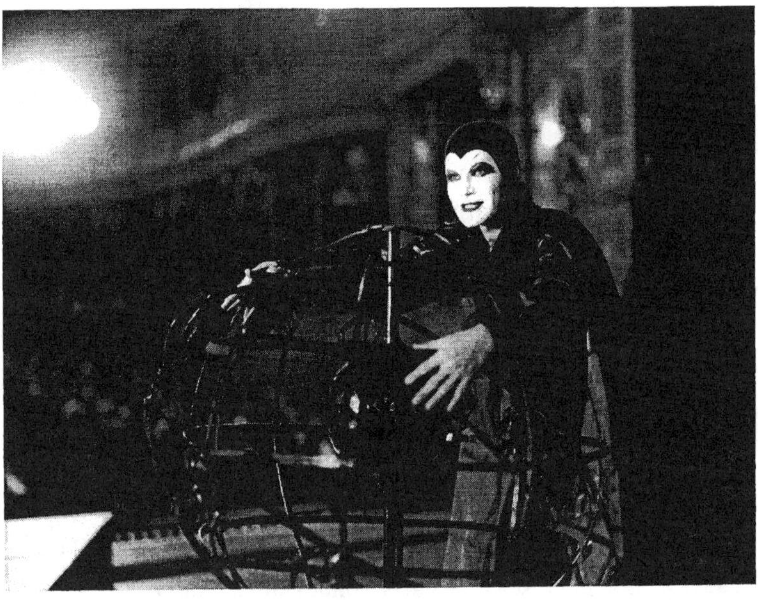

Klaus Maria Brandauer in *Mephisto*

refuse the blandishments of the Nazis and is gradually sucked into placing that talent in the service of their dictatorship, the film eschews laboured parallels or hidden allusions to deliver a message, in readily understandable language and with great aplomb, that is of much wider validity, as are the subsequent *Redl ezredes* (Colonel Redl, 1984) and *Hanussen* (1988). These three internationally acclaimed films can be regarded as constituting a trilogy. The latter two likewise concern the individual's relationship to authority, how talented men like Redl, the Austro–Hungarian Monarchy's spy chief, or Hanussen, an illusionist with power to sway the masses in Germany of the thirties, surrenders himself into the clutches of the authorities: all three films are anatomisations of self-delusion. (Lajos Koltai (1946), the cameraman in most of Szabó's films, himself gained an international reputation and has since often worked in Hollywood.)

Among Szabó's other contemporaries who were by now approaching middle age, Zsolt Kézdi-Kovács was able to project the stagnancy of the 1970s with great evocative potency, whilst the work of Pál Sándor (1939) and János Rózsa was also significant. In *Vasárnapi szülők* (Sunday Parents, 1979) and what may be regarded as its follow-up, *Kabala* (Mascot, 1981), Rózsa portrayed the (sometimes) tragedy-ridden lives of neglected youths with empathy and lyrical sensitivity. Pál Sándor's *Régi idők focija* (Football in the Good Old Days, 1974) recreated the fantasy, dreams and visions of writer Iván Mándy's world in a film that confirmed an original and versatile talent with its homage to classical slapstick comedy and diverse other stylistic layers.

Representation of the pre-1956 period of Stalinist dictatorship had already been permissible for some time, albeit within certain limits, but by the early 1980s those boundaries seemed to be stretched a bit further. Ferenc Kósa set the story of one of his films in the turbulent spring of '56. Pál Sándor, following his *Szabadíts meg a gonosztól* (Deliver Us From Evil, 1978), a slightly surreal tragicomedy that takes place in 1944, turned to the subject of 1956 in *Szerencsés Dániel* (Daniel Takes a Train, 1982), which is about a whole trainload of people fleeing from Hungary to the West in the immediate aftermath of the suppression of the revolution. With the wider range of permissible subject matter it also became easier to pass between film genres. Ferenc András (1942) demonstrated equal fluency with the latest trends in social satire, as in *Veri az ördög a feleségét* (It's Rain and Shine Together, 1977), the crime thriller, as in *Dögkeselyű* (Vulture, 1982), and moral scrutiny of an age, as in *A nagy generáció* (The Great Generation, 1986). The latter created quite a stir when it came out, and the term 'the great generation' has been used ever since, with a mixture of respect and irony, to refer to the '68 generation of malcontent but not rebellious youngsters, many of whom defected to the West after 1968–69 only to return some twenty years later, boasting of their achievements, real or bogus, to a Hungary that had changed in many ways whilst they were away.

As the 'doldrums' of the late Kádár regime approached, the waves made by Solidarity's struggle against Poland's military dictatorship and Hungary's worsening economic situation strained the general political climate—and cultural life particularly. In *Egymásra nézve* (Another Way, 1982), the story of a tragically ending lesbian relationship, Károly Makk sought, for the first time in Hungarian cinema, to give an unvarnished, illusion-free portrait of the grim years after 1956. The 'period of reprisals' directly following 1956 was as much of a taboo subject as was the revolution itself. (Zsolt Kézdi-Kovács, in *Kiáltás és kiáltás* (Cry and Cry Again, 1987), was later to tackle this subject.)

Another Way was followed into Hungarian cinemas, within the space of a few weeks, by *Megáll az idő* (Time Stands Still, 1981). Made by young theatre and film director Péter Gothár (1947) as a follow-up to a psychological drama, this treated the early 1960s in a new and unconventional manner. The story of *Time Stands Still* takes the final days of the 1956 revolution as its point of departure: "So, we'll be making our life here, then," says one of the adult protagonists. The real main characters, however, are the adolescent children of these grown-ups, who recognise the ubiquitous lie even in the 'thaw' and greater social openness of the sixties.

In its own way, the Hungarian cinema thereby helped to pave the way for—and even, to some extent, anticipate—the 1989–90 change in régime. For this reason Gothár's *Time Stands Still*, his most successful film both domestically and internationally, can also be considered as marking an epoch: after taboo subjects like the Stalinist dictatorship or 1956, the taboo on treating the post-revolutionary retribution likewise appeared to be toppling, though the process was elusive, uncertain and faltering. In any event, film artists won a greater measure of freedom of expression than, for instance, writers. Jancsó's visionary *Szörnyek évadja* (Season of Monsters, 1986), which probed the alternatives of history by posing questions of philosophical absurdity, followed by his *Jézus Krisztus horoszkópja* (Jesus Christ's Horoscope, 1988), a work that makes innovative use of video images within film, again attracted more than usual attention.

A number of new young directors emerged during this period, most notably Géza Bereményi (1946). Already known for his work as the screenwriter of *Time Stands Still*, he directed *Eldorádó* (The Midas Touch, 1988), winner of a European 'Felix', which tracks the path of blood and gold that its protagonists tread over the period from 1945 to 1956.

The sprawling talent of Gábor Bódy (1946–85) experimented in the avant-garde tradition in early films he made at the Béla Balázs Film Studio, carving out a special niche for himself with *Amerikai anzix* (American Fragment (or Torso), 1975) before going on to create a cult film for an entire generation with *Nárcisz és Psyché* (Narcissus and Psyche, 1980), a two-part historical fantasy based on the work of the then still living poet

Sándor Weöres. Two years after the premiere of *Kutya éji dala* (Dog's Night Song, 1983), still bubbling with ideas, Bódy was to take his own life.

Nearly all the works of another uncompromising young director, András Jeles (1945), no matter how clever their form and symbolism, ran into a brick wall of censorship. Even at this late stage, his debut, *A kis Valentino* (The Little Valentino, 1980), created an immediate stir with the stylistic bravura, unaccustomed in a Hungarian film, of its tragicomic, dispassionate, yet at the same time gripping and hyper-real reconstruction of a criminal case. Nor are Jeles's films of the mid-eighties—*Angyali üdvözlet* (The Annunciation, 1983) and *Álombrigád* (Dream Brigade, 1983)—any easier. The former is a cinematographic transcription (certainly no straightforward adaptation) of Imre Madách's classic philosophical drama, *The Tragedy of Man*, with child actors, the latter a dramatically complex, surreal parody of Soviet 'dramas of production'—the last film that the communist régime banned from screening.

Béla Tarr, having gained recognition with his documentary fictions, branched off in a new direction in *Őszi almanach* (Autumn Almanach, 1984 and *Kárhozat* (Damnation, 1987)—tense and slow-moving fictions based on strongly literary texts, whose craftsmanlike idiom further enhances the stylisation of the literary text. Only now, from the vantage point of a decade and a half on, is Tarr's work starting to be widely hailed in the film world.

Ferenc Grunwalsky (1943), whose significance was also sadly long underrated, directed *Egy teljes nap* (A Full Day, 1988) and *Kicsi, de nagyon erős* (Little But Tough, 1989)—tightly composed psychological chamber pieces about fallen people and criminals, remarkable for their lighting effects, in which his extraordinary talent for close-ups allow him almost to inhabit the very souls of the characters. Grunwalsky is an original talent as a cinematographer in his own right, becoming virtually a co-director with his work on

György Szomjas's more recent films, such as the highly successful *Roncsfilm* (Junk Movie, 1991), a picaresque suburban tragicomedy.

Szürkület (Twilight, 1989), by the unclassifiable György Fehér (1939–2002), is a minutely drawn-out and immensely tension-building chronicle of a psychological crime that stands as a kind of apotheosis of the art of black-and-white film. The infinite pains that Fehér took over his work meant that it was almost ten years before the next film appeared, but *Szenvedély* (Passion, 1998), a free adaptation of James Caan's much-filmed detective story *The Postman Always Rings Twice*, corroborates that he was in the front rank of European art-film makers.

The *oeuvre* of Miklós Erdély (1928–86), a prominent figure in Hungarian neo-avant-garde art circles, is especially noteworthy. Although not one of his films gained commercial release, *Partita* (1974) and the trilogy of *Álommásolatok* (Dream Copies, 1977), *Verzió* (Version, 1979–81) and *Tavaszi kivégzés* (Spring Execution, 1985) showed him to be one of the most original experimenters in Hungarian cinema. The public was only allowed to view them after his death but the films already had a strong influence on several other artists during his lifetime. He certainly influenced Gábor Bódy, for example, and also, if not so directly, Péter Forgács's (1950) documentary experiments and possibly the works of András Szirtes (1951), a number of whose personal or diary-type experimental films can be regarded as avant-garde 'journeys' into the fundamentals of visuality, formal idiom, film imagery and the cinematic approach.

Among the most important events in Hungarian cinema during the latter half of the '80s, prior to the change in régime, was Márta Mészáros's autobiographical *Diary* trilogy. The huge success of *Napló gyermekeimnek* (Diary For My Children, 1983), the first part, was duly followed by *Napló szerelmeimnek* (Diary For My Loves, 1987). Censorship delayed the release of the third part, *Napló apámnak, anyámnak* (Diary For My Father And Mother,

1990), which deals with the '56 revolution, but that only intensified public interest when it was finally shown.

By the end of the decade, 1956 had become a subject that was no longer a matter of political courage so much as a point of honour, an artistic challenge, for many film directors. Filmed on the heels of *A mérkőzés* (The Match, 1982), Ferenc Kósa's *A másik ember* (The Other Person, 1987), a hotly debated film set in the spring of 1956, was able to show, for the first time, unfaked footage of the actual street fighting and day-to-day life during the revolution, and placing this in a wider historical perspective points up its moral lessons above all.

The greatest critical acclaim nevertheless during the years preceding the change in régime was rightly reserved for several key long documentaries. Judit Ember's long-banned *Pócspetri* (1982) presents the recollections of survivors of a notorious incident where dozens of villagers were arrested and tortured, and several subsequently received long prison sentences in a staged trial, following the shooting of a Communist policeman with his own weapon during a protest against the nationalisation of church-run schools in June 1948. It was also only then that a go-ahead was given for TV transmission of all 25 episodes of *Pergőtűz* (Barrage, 1982), a document of stunning human and historical perspectives in which Sándor Sára questioned survivors of the Hungarian Second Army's bloody defeat on the Don Bend in 1943.

Recsk—Egy magyar kényszermunkatábor (Recsk, a Hungarian Forced Labour Camp, 1989), a hard-hitting and heart-rending documentary by Lívia Gyarmathy and Géza Böszörményi, which uncovered many new facets of the country's infamous Gulag, was one of the most significant of all Hungarian documentaries. That same exciting Film Week also fêted Ildikó Enyedi (1955), previously known only within a narrow circle of film professionals, for the fresh narrative voice that she brought to her first feature film, *Az én XX. századom* (My Twentieth Century, 1989), an original work blending playfulness and philosophy that has no precedent in Hungarian cinema. Having won her an award at the Cannes Film Festival, it quickly conquered art-movie houses abroad as well as in Hungary. There were great expectations for her next film, *A bűvös vadász* (Magic Hunter, 1994), and even more so for *Simon mágus* (Simon Magus, 1998), which sets an apocryphal biblical figure down in the Paris of our times.

Prospect: a New Beginning

Ildikó Enyedi brings us to the present day. The 1989–90 change of régime battered and only gradually re-structured the film industry. Privatisation of MAFILM, the Hungarian State Film Production Company, was stalled; fewer and fewer full-length feature films were produced. In the meantime, distribution channels underwent a complete change, with American mass entertainment films squeezing out Hungarian and European art films into a small network of specialist art-movie outlets.

It is harder, of course, to assess the latest Hungarian films and trends than those of the recent past, for the latter already have discernible and codifiable common features whilst that cannot yet be said of today's films. These are coming from established as well as completely new directors, so the main guide has to be the individual creative output rather than any hypothetical stylistic trends that may be defined at some point in the future. Here Péter Gothár is the name that springs most readily to mind. Most of his work during the 1990s attracted great attention and was garlanded with prizes, notably *A részleg* (The Outpost, 1994) and *Haggyállógva Vászka* (Let-me-hang Vaska, 1995), a Soviet 'lager tale' of indescribable genre, set in Leningrad/St. Petersburg and employing fantastic elements.

Conspicuous amongst the recent output of documentary specialists is a multiple-part series by Pál Schiffer (1939–2001) which follows the lives of a group of now unemployed former workers of a defunct state enterprise as they come to grips with an unfamiliar and, for them, incomprehensible new set of economic conditions. A somewhat similar intent is evident in the films of Tamás Almási (1948), which interrogate unemployed workers in an industrial city in eastern Hungary as they unsurely seek to put their lives together anew. Considerable interest was roused by Péter Forgács with a very different sort of documentary project, under the umbrella title *Privát Magyarország* (Private Hungary, 1991–97), in which he restored and, as it were, rewrote pre-war and, in some cases, post-1945 footage originally shot by amateurs. Whether these are regarded as fictions that recreate an era or documents that evoke an era, or perhaps both, he has devised and developed a genre without parallel in international cinema.

Ferenc Grunwalsky's most recent films—*Goldberg-variációk* (*Goldberg Variations*, 1991), *Utrius* (1993) and *Visszatérés* (Homecoming, 1998)—are important achievements of the latest Hungarian cinema, without obvious kinship: with their consummately calculated presentations of human faces in close-up, their muteness and taciturnity, the plays of light on them, they provide savage illuminations of latent and overt aggressions, the 'new savagery' of unbridled capitalism in Hungarian society.

Bolse Vita (1995), the début film of Ibolya Fekete (1951), reflects the evaporation of euphoria over the change in régime, a bitter yet not hopeless vision of populations adrift in East Europe in the wake of its political earthquake which picks out the faces of a few of the losers amidst the mass of migrants. It has been awarded more international prizes than any other Hungarian film of the past five years. Fekete's most recent film, *Chico* (2001), which tackles the problem of international terrorism, was awarded a prestigious prize at the Karlovy Vary Film Festival.

János Szász (1958) won recognition both at home and abroad with the extraordinary density and tension of his modern reworking of Büchner's *Woyzeck* (1993), filmed in exquisite black-and-white, whilst his subsequent *A Witman-fiúk* (The Witman Boys, 1997), a tense fin-de-siècle drama, also stands out for a psychological authenticity akin to that of *Woyzeck*. Most recently, Szász was involved as one of the directors engaged on Steven Spielberg's Shoah Foundation series, which, as James Moll's *The Last Days* (1998), records interviews with Hungarian Holocaust survivors. Stylistically close in certain respects to Szász's fictional films is the work of Attila Janisch (1957), who first came to attention with a psychological crime thriller and, more recently, for a somewhat Borgesian film that employs fantastic elements in a blend of reality, imagination and half-dream. Another member of this generation of fortysomething directors, Árpád Sopsits (1952), made a critically well received debut with a black-and-white film and confirmed his talent with *Torzók* (Abandoned, 2001), premiered at the 2001 Budapest Film Festival, whose child hero is captive in a cruelly run institution that specialises in recriminatory punishments.

Frigyes Gödrös (1942), who graduated from the Budapest Film School in the late 1960s, was for decades a mere 'up-and-coming' talent until he suddenly arrived with *Glamour* (1998), his second feature film. In telling the story of a part-Jewish, part-German, middle-class Budapest family, this largely autobiographically inspired epic employs an intriguing narrative frame as well as considerable self-irony, coupled with respect for its protagonists. Both the shooting of the film and its premiere virtually coincided with the European and world premieres of István Szabó's somewhat similar *A napfény íze* (Sunshine, 1999), promoting a productive and occasionally lively debate in both the American and Hungarian specialist magazines and even dailies.

Both critics and fellow professionals have had high expectations of Béla Tarr from the outset, and were looking to be surprised by him. To start with, Tarr explored the possibilities of fictional documentaries, but he soon changed tack, becoming what one might call—if the epithet were not restricting in its import—a 'researcher' of visual imagery and cinematic rhythm, of a slowly unfolding then onward-plunging film time. The expectations were proved to have been justified. The seven and a half hours long, tautly measured, black-and-white *Sátántangó* (Satan's Tango, 1994), based on a 1985 novel of that title by László Krasznahorkai, is the most daring achievement of Hungarian cinema in its uncovering of the possibilities of cinematic narrative, film time and rhythm. *Werckmeister harmóniák* (Werckmeister Harmonies, 2001), Tarr's most recent work, likewise based on a Krasznahorkai novel, *Az ellenállás melankóliája* (published in English as *The Melancholy of Resistance*), is of similar intellectual profundity, slow rhythm and extraordinary tension. Like the earlier film, it projects within an apocalyptic landscape *Satan's Tango* in a rural backwater, *Werckmeister Harmonies* in the confines of a small town—a story about the impossibility of human redemption that for all its difficulties of interpretation is nevertheless crystal-clear in its harrowing morality.

One of the most interesting developments in Hungarian cinema at the turn of the new century has certainly been Miklós Jancsó's artistic rebirth. The films he produced in the late '80s and early '90s react with a journalistic reflex to the socio-political challenges of post-communist Hungary—fascinating, passionate works, shot through with bitterness. *Isten hátrafelé megy* (God Walks Backwards, 1990) was a sinister vision of a new military take-over by the Russian army, whereas *Kék-Duna keringő* (Blue Danube Waltz, 1991) maps out the venal rule of a right-wing Hungarian dictatorship with reference to events that are in part purely fictive, in part real. For all their merits, these films were, paradoxically, at once clichéd, self-repetitious, drowning in symbolism and, at the same time, too labouredly 'up-to-the-minute', sensationalistic, overcharged with journalistic public crusading fervour—a stylistic frame in which the other function of fantasy sits rather uncomfortably.

Anyone (perhaps including Jancsó himself) who thought that this marked the final flicker in the great span of his career could not have been more mistaken. The film he shot towards the end of the nineties, *Nekem lámpást adott kezembe az Úr—Pesten* (The Lord's Lantern in Budapest, 1998), was a major departure from all his previous works and had an unexpectedly favourable reception from critics and public alike. Talk of Jancsó's 'rebirth' was in this case fully justified. Instead of the sort of symbolically freighted pseudo-historical parable that had come to be expected of him, Hungary's supreme film-maker came up with an irresistibly funny slapstick farce which has a pair of gravediggers as its main protagonists, playing now street cleaners, now bank robbers, even financiers setting up a new bank, untouchable by Mafia bullets as they blithely circumvent the laws of reality with eternal clown's grins on their faces. Crammed with allusions to Hungary's political and ordinary daily life, self-avowedly 'daft', and drawing on a veritable compendium of the devices of slapstick, artistic subterfuge and even vulgar clowning, this provocative film was awarded the foreign critics' prize at the Budapest Film Festival. In one of its bittersweet scenes Jancsó himself makes an appearance, together with Gyula Hernádi, his long-time screenwriter, to bid farewell to his career, to the twentieth century, and maybe to life itself; however, the film's success inspired Jancsó, aided by Hernádi and with Grunwalsky as his cameraman, to go on and shoot two follow-up films in similar vein—*Az anyád! A szúnyogok...* (Damn You! The Mosquitoes..., 1999) and *Utolsó vacsora az Arabs szürkénél* (Last Supper at the Arabian Grey, 2000)—which display the same virtues (and the same two clown-actors in the main roles) and so round out the work into a trilogy. At 80, Jancsó is still the youngest at heart of Hungarian directors, who has managed the trick of keeping the loyalty of at least part of his old fan base whilst winning a younger generation of cinemagoers that sees and thinks about things in a very different way.

Winning over the public is now a bigger challenge for Hungarian directors than at any time during past decades. In the 1990s Péter Tímár (1950) was supreme in being able to attract audiences with comedies that have an idiom and technique all of their own. *Csinibaba* (Dollybirds, 1996) is a splendid satire on the sixties, spiked with popular hits and film clips, whilst *6:3* (1998) whisks its street cleaner main character (and with him the viewer) back to the historic afternoon in 1953 when the Hungarian soccer team, captained by 'Kiddo' Puskás, became the first foreign side to beat the English on their hallowed home turf of Wembley Stadium in London. A satirical portrait of that era, it wrings much wry and disconcerting humour from this basic premise. Another noteworthy director is Gábor Koltay (1950), whose heavily state-funded *Honfoglalás* (Conquest, 1996) and *Sacra corona* (2001) provide screen spectaculars of two passages in Hungary's history shown in a heroic, glorious and pathos-filled light.

Judging from the experience of recent years, it seems Hollywood-style mass entertainment films like *A miniszter félrelép* (Out of Step, 1998) by Róbert Koltai (1943), *Sztracsatella* (Stracciatella, 1996) and a remake of *Hippolyt, the Lackey* (2000), both directed by András Kern (1948) have some limited chance of success. Even more important, though, is that Hungarian documentary and feature films of artistic merit, which finally came of age during the sixties, can also still survive to shape the intellect and emotions in the new century.

It is a new and noteworthy development that in 2001 two particularly lavish productions were embarked on that without massive state support could never have come to fruition within the standard domestic film-financing system. One is Géza Bereményi's *Hídember* (The Bridge Man), a historical film dealing with the tragic and yet uplifting life of Count István Széchenyi, the leading figure of Hungarian economic and political reforms during the first half of the nineteenth century; the other, a spectacular film adaptation of the opera that Erkel based on József Katona's tragedy *Bánk bán*, which is to be directed by Csaba Káel (1961), previously known mainly for his work on commercials.

These lines are being written in the weeks after the 2001 Budapest Film Festival. There the films that provoked the greatest interest were not mass-entertainment products but demanding works by three new young directors—Szabolcs Hajdu (1974), Kornél Mundruczó (1975) and Ferenc Török (1970)—that offered new ways of looking at film and at the world. The Hungarian cinema of the future not only made its debut through this trio but also, if one may be permitted a touch of pathos, clamorously smashed open the gates of the new millennium.

Notes

1. Béla Balázs, *Der sichtbare Mensch, oder die Kultur des Films* (Wien–Leipzig, 1924), p. 75.
2. Guido Aristarco, *A filmművészetek története 1–2* (The History of Film Theories. Vols. I–II) (Budapest, 1962), p. 92.
3. István Nemeskürty, *A magyar film története (1912-1963)* (The History of Hungarian Film (1912-1963)) (Budapest, 1965), p. 102.
4. *Ibid.* p. 116.
5. *Ibid.* p. 118.
6. *Ibid.* p. 123.
7. *Idem, A képpé varázsolt idő* (Time Magicked into Pictures) (Budapest, 1983), pp. 467–469.

László Beke

VISUAL ARTS AND VISUAL ARTISTS, 1848–2000

There lingers to this day a romantic notion of the artist as a tormented genius whose talents go unrecognised by his contemporaries but who, setting his face against society, trusting only in posterity, ekes out a lonely, impoverished existence painting or sculpting masterpieces that he is forced to sell for peanuts, eventually to end up lining the pockets of canny art collectors three or four generations later. Van Gogh's 'passionate life' is archetypal. In Hungary too, though such perceptions had become discernible by the beginnings of the period surveyed here, few artistic careers can actually be said to exemplify this scenario. Leaving aside the figures of Mihály Munkácsy, who started off as a joiner's apprentice but went on to become a lion of the Paris Salon, or the eccentric but mentally unbalanced prophet Tivadar Csontváry Kosztka, or the bohemian-hearted aristocrat László Mednyánszky, it would be hard to characterise the past century and a half of Hungarian art history as one of heroic individual struggles against adversity. Inclined as Hungarians still are to play up the role of 'misunderstood geniuses' before the wider world, it seems more apt to analyse the historical and cultural trends in this field as more sober and 'public-friendly'.

From 1849 to the Compromise of 1867

The ferment of Enlightenment ideas during the final decades of the eighteenth century spawned the classicist and romantic aesthetics that were the twin vehicles for the aspirations for independent nationhood that were manifested in Hungary from the 1830s onwards. Suggestive examples of this parallelism of conceptual approach and taste in the visual arts can also be traced in the capital's architecture of this era: Mihály Pollack's masterwork, the National Museum with its archaising style, constructed between 1837 and 1847, from whose steps the poet Sándor Petőfi was to recite his *National Song* when the revolt against Habsburg domination broke out in Budapest on 15th March 1848, followed two decades later by Frigyes Feszl's Vigadó (Redoute, or Ballroom, Building; 1859–65), which, in a spirit of romantic wish-fulfilment, is posited on a belief that the origins of 'Hungarian' style are to be located in the Orient. The blend of patriotic poignancy and cool, classicist restraint displayed by the *Pásztorlányka* (Shepherd Girl, 1822) of István Ferenczy (1792–1858), executed whilst he was working as an assistant to the Danish sculptor Bertel Thorvaldsen in Rome, was designed to mark the beginning not only of 'fine arts' but also of the visual arts in Hungary. Ancient Greek in inspiration, this sculpture alludes to the antique myth of the shepherdess who invents figurative art by the gesture of tracing the shadow of her departing lover in the dust. Ferenczy's intention was that his public would read 'Hungarian' for 'antique' but he did not fully bring off his plan to found a national art. A projected bronze equestrian statue of *King Matthias*, had he been able to secure the finance for it, might well have turned out to be the first truly Hungarian statue of this kind, but it was never realised. As it was, it was his pupil, Miklós Izsó (1831–75), who around 1870 grasped the real function of sculpture in the brilliant dynamism—the 'romantic' subject, if you will—of his dancing peasant figures and thereby genuinely succeeded in formulating a new and characteristically Hungarian fund of motifs.

In order to speak with any justification of 'progress' in the arts, one needs to be able to point to distinct changes in their 'infrastructure'. Classicism in Hungary was born with the rise of the middle classes. The urbanisation of the then separate municipality of Pest, on the left (eastern) bank of the Danube, was subject to an integrated city development plan (drawn up by János Hild in 1805) and a Municipal Beautification Commission (established in 1808). The entire institutional basis of the arts underwent major changes. The collection of the National Museum was established in 1802. The Hungarian Academy of Sciences, founded in 1825, had its emblem, the allegory of Science, painted by the Austrian Johann Nepomuk Ender and even went as far as electing artists as academicians, the first being István Ferenczy then, after a prolonged delay, Miklós Barabás (1810–98) and Károly Markó Sr. (1791–1869). One curious aspect of the Academy's history is that it managed virtually to replicate a split within the *belle arti* between the four greater and three lesser arts, which goes back to classical antiquity: the 'fine arts', including art theory and history of art (along with belles-lettres and poetry), were assigned as provinces of the Academy's Department of Linguistics, whereas aesthetics was the domain of the Department of Philosophy. (Only in 1862 was a competition eventually announced for the design of the building that the Academy currently occupies in Budapest.) In 1839 an Art Society was founded, with Ágoston Trefort as its president, and remained in existence for thirty years. For training in art, though, students had traditionally had to pursue their studies abroad—in Rome, Vienna, Paris, Düsseldorf or Munich—but modest opportunities also began to open up to obtain instruction under the guidance of certain masters in Pest itself. Jacopo Marastoni (1804– 60), a painter of Venetian origin, started a private school in 1846, teaching a variety of graphic and printing techniques including lithography, then so important for the newspaper press (he was, incidentally, also one of the very first photographers in Hungary).

Romanticism was somewhat tardy to arrive in Hungary, as compared with other parts of Europe. Undoubtedly, this was due to the break in momentum that came with defeat in the country's War of Independence in 1849. In the pre-revolutionary Reform Era there had been painters who tried out the fashionable 'Italianate' or

Biedermeyer approaches to depicting parts of the Hungarian landscape such as Lake Balaton or the empty *puszta* of the Great Hungarian Plain. The bulk of the work by Károly Markó Sr. (as with István Ferenczy) was done in Rome, but his *Visegrád* (1826–30) offered a slightly classicising and yet strikingly fresh interpretation. Others catered to the demand for portraits from the nobility and the bourgeoisie, with Miklós Barabás, by far the most talented painter of the mid-century and renowned for his visual memory, capturing all the celebrities of the day in oils. He made relatively few genre paintings, and these—*Vásárba menő román család* (Romanian Family Going to Market, 1844) is typical—tended to be more in a sentimental Biedermeyer manner than an expressive-romantic or full-blooded realist style.

Károly Kisfaludy (1788–1830), who was first and foremost a writer and poet, was the first to experiment with Romanticism in painting, following in the footsteps of English and German prototypes. Painters were occasionally carried away by the sense of boundless space instilled by the Great Plain or the spectacular drama of one of the huge storms that occasionally brew up on it, as in *Vihar a pusztán* (Storm on the Puszta) by Károly Lotz (1833–1904). Though shipwrecks and floods were other popular subjects, as in *Mentőcsolnak* (Lifeboat, 1847) by Mihály Zichy (1827–1906), painters reserved their real heights of romantic passion for the pains of defeat in the War of Independence. They painted grandiose historic tableaux whose nominal subjects—the defensive war against the Turks or the clashes of Hungary's *kuruc* armies against the Habsburgs in the war of liberation a century and a half previously—made it very clear to the beholder that more recent events were implied.

Industrialisation, the advance of the press then, later on, the graphic and technological processes prompted by the emergence of photography paved the way for changes not only in methods of execution but also in how things were viewed. The spread of photography impelled the visual arts to take stock of themselves. The finest portrait of Sándor Petőfi—the poet of the 'blazing soul' who died young on the battlefield towards the end of the 1848–49 War of Independence—is a daguerreotype by an unknown photographer. Before destroying the plaster model for his *King Matthias* statue, István Ferenczy had Ádám Gola make a photographic record of it. And it was a Hungarian priest and teacher, József Petzval (1807–91), who constructed one of the first objectives suitable for portrait photography, which was incorporated by the Austrian firm Voigtländer into the design of its renowned wooden-cased cameras.

From 1867 to World War I

The 1848–49 War of Independence is conventionally regarded as one of the most glorious chapters of Hungary's history, epitomising its centuries-long quest for national identity, whilst the two decades that followed the defeat are seen as a period of melancholic yet pragmatic recognition that accommodation with the Habsburg power was the necessary pledge for modern capitalist development. The 1867 'Compromise' had a positive effect on the Hungarian economy; following the unification of the hitherto separate municipalities of Pest, Buda and Óbuda in 1872, the centre of the country's capital built up on both banks of the Danube; the use of gas and electricity for lighting became widespread (with knock-on effects for light and colour values in painting); the railway became a means of mass transport, while communications were enhanced by the introduction of the telephone (and ancillary pre-radio services such as Tivadar Puskás's system for telediffusing news bulletins and live music performances). Those who flocked to the City Park exhibition site that was the focal point of the 1896 celebrations to mark the millennial anniversary of the arrival of Magyar tribes in the Carpathian Basin could do so along the elegant, historicising thoroughfare of the Radial (today's Andrássy) Avenue, below which ran the brand-new underground tram service, running out from Vörösmarty Square in the centre of Pest and crossing the Grand Boulevard (Nagykörút) with its eclectic rows of apartment blocks, to terminate at the exhibition site. By the early years of the twentieth century the Millennial Monument, a broken semicircle of two colonnades of monumental statues by György Zala (1858–1937), had risen in the centre of Heroes' Square, flanked by two new buildings, the Exhibition Hall (Műcsarnok) on one side and the Museum of Fine Arts on the other. (At the core of the latter's permanent collection was a large group of old masters donated to the nation by the Esterházy family, which had previously been temporarily displayed (from 1865) in the Hungarian Academy of Sciences building and then (from 1871) the National Gallery.) Adding to this were the contrived popularism and eclecticism of the clutch of representative Hungarian architectural styles that make up Vajdahunyad Castle (1896), originally designed by architect Ignácz Alpár as a temporary centrepiece for the exhibition site but later rebuilt as a permanent structure, which can be seen in retrospect as a precursor of pop art montage and post-modern referentiality.

Various organisations to safeguard artists' interests were by then also gaining in strength. Miklós Barabás, Sándor Brodszky (1819–1901) and Károly Telepy (1828–1906) quit the Art Society to set up a National Society of Fine Arts in 1859; their first exhibition opened in 1863, and from 1877 it had its own premises at No. 69 Andrássy Avenue before taking over the new Exhibition Hall in Heroes' Square in 1896. A parallel Hungarian Society of Applied Arts was formed in 1885. A Royal Hungarian School of Design and Art Teacher Training opened its doors in Budapest in 1871, with Gusztáv Keleti (1834–1902), himself a painter, as principal. (This was the predecessor of today's Academy of Fine Arts, now located at No. 69 Andrássy Avenue.) A National Committee of Monuments assumed responsibility for the upkeep of the country's architectural heritage in 1881. The final step in setting up a nucleus of institutions that, in essence, still defines Hungary's art life to present-day was marked by the construction of the Budapest Museum of Applied Arts (designed by Ödön Lechner, 1891–96).

Hungarian art at the turn of the century was shaped, on the one hand, by the Millennial Exhibition, with its ancillary facilities and programmes, and on the other, by the efforts of the Museum of Applied Arts. Official 'image-boosting' encouraged the artistic expression of grandiose nationalistic ideas—most symbolically, the separate pavilion built to display *A magyarok bejövetele* (The Arrival of the Magyars, 1895), a vast cyclorama painted by Árpád Feszty (1856–1914)—particularly if this chimed with the approved canon of historicist, academicist and eclecticist styles. No matter that technology by then was showing that the way ahead lay with new media, from the 'peep-show' kinetoscope to motion pictures, from the Dante's *Inferno* type of panorama to a diverse range of major photographic enterprises. Ödön Lechner, by contrast, typified the Hungarianised version of the 'Secessionist', or Art Nouveau, movement which was to prove forward-looking not just within architecture. Having first experimented with the adaptation of oriental motifs, Lechner eventually hit on the style with which he built up a school of disciples, one distinctive element of which was the use on roofs and facades of colourful pyrogranite and glazed ceramic tiles produced by the Zsolnay factory in Pécs. His followers, most notably the 'Fiatalok' ('Young Ones') group of Károly Kós, Ede Thoroczkay Wigand, Lajos Kozma, István Medgyaszay and others, did not aim simply to transplant the Viennese or German *Sezession* but consciously strove to create a 'Hungarian' architectural vernacular. To that end, they embarked on their own collecting trips to Hungarian villages to search for ideas, just as Zoltán Kodály and Béla Bartók were to do in the area of folk music. Amongst Lechner's followers, it was Béla

Lajta who, through his heavy forms and pre-functionalistic solutions (as with his design for a school in Vas Street, Budapest), who provided a bridging link between the modernist aspirations of the *fin-de-siècle* and the inter-war years.

In painting, following Hungary's defeat in 1849 the endeavours of Bálint Kiss (1802–68), Soma Orlai Petrich (1822–80), Mihály Kovács (1818–92) and other artists of the Reform Era found worthy successors in the creators of monumental historical tableaux. Viktor Madarász (1830–1917) and Bertalan Székely (1835–1910) perceived subjects from the country's tragic history as symbolically applicable to the present. Thus Madarász' *Dózsa népe* (Dózsa's People, 1868) or Székely's *Egri nők* (The Women of Eger, 1867) were both reminders of the self-sacrificing heroism of the recent past. Mór Than (1828–99) and Sándor Wagner (1838–1919), being driven by similar ideas, equally ruled themselves out of the running to be considered representative official painters of the period. This role fell to Károly Lotz and Gyula Benczúr, the latter, for instance, with *Budavár visszavétele 1686-ban* (The Re-capture of Buda Castle in 1686, 1896). Fortunately, though, there was another, more progressive trend in the second half of the nineteenth century: the realism of Mihály Munkácsy. Munkácsy (1844–1900) did in fact produce several genre pictures invoking

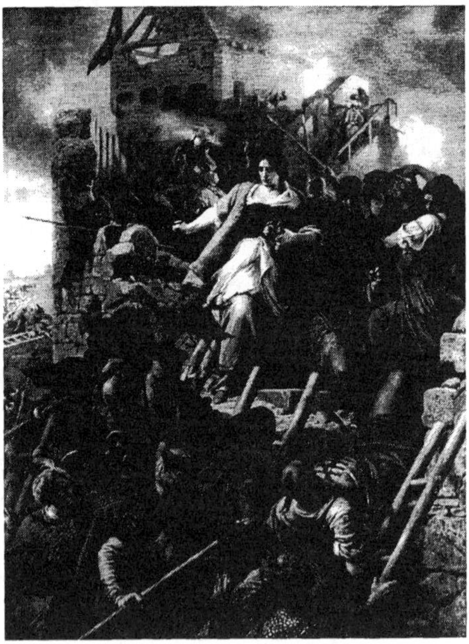

Bertalan Székely: *The Women of Eger*

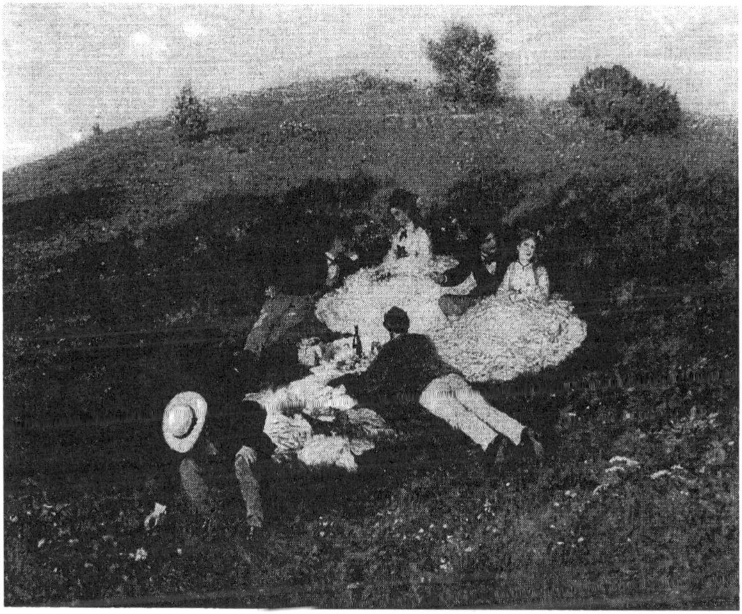

Pál Szinyei Merse: *Picnic in May*

the War of Independence but the compositions were dominated by heavy, sombre tones. At the peak of his career, by contrast, though he had inklings of the significance of impressionist *plein air* techniques, he turned out mainly studio portraits for his adoring public in Paris. László Paál (1846–79), the bulk of whose output was accomplished at Barbizon, likewise came close to realising the importance of colours in sunlight. The final step in that direction was not taken by him, however, but by Pál Szinyei Merse (1845–1920). The significance of Szinyei's *Majális* (Picnic in May, 1873), painted in Munich, with its Courbetesque shrill greens and mood akin to that of the impressionists Manet and Monet had to wait until the turn of the century before its significance was recognised by the generation of young artists who were to usher in the modern era of Hungarian art.

Modern Hungarian painting dates its start from the artists' colony that Simon Hollósy (1857–1918), along with a group of painter friends and students at his private school in Munich, established at Nagybánya (Baia Mare, Romania) on their return to Hungary in 1896. The arrival of the Nagybánya group, which included Károly Ferenczy (1862–1917), István Réti (1872–1945), Béla Iványi-Grünwald (1867–1940), János Thorma (1870–1937) and István Csók (1865–1961), was at one and the same time a 'retreat' from the hold of academicism and metropolitan life into a natural environment that supplied impressionist atmosphere and equally served to demonstrate a form of plebeian solidarity with the rural populace. A seminal work here was Ferenczy's *A hegyi beszéd* (Sermon on the Mount, 1896), which catches a landscape plunged in a rapt hush. This school had its continuation, a few years later, in full-blooded *plein air* and a separate 'urban' school near-photographic fidelity originating in the 'painting of the poor'. Painter Réti and art historian Károly Lyka managed to rescue the Nagybánya tradition after the First World War by incorporating it into the training at the Academy of Fine Arts, where they both taught, and this in turn inspired the Gödöllő colony of Sándor Nagy (1869–1950), Aladár Körösfői-Kriesch (1863–1920), Jenő Remsey (1885–1989) and others to try and set up a Secessionist commune that cultivated all the arts. An important separate group of Great Plain painters—János Tornyai (1869–1936), Adolf Fényes (1867–1945) and József Koszta (1861–1949) among them—took their lead from Munkácsy and aimed explicitly at capturing the peasant *couleur locale*, though the last-named was also receptive to brooding views.

Despite the artistic will to create communities, the greatest Hungarian painters at the turn of the century were all solitary figures. Among these was József Rippl-Rónai (1861–1927), who was a member of the Nabis in France at the end of the nineteenth century, and Mihály Zichy, the success of whose *Life-boat* led to his appointment as official painter to the Russian imperial court at St. Petersburg, with a spell of time in Georgia. After a five-year interlude in Paris, painting huge canvases such as an allegory dedicated to freemasonry and producing several big engravings for literary works rather in the style of Gustave Doré, he went back to Russia in 1879, producing illustrations for, amongst other things, printed editions of János Arany's ballad *Hídavatás* (Inauguration of the New Bridge, 1892) and Imre Madách's play *Az ember tragédiája* (The Tragedy of Man, 1887). Baron László Mednyánszky (1852–1919) left behind a large *oeuvre* of expressive landscapes but was drawn to the world of tramps and Oriental mysticism. The symbolist painter Lajos Gulácsy (1882–1932) was another who gave himself up to his dreams, going about in Renaissance garb and imagining himself to be living at various historical periods and in his own invented realm of Na'Conxipan; *Az Ópiumszívó álma* (The Opium Smoker's Dream, 1913–16) was typical of his work. Above all, there was Tivadar Csontváry-Kosztka (1853–1919), the great eccentric of Hungarian art history and yet, for all his eccentricity, a singularly prophetic apostle. According to his autobiography, Csontváry owed his creativity to a mystic hallucination. He was obsessed by a quest for 'the great motif' that would help him to discover what he called the 'Sun's route' (his idiosyncratic, near-mystical interpretation of *plein air*)—and there can be little doubt he achieved that in a number of his monumental paintings such as *A taorminai görög színház romjai* (Ruins of the Greek Amphitheatre at Taormina, 1904–05), *Baalbek* (1906), and *Magányos cédrus* (Solitary Cedar, 1907). Csontváry had an odd notion of Hungarianness in which esoteric doctrines were linked in his mind with myths about the Hortobágy region of the Hungarian *puszta* (as revealed by a film show), the Lebanese origins of the Magyar people, and the cedar of Lebanon—a potent brew of ideas that eventually unhinged him completely. It is one of the great sorrows of Hungarian culture that, beyond the recognition accorded by André Breton, some other surrealists and a few specialists, the work of this 'Hungarian Rousseau' is still barely known to the wider world.

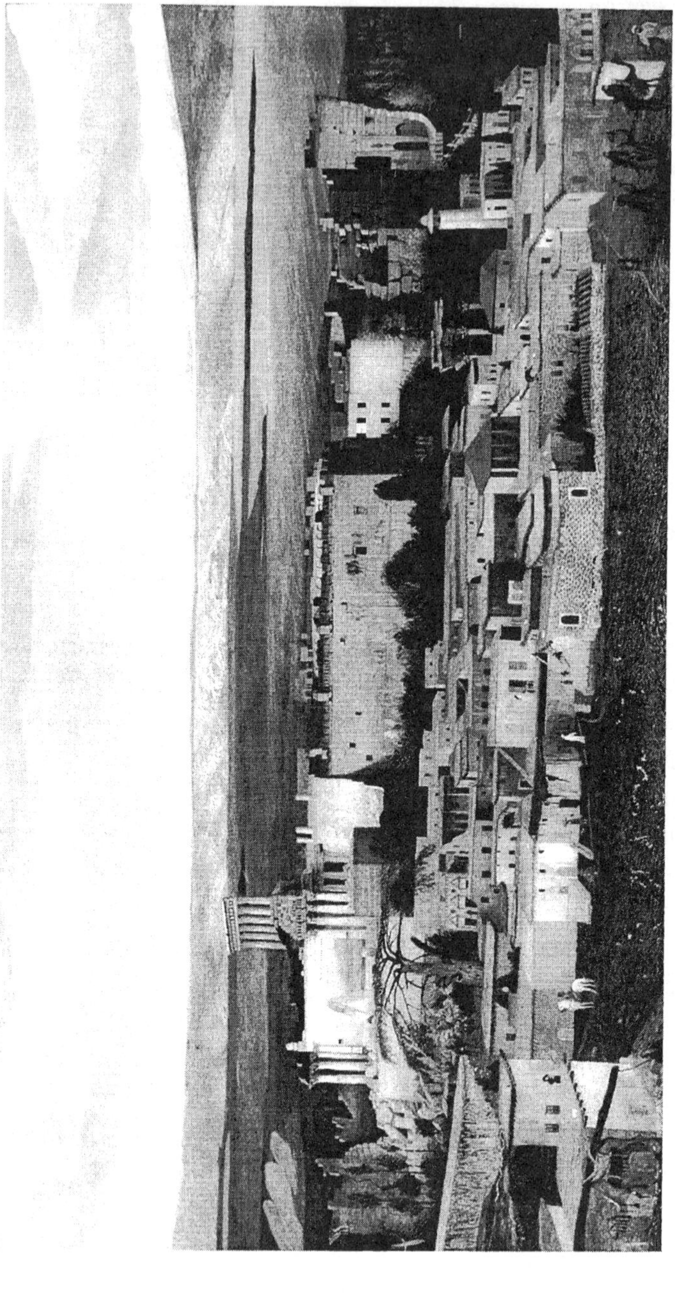

Tivadar Csontváry Kosztka: *Baalbek*

In speaking about painting at the turn of the century it is impossible to avoid some appraisal of the impact of photography. The latter half of the nineteenth century witnessed a continuous series of innovations: József Chudy's optical telegraph, Jenő Gothard's astronomical and X-ray pictures, István Kruspér's measurements, Antal Simonyi's very early snapshots (1855), Ányos Jedlik's experiments with photography and so on. Arguments over whether the new image-making process would eventually make portraitists redundant, and was it permissible for an artist to use a camera, had already been raging back in the 1850s and 60s, with Miklós Barabás and Bertalan Székely amongst the contributors to the debate. For a short period during the 1860s, when he was financially pressed, Barabás took up photographic portraiture, as did József Borsos (1821–83), a contemporary portraitist and genre painter (e.g *Lányok bál után*; Girls after the Ball, 1850). Neither became an epoch-making photographer. As far as the subject matter goes, by contrast, the Latin American travel photographs of Pál Rosti (1830–64?) and the documentary photographs of the Crimean War taken by Károly Szatmáry-Papp (1812?–87)—irrespective of whether the latter is claimed to be a Romanian or Hungarian—broke new ground. In their footsteps came the first professional photographers, each specialising in a particular subject: Károly Divald (1830–97) in landscape photography, György Klösz (1844–1913) and József Erdélyi (1866–1934) in views of the capital (or the Millennial Exhibition itself), and József Plohn in portraits of veterans from the War of Independence. By around the turn of the century, the arguments had moved on: it was photographers who were battling to gain acceptance for the new medium as an art form in its own right and, in the process, turning to so-called 'pictorialist' ways of looking at their subjects and 'fine' rendering of surfaces, with József Pécsi (1889–1956), Dénes Rónai (1875–1964) and Aladár Székely (1870–1940) in the vanguard—the latter famous for his portraits of the poet Endre Ady.

Whereas literary modernism can be said unequivocally to have set its foot in Hungary at the very opening of the twentieth century with the publication of Ady's poetry and the advocacy of French symbolism by the literary periodical *Nyugat*, it is harder to discern clear parallels in the field of visual arts in relation to specific individuals or trends. For a better understanding of connections, mention should be made of the influential intermediary role played by one particular group of intellectuals at this time, the so-called Sunday Circle, virtually all of whose members achieved later renown as thinkers and scholars. Among them were Leo Popper (1902–94), Karl Mannheim (1893–1947), the subsequently prominent Marxist thinker György Lukács (1885–1971), a clutch of art historians with a penchant for the history of ideas, several of whom—Lajos Fülep (1885–1970), Charles de Tolnay (1899–1981) and Johannes Wilde (1891–1970)—became left-wing art sociologists, others—Arnold Hauser (1892–1978) and Frederic Antal (1887–1954)—sociologically slanted art historians, as well as Béla Balázs, a poet and later film theoretician, and the painter Anna Lesznai (1885–1966), with composer Béla Bartók a frequent guest at their gatherings. This group was at the forefront of picking up new European trends in the visual arts, with Lukács, Fülep and later Tolnay taking the lead in affirming their beliefs though their writings.

It was the appreciation of Cézanne and cubism, together with Matisse and the Fauves that provoked the emergence of the first avant-garde movement in Hungarian painting. Several Nagybánya artists—the so-called 'neos', amongst whom were Dezső Czigány (1883–1937), Vilmos Perlrott-Csaba (1880–1955) and Béla Czóbel (1883–1976)—were by then already turning towards Paris, but a theoretically underpinned programme—'a declaration of war against all impressionism, all sensation and mood, all disorderliness and the negation of values', as Lukács described it—was only articulated with the formation of a group who called themselves 'The Eight' in 1909. Róbert Berény (1887–1953), Dezső Czigány,

Béla Czóbel, Károly Kernstok (1873–1940), Ödön Márffy (1878–1959), Dezső Orbán (1884–1987), Bertalan Pór (1880–1964) and Lajos Tihanyi (1885–1938) treated visual elements in a cubist or constructionist manner, preferring strident 'primary' colours to a more subdued, blended palette.

The first to part with *fin-de-siècle* academicism in sculpture was the Art Nouveau–symbolist artist Fülöp Ö. Beck (1873–1945), who was also an important goldsmith. However, he—no more than any other prominent sculptor in the first half of the century—was unable to jettison entirely the spirit of classicism. His brother, Vilmos Fémes-Beck (1885–1918), along with Márk Vedres (1870–1961), maintained close contacts with painters of The Eight. The work of József Csáky (1888–1971), a 'genuine' cubist sculptor of Hungarian origin who worked in Paris from 1908 onwards, has curiosity value (little of it has survived). Paris was likewise where painter Alfréd Réth (1884–1966) exhibited with Metzinger and other cubists.

It was partly from within the ranks of The Eight that Lajos Kassák (1887–1967) attracted recruits for the 'activism' that he began promoting some years later through his periodicals *A Tett* (The Deed, 1915–16) and *Ma* (Today, 1916–19), the most substantial painters amongst whom were Sándor Bortnyik (1893–1976), János Mattis Teutsch (1884–1960), József Nemes Lampérth (1891–1924), Béla Uitz (1887–1972) and László Moholy-Nagy (1895–1946). Kassák, now considered a legendary figure of the Hungarian avant-garde, was a poet, writer, artist and theoretician but, first and foremost, an impresario, publisher, socialist and cosmopolitan, who even during World War I contrived to stay in touch with avant-garde groups across the continent. He wrote Dadaist poems but soon became a constructivist, coming up with an engaging dialectic slogan to contrive a link between these fundamentally antithetical trends: 'Destroy in order to construct, construct in order to win!'. He composed pictorial poems and pasted together photographic collages, whilst also devising, together with Béla Uitz, the compositional rules and ideology for a new genre of pictorial art on the Soviet Russian constructivist pattern. The Transylvanian Saxon János Mattis Teutsch was important to the group in that by the 1920s he had worked out a way of reconciling his German expressionist and abstract styles of painting with a vaguely leftist ideology and pedagogical theory. Moholy-Nagy at this stage of his career was still a fledging 'cubo-expressionist' and only became a leading figure of international constructivism after leaving Hungary.

These various endeavours all fed into the brief life of the Hungarian Soviet Republic in 1919. Within the space of a few months a constructivist decoration was installed in Heroes' Square (covering up György Zala's group of statues) to mark the 1st May celebrations, and a start was made on nationalising several major art collections, setting up a register of artists, and putting together a portfolio of revolutionary red–white–black posters.

The Inter-War Years Within and Outside Hungary

With World War I barely over, the artistic community was rent first by the Soviet-style republic then by the 1920 Treaty of Trianon, with its attendant massive losses to Hungary not only of territory and population but also of artistic heritage (a break that has customarily been flagged in scholarly discourse by making a distinction between the art of 'Greater' or 'historical' Hungary and that of mere 'Hungary'). In addition, virtually anyone who had been publicly committed to the revolution—and a large proportion of artists and intellectuals did espouse left-wing views—was forced to emigrate. Even during the Commune there had been a parting of ways between the Bolshevik leader, Béla Kun, and Kassák, a socialist who professed support for world revolution, when Kun demanded that the latter engage in political propaganda. By late 1920, Kassák,

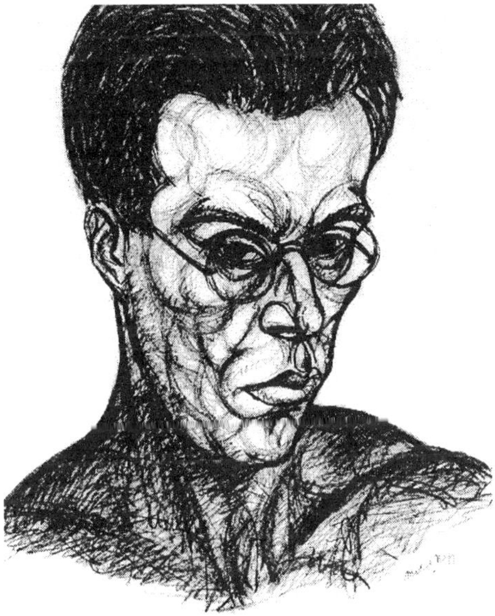

László Moholy-Nagy: *Self-portrait*

Béla Uitz and Sándor Bortnyik had tacitly accepted the ideas of international constructivism in their own manifesto on *Bild-architektur*, or 'pictorial architecture'—a radicalisation that also accounts for why so many Hungarians, including Moholy-Nagy, Farkas Molnár (1897–1945), Marcel Breuer (1902–81), and Andor Weininger (1899–1986), were prominent as teachers and students in the Weimar Bauhaus school.

The essence of *Bildarchitektur* was to produce a construction on the plane of the paper or canvas that did not depict anything but, precisely through that autonomy, could be interpreted as the model of a new society founded on a new ethics. There was likewise the constructivist motive behind the introduction of the new genre of photographic montage into Hungarian art. As a publisher, Kassák put out an important anthology *Buch neuer Künstler* (Vienna, 1922), co-edited with Moholy-Nagy. Having literally walked half the breadth and length of Europe, he returned to Hungary in 1926, publishing volumes of his own poetry, such as *Tisztaság könyve* (The Book of Purity, 1926), and generally resuming the role of a link between West and East or, to be more exact, the avant-garde émigrés and those who stayed in Hungary.

In Germany the torch passed from Kassák to László Moholy-Nagy, arguably the most influential of all twentieth-century artists of Hungarian birth working abroad. His first exhibition was put on jointly with László Péri (1889–1967) at the Der Sturm gallery in Berlin. (Péri made a name through his invention of reliefs constructed from tinted concrete shapes. On moving to Britain in the 1930s, he adopted the name Peter Peri, became a Mormon, and was the model for the protagonist of John Berger's novel *A Painter of Our Time*, 1958). The gallery, run by Herwarth Walden, though nominally an expressionist stronghold, was actually becoming a platform for all that was new, including the works of Hungarian artists such as the post-cubist Béla Kádár (1877–1956), the later futurist Hugó Scheiber (1873–1950), a café-painter of exceptional virtuosity, as well as the then genuinely expressionist Aurél Bernáth (1887–1953).

Moholy-Nagy soon became director of the Bauhaus metal workshop. Basically a constructivist, he had the shining virtue of being open to all novelty and being able to contextualise art within a modern scientific world view. He made his explorations of the photogram—a print of film exposed to light without use of a camera—a focus of both his theory and practice, from which he drew searching conclusions for painting, three-dimensional art and film as well as the pure play of light. Among his most important works (early pinnacles of international kinetic art) was the so-called *Light–Space Modulator*, or *Light Requisite* for short (1921–30); *Lichtspiel, schwarz weiss grau* (1930), an abstract motion picture based on the use of these devices was a characteristic work in his film oeuvre. A special Moholy-Nagy issue of the Brno-published *Telehor* (1936) got its title from a device for the wireless transmission of photographs that had been patented by the inventor Dénes Mihály, a pioneer of talking-picture technology in the late 1910s who went on to design one of the prototypes of television during the 1920s. Moholy-Nagy's principle of transmitting geometric compositions ('telephone images') and his numerous innovations in the field of design (e.g. the form of the Parker fountain pen) were similarly revelatory from an aesthetic as well as practical point of view. As a pedagogue, Moholy-Nagy advocated that everyone has talent; he played a crucial role in drawing up the Bauhaus curriculum at Weimar (and later also Dessau and Berlin), and his theoretical magnum opus, *Vision in Motion* (1947), was published as a course-work for the Chicago Institute of Design, which he himself had founded. Other major Hungarian-born Bauhaus artists, such as Farkas Molnár, Alfréd Forbát (1897–1972) or Marcel Breuer, inventor of the tubular chair, were better known for their achievements in architecture.

Berlin was the most important centre of the avant-garde during the inter-war period, certainly during the 1920s, but it was not the only destination for Hungarians. István Beöthy (Étienne Béöthy; 1897–1961), creator of an intriguing system of proportions, the 'golden series', was attracted by the international scene of the École de Paris, as was Henri Nouveau (Henrik Neugeboren; 1901–1959), a devotee of 'musicalism', and several now world-famous photographers. Vilmos Huszár (1884–1960), on the other hand, attracted international attention as a member of the De Stijl group in the Netherlands with his semi-constructivist, semi-Dadaist 'mechanical stage' automatons.

As for those who chose to stay in Hungary, the 'Italian orientation' of official cultural policy during this period is best reflected by the so-called 'Rome School'. Through the influential art historian Tibor Gerevich, curator of the Accademia Ungherese on the Via Giulia, many had an opportunity to work there on state stipends and, in some cases, to participate in the Venice Bienniale. Numerous Hungarian painters and sculptors thereby came into contact with Italian neo-classicism, and its 'novocentismo' core in particular, most notably Vilmos Aba-Novák (1894–1941), best known for his fairground and circus scenes and state-sponsored frescoes, István

Szőnyi (1894–1960), Károly Patkó (1895–1941), Ernő Jeges (1898–1956), Jenő Medveczky (1902–69), Pál Pátzay (1896–1979), creator of the equestrian *Hussar Monument* at Székesfehérvár (1939), László Mészáros (1905–45), Tibor Vilt (1905–83), with his figure of *Istenes Szent János* (St. John the Divine, 1937), Béla Kontuly (1904–83), and Pál Molnár C. (1894–1981). The Rome School, however, was but one facet of the conservative domestic tendencies of the period, and neo-classicism but one of the approved aesthetic postures: amongst others there was the German 'Neue Sachlichkeit' adopted by Kontuly, or the 'Art Deco' style of the decorative urban paintings of János Vaszary (1867–1939) and small sculptures of Elza Kövesházi Kalmár (1876–1956). Several painters, taking a strong, Rembrandt-inspired chiaroscuro as their point of departure, developed a highly plastic technique which István Szőnyi carried over into his engravings, creating a veritable school that drew in the likes of Aba-Novák, Patkó and Erzsébet Korb (1899–1925).

A good idea of the problems that Hungarian painters faced at this time may be conveyed by noting that Szőnyi, a key artist amongst those just listed, was also one of the Gresham Circle, a group of artists, critics and collectors who frequented a well-known Budapest coffee-house of that name and upheld a by then distinctly outdated post-impressionist aesthetic (often referred to in Hungary as 'post-Nagybánya' or 'scenic' painting). This type of conservatism persisted even as late as the 1960s, especially in the views propagated by Aurél Bernáth and Pál Pátzay in their positions as teachers at the Academy of Fine Arts. Other artists often seen at the Gresham table were Róbert Berény (1887–1953), József Egry (1883–1951), a painter of moody landscapes—indeed, visions—of Lake Balaton, the sculptor Béni Ferenczy (1890–1967), and the latter's sister, Noémi Ferenczy (1890–1957), who restored tapestry-making to the rank of high art. As for classicism, Ferenc Medgyessy (1881–1958), one of the most influential

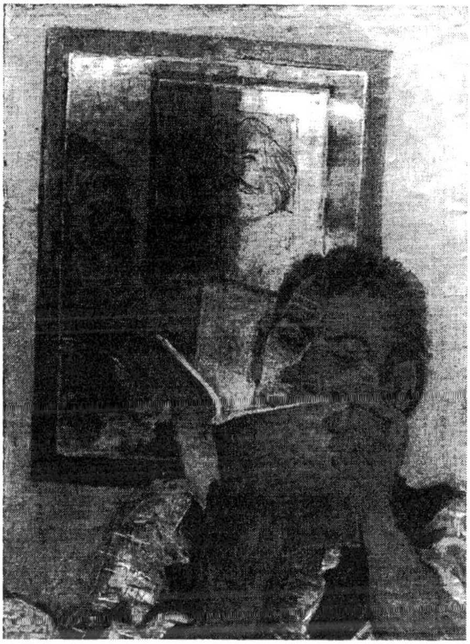

Gyula Derkovits: *Three Generations*

sculptors of the inter-war years, endorsed a rural–peasant classicism that was not averse to the grotesque and a crude humour.

Like Szőnyi, Gyula Derkovits (1894–1934), another major painter of the period, was initially inspired by Rembrandtian chiaroscuro, but his drawings later gained a sharply critical edge close to that of George Grosz, whilst his paintings display an incredibly refined pictorial world, at once powerfully monumental and delicately lyrical, replete with silvery tones, accidental lights and space-dissolving reflections, combining colour harmonies of Bonnard-like refinement with expressionistic forms. Derkovits had no followers, yet his reception was all the more problematical for posterity, given that he was a genuinely proletarian artist, with impeccable leftist credentials, who nevertheless catered for a grand bourgeois clientele. Having been evicted from his home, he all but starved to death, and after 1945 (like the poet Attila József) was the subject of constant debate as whether he should be considered an example of the socialist artist *par excellence* or a formalist who mimicked Western trends.

Lajos Vajda (1908–41), dying tragically in a TB sanatorium after a forced labour battalion, found an even greater post-war echo than Derkovits. He deliberately tried to synthesise the worlds of thought and form of both East and West, Christianity and Jewry,

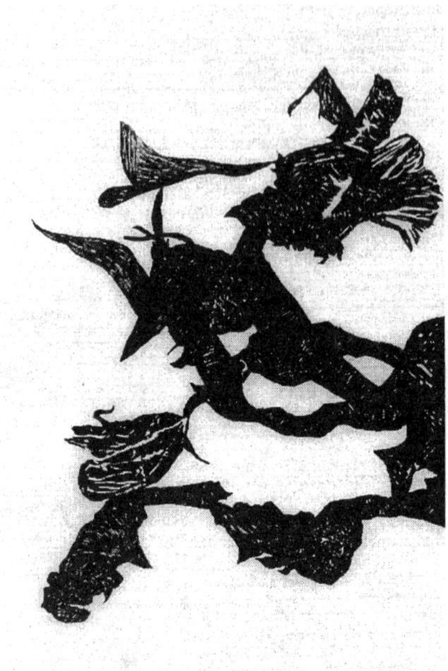

Lajos Vajda: *Vegetation*

constructivism and surrealism, painting icon-like portraits with a sacral content but also reminiscent of multiply exposed photographs. While living in Paris, Vajda came under the influence of Soviet Russian avant-garde films, and he populated his later pictures with monsters (1930–33). In a letter written in 1936, when he was working in the artists' community of Szentendre, just outside Budapest, he notes that he and Dezső Korniss (1908–84), a fellow painter with whom he went out collecting ethnographic motifs, made an efficient team because Korniss was from Transylvania and had soaked up the De Stijl tradition during a stay in the Netherlands. These two were repositories of the characteristically Hungarian intellectual endeavour to seek a mediatory role for Hungarian culture between East and West, between folk tradition and modernity. Also belonging to this tradition are Sándor Trauner (1906–93), later to become a renowned stage designer, and György Kepes (1906–2001), who, both in his books and as a director at the Massachusetts Institute of Technology, continued Moholy-Nagy's work in becoming a pioneer of the 'art & technology' movement. Mention should also be made of the painter Béla Veszelszky (1905–77), who, together with philosophers Lajos Szabó and Béla Tábor and, more indirectly, the writer Béla Hamvas, was later to exercise considerable influence on the visual arts of the post-war decades especially.

Károly Kincses, historian of photography, gave the extraordinarily apt subtitle of *Aki elment, aki maradt* (Those Who Left and Those Who Stayed) to a recent album. Between the two World Wars, Hungary gave the world numerous outstanding photographers, including several acknowledged giants of the genre, nearly all of whom achieved their fame abroad—in France, Germany or the USA. Apart from the already mentioned Moholy-Nagy, these figures include André Kertész (1894–1985), who immortalised himself with the unmatched mastery of captured time of his genre photographs and with his *Distortion* series of nude figures. Then there are fashion photographer Martin Munkácsy (1896–1963); Brassaï, the name adopted by Gyula Halász Jr. (1899–1984), unveiler of *Paris de nuit* (1933), friend of Picasso, discoverer of graffiti as an art genre; and Robert Capa (1913–54), the ace war photographer through his classic *Death of a Militia Man* (1936) and his photo reportage on the D Day landings (*Omaha Beach, 6 June 1944*).

One cannot omit mention of 'those who stayed'. Among them was József Pécsi (1889–1956), a highly cultured studio photographer, close to the Gresham Circle, and the first person anywhere to produce a book on photography in advertising. There were also the two seemingly opposed trends of 'glamour' photography, as pursued by Jenő Dulovits (1903–72), Ernő Vadas (1899–1962), F. G. Haller (1898–1954) and others, which conceived the landscape and life of rural Hungary as an idyllic beauty—the 'Hungarian style' as amateur photography clubs across the world came to call it—and the social reportage of Lajos Kassák's Munka (Work) Group, more attuned to the naturalism of poverty. In hindsight, however, the two camps may have been closer to one another in their treatment of light and shade than was thought at the time.

From 1945 to 1956

For three or four years between the end of World War II and the full communist take-over in Hungary there was an opportunity for stocktaking and mourning to unite with an impulse to start afresh; in culture and the arts, new community ideals were responsive to notions of links with international experimentation—all 'formalisms' against which communist ideology was to erect administrative barriers.

The European School, whose founders included men of letters, doctors and scientists as well as artists, resuscitated the legacy of Lajos Vajda in embracing avant-gardists of all stripes, especially abstract artists and surrealists. Emblematic not only of the European School but of the post-war situation as a whole is *Dunavölgyi népek kórusa* (Choir of the Danubian Nations, 1946), a statuette by Dezső Bokros Birman (1889–1965) which seeks, despite the crabbed expressiveness of its modelling and despondent passivity of its figures, to nurture some hope in the midst of hopelessness. Another of the group's typical attempts to give international surrealism a domestic currency was to link it to archaic (and equally folkloristic) roots like the masked carnival procession—relic of an originally Croatian ethnic community—at Mohács on Shrove Sunday. This is palpable in the paintings of János Martinszky (1909–49), who worked in Mohács itself, and, to some extent, in the semi-abstract, semi-surreal paintings of Ferenc Martyn (1899–1986), based in Pécs. Taking a different track, Dezső Korniss, who had worked with Vajda in collecting folk-art motifs before the war, in a series of miniatures entitled *Metamorfózisok* (Metamorphoses, 1946), now set about producing an endless variety of head shapes to arrive at an archetypal abstract formulation of a universal collective subconscious. The surrealist line was further bolstered by the fairy-tale puppet world created by Margit Anna (1913–91), who also preserved the legacy of her husband, Imre Ámos (1907–44), a painter who drew on the mystical elements of Hassidic folklore and had died probably in a concentration camp, whilst Júlia Vajda (1913–82) perpetuated the artistic vision of her husband, Lajos Vajda, by translating, crystallising it into the abstractly organic, map-like surfaces of her photomontages. Contributing to this were also sculptor József Jakovits (1909–94), with his organic-abstract monsters, and Endre Bálint (1914–86), who during the 1960s and 1970s transmitted the 'ideals' of Szentendre—the picturesque small town on the Danube, just north of Budapest, dotted with Serb Orthodox and late Baroque churches and home for an ever-growing colony of artists—to Lili Ország (1926–78) and a host of other young painters.

Ideological controversy within the European School, however, soon prompted fissures, with a group of 'abstract artists'—consisting notably of Tihamér Gyarmathy (1915), Béla Fekete Nagy (1905–83), Tamás Lossonczy (1904), Sameer Makarius (1924), and Ferenc Martyn—breaking away as early as 1946 to mount two shows and set up a 'Gallery at the 4 Points of the Compass'. This 'secession' signified a manifest radicalisation towards international non-figurative trends and new connections between science and art. Ernő Kállai, the group's chief theoretician, put his authority behind the Moholy-Nagy–Bauhaus–György Kepes line to organise an exhibition, under the title *Új világkép* (New World Image, 1947), which displayed paintings and statues alongside photographs of microscopic details, machines and other engineering constructions.

Apart from Kállai, Lajos Kassák (again active both as writer and publisher), and the theoreticians of the European School (Árpád Mezei, Imre Pán), another figure of crucial importance was Béla Hamvas, who, with his wife Katalin Kemény, co-wrote *Forradalom a művészetben* (Revolution in Art, 1947), which became the *ars poetica* of virtually every major artist of that period. Hamvas was among the first to have his work condemned and banned by the new communist ideologists (specifically by philosopher György Lukács, again installed in the role of 'cultural commissar' that he had held in the 1919 Commune), and over the next three decades or more it was 'forbidden fruit' that circulated solely in *samizdat* form. These pre-Platonist, gnostic and esoteric (or simply 'poetic') enquiries across the entire spectrum of cultural history did, however, begin to resurface from the 1980s onward to exercise considerable influence over younger generations of artists and intellectuals alike.

From 1949, the year of the political turn, cultural policy bowed to Moscow in setting up the battle stations of socialist realism in every branch of art. Undoubtedly much documentation about the oddities of Hungary's version of the Stalinist 'cult of personality' is still waiting to resurface, but two extreme examples are worth citing here. One is the 'popular' form, Mrs. Anni Feleki's painting of *A füttyös kalauznő* (Tram Conductress with Whistle, 1950), which depicts a chastely attractive and blithe 'heroine of work' as she stands, flooded in sunlight, beside her tram, and requires no comment (it was chosen, with obviously ironic intent, as the dust-sheet illustration for the book entitled *Esterházy-kalauz* (Esterházy Guide, 1991), which records conversations that the US-based literary historian Marianna D. Birnbaum held with author Péter Esterházy). The other is the monumental, 8-meter high statue of Stalin (1951), designed by Sándor Mikus, that was erected behind the grandstand on Budapest's 'parade route', leading from Heroes' Square, from which party leaders and other dignitaries used to view the country's main May Day parades. Its toppling and demolition on October 23rd 1956, the first night of the popular revolt against the Soviet occupation and totalitarian system, remains to this day an eloquently symbolic action.

Public sculpture, arguably the most controversial of all artistic genres throughout the twentieth century, indeed produced more than its share of extremes and transitional forms during this particular period. On the one hand, there are classicising and heroising monuments such as the *Felszabadulási emlékmű* (Liberation Monument, 1947) on Budapest's Gellért Hill, the work of Zsigmond Kisfaludy Strobl (1884–1975), which had actually been commissioned before the war; on the other, cautiously abstract works, such as the *Zero Kilometre* stone produced by Miklós Borsos (1906–90), set in Adam Clark Square between the Chain Bridge and the entrance to the tunnel under Castle Hill, which marks the point from which road distances are measured, or *Fóka* (Seal) by Lajos Barta (1899–1986), which is disguised as a piece of playground plastic art on a housing estate. Between these extremes are suitably realistic but mildly expressionist statues, such as *Martinász* (Smelter, 1953), commissioned from József Somogyi (1916–93) by the town of Dunaújváros (then called Sztálinváros), or the *Partizánemlékmű* (Partisans' Memorial) by Jenő Kerényi (1908–75), set up in the northern town of Sátoraljaújhely. It will surely prove to have been one of the more enlightened decisions taken during the 1989–90 change of régime that the most flagrant of these socialist mementoes should be preserved, rather than destroyed, and put on display as documents of the era, in due critical and somewhat ironical fashion, in the Budafok Statue Park, on the outskirts of Budapest.

From 1956 to the late 1970s

As far as art is concerned, the period that led from the suppression of the 1956 revolution, with its subsequent reprisals, through three decades of 'soft dictatorship' to the change of political régime can be labelled as one that encompasses the unfolding and triumph of the neo-avant-garde and the eventual debouching of the entire venture into postmodernism. Put simply, the forty or so years can be summarised as a fight between the 'official' and the 'unofficial'. The entire institutional system of culture and art came under the control of 'party direction'. The 'New Economic Mechanism' on which Hungary embarked between 1968 and 1972, coupled with the liberalising impact of New Left movements in the West, led to a relaxation of Marxist doctrine, with the simultaneous introduction of the celebrated, though distinctly inconsistent, official policy of distinguishing between work that was officially 'touted', merely 'tolerated' or simply 'taboo' (the 'Three Ts' approach as it was dubbed). Though another spell of retrenchment and reprisals ensued after 1972, this was also the period that saw the emergence of a 'democratic opposition'—occasionally in explicit collaboration with the avant-garde of the art world—that was eventually to play a key role in the political transition of 1989–90. There is no need to say much about the artists who were promoted by the 'Three Ts' policy, for very few actually pursued strict socialist realist dogma; indeed, the then current jargon term of 'soc-realist' would be a more apt description, since it was more a case of individuals from the pre-war 'post-Nagybánya' Gresham Circle—principally Aurél Bernáth, Endre Domanovszky (1907–74), and Pál Pátzay—striving, through making concessions, big and small, to socialist ideology, to retain positions in the art educational establishment and the allocation of commissions for public monuments. Representatives of the new avant-garde, by contrast, built up illegal or semi-legal redoubts in cultural centres, university clubs, student hostels, the meeting halls of state enterprises and, last but not least, private homes. The outstanding events of the late 1960s and early 70s were two 'Iparterv' shows (the name comes from the industrial construction company which supplied the dining-hall venue) and one-off exhibitions put on under the banners 'Szürenon (Surrenon)' and 'R'.

In terms of style, taste and fashion trends, one can discern a continuing phase of late surrealism, a product of the encounter during the mid-'60s of the post-Nagybánya approach taught at the Academy with Western pop art, ideas that were fostered about the 'American way of life' and much else. Tibor Csernus (1927), one of Aurél Bernáth's most talented students, was the prime mover of an approach that one critic appositely dubbed 'sur-naturalism'—a style in which a wry sociological interest in derelict industrial sites was married to a virtuoso handling of surfaces, rich in detail and technical trickery, and thus no stranger to the machines, streamlined objects and other vocabulary of pop art. That in turn helps to understand the fairly rapid switch to hyper-realism that was seen in the work of László Lakner (1936), László Méhes (1944), Gyula Konkoly (1941) and, during a spell in Paris, Csernus himself (who then moved on to a 'Caravaggist' phase and eventually to a more generalised treatment of light and shade that seeks to measure up to the master realists in the history of painting). Méhes made his hyper-realistic paintings by employing special screen-print technique to achieve a highly translucent finish, whereas Lakner kept several options open: besides paintings based on photographs, he also produced conceptual works and objects, and later, when 'new painting' emerged, returned to pure gestural painting.

László Gyémánt (1935), Ignác Kokas (1926), Ákos Szabó (1936) and another ten or so painters still hold to a sur-naturalist direction, whilst Sándor Altorjai (1933–79) soon progressed beyond this to a world of montages, collages and '*gyagya*ist' performance art (meaning 'crackpot', and obviously deriving from Dada, the term was coined by poet, media artist and theoretician Miklós Erdély (1928–86)). Lili Ország, having started off as a committed surrealist, arrived at a sovereign visual world in which various surface effects, such as the imprints of printed circuits, are combined with ancient letterings to invoke historic memories. The virtuoso graphic artist János Major (1934) fashioned a self-lacerating, morbidly humorous Jewish iconography, which (also combined with photos) he developed into an ironic conceptual art, while Béla Kondor (1931–72), perhaps the broadest-thinking of the painters of this period, blended his own visions with phantasmagoria that reached back to Rembrandt, Blake and Hieronymus

Bosch. (The truculence that Kondor displayed towards the authorities, his deviant lifestyle and early death, all helped fuel a negative Hungarian mythology of the self-destructive talent.) Others found it equally hard to shake the hold of surrealism, out of which arose the 'szürenon' line (for 'sur et non', implying both an embracing of surrealism and its repudiation) taken by Attila Csáji (1939) and his associates, such as Sándor Csutoros (1942–89) and Gábor Karátson (1935). With Paris-based Simon Hantaï (1922) this amounted to transcending at the same time as he preserved the formal repertoire of André Breton and, more particularly, Max Ernst.

Another trend arrived at pop art in the pursuit of radical lifestyle models rather than via the (roundabout) route of the visual arts. The happening and Fluxus-inspired aesthetic reached Hungary in 1965–66, along with British-led fashion trends, the hippie movement, and the successive waves of pop and rock music, though Miklós Erdély retrospectively designated the spontaneous street collections that appeared in late October 1956 for victims of the revolution a work of art—which he titled *Őrizetlen pénz* (Unguarded Money), in effect conferring on the money the status of a peculiar kind of *objet trouvé*. The first genuine happening, organised by Tamás Szentjóby (1944) and Gábor Altorjay (1946) under the banner *The Lunch—In Memoriam Batu Khan*, was mounted in 1966. Both artists went on to devise numerous objects, montages, pop poems and, later on, conceptual artworks.

The parallelism between happening and actionism, on the one hand, and 'alternative' or 'experimental' theatre on the other is noteworthy. Whilst 'street theatre' groups in the West were promoting a hippie chic, New Leftism, and the views of students' movements, in Hungary that role was played by the 'living-room theatre' of the Lajos Kassák Studio under Péter Halász (1943), which, after being banned, was reborn as the later famous Squat Theatre when Halász emigrated to New York in 1975, and the István Kovács Studio, headed by László Najmányi (1946), which increasingly took on a punk and new wave edge. It is all the more paradoxical, from an institutional point of view, that the similarly radical New Musical Studio in Budapest was all the while sponsoring regular performances of post-Cageian and 'minimalist' music under the auspices of KISZ, the official League of Young Communists. The best known of the country's 'alternative' centres during the 1970s were the Balatonboglár Chapel Exhibitions (in reality the private studio of sculptor György Galántai (1941)), and, from 1973 until the late 70s, the Young Artists' Club in Budapest, whereafter Galántai and his Artpool once more took over that role.

From the second half of the 1960s, under the pressure of a uniform cultural policy imposed by authorities, several avant-garde generations co-existed more or less peacefully. Older abstract painters, such as Korniss, Bálint, Kassák, Martyn, Gyarmathy, Károly Tamkó Sirató (1905–80), Mihály Schéner (1923) and Jenő Barcsay (1900–88), younger pioneers such as Kondor and Lili Ország, and advocates of the latest trends, were all keeping an eye on each other's efforts. Though it would be tedious to try and list who influenced whom, or who attracted the bigger group of disciples, it would be instructive to compare the gravitational pull of the constructivists Korniss or Kassák, or consider the isolated position of Barcsay, who throughout his career attempted to synthesise the ideals of the Szentendre school, composed monumental wall mosaics, and in his later years arrived at a kind of purely geometric yet lyrical abstraction, but who exercised his greatest influence on innumerable pupils via his book *Művészeti anatómia* (Anatomy for Artists, 1953), a product of the soc-realist era which was published in many countries.

During the mid-'60s neo-constructivism was a particularly potent organizing force but it should be noted that many of the artists who flocked to it—Gábor Attalai (1934), Imre Bak (1939), István Nádler (1938), Krisztián Frey (1927–97), etc.—had a background in French-style lyrical abstraction, tachisme (action painting) and informal art. This alignment gained considerable support from the Zugló Circle, a group of self-taught young painters led by Sándor Molnár (1936), as well as from the always well-oriented Ilona Keserü (1933) and older masters like Veszelszky, Schéner or Menyhért Tóth (1904–80) who held aloof from all specific groupings.

At this juncture, a veritable welter of abstractionist and geometric-constructivist trends was on parade whilst the artists who are currently active evolved their markedly individualistic styles. Both János Fajó (1937), a follower of Kassák, and Bak, who nurtured contacts with the Zero group in West Germany, worked towards 'hard edge' and 'shaped canvas' approaches; Károly Halász (1946), a member of the Pécs workshop, temporarily fell under the influence of Victor Vasarely (Győző Vásárhelyi, 1908–97); István Haraszty (1934), the most significant Hungarian sculptor of mobiles, established an intellectual partnership with Nicolas Schoeffer (Miklós Schöffer, 1912–92), a kinetist working in Paris; Tibor Gáyor (1929), András Mengyán (1945) and Attila Kovács (1938), the latter by then living in Cologne, explored systems based on principles of mathematical logic; Tamás Hencze (1938) experimented with the deceptive brushwork and patterns created by the use of the paint-roller; whilst Ferenc Lantos (1929) and Ferenc Ficzek (1947–87) in Pécs, Pál Deim (1932) in Szentendre, and Gizella Rákóczy (1947) in Budapest each created their own diverse worlds of geometric forms.

By then it had become impossible to pretend any longer to the powers-that-be that more than a few Hungarian émigrés had in the meantime become internationally acclaimed artists. In France alone, besides Vasarely and Schoeffer, with his 'cybernetic towers', there were the likes of Marta Pan (1923), with her masterly swimming and moving abstract sculptures, Vera Molnar (1924), a pioneer of computer-aided graphics, Simon Hantaï, whose 'pliage' creations took him into analytical painting, or the Transylvanian-born, organic abstractionist sculptor Étienne Hajdu (1907–96), whilst Ervin Patkai and Pierre Szekely (1923–2001) contributed various stone-dressing processes to the field of monumental sculpture, and Szekely's wife, Vera Szekely reigned supreme with her installations of textiles, wood and metal. Zoltan Kemeny (1941), the greater part of whose creative life has been spent in Switzerland, is regarded as one of the pioneers of informal sculpture.

Turning to domestic sculpture, by the late 1960s Tibor Vilt (1905–83) had already seized on the tectonic form of 'vise-jaws' as a distinctive central problem of his works, while Rezső Berczeller (1912–92), a close associate, toiled away at how to meld the human figure with machinery or scaffolding. Towards the end of his life he composed increasingly ecstatic groups of statues: winged altarpieces on the medieval pattern, a procession of crippled figures, a woman in labour, angels descending from the ceiling. Erzsébet Schaár (1908–75), Vilt's wife, progressed during the 1960s from an expressivist treatment of human suffering to using small sculptures to characterise the relationship between figure and space, and eventually to creating a strange amalgam of pop art and funeral memorials with her life-size, snow-white figures. The most successful sculptor of this generation, in terms of the sheer numbers of commissions he won, was Imre Varga (1923), who constantly pitted his undeniable proficiencies in portraiture against the latest fashions-a crown that, since the change in régime, has now passed jointly to Pál Kő (1941) and Miklós Melocco (1935), with the somewhat trivial, populist touch and the mannered pathos that they respectively bring to their equally virtuosic (at worst, highly accomplished) responses to official commissions.

The new generation that was emerging in the latter half of the '60s again turned to non-figurative approaches. Tibor Csiky (1932–89), with his simplified wood and metal structures, was Hungary's first minimalist sculptor. György Jovánovics (1939), after an initial pop-art-style figurative period, increasingly sought a form of 'facultative' geometry, whereas the 'Pseudo' art of Gyula Pauer (1941) is based on illusionist manipulations of rumpled surfaces. The improvised constructions of thin wooden sticks built by Vienna-based János Megyik (1938) had their analogies to projective geometry and eventually developed into a modelling of the actual process of mapping. Ádám Farkas (1944), for a while, worked out a variety of topological problems in his statues for public spaces, and Dóra Maurer (1937) tackled systematic series of geometric paintings before ending up with three-dimensional solutions. The ironic designs of the machines assembled by István Haraszty made him the most important sculptor of mobiles and kinetic artist. His own studio-home, built on an absurdly small area, was at the same time the embodiment of an extended environment for his sculptural ambitions.

Alongside neo-constructivists (Bak, Hencze, Nádler, Fajó), pop artists and hyper-realists (Lakner, Méhes), the two semi-legal 'Iparterv' exhibitions put on in the late '60s also displayed early works of the then nascent conceptual art movement, which went on to become the dominant trend of the '70s. What, then, is conceptual art? A dictionary definition would have to take the activities of pioneering American practitioners such as Henry Flynt, Joseph Kosuth or the Art & Language Group as a starting point and embrace any logical operation that hinges on the notion of art or, put more simply, an acceptance of the notion that concepts and ideas can constitute a work of art just as much as materials. Strict linguistic conceptualism is quite rare in Hungary, except in certain works by Péter Türk (1943); more important has been a recognition of the freedom, and hence social significance, of the genre—in practice the 'concept' or idea—and of the fact that *anything can be art*. One of the most consistent practitioners, former painter Endre Tót (1937), used typewritten rows of 0s as manifestations of nothingness, wrote letters (mail art), designed stamps, sent 'evergreen ideas' to his friends, and was glad of anything ('I am glad if…'), especially the most banal phenomena. The diverse projects of such artists did not necessarily have to be realised (Attalai, Bak), but practitioners in many branches of art felt compelled to introspection on the media they were employing (e.g. Dóra Maurer's etchings), and photography became an essential tool for documenting or expressing the concepts. Important early figures, in addition to the aforementioned, included Gyula Gulyás (1944), Gyula Konkoly, János Vető (1953), Gusztáv Hámos (1955), Tamás Szentjóby, Gyula Pauer, Tibor Csiky, Sándor Pinczehelyi (1946), and András Baranyay (1938).

The first comprehensive exploration of the relationship between photography and visual art was the *Expozíció. Fotó/ művészet* (Exposure. Photo/Art) show curated by Dóra Maurer and László Beke at the Lajos Hatvani Museum, Hatvan, in 1976. Besides photography, other technologies, such as photocopying, the telephone, and especially film and video, gained in importance. In the early '70s a number of artists, Dóra Maurer and Miklós Erdély prominent amongst them, embarked on projects at the Béla Balázs Studio to explore the expressive possibilities of film in an enterprise that initially went under the umbrella *Film Language Series*, the leading figure in which was Gábor Bódy (1946–85), director of a dozen or so experimental, feature and videotape films, producer of INFERMENTAL, the world's first videotape periodical, which ran for twelve issues, and contributor to several international videotape anthologies.

Indisputably the most influential artist of that period, however, was architect–polymath Miklós Erdély—of pivotal significance for his poems, his theoretical writings, and his late paintings alike. For him too conceptual art had a liberating effect on his objects, projects like *Szolidaritási akció* (Solidarity Action; a process which enabled any chosen message to be disseminated by word of mouth to the entire human race in thirty-odd steps), and, of course, his concepts. Erdély's films, especially the trilogy of *Álommásolatok* (Dream Copies, 1977), *Verzió* (Version, 1979–81) and *Tavaszi kivégzés* (Spring Execution, 1985), were concerned with dreams, repetition, and his ambiguous attitudes to his own Jewishness. His largest extant installation, *Haditítok* (Military Secret), was completed for the year marked by Orwell's *Nineteen Eighty-Four*. He is also credited with 'inventing' the idea of *indigó* (i.e. carbon-paper) graphics, which allows an original and multiple reproductions to be produced simultaneously and later also gave its name to *In*ter*Di*szciplináris *Go*ndolkodás (Interdisciplinary Thinking), Erdély's concept of a private art school or self-educating group.

Performance art made its first appearance in Hungary, as a continuation of happening and body art, during the second half of the 1970s. Its most significant practitioner was Tibor Hajas (1946–80), whose brief career began in poetry and conceptual pieces of art. Hajas combined his performances with the use of photographs, being of the view that the essence of his works was the unrepeatable moment; an effect of the moment was thus often underscored by the dazzle of a flashlight in the dark. For artists in Hungary, his death marked the end of an era: the avant-garde yielded place to the post-modern and to individual mythologies. This latter trend assumed a life strategy for certain artists, such as János Szirtes (1934), El Kazovszkij (1950), Imre Bukta (1952), Géza Samu (1947–90) and István Kovács (1964) who, among other things, were also involved in the performance scene.

The Last Two Decades

Looking back on events of the recent past, developments in Hungarian art need to be viewed in relation to two major turning points. The first was bound up with the emergence of 'new painting' around 1980 and amounted to a decisive change within art itself; the second, the change in régime of 1989–90, seemingly had nothing to do with art.

'New painting' was the most characteristic symptom of a bigger paradigm shift which also took in postmodernism, the trans-avant-garde, neo-expressionism, the new sensibility, new wave and a good few other attendant phenomena. Postmodernism in architecture proclaimed a return to symbolic forms and a 'radical eclecticism' in its citation of the past as a reaction to the previous rigors of functionalism (modernism). That eclecticism was also particularly welcome to visual artists, above all those who had grown tired of the purism and radicalism of the earlier avant-garde, including the extremes of conceptualism within that. Painting, having been more or less given up for dead, made a return (especially in its 'passionate', expressive and 'informal' yet decorative forms), and shortly afterwards so too did sculpture. Theoretical support for the implications of this revolution was forthcoming from art historian Lóránd Hegyi, who later held the directorship of the Museum of Modern Art in Vienna for a ten-year stint, and from the painter Ákos Birkás (1941). On his return from a study trip in the US, Birkás painted a host of stridently colourful, hectically brushed semi-abstracts before eventually hitting on the oval as the basic form that he has stubbornly clung to ever since. The oval can be a head, a face, a flower, a loaf of bread, an egg, or an icon; it can have cosmic, symbolic or religious implications; it can be small or large and have one or two parts; most recently, it can even re-emerge as a human face, rendered with photorealistic fidelity, or—as in his *kopfraumkörperbilder* (1998)—as a montage-like combination of two faces. Though he has long worked in a string of cities across Europe, Birkás has retained his Hungarian citizenship and permanent base and thus, sociologically speaking, is a prototypical 'nomad' of the country's political transition whose Hungarian roots can hardly be denied though it would be harder still to put one's finger on any specific 'national characteristic'. László Fehér (1953), by contrast, has worked out a totally simplified, appealing and 'readily intelligible' pictorial vocabulary and is widely considered

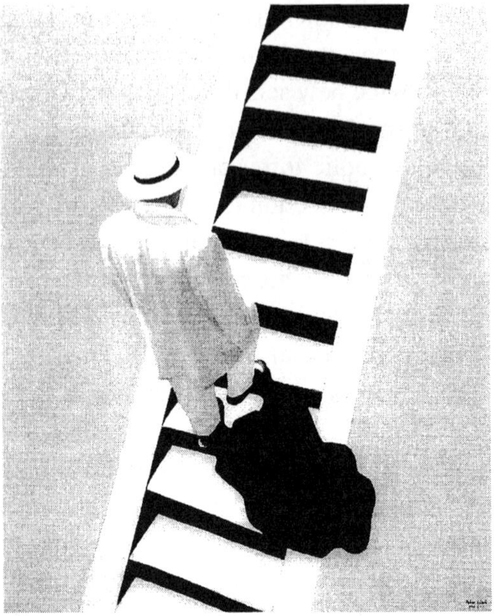

László Fehér:*On the Stairs*

to speak for the childishly naive little man cast adrift in the world. His human figures, reminiscent of strip-cartoon characters, are drawn in simple, well-judged outlines on evenly coloured backgrounds of (depending on the period) yellow, pink, silver or, most recently, black, with each picture highlighting a single object whose detail is picked out with photographic pedantry. A special attraction of this world is that its figures sometimes step out from the surface in the form of life-size iron-plate silhouettes. The secret of the success of Fehér's reduc-tionism is his feel for the appropriate forms to meet the appropriate needs at the appropriate moment. Arguably Hungary's best-known artist at the start of the twenty-first century, one obvious reason for his favourable reception abroad is his clearly identifiable acceptance of Jewish traditions.

Imre Bak can be taken as a third prototype of the new sensibility, distinctive for not breaking with the pedantic strictness of his constructivist forms even as he revelled in a freshly liberated, frenetically expressive approach. Even after the switch, his paintings retained a machine-like homogeneity of colour finishes and razor-sharp demarcation of shapes, and to that extent found common ground with 'neo-geo', one of the interesting contemporary trends at the time. Nevertheless, his canvasses acquired a decorativeness and gaiety, if only because (not least under the influence of architectural eclecticism) he began to populate the surfaces of his pictures with motifs from diverse painters of diverse periods. In recent years, he has returned again to a kind of constructivism insofar as his large, colourful, geometric forms resemble the facades of buildings.

In hindsight, the emergence of 'new painting' looks very simple and reasonable, but the reality is that behind it lay a complex alternation of generations. Hungary too was hit by the impact of an international movement that had been initiated by the younger generation, but in such a way that (as in Germany) certain older artists who had been languishing in obscurity again came to the fore. In Hungary, it was these older artists, members of an earlier avant-garde, who took the initiative, and for some this switch (or return) to painting even presented a genuine ethical crisis. Endre Tót, for example, had not taken a brush in his hand for nearly a decade, and when he did so his works were explicitly conceptual paintings. Miklós Erdély too may have started to paint again, and even regarded 'painterly' painting as his ideal, but he remained a conceptualist to his very marrow. Similar in age, Károly Klimó (1936), by contrast, had been a painter all along but only now, in exploring an 'informal' direction, did he truly start to blossom (finding an intriguing parallel in the work of Michael Horsley (1941), a British painter who moved to Hungary for a few years). Much the same opening out and shift towards the 'informal' style was seen with others of this generation, all born around 1940, including István Nádler, Krisztián Frey and László Lakner. The expressivist–informal movement also gave new impetus to the work of other die-hard painters like Ilona Keserü, one of the most 'colour-conscious' Hungarian masters, who was then setting up a master's school in Pécs, or Gyula Konkoly, returning from Paris, whose imposing persiflages reanimated some of the historical greats of painting (Monet, for instance).

The next Hungarian generation, which, on the international model, would have been expected to be the standard-bearers of neo-expressionism, instead seesawed at first between the avant-garde and trans-avant-garde. Even whilst still at art school, they had announced a number of collective actions, and then, with help from art historian Zsuzsa Simon, they had set up the country's first alternative gallery (under the name 'Office', then 'Rabinec' and finally 'Rabinext'). Out of this came artists like Zsigmond Károlyi (1952), a painter who also used photographs in his work and who later, by then as a teacher at the Academy, ushered in the monochrome movement; András Halász (1946), who, having started off as a painter, made a detour via photography and performance art before returning to painting in New York; Orsolya Drozdik (1946), who became a feminist then a designer of fake medical installations and also gained a growing international reputation in New York; or Károly Kelemen (1948), who, after dabbling with photography, aligned himself to the Italian trans-avant-garde and for a long time painted cubist teddy-bears. András Koncz (1953) instead went more after German examples; Ernő Tolvaly remains to this day more a conceptualist, even though he has never stopped painting; whilst László Méhes, having abandoned the hyperrealism of his student days, converted to a more cheerful, graffiti-inspired painting in partnership with self-taught photographer and rock musician János Vető (who had previously teamed up with alternative–conceptualist photographer Tibor Hajas).

This latter pair brings us to the subject of the interplay between music and painting in the early 1980s. With rock music having established itself in Hungary during the '60s and 70s, a brief flirtation with intellectual punk (involving the likes of Gergely Molnár and László Najmányi) in the late 70s was superseded by Hungarian versions of new wave, no wave and associated trends from bands like URH, Európa Kiadó, Balaton and others. János Vető was a founding member of the group Trabant, while the neo-dadaist shows put on by the Albert Einstein Emlékbizottság, or 'Albert Einstein Memorial Committee' (known as A.E. Bizottság for short), were principally the work of a trio of painters: András Wahorn (1953), László fe Lugossy (1947) and István ef Zámbó (1950). This was not a far cry from the outright performance art devised by the likes of El Kazovszkij (1948) or János Szirtes, who at this stage was best known for pitting tribal cultures against urban civilisation and whose paintings sometimes acted as drafts, sometimes as accessories, for his performances, or again the duo of András Böröcz (1956) and László László Révész (1957). Böröcz had actually started out as a sculptor and later on, in New York, carved miniature pieces of plastic art from pencils, though his magnum opus is the installation *Akasztottak* (Hanged, 1969–90), now in the Museum of Fine Arts, Budapest, whereas Révész (his doubled first name is a 'tribute' to a fraudulent spiritualist medium) produced grotesque pictures and computer animations. As a duo, 'Böröcz and Révész' were a sort of Amos and Andy, or Laurel and Hardy, their performances unfolding either as absurdist skits or music-hall numbers, with neo-figurative painter Gábor Roskó (1958) a sidekick providing accompaniments on clarinet and other woodwind instruments. The situation was further complicated by the fact that some of the above-mentioned artists (Birkás, Fehér, Böröcz, Roskó), along with János Sugár (1958), György Fazekas (1948) and others, as art students or later, were regular participants in Miklós Erdély's conceptualist- and actionist-inspired 'exercises in creativity' under the 'INDIGO' banner.

By the time the first major exhibition of the 'new sensibility' was mounted at the Ernst Museum, Budapest, in 1984, therefore, few of the participants came without an avant-garde 'ballast' or 'unburdened' by connections with other branches of art. The few who might be called true 'innocents' were to be found amongst the very youngest, painting purely for the sake of painting, such as Zoltán Ádám (1959), András Bernát (1957), József Bullás (1958),

István Mazzag (1958), László Mulasics (1954) and Tamás Soós (1955). Their counterparts in sculpture were Gyula Gulyás, with his frivolous, coloured plaster portraits, György Chesslay (Cseszlai, 1957), working with cubist forms, or Attila Mata (1953), with his painted blocks of split wood.

With 'new painting' the administrators of Hungary's official cultural policy, having long laboured under the stress of trying to deal with the constant provocations of the avant-garde, could feel themselves entitled to heave a little sigh of relief, as this cheerfully abstract (or non-abstract) approach posed no obvious threat to either the socialist order or the status of Soviet troops stationed in Hungary. Art critics, however, had a hard time trying to work out what to make of the sudden 'consolidation'. The title 'In Quotation Marks' bestowed on an exhibition at the King (and now again St.) Stephen Museum in Székesfehérvár was laconically emblematic of the period. It was indeed truly difficult to know where to slot artists who did not unambiguously belong either to the earlier avant-garde or to new painting but to both one and the other, who were both 'abstract' and 'figurative'. Among them were not only Roskó and Péter Ujházi (1940), but also Árpád Szabados (1944), now dean of the University of Fine Arts, Budapest, who studied children's drawings and painted neo-primitive, psychoanalytical pictures; Géza Samu (1947–1990), who incorporated the implements and objects of rural life into installations that displayed radical approaches to nature; the 'individual mythologist' Imre Bukta (1952), an observer of the transformation of agriculture and well versed in every genre; Frigyes König (1955), whose private passion for collecting cabinets of *objets d'art* and curios was a basis for his meticulous graphics and paintings; Zoltán Érmezei (1955–91) and János Rauschenberger (1955), who painted one another's mirror images; András Lengyel (1952), who constructed book objects and painted books; and Márton Barabás (1952), who created objects, installations and paintings out of the component parts of pianos. Then too there were groups such as the *Hejettes Szomlyazók* (roughly 'Surrigate Thursters': the deliberate misspelling of their name was a signal of their aspiration to reform the rules of both spoken and written language), whose grotesque installations, based as they were on well-known paintings and events of Hungarian history, by their very nature placed everything 'in quotation marks'.

As the requisite of an artistic event, an installation is related to performance, in regard to its origins (in Hungary as elsewhere), in the same way as the environment, as natural surrounding, had been to a happening. As applied to an autonomous spatial construction, articulation of space, or apparatus, it is a technical term that even nowadays springs more readily than any other to the lips of young artists. The 'classical' installation artists include Tamás Trombitás (1952), applying neon tubes to forms produced courtesy of the assistance of iron works; Ilona Lovas (1948) and her spatial arrangements of the man-sized spindle shapes formed by cattle intestines; or Tamás Körösényi (1953), installer of three-dimensional forms reminiscent of the blotches on military combat fatigues. Mention must be made here of an architecture-related trend that could be considered a truly Hungarian speciality.

In part due to the difficulty of gaining serious commissions in their own field, a number of young architects started to take an interest in temporary projects like stage décor, film set design, interior design, furniture design, etc. Gábor Bachman (1952), László Rajk (1949), Attila (Ferenczffy) Kovács (1951) and their associates began to gain expertise in a huge diversity of genres and create in-between genres. With their eclectic mixing of constructivism, futurism, pop art and soc-realism (in an ironic context), to say nothing of new wave, these artists were attuned to the post-modern spirit of the age. Some of them, Bachman and Rajk in particular, were the creators of a Hungarian version of deconstruction. A memorable peak of Bachman and Rajk's activity was their design for the biers in the colonnaded main hall of the Exhibition Hall and in

Heroes' Square, Budapest, where the remains of Imre Nagy and other executed leaders of the 1956 Revolution, were ceremonially laid prior to their symbolic re-burial on 16th June 1989—the event that can be said to mark the beginning of the end for the country's communist régime. From his early work as an interior designer for night-clubs and set designer, Bachman arrived at a creative way of treating architect's paper models as statues, which in turn were the basis for preparing technical drawings that were then used as blueprints to construct a building—all of which fed in as subjects for his paintings. When put on display for several months at the 1996 Venice International Architectural Bienniale, Bachman's work was looked on as a full-scale model for a utopian building. A similar path was followed by the somewhat younger Brettschneider group, whose members included Tibor Szalai (1958–98), designing huge paper installations and also the furniture-like instruments on which they performed their 'Brettschneider music', thereby linking themselves, knowingly or not, to the sound installations of György Galántai, Viktor Lois (1950), and Rezső Móder (1954).

In the meantime, media art was catching on, with video and, later on, computer art followed in the footsteps of experimental film. Particularly noteworthy here have been the film oeuvre of Zagreb-based László (Ivan Ladislav) Galeta (1947), planned with mathematical precision and pregnant with symbolism, along with several installations of Gusztáv Hámos (1955) and Péter Forgács (1950)—among others, the latter's *Privát Magyarország* (Private Hungary) series, which utilises old amateur family films as its raw materials—and the work of Gábor Császári (1959), who devised a way of using running track photo-finish apparatus as a motion picture camera. Even in the field of 'traditional' photography the work of some practitioners, such as Lenke Szilágyi (1959) or Gábor Kerekes (1945), has entered the history of graphic art, or vice-versa, in the case of Károly Schmal (1942) and Péter Gémes (1951–96).

Perhaps paradoxically, the change of régime in Hungary was not attended by any sea-change in art—at least not immediately or in any direct manner. Rather, the converse was true: the intellectual battles that artists had fought for freedom of expression and new forms had helped pave the way for social and political change. What was striking about the public visual face of 1989–90 was the sheer multiplicity and diversity of political posters, especially during the period of the first free parliamentary elections since 1945. One particularly popular bull's eye was *'Tovarishchi, koniets!'* ('That's It, Comrades!', 1989), designed by István Orosz (1951). Contrary to all expectations, however, the transition was *not* accompanied by a wholesale de-struction of communist-era images. Apart from the already mentioned establishment of the Budafok Statue Park, Tamás Szentjóby's idea of covering the Liberation Monument with a Ku-Klux-Klan-style hooded cloak to embody the 'statue of the spir-it of freedom' lingers in the mind as an original individual gesture.

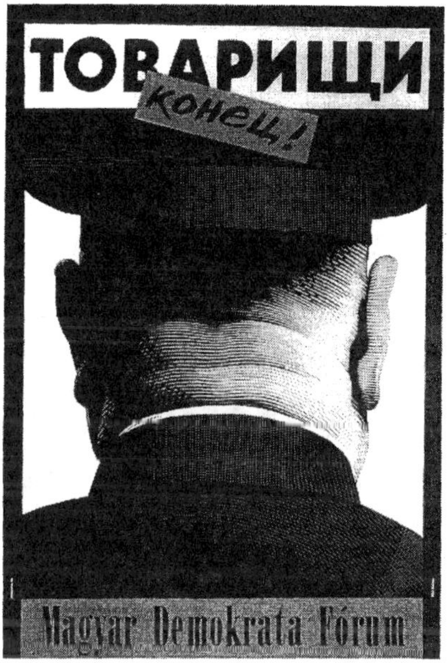

István Orosz:
'Tovarishchi, koniets!'

The first significant new public statue (and arguably the finest erected in Hungary since 1945) was György Jovánovics's *1956 Monument* (1992) for Plot 301 in the Rákoskeresztúr Cemetery, Budapest, where Imre Nagy and his associates had been anony-mously buried after their execution. That does not necessarily apply to another monument, in the form of a traditional arch of the

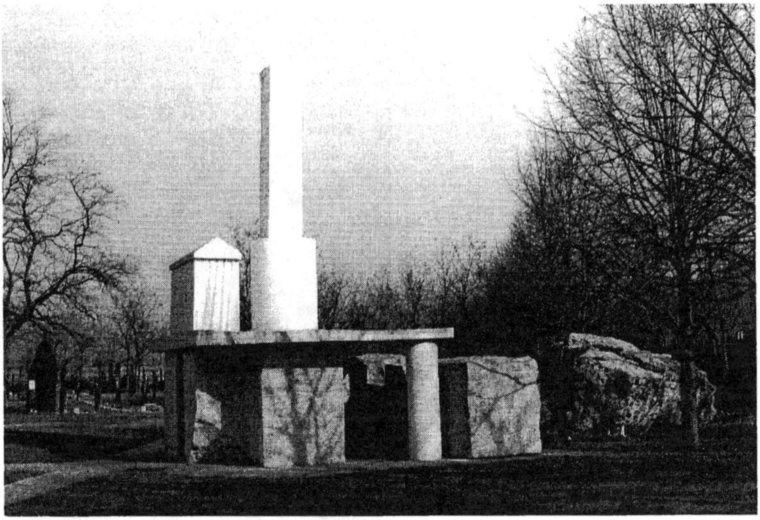

György Jovánovics: *1956 Monument*

Transylvanian Székely people, which has since appeared on the same site (as conservative tribute to the memory of 1956), or indeed the hundred-plus grave headboards that were set up by Inconnu, a far from nationalistic group of artists who had organised several political protests during the final decade of the Kádár régime, which was also their way of announcing that the group's *raison d'être* had come to an end.

There has been little talk of stylistic shifts since 1990, but the major changes that have been seen in the art world's infrastructure, institutions and sociological structure have been all the more striking. Apart from the lapse (or at least tempering) of centralised controls, the most crucial change has been in its financial base, with the emergence of private commerce (gallery network, auction houses) and corporate sponsorship and the transformation of government funding. A non-profit foundation was set up and was in existence for almost a decade to run Budapest's Art Expo fair. Censorship was abolished; publishing houses and periodicals have been revitalised, and new forums have emerged. Out of the two centralised, party-controlled organisations that once embraced all the artists—the Arts Fund of the People's Republic of Hungary and the Association of Hungarian Artists and Industrial Designers—have arisen four or five larger and some thirty smaller successor associations to protect artists' interests. Art schools at the higher education level have undergone several reforms and new scholarship opportunities have been made available for the young (and also mature students). Many émigré artists have returned to Hungary. The (art) world has been reassessed, and within that the role of Central and South-Eastern Europe as a whole, Hungary included.

Outlook at Start of the New Millennium: the Problem of Interdisciplinarity

From the earliest histories of art onwards, the hardest part of any serious survey is ending it at the very point where the past is about to tip into the future. Were one to undertake a comparison of the closing sections of the histories of art written at various times, the historical trajectory that would emerge might, perhaps, be very different from the accustomed picture: different trends and different names would be thrown up by accidents of the moment. At this moment of writing, the only assertion that the author of this paper can make with some certainty is that Hungarian art currently boasts *no* global stars (in the universe of art the stars—the 'painter-princes' of old—operate in a different way from those in the worlds of music, film or sport). Nonetheless, even in the Hungary of today there are defining circumstances, dominant trends, new phenomena and, indeed, notable young creative talents who are finding their place in the context of the world's artistic life.

What are the main characteristics of this global situation? As the model of the information society is being turned into reality, communication is accelerating; more and more scientists and artists use the computer. Hungarian scientists are working on the extraordinarily intricate problems that are posed by genetic surgery, for instance, or the creation of an artificial retina. Some artists, on the other hand, see new means of expression in the Internet or in computer print-outs, whilst with their complex installations Tamás Waliczky (1959) and Attila Csörgő (1965)—two of the most familiar Hungarian faces at international art festivals—have opened up genuinely new possibilities of creating imagery and innovative conceptual models. Not long ago, the C3 (Center for Culture & Communication) Foundation in Budapest presented several intriguing interactive computer animations, amongst others *Demeduzátor* (De-Medusator, 1998–99) by Zoltán Szegedy-Maszák (1969) and *Kultúra Kettőssége Kalendárium* (Almanac of Cultural Duplicity, 2001) by Márton Fernezelyi (1972), Róbert Langh (1967) and Miklós Szalay (1972).

While the symbolism of the year 2000, which also marked the millennium of the foundation of the Hungarian state, has thrown up new problems and opportunities for the ways the state and nation choose to project themselves, the artist seeking to navigate his or her way amongst the welter of international trends is paying attention to such global phenomena as multiculturalism and post-colonialism, or ethnic and other minority identities, with a whole group of younger female artists concerning themselves with the social issues of gender. Conspicuous among the artists receptive to social problems are Róza El-Hassan (1966), with her 'overpopulation' project, and Miklós Erhardt (1966), who has been working with the Scottish artist Dominic Hislop (1967) to provide homeless people in Budapest with cameras and encouraging them to supply their own comments on their photos. The success of the project is a reminder that even a seemingly hackneyed avant-garde idea like Warhol's 'five minutes of fame' may contain hitherto unexploited aesthetic reserves that have yet to be mined during the new millennium.

Péter Halász

TONES REVERBERATING AROUND THE WORLD

Introduction: European music versus national music?

Thinking about what gives Hungarian music its distinctiveness—whether it was something that the eastern periphery of Europe managed to retrieve from western Christian musical developments of the day or, rather, something that Hungarian musicians added to the orchestral palette of European music as their own independent contribution—has undergone such a big change in recent decades as to require a radical reassessment of what is meant by the very term 'Hungarian music'.

In the early 1930s Bence Szabolcsi, one of the greatest Hungarian scholars of musical historiography, inclined to the former theory, downplaying such secondary phenomena of musical life in Hungary as its adaptation and elaboration of medieval Gregorian chant, and even the genres of religious and instrumental music derived from southern German, Austrian and Italian models that took root in the eighteenth century. Instead, he emphasised such peculiarly Hungarian musical features as the national art-music style that evolved around the turn of the eighteenth into the nineteenth century from the folk traditions of the *verbunkos* recruiting dance, or the folk tradition-based modernism that was associated with Béla Bartók and Zoltán Kodály in the early twentieth century.

Half a century later, during the 1980s, László Dobszay stood views of what was important about this music's evolution on their head: for him, it was precisely Hungary's ability to integrate into the history of European music, the way it was able to accept European musical models and refashion them to local tastes that should be regarded as one of the most crucial factors of the country's musical history. Accordingly, as Dobszay saw it, even the strong tradition of Gregorian chant, extending well past the Middle Ages, is to be regarded as a distinctively Hungarian tradition, which may well have been preserved due to the lack of skilled church choirs and the absence of late-medieval and Renaissance polyphonic music. Like Szabolcsi, Dobszay naturally attached equally great importance to the verbunkos as the starting point for Hungary's art-music aspirations during the nineteenth century, but he sees the chief puzzle as being how it was possible to place this style, suitable for composing essentially lyrical or small instrumental forms, in the service of such typical European musical genres as opera or symphonic music.

Where Szabolcsi and Dobszay are in fullest accord, however, is on the big part that folk music—the music of peasant strata that remained in a condition of relative changelessness for centuries on end—played over and over again in the history of music in Hungary, as in the other countries of Eastern Europe. Even though it was not, of course, fixed in notated form, through its very widespreadness in society this musical culture was ever-present to serve both as a potential source for renewal of the national musical culture and as a kind of counterexample in any comparisons between national aspirations and the adoption of the products of European musical culture.

The Foundations: New Settlers and the Verbunkos

Following the Turkish retreat from Hungary in the early eighteenth century, after almost a century and a half of occupation of much of its territory, the country's Habsburg rulers saw fit to secure its economic, social and cultural 'resuscitation' mainly by encouraging an influx of German- and Slavic-speaking settlers. Settling predominantly in the towns, large and small, and thereby founders of a gradually emergent Hungarian middle class, they naturally brought their own musical cultures with themselves.

The need to transplant and domesticate European trends in music must have arisen, first and foremost, in the entourages of church dignitaries and the aristocracy in episcopal towns and aristocratic residences over the course of the eighteenth century. As part of this process, at Nagyvárad (Oradea, Romania) and Pozsony (Bratislava, Slovakia) orchestras came into existence and functioned for shorter or longer periods during the 1760s and '80s. Most significant of all was the orchestra maintained by the princely branch of the Esterházy family, dividing its time between the palaces of Esterháza and Eisenstadt and also supplying the accompaniment for regular opera performances. Orchestras like this, being big consumers of the contemporary musical repertoire, were magnets for composers from Vienna and Bohemia, as a result of which numerous figures in the musical history of Austria, and indeed Europe at large, spent shorter or longer periods on Hungarian territory. The most considerable of them was Joseph Haydn, who was in the service of the Esterházys for three decades from the early 1760s and thus, as he himself wrote in a letter, his creation of 'Viennese Classical style' was achieved whilst "left to my own devices, in the seclusion of a backwater plain," through a series of symphonies and string quartets that are popular to the present day and are held to be among the most important musical pieces of the last third of the eighteenth century. Although Haydn's style can be said to be universal—despite the characteristically Austrian minuet-ländler and the less well known Italian operas—more than a few passages show the influence on his music of the Hungarian (or, for that matter, Gypsy) musicians who would have played light music for entertainment in the Esterháza region. The final movements of the *Piano Concerto in D major* and the *Piano Trio in G major*, both bearing the marking *all'ongherese* (in the Hungarian style), are typical examples.

Alongside the immigrants, there were also noteworthy Hungarian-born composers among the musicians of that period, the most important of whom was Benedek Istvánffy (1733–78). His career took him through various Transdanubian towns, which brought him into contact with the Széchényi family, who were to play a major role in advancing national culture during the first half of the nineteenth century, and with the Esterházy orchestra at Eisenstadt, perhaps even with Haydn himself. Eventually, though, he settled down in Győr in 1766, where he became the choirmaster at the local cathedral. Only a dozen or so of his compositions of various lengths, including two masses, have survived in manuscript form, yet even that slight oeuvre reveals Istvánffy to have been a talented emulator of the late Baroque and early Classical styles that were flowing out of Italy through Vienna to Hungary. Handling his forces with great assurance and a high degree of artistic imagination, he turned the diocese of Győr into one of the country's most important musical centres.

Since the term all'ongherese has already been mentioned in relation to Haydn, it is worth looking at how this style arose and what that signifies. The first traces of 'Hungarian dances' can be found in sixteenth-century collections of dance music printed in Germany, Italy and France. From then on all'ongherese became just as much a distinctive, exotic form of instrumental music in Western Europe as the Polish *polacca* or, if it comes to that, the English 'hornpipe' and 'jig' (or 'gigue').

During the second half of the eighteenth century, it was not just Haydn, working as he was on Hungarian soil, who became receptive to this style, but other Austrian composers too. One such was Mozart, who employed it in the finale of his *Concerto for Violin in A major, K.219* and his *Concerto in C major for Flute and Harp, K.299*; indeed, it is barely distinguishable from the Janissary music of *Entführung aus dem Serail* or the *Rondo alla turca* finale of the *Piano Sonata in A major, K.331*. The so-called 'Hungarian style' was also very popular with German and Austrian composers in the nineteenth century—not just Brahms in his *Hungarian Dances* but Beethoven, who inserted a characteristically Hungarian tune in his *Symphony No. 3* (Eroica).

The exact route by which Hungarian music—or the verbunkos, to use the modern collective term—came to the attention of composers of the day is hard to trace. (The term itself comes from the German 'Werbung' for the music played in the villages on army recruiting drives.) At all events, during the early decades of the nineteenth century the verbunkos came to be the most characteristic musical idiom in Hungarian art-music, for the ideals that it was perceived to represent allowed it to fulfil the role of a national music. For Hungarians in the nineteenth century the verbunkos did not just revive memories of recruitment in the era of the War of Independence of the early eighteenth century, the *kuruc* era of Prince Ferenc Rákóczi II, but was also perceived more generally as an embodiment of Hungarian identity and national character.

The performers of verbunkos music during the early nineteenth century were mostly Gypsy bands supplying dance music for the gentry class. Conspicuous among them was the band put together by János Bihari (1764–1827), around 1801, in the city of Pest, typically a five-piece ensemble of two violins, viola, double bass and cimbalom. Bihari was considered the most famous of all performers, his skills as a violinist in demand even as far away as Vienna; his admirers included the leading writers and politicians of the day, such as Dániel Berzsenyi, Count István Széchenyi and Sándor Kisfaludy. Born in the northern uplands of the country (now Slovakia), Bihari could not read music, so he himself left no written record of any of his own pieces; the extant versions of Bihari's music were notated by schooled musicians. Although the ternary slow–quick–ornamented scheme of the verbunkos was already established by the time Bihari embarked on his career, it was in his dance adaptations that it gained the characteristic form that it retained throughout the nineteenth century, which Ferenc Liszt was to follow in his *Hungarian Rhapsodies* and other composers in their Hungarian-style compositions. It is a measure of Bihari's influence that his intonation remained something of an ideal for Hungarian music throughout the nineteenth century and, indeed, for Gypsy performers to the present day. The other great achievement of Bihari's art was that his verbunkos art-music also found its way to the peasantry and in the process was soon transformed into a kind of folk music. It also revived a tradition of kuruc songs that goes back to the early eighteenth century and thereby contributed to the forging of diverse musical traditions into a single, seemingly ancient tradition, which was the initial step in the creation of a new genre of distinctive national music. Apart from his own compositions, Bihari often played the *Rákóczi-nóta* (Rákóczi Song) and *Rákóczi-induló* (Rákóczi March), as well as pieces composed by János Lavotta and Antal Csermák.

The performance style and works of János Lavotta (1764–1820) and Antal Csermák (ca. 1774–1822), alongside Bihari, played a crucial part in Hungarian music during the first decades of the nineteenth century. Both were trained musicians and so notated their own compositions. Their historical importance lies in the fact that they were the first to compose dance suites for small ensembles, thereby laying the foundations for chamber music in Hungary. Csermák's 1809 string quartet, with the programmatic title of *Az intézett veszedelem avagy a Haza szeretete* (The Peril Addressed, or Love of One's Country), which was a reaction to Napoleon's incursion into Hungary, was composed in the verbunkos style but its orchestration is Classical. The reason why those initial attempts at chamber music had no successors in this form is rooted in the characteristics of the verbunkos itself, as these short forms, designed as dances, are broken up by frequent cadences and did not lend themselves readily to more extended art-music compositions. They were all the better matched to the needs of dance compositions for the salon, and the most prominent exponent of this genre in the 1830s and '40s was Márk Rózsavölgyi (1789–1848). It was through Rózsavölgyi and his contemporaries that impulse came to convert the 'lively' section of the verbunkos into the *csárdás*, refine the melodies and adapt the dances to westernised patterns of social dancing (one example was the creation of the *körmagyar* (Hungarian round dance) as a sort of Hungarian equivalent of the quadrille). Rózsavölgyi's most successful pieces, with their well-crafted structures, the closed forms created by repeated dance tunes, and the unaffected elegance of their salon ambience, were not just successors of Bihari's art but also marked the culmination of the mature period of the verbunkos.

Ferenc Liszt, the Virtuoso Who Spellbound the World

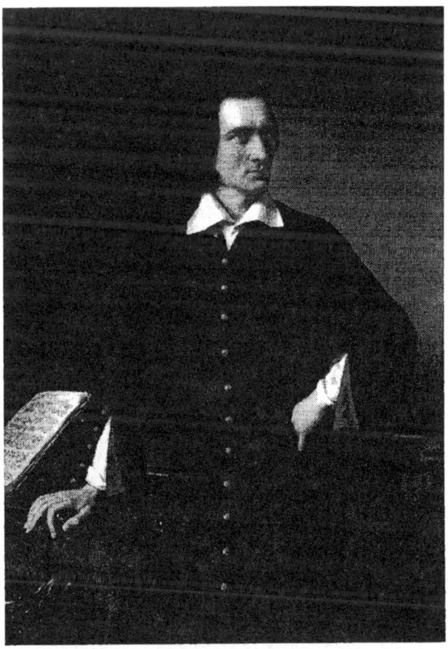
Ferenc Liszt

Well, is he a Hungarian or not? Few questions in the history of music are as barren as those that delve into the origins and national allegiances of Ferenc Liszt (1811–86)—or Franz, to give him the German equivalent of his Christian name by which he is most widely known. Yet the facts are plain enough: Liszt was born into a German-speaking family in Doborján on what was then Hungarian territory. (It is a minor complication, in retrospect, that the small village is now called Raiding-Unterfrauenhaid and lies in Austria's Burgenland.) Ferenc's father, Ádám, was an amateur cellist in the Esterházy orchestra (reputedly when Haydn was Kapellmeister), so it is hardly surprising that the son's outstanding talent as a pianist should quickly have been noticed.

When Ferenc Liszt left Hungary in 1819 to study and to give concerts, first in Vienna then, from 1823, in Paris, he cannot have anticipated how long that absence from his homeland would last. The character and personality were shaped, above all, by the Paris years, the vivacious intellectual life around the young man from 1830 onwards, the friendships with Berlioz, Chopin and Paganini as well as Victor Hugo, Dumas, George Sand, Delacroix, and the Abbé Lamennais, the world of salons and boudoirs, opera houses and concert halls. The incredibly gifted pianist and experimentally-minded fledgling composer's thinking was imbued by the civil democratic ideals of France's July 1830 Revolution and the transcendentalism of German and French Romanticism. This was a world not only geographically remote from Hungary but intellectually too an infinity away from an undoubtedly backward country that was still largely stuck in a pre-Enlightenment time warp. Yet there were some shared features: one of the guiding principles of the Reform Era in Hungary, the start of which was marked by the Diet of 1825, was the liberation movement against Austrian absolutism, which could count on the sympathy of western Europe's progressive middle classes and was to culminate in the 1848–49 Revolution and War of Independence.

The adult Liszt's reunion with his home country was not to take place until the late 1830s. Liszt suddenly discovered his Hungarian identity on hearing news of the tragedy that had befallen Pest with the bursting of the River Danube's banks in March 1838. He hastened to Vienna and there, in the capital of the Habsburg Empire, put on a series of benefit concerts to raise money from the affluent public. When he finally set foot again on

Hungarian soil in December 1839, long preceded by his reputation, the tumultuous reception from Hungarian audiences knew no bounds. At a concert he gave on the stage of the Hungarian (later National) Theatre in Pest, which had opened two years previously, representatives of the aristocracy, in the typically combative spirit of that age, presented him with a ceremonial sword, for which Liszt, having no Hungarian, gave a short speech of thanks in French. The star only further increased his already huge popularity by rounding off his customary program of paraphrases of popular operas and improvisations, with due regard to the sentiments of his public, with a performance of his arrangement of the *Rákóczi induló*—a tune whose origin is obscure (though it most likely reached Hungary through Turkish hands) but at that time was generally regarded as deriving from the kuruc era, and thus as a musical symbol of national independence. (Hector Berlioz, having heard the melody during his visit to Hungary in 1846, slotted his own orchestral arrangement of it into his oratorio *La damnation de Faust*.)

Thereafter Liszt was never again to lose his contact with Hungary, although he never relinquished his residences abroad to settle down there for good. At first the visits were infrequent, but his formal avowal of his Hungarianness was all the more explicit. There are innumerable attestations to the fact that he felt and declared himself to be Hungarian, despite having spent the greater part of his life outside Hungary, with an education and culture dominated by French and German influences and, worldly spirit that he was, often found the provincialism of his native land hard to take. Over the next few years he toured across Europe several times, from St. Petersburg to Lisbon, and the next time his concert schedule was to bring him to Hungary was not until 1846. He then gave up this peripatetic existence in 1847 to settle down at Weimar and turn the hearth of German Classical poetry, the former residence of Goethe and Schiller, into a centre of German musical life. The following decade was the period in which the Lisztian oeuvre was crystallised with the composition of a series of symphonic poems (e.g. *Tasso*, *Mazeppa*, *Les Preludes*)—one being *Hungaria* (1854), in which Liszt painted a musical portrait of his homeland, employing typical verbunkos idioms and original Hungarian melodies.

In order to understand Liszt's relationship with this music, it is instructive to quote his own reminiscence, as set down in his *Des Bohémiens et de leur musique en Hongrie* (1859), of hearing János Bihari perform when he was a child: "What a colossal magic he exercised when he took his violin in his hand, playing for hours on end as if oblivious to the passage of time, enticing out of it cascades of notes that now rushed on with a wild tumbling, now trickled as over a moss-bed of almost silky smoothness. The effect of the notes of that magic violin was like the distillate of a spirituous essence."[1]

Even at that tender age, what undoubtedly fascinated Liszt, who was of course himself later to dazzle audiences with his own improvisations, was the ability of a performer's soaring imagination, the force of individual expression, to make a life-long impression. When he re-encountered this type of music during his visits to Hungary as an adult, it was to affect both his repertoire as a pianist and his compositional output. Already during his first visit in Hungary, in 1840, he formed the idea of collecting Hungarian music and popularising it through his own piano piece arrangements. He made a start by himself jotting down tunes he heard in restaurants and at balls, and later on he tried to get hold of as many arrangements made by others as he could; indeed, romantic that he was, at one stage he even conceived the plan of tramping around the country's more out-of-the-way villages on foot to collect hitherto unknown songs.

It was this collected material that was the basis for the books of *Magyar Dalok* or *Ungarische Nationalmelodien* that he published from 1840 onwards, which during the Weimar years were transformed into the *Hungarian Rhapsodies* Nos. 1–15. The main reason why these compositions stand out among other Hungarian-style pieces of the era lies in the highly sophisticated western European schooling that Liszt brought to bear on his sources. As in his paraphrases of popular operas of the day, Liszt strove in the rhapsodies to string the successively presented themes into a single musical chain with bridging passages that comprise variations of the highest pianist virtuosity. What is typically Hungarian, though, is the binary slow–lively structure that he picked up from Gypsy musicians.

Even as he was composing these, Liszt was struck by the notion that it was not so much the nature of the individual pieces that mattered as the completeness of the collection, which he regarded as forming a sort of musical folk epic. That was all very well, but which folk are we talking about? This brings us to the heart of the controversy about the Lisztian musical concept of the *Hungarian Rhapsodies*. Taking as a point of departure the fact that what Liszt, like his contemporaries, called 'Hungarian' style, with its well-known minor key changes, elementally simple rhythm and almost heel-clicking plucked cadences—was known to them primarily from music performed by Gypsy bands, it follows that the folk in question are Gypsies and not Magyars. As he expounded in his above-mentioned little book, to the outrage of the Hungarian public, the verbunkos and its derivatives had nothing to do with the ancient music of the Magyars but was in reality a music that the Gypsies (or Romanies), originally from the East, had brought with them. In Liszt's view, the Hungarian music that had been unearthed with such difficulty was nothing more than a foreign import. Understandably, the book provoked huge controversy in Hungary as many people thought Liszt was stabbing the nation in the back by denying them a music they felt was their own. Small wonder that for a long time after the book's appearance Liszt was the target of harsh attacks by Hungarian journalists who felt their most cherished sensibilities had been besmirched and, not entirely without reason, demanded that he return their national music to them. From a modern perspective, the music is in fact a dance music of Central European origin, with recognisable Hungarian as well as Romanian, Slavic and other elements, the chief exponents of which were Gypsy musicians. It is equally unjustified to extol it as entirely Hungarian as to dismiss it as entirely non-Hungarian; it should be accepted for what it is, in its own historical context, in its own musical diversity, as a typical hybrid genre of its era. And whilst it is equally an error to suppose that it has anything to do with the Gypsies' own indigenous music, Liszt was nevertheless right to assert that the uniformity of the verbunkos was determined by its Gypsy performers, insofar as the way the music is played, its ornaments and its frequently improvised arrangements must all have had their origins in a non-notated tradition of performing. (Rather akin to jazz performing skills, this tradition can still be observed today in the way that Gypsy bands, now largely restricted to engagements in restaurants, treat tunes from operettas and musicals and even modern hit-parade songs in the same way as they treated popular melodies in the nineteenth century.)

During the 1850s Liszt became increasingly active in Hungary's music-life with a string of commissions to compose major liturgical works, such as the *Esztergomi mise* (Missa Solemnis, 1856), for the consecration of Esztergom Cathedral, or the *Koronázási mise* (Hungarian Coronation Mass, 1867), for the Buda coronation of Emperor Francis Joseph as King of Hungary following the Compromise. From 1869 Liszt spent more and more time in Hungary as he entered the period of what he called his 'vie trifourqué', or threefold life, dividing his year between spring in Rome, summer and autumn at Weimar, and winter usually in Budapest. His main activity was as a piano teacher, with an international entourage of students in tow as he made the round of his residences. Although he had given up his career as a professional performer as early as 1847, he continued to appear regularly at charity concerts. The weight of his authority played a key part in the establishment of the Budapest Academy of Music, and he rendered an inestimable service for the high standards that were to be cultivated in later Hungarian musical life by becoming its first director in 1875.

A Hungarian tone gained increasing expression in the compositions of his final years, albeit very different in style from that of his earlier output. The enthusiastic, optimistic sparkle of the mid-century Liszt receded ever further as the greyness of the bourgeois world of the 1870s and '80s left a decided stamp of resignation on the strange works of his late period. His musical horizons were no longer shaped by the grand virtuosity of his concerts' style; the new piano pieces were more bittersweet and introverted, often formidable and aphoristically pessimistic, as in his *Ungarische Bildnisse* (seven musical portraits of the poets Petőfi and Vörösmarty and fellow composer Mihály Mosonyi as well as a clutch of politicians: István Széchenyi, Ferenc Deák, László Teleki and József Eötvös), the *Hungarian Rhapsodies* Nos. 16–19, or the *Csárdás macabre* and *Csárdás obstinée*. This is Hungarian music that no longer has flesh, blood and spirit, just a rattling skeleton dancing a frightening dance of death amidst the imposing, newly built palaces of Budapest's rapidly proliferating prosperous streets. There are many who discern in the works of Liszt's old age a prototype for the harsh, dissonant music of the catastrophe-ridden twentieth century, the music of Bartók.

Ferenc Erkel: Opera Like Anywhere Else?

If Ferenc Liszt signified an extreme advance guard of 'Hungarian nationhood', his almost exact contemporary Ferenc Erkel (1810–93) was its rear guard. The creator of Hungary's national opera and composer of its national anthem, not to mention conductor for close to half a century of the orchestra at Budapest's National Theatre and founder of the Philharmonic Society Orchestra, the country's first professional concert orchestra—those combined achievements make Erkel the most prominent Hungarian-domiciled musical figure of the nineteenth century.

Born in the south-eastern corner of the Great Hungarian Plain, the Alföld, rather than in the western part of the country, Erkel made his name, after finishing his studies in Pozsony (Bratislava, Slovakia), as a piano virtuoso, initially earning his living as a music tutor to the Transylvanian aristocracy of Nagyvárad (Oradea) and Kolozsvár (Cluj) before moving to the capital in 1835. Appointment to the musical directorship of the new National Theatre in 1838 was to place him at the helm of the country's operatic culture over the next half century.

The National Theatre put on a program of alternating performances of spoken and musical works the common feature of which was that these were intended, first and foremost, to nurture Hungarian language and culture. Of course, the available Hungarian repertoire, especially in the operatic vein, was not broad enough to restrict the selection to home-grown works, so the backbone of the opera seasons was made up of the then fashionable works of Italian (Rossini, Donizetti, Bellini) and French (Auber, Meyerbeer) composers, though the librettos were translated into Hungarian. Critics, however, constantly (and quite rightly) deplored the lack of original Hungarian opera, represented by a handful of undemanding works by mediocre composers, which explains the huge anticipation that attended the premiere of Erkel's first opera, *Bátori Mária*, in 1840. With a story that takes place in medieval Hungary, the drama runs virtually the entire gamut of Romantic operatic stock devices, from the ecclesiastical chants of its wedding scene through a chorus of hunters and a drinking song to a grand love duet. In setting this to music, Erkel fell back mainly on the stylistic model of Italian opera, though pepping it up every now and then with a tang of verbunkos-based Hungarian music. Critics of the time were oddly divided, with the Hungarian-language newspapers finding the music foreign whereas the German-language newspapers of Pest emphasised its Hungarian characteristics—a contrast that is instructive in that it underscores how imprecise the notion of a 'national music' is, and how the ideals that audiences cherish are never fully met by reality.

Erkel achieved enduring success with his next opera, *Hunyadi László*, which was first staged in 1844. As was to be the case in virtually all of his operas, this too has a historical subject—here events from the fifteenth century that would have had clear political resonance for a nineteenth-century public. The clash between Hunyadi, an epitome of the Hungarian aristocracy, and the treacherous Habsburg king, Ladislas V, accurately captured the nation's opposition to its foreign ruling dynasty during the 1840s, and the fact that the struggle culminates in Hunyadi's downfall and eventual execution, due to false accusations made against him, endowed it with a tragic pathos. Though the fashionable Western-style set pieces are still present, as in *Bátori Mária*, what is new is the greater attention to dramatic continuity, with recurrent motifs to characterise individuals and situations, and the Hungarian-flavoured elements integral to the dramatic essence. The real crowd-pleasers, though, were self-sufficient numbers like *Meghalt a cselszövő...* (Dead is the intriguer...), the revolutionary-spirited chorus that follows the murder of Ulrik Czillei, who has been scheming to ruin Hunyadi, or the florid coloratura of the 'La Grange aria' or, for that matter, the imitation of the old Hungarian dance of the *palotás*. Contemporaries justly recognised this conquest of the operatic genre as an event of truly historic importance, with one critic writing after the premiere that "Erkel has elevated the wonderful melodies of the *csárdás* to European rank. If Hungarian music has a future, then our nation will be grateful to the composer of *Hunyadi László* for laying its foundations."

It was also in 1844 that Erkel composed the entry that won a competition for the best setting of Ferenc Kölcsey's famous poem *Himnusz*, which has been the country's national anthem ever since—not a military march but a hushed supplication to providence that Hungary's stormy history may see the nation's fate take a turn for the better in the future.

What was to ensue, however, was Hungary's 1848 Revolution, and the disaster of its eventual suppression in the War of Independence that was fought against Austria, followed by a decade and more of secret police supervision and censorship sensitive to the slightest expression of Hungarian national identity. Erkel held his own council during this depressing period, but when he finally broke that silence it was to offer his public a more mature masterpiece that had in fact been written in the aftermath of the defeat but been kept in the bottom drawer until it was possible to stage it. *Bánk bán*, is based on the play by József Katona that the audience had demanded be staged on 15th March 1848, the day the revolution broke out in Pest. Taking the scenarios of Verdi's grand operas as its models, *Bánk bán* tightly interweaves politics and the dramas of private life, once again bringing a story of conflict between the Hungarians and their foreign oppressors to the stage. Set in the early thirteenth century, it is about the personal conflict of Bánk, a proud aristocrat who is acting as the country's governor whilst the king is away crusading in the Holy Land, as he is torn between loyalty to his sovereign and outrage at the immorality of the non-Hungarian queen's entourage. The final straw for him is the seduction of his own wife, Melinda, by the queen's younger brother, which provokes him to take the law into his hands and kill the queen.

Erkel deliberately sought to interlace the various musical styles in a happy conjuncture of French, Italian and Hungarian styles, with the latter never being allowed to gain sole ascendancy, as his aim was to employ the various stylistic idioms within the international operatic language of the day to characterise the individual protagonists. The passionate duet of the murder scene, the farewell aria of Melinda, who is driven insane by the disgrace she has suffered, or the *Bordal* (Drinking Song) of the rebel leader Petúr, are all musical-dramatic peaks, whereas the patriotic declaration of Bánk's romanza *Hazám, hazám* (My land, my native land) is a lyrical supplication similar to the *Himnusz*. The figure of Tiborc sneaking into the palace is sociologically a particularly intriguing element, for this represents the first time a Hungarian peasant was allowed to appear on stage alongside contending and suffering noblemen.

Erkel composed another five operas over the next 25 years, the last of which, *István király* (King Stephen), was first staged in 1885 by the newly opened Royal Opera House in Budapest. He gradually transformed his composing style, partially embracing a Wagnerian-style recitative, though by then he was increasingly reliant on the assistance given by his musician son so the late works should be seen rather as the products of a kind of family workshop than as the fully integrated manifestations of an individual creator. That may go some way to explain why these works, although respectfully received, never achieved real popularity.

True success likewise eluded Mihály Mosonyi (1815–70). His career was characteristic of the assimilating power of the idea of Hungarian national identity. Born into an Austrian family (his original name was Michael Brand), he worked initially as a music teacher in Pozsony and in southern Hungary then took up residence in Pest in 1842, where he gained attention with his symphonic and chamber works. He tried to find his own musical voice by emulating Schumann in his instrumental music and Wagner in an unperformed opera (*Kaiser Max*, 1857). He then broke completely with his former ideals in 1859 to start labouring on a new art music that was to be constructed from purely Hungarian elements (it was at this point that he changed his German name for a Hungarian one).

His main work, the opera *Szép Ilonka* (Belle Hélène), was premiered in December 1861, a few months after *Bánk bán*. Though Mosonyi's aim, unlike that of Erkel, was to draw solely on authentic Hungarian sources, it resulted in a music that was so relentlessly in the folk style that its sheer lack of variety hindered the development of any dramatic plot. Mosonyi's oeuvre, therefore, faithfully reflects the hardest task that every 'national' composer of his day had to face: how were elements of the national music to be fitted smoothly into the framework of a high-art genre like opera.

Lightly Kissed by the Muse: Hungarian Operettas

For all the efforts made to place Hungarian national opera at the centre of the cultural policy agenda during the nineteenth century, the wider domestic public and many foreigners were more inclined to recognise and esteem Hungarian music of another kind: the operetta. The reasons behind this hark back to the same period as national opera, when the national movement of the 1840s promoted the rise of another distinctive genre: that of the *népszínmű*, or 'folk play', a sort of rural musical melodrama in which the spoken dialogue is interspersed with set-pieces of Hungarian-style dancing and singing. Over and above its musical style, the népszínmű derived its national character from a plot line that allowed the stage to be populated with idyllically drawn folksy characters and set pieces. Composers of the songs for these works would occasionally adapt a genuine folk melody but more often they would draw on already widely familiar folk-style art-songs, or 'magyar nóta', or else produce new compositions in that vein, which popular performers would disseminate the length and breadth of the land so that they, in turn, would be perpetuated within the general public as 'nóta', by now without any knowledge of who had composed them.

As with national opera, the birthplace of the first such folk plays was the National Theatre. From 1843 through to the 1870s the National Theatre then, from its opening in 1875 and into the early years of the twentieth century, the Népszínház (Popular Theatre) mounted an unbroken string of such productions in Budapest, and that example was copied by a host of provincial theatres. There is little sense in listing the names of those who created such works, ephemeral as they were in artistic merit and life-span. They were hugely influential all the same, with a wide cross-section of the Hungarian public for a long time identifying with the relatively undemanding genre, which no doubt accounts for its success. Yet, like every genre of popular entertainment, this one too was at considerable risk of obsolescence, and although individual songs remained popular, in many cases for decades to come, the genre itself had essentially gone out of fashion by the end of the nineteenth century.

Possibly the very last of the dozens of népszínmű to reach the stage over the course of this half a century was *János vitéz* (John the Valiant) by Pongrác Kacsoh (1873–1923), first performed in 1904, but it alone has demonstrated true durability, enjoying success to the present day. There are two reasons why it has stayed in the repertoire: for one, its subject is deeply rooted in the nation's experience of literature, being based on Sándor Petőfi's epic poem of that title, and second, the musical sensitivity and basic compositional flair that Kacsoh displayed in fusing a Hungarian feel to operetta sentimentality. Although the piece was originally premiered at the Népszínház, eventually even the élitist Budapest Opera House was compelled to bow to the unbroken tide of its popularity: Kacsoh's work has been almost permanently in the company's repertoire since 1931.

From the 1860s onwards, the newly imported genre of operetta was also to target the audience for the népszínmű. In the hindsight of the resounding success that we know operettas were to achieve in the twentieth century, it is surprising that their route to initial acceptance in Hungary was anything but plain sailing. Offenbach, the most prominent composer of the operetta's first golden age, failed to exert much impact in Hungary. The sophistication of his music, often rooted in the parody of more serious genres, assumed a cultural grounding that Hungarian audiences did not possess. The neurotic cavalcade of Johann Strauss Jr.'s *Die Fledermaus*, with its intimations of *fin-de-siècle* Vienna, likewise cannot be said to have been notably well received when it was shown in Budapest in 1882. It was a different story with *Der Zigeunerbaron*, however, with its libretto based on a Mór Jókai novella, in which, from a Westerner's standpoint, the Gypsy flavour passes for Hungarian *couleur locale*, forming just one exotic element in a parade of the many cultural facets of the Austro–Hungarian empire. Strauss was so concerned to represent the unity of the dual Monarchy that he features Viennese waltz and Hungarian csárdás side by side in the Act 2 finale; indeed, his effort to win over the Budapest public went as far as incorporating the sure-fire patriotic appeal of the *Rákóczi March* into the music.

Franz (Ferenc) Lehár (1870–1948) and Emmerich (Imre) Kálmán (1882–1953), two great Hungarian-born composers of operetta during the first decade of the twentieth century were at least as much Viennese as they were Hungarian. The amalgam that is identifiable both in their choice of subjects and the character of their music was primarily a product of the unifying spirit of the Austro–Hungarian Monarchy, or in other words an attempt to harmonise the barely reconcilable diversity of its cultural legacies. At the same time, there is a conspicuous difference in the way the two handle precisely the Hungarian colouring.

Franz Lehár was decidedly a citizen of the Austro–Hungarian Monarchy. His father was of Czech descent and as an army bandsman had been posted to Hungary, where he met Lehár's mother, who was German by name but spoke Hungarian as her mother tongue. The reason why Lehár won his success with German-language operettas, and first and foremost in Vienna, is to a large

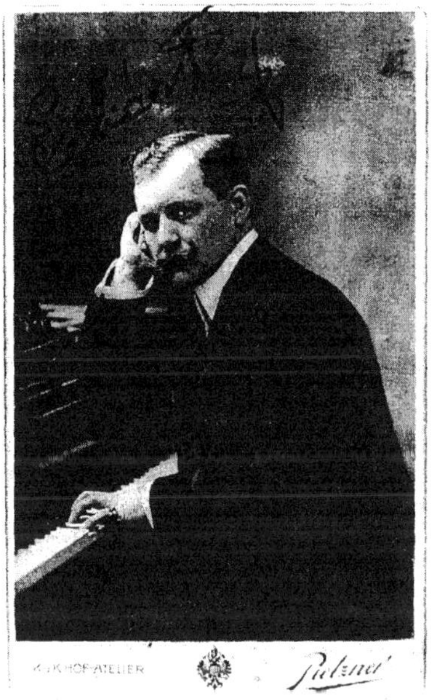

Franz Lehár

extent bound up with the attractions of the imperial capital. His first hit was with *Die lustige Witwe* (1905), in which the operetta's potential for current political comment could be fully exploited in the light of the ongoing wars in the Balkans. As later scholars were to realise, *The Merry Widow*, in its own delicately spiced manner, touched upon such highly contentious issues of the day as universal suffrage and women's rights, though of course its enduring success, over close to a century now, can be put down not to that but simply to Lehár's ingratiating, ravishing melodiousness, his syrupy yet tasteful style. His later works were proof enough that the basic clichés of the world of operetta, like those of Hollywood, can readily be stretched to fit plots transposed to the unlikeliest of settings. Whether the milieu happens to be aristocratic (*Der Graf von Luxemburg*, 1909) or a staidly portrayed working class (*Eva*, 1911), Spanish Gypsy (*Frasquita*, 1922) or Chinese (*Das Land des Lächelns*, 1930), the pieces always concern the same dreams and desires. In Lehár's world the Hungarian flavour, at its strongest in *Zigeunerliebe* (1910), is never more than just one of a multiplicity of exotic seasonings.

Not so with Emmerich (Imre) Kálmán, who was much more systematic and consistent in his efforts to bring out a sense of the Hungarian locale and lifestyle in staging his operettas. Part of the reason for that may have been that he was a student at the Budapest

Academy of Music at the same time as Bartók and Kodály. His first success, *Tatárjárás* (The Gay Hussars)—presented in Budapest in 1908 and the following year in Vienna, in a revised version, retitled *Herbstmanöver*—was a caricature of the army applicable to the whole of the Monarchy. From then on, Vienna took priority in being given the first chance to see such works as *Die Csárdásfürstin* (1915) and *Gräfin Mariza* (1924). In these, as in his many other operettas, Kálmán always took pains to include scenes, and hence music, of a Hungarian character, so that their staging across the world, from New York to Moscow, played a significant role in broadening knowledge of Hungarian Gypsy music and in reinforcing the close association of Hungarian music with 'Gypsy romanticism' that has lived on in the general consciousness in many parts of the world long after the decline of operetta.

Viktor Jacobi (1883–1921) and Albert Szirmai (1880–1967) were of the same generation as Kálmán, and good friends during his Budapest days, in the early phase of his career. Jacobi's biggest hits were *Leányvásár* (Girls' Fair, 1911) and *Sybill* (1914), whilst for Szirmai it was *Mágnás Miska* (Max the Magnate, 1916). Both left Hungary after the First World War in order to escape the country's economic slump, crossing to the United States in search of opportunities for work. Jenő Huszka (1875–1960), by contrast, stayed at home to the end of his life, which may be why he is not so well known abroad as his lesser contemporaries, though many consider *Bob herceg* (1902) to be one of the cornerstones of the Hungarian-language operetta repertoire (it was put on in the USA under the title *Prince Vagabond*). In this, as in later works, Huszka showed he was at least as receptive to the English operetta tradition of Gilbert and Sullivan as he was to the influence of his native musical heritage.

Over the course of the twentieth century many Hungarian composers were to leave the country and make a name for themselves in the field of light music and, through the new mass-entertainment medium of cinema, that of film music in particular. During the 1940s and '50s virtually every Hollywood spectacular seemed to have a musical score by Miklós Rózsa (1907–95), a three-times American Academy Oscar winner whose credits included *The Thief of Baghdad*, *The Jungle Book*, *Quo Vadis* and *Ben Hur*. The title *Double Life* that he gave to his autobiography, however, was meant as a clear hint that behind the front of a master craftsman of music designed for broad popular appeal there lurked a serious composer who produced a body of work very much in the stamp of Kodály, his former mentor. Joseph Kosma (born József Kozma, 1905–69) chose to settle in Paris in 1933, and his scores became just as integral to French films of the same era, though he was perhaps even better known as a composer of countless widely acclaimed chansons performed by the likes of Edith Piaf, Juliette Gréco and Yves Montand.

Béla Bartók: East and West Shake Hands

A Hungarian encyclopaedia that appeared in 1982 pronounced: "Béla Bartók is not only the greatest figure in Hungary's music history, he is also one of the greats of Hungarian culture generally, and beyond a shadow of doubt he embodies the most significant contribution that Hungarian culture has made to date to world civilisation." Although such superlatives are always difficult to substantiate, at least it seems reasonable to suppose that music's ability to communicate universally, above and beyond language, have enabled Bartók's oeuvre to reach foreign audiences much more readily than, say, the great works of Hungarian literature. Equally, since it is a body of artistic creative work, and so more closely linked in the public mind with the figure of its creator than would be the case with scientific or technological achievements, it is Bartók's name that has become inscribed in the general consciousness rather than that of any of the great scientists who were his contemporaries.

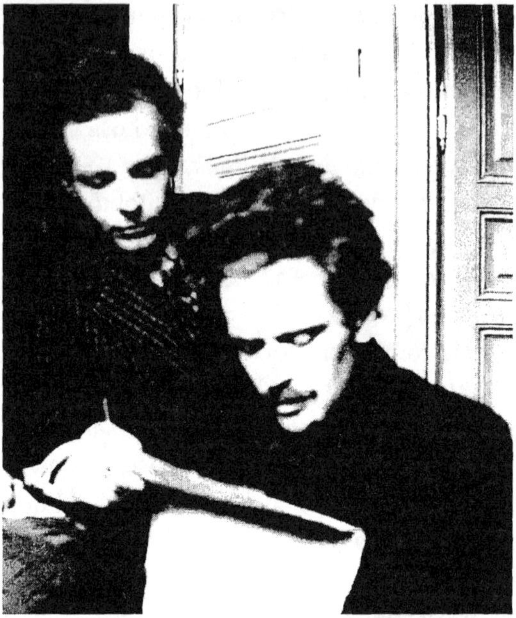

Béla Bartók and Zoltán Kodály

In his early years, Béla Bartók (1881–1945) lived still very much under the mark of the nineteenth century's dualism. He grew up in a bilingual German–Hungarian family, gaining most of his schooling in Pozsony (Bratislava, Slovakia), where his piano teacher was one of Ferenc Erkel's sons. At the Budapest Academy of Music his composition teacher was Hans Koessler, who hailed from Germany and gave him a thorough grounding in German musical style, above all the Beethoven and Brahms tradition. The young Bartók, however, had set his sights on a renewal of the national music, and in 1903 supposed that the route to this lay through Richard Strauss, then deemed an ultra-modernist. This was the genesis of the symphonic poem *Kossuth*, composed in commemoration of the leader of the 1848–49 War of Independence.

A continuance of nineteenth century music could only be an illusion, however, and Bartók soon put that idea behind him. The mainstay that enabled him to do this, one that was to sustain him for the rest of his life, was his discovery of Hungarian folk music. Almost by accident, on hearing the singing of a chambermaid, it dawned on him that Hungary boasted a musical style that had remained essentially untouched by art-music. He soon set out on a series of song-collecting trips, mostly in the villages of Transylvania, arming himself as early as 1906 with one of Edison's wax-cylinder phonographs. By then he was working with Zoltán Kodály (1882–1967), whom he had met at one of the salons that were fashionable around the turn of the century in Budapest's affluent middle-class circles—one hosted by Emma Sándor, a patroness of talented young musicians (she was later to become Kodály's first wife). That alliance in the field of folk-song collection was to lead to a close collaboration and subsequent division of labours which was all the simpler when Bartók's attention turned fairly soon, and decisively, towards the folk music of Hungary's ethnic minorities, especially that of the Slovaks (from 1907) and Romanians (from 1909). He did not do that purely for its own sake but out of a sense that a knowledge of their song materials was a prerequisite to a better understanding of what was genuinely distinctive about Hungarian folk music: this was the only way of clarifying the interactions of the songs and musical styles of communities that co-existed in close proximity with one another.

The two composers-cum-ethnomusicologists (as we would now call them) uncovered previously totally unrecognised strata of a wealth of folk music that had been preserved largely due to the sequestered lifestyles of these peasant communities; indeed, they radically altered perceptions of what the notion of folk music entails. They discerned the most ancient musical 'idioms' of the peoples of East Europe, including Hungarians, in a body of music displaying numerous 'archaic' features (pentatonality, modality, a generally isosyllabic four-line song structure, etc.) that was current in the lower strata of society, rather than in the treasury of popular nóta—the kind that Liszt, among others, had chosen to arrange—which existed on the fringes of nineteenth-century art-music. Sociologically speaking, that required them to transgress the boundaries of who was considered as constituting the nation, which up till then had been seen as an exclusive preserve of the educated and/or affluent classes. Nowadays it has come to seem almost a self-evident truth that peasant communities largely cut off from external influences would be far more conservative in perpetuating the ancient music of the Hungarians than the more mobile and intermixed communities of the towns. Music played an integral part in the whole culture of the peasant community, and its ‘unconscious’ (or, more properly, formally untrained) cultivation must have seemed like a paradisiacal state to ‘specialist musicians’ who felt themselves to be isolated in the stratified bourgeois society of the metropolis.

What relevance did this have for composition? Above all, it yielded a completely new solution to the general desideratum of how to create a national art-music. This solution was new if only on account of the ‘novelties’ that inhered in the musical characteristics of the unexpectedly revealed material, but new also in the opportunities it offered to re-think the options that could be explored by art-music, ranging from straight arrangements of the folk songs to bigger forms and works of several movements. The immediate products of the collecting work were, firstly, scholarly publications of the song melodies and texts and, second, folk-song arrangements. The volume of *Magyar népdalok* (Twenty Hungarian Folk Songs, 1906), ten each arranged by Bartók and Kodály, was a declaration of intent. The series of songs was aimed at amateur pianists, with the intention of introducing and attracting them to genuine folk songs, which is why the arrangements of these originally monophonic songs, though departing from art-music conventions in their respect for the features of the tunes, were given a European guise. The experiment, however, went largely unnoticed. It soon became clear that other approaches would have to be tried. From the *Gyermekeknek* (For Children, 1908–09) album onwards, Bartók’s didactic works, high as the aesthetic values they incorporate may be, were directed specifically at children beginning their piano studies. For the wider public he sought to present genuine compositions, whether those be arrangements of folk songs or works that did not draw on specific folk-derived materials; in any event the application of their experience of folk music was to define the approach that both he and Kodály took to composition.

If publication of the arrangements in *Twenty Hungarian Folk Songs* can be called a milestone in musical history, the new possibilities of expression realised in the many cycles of highly experimental piano pieces over the ensuing years were even more important compositionally speaking. Emblematic of these—even more so than the *Allegro barbaro* (1911), which became particularly well known on account of its title—were the *Tizennégy bagatell* (Fourteen Bagatelles, 1908), with which Bartók created a considerable stir when he had the chance to introduce them in Berlin. The next big step in terms of communicating the new ideas in a highly refined art-music form was the *String Quartet No. 1* (1908–09) which, although it also drew on the doleful tonalities of a Wagnerian harmonic world and the modally saturated progressions of Debussy's music, presented a new Hungarian style resting firmly on the recently discovered cache of folk song.

During the spring of 1910 Bartók and Kodály introduced their new style at a programmatic series of soirées in Budapest but this gained attention only within a close circle of music lovers. The new compositional endeavours were to meet with continued rebuffs from the country's official musical establishment. The most significant of these befell Bartók, whose one-act opera *A Kékszakállú herceg vára* (Duke Bluebeard's Castle, 1911) was repeatedly rejected before it was finally premiered seven years later. Its libretto, a mystery play by Béla Balázs, is a Secessionist-symbolist stylisation of the Hungarian folk ballad in verse that sustains its distinctive folk-inspired tropes throughout, though the story-line of an encounter between a man and a woman and the impossibility of mutual understanding is not at all folk-inspired but reflects typical concerns of a turn-of-the-century era still very much in the hold of

Nietzsche and Strindberg. Bartók's opera was the first work to achieve a synthesis of western European large-scale form, a declamation based on the parlando rhythms of folk song, and the dramatic style of a new harmonic world.

The commission received for the ballet *A fából faragott királyfi* (The Wooden Prince, 1916) was the first opportunity Bartók had to give orchestral (in point of fact, symphonic) expression to his new style. The 1917 Budapest Opera House season in which the work was performed represented the first occasion on which his concept of the new Hungarian music received a proper hearing; indeed, it paved the way for the house to stage *Duke Bluebeard's Castle*, the premiere of which took place in the dying days of the Austro–Hungarian Monarchy, on 24th May, 1918. By then, however, Bartók had already progressed beyond those earlier stage works and was busy with the score for *A csodálatos mandarin* (The Miraculous Mandarin, 1919) a pantomime ballet based on a scenario by Menyhért Lengyel. Like the previous two works, it too examines aspects of gender relationships, in this case the encounter of a city prostitute with a mysterious stranger, the Mandarin, who embodies the primordial force of nature and a desire stronger than death itself. In the sensitivity of his characterisation of the action taking place on the stage and his radically modern orchestral tonalities there is no question that Bartók here produced a large-scale score to vie with Stravinsky's *Rite of Spring*. The ballet remained unperformed until 1926, when its premiere was mounted in Cologne, though further performances were promptly banned by Konrad Adenauer, the city's Christian-Socialist mayor, on account of its 'immorality'. It was not to be staged in Hungary until the autumn of 1945, during the country's brief period of democratisation under the shadow of Soviet occupation.

The troubled political aftermath of the First World War, and above all the dismemberment of historical Hungary under the Trianon Peace Treaty, totally ruled out further folk-song collecting amongst the ethnic minority or mixed communities of the territories that had passed to the successor states, which prompted Bartók to restrict his ethnomusicological efforts henceforth to systematic classification of the material already in his possession. Folk song for him was thus no longer a living reality but the legacy of a now vanished Golden Age that was to be preserved in the soul, rescued for posterity, and provide, as it were, inner sustenance for the new compositions that were to be written. The orchestral *Táncszvit* (Dance Suite) of 1923 is typical, with its programmatically avowed survey of aspects of various folk musics, the 'fraternisation of peoples' implied by the Romanian, Slovakian and Arabic character of its successive movements instilling it with a supremely idealistic world-view in the midst of Central Europe's nationalistic tensions at the time, though admittedly Hungarian folk music is given a preeminent role throughout in the cavalcade of nations, with its memorial appearing in the form of the ritornello passages that link the five movements.

As a gesture of personal and social disillusionment, Bartók virtually withdrew from the domestic musical scene in the years immediately after 1920. From then on the first performances of most of his major compositions took place outside Hungary. Whilst he put ever-greater weight on emphasising the fundamental inspirational role that folk music played in his own work, he sought his standards and points of reference amongst leading European contemporaries, his name becoming increasingly bracketed with Schoenberg and Stravinsky rather than Kodály. It was, in part, precisely this orientation towards Europe that led to his change to a neoclassical style with the 1926 piano pieces (*Szonáta* (Sonata), *Szabadban* (Out of Doors)), which, even as they mercilessly piled up on the surface the new dissonances that struck many as off-putting, underneath established even closer links with the supreme creations of Europe's musical tradition, Beethoven's oeuvre in particular. What the string quartets of this era—*String Quartets No. 3*

(1927), *No. 4* (1928) and *No. 5* (1935)—and the *Piano Concertos No. 1* (1926) and *No. 2* (1931) have in common is a perfection of musical structure and a consummate distillation of style and form into something unmistakably individual. The same period gave rise to the series of 153 piano pieces that comprise *Mikrokozmosz* (Mikrokosmos, 1926–39), which, on the one hand, provides an introduction for children to twentieth-century music and, on the other, can be regarded as an anthology of Bartók's musical vocabulary. The most personal of his compositions during this period was the *Cantata profana* (1930), set to the Hungarian translation of a Romanian folk ballad text, in which the dramaturgy of a Baroque passion is fused with an intonation that lies close to nature and to the peoples of East Europe. The work, the closing line of which ("Csak tiszta forrásból"—"From purest springs alone") was to become a dictum that epitomised the ambitions of the new Hungarian music, is a declaration of definitive withdrawal on the part of an individual who is turning his back on modern society at least as much as it is a confession of community with folk culture.

The trio of works that were written during the second half of the 1930s as commissions for the Basel Chamber Orchestra and its patron-conductor, Paul Sacher—*Zene húroshangszerekre, ütőkre és celestára* (Music for Strings, Percussion and Celesta, 1936), *Szonáta két zongorára és ütőhangszerekre* (Sonata for 2 Pianos and Percussion, 1937), and *Divertimentó* (Divertimento for strings, 1939)—brought a further purification of style and encyclopaedic summation. Bartók's melodic world acquired an exquisite polish (*Hegedűverseny*, Violin Concerto, 1937–38), and he again took on board the popular art-music, including the verbunkos, that he had once rejected so vehemently (*Kontrasztok hegedűre és klarinétra*, Contrasts for violin, clarinet and piano, 1938). It was not directly personal reasons but concern and anger at the expansion of Nazi Germany and the conduct of Hungary's political leaders that prompted Bartók's decision, which shocked many, to leave the country for America following a farewell concert on 8th October 1940. The five years he spent as an émigré in New York were overshadowed by constant financial worries and depression over his ultimately terminal illness. Yet this was also the period during which he composed what were to become his most popular works, the *Concerto zenekarra* (Concerto for Orchestra, 1943) and *Piano Concerto No. 3* (1945), which showed a further relaxation of his harmonic rigor that many attribute to an influence of contemporary American music.

The immense significance that Bartók's figure and life work have had for twentieth-century Hungarian culture is almost impossible to put into words. The indelibly Hungarian roots of his music, his ideals of a Central European political identity, his European perspective and his uncompromising commitment to modernity offer such an impeccable and valid accomplishment of the linkage of a distinctly Hungarian culture to world culture as to set an example not merely to the country's musicians but to its entire intelligentsia. Many works of fine art (Dezső Korniss and Endre Bálint among others) and volumes of poetry (Lajos Kassák, Attila József, Gyula Illyés, Sándor Weöres, Ferenc Juhász) are testimony to that influence.

Zoltán Kodály, Music Teacher to the Nation

In Hungary, far more than elsewhere, it is usual to mention the name of Zoltán Kodály in the same breath as that of Bartók: Bartók and Kodály go together like Bach and Händel, or Debussy and Ravel—binary stars that, for all their differences, draw attention more to their affinities. Congruences are plentiful enough too: born within some twenty months of one another, graduating from the Budapest Academy of Music in the same year, the joint work on discovering, collecting and scientifically working up Hungarian folk music, the renewal of Hungarian music on the basis of folk music—to list just the most obvious. Any open-minded assessment cannot ignore the differences, however, and particularly not the rather sharply divergent paths their lives were to take after 1920. This was when Bartók became recognised as a leading figure in European music, a name uttered like a slogan along with those of Schoenberg and Stravinsky, whereas Kodály, despite signal successes outside Hungary, became an increasingly influential figure in Hungarian cultural life—in certain respects, even more so than Bartók himself. Outside Hungary, knowledge of this side of Kodály's activity and influence is only sketchy, specifically through the musical pedagogical method that is named after him, yet that method, which since the 1960s has spawned whole schools on the North American and Asian continents, was originally part of a much broader concept than just a new apprehension of the young child's relationship to music. In Kodály's multifaceted philosophy the central role that he allotted to singing and relative solmization, along with its favouring of choral singing, were a great deal more than pedagogical devices, for they were founded on a program he envisaged for musically educating a whole nation and, indeed, for using music as a tool to shape general education.

In the early years of the twentieth century Kodály faced up to the reality that Hungarian culture was in practice divided into three. One group, doubtless the smallest, comprised sections of the intelligentsia and middle classes interested in serious art-music, that narrow circle of concert-goers and opera lovers who demanded the best possible performances of the international repertoire and tended to deprecate Hungarian musical initiatives, which were provincial and mediocre in their eyes, compared with the standards of a Beethoven, Brahms or Wagner. A middle group was made up of the petty bourgeoisie, the provincial intelligentsia and gentry, who had no call for the products of high culture and as far as music goes were satisfied with the dance music played by Gypsy bands at local balls and other festivities, popular art-songs of the nóta-type, supposedly national in character but increasingly shoddy ditties that were turned out by the dozens, or the German-based repertoire sung by amateur male-voice choirs (*dalárdák*). At the very bottom were the mass of the populace: the illiterate rural strata, mostly peasants, who whilst having little or no contact with high culture possessed their own truly ancient folk song culture, which more affluent classes chose to ignore.

Kodály summed up the situation with a much-quoted quip: "The educated are not Hungarian enough, and the Hungarians are not educated enough." As he saw it, the solution was to elevate Hungarian national character into high culture, and to some degree he subordinated all that he did to this principle, including his own compositional activity, thereby in part renouncing the creative path that his talents suggested was open to him. Instead, he chose self-sacrificingly to put fulfilment of personal ambitions to one side and undertake a task that would bring profit to the nation as a whole. Those aspirations went beyond musical culture, aimed as they were at laying the foundations of a uniform national culture that crossed the bounds of social strata through the vehicle of music—a grandiose project that, at all events, has a touch of pathos about it. Equally, the grounds on which the mediation of a vision of a uniform nation, as Kodály conceived it, might be justified in the middle of the twentieth century, are questionable. Some decades on from his death, it is fair to say that the movement he initiated truly did make a huge contribution to advancing Hungary's musical culture, making it a resource that is accessible to all and generally enhancing its prestige; at the same time, a uniform national musical culture of the kind that Kodály envisaged remained an illusion, indeed never could have become a reality.

Kodály, unlike Bartók, had not only graduated from the Academy of Music but also had an arts degree from the University of Budapest, a qualification that was to serve both his scientific work and his teaching activity. In the years before the First World War he was less in the public eye than Bartók. His chamber music

(*String Quartets Nos. 1* and *2*, *Cello Sonata*) and songs based on folk texts or on poems by Dániel Berzsenyi, Endre Ady, Béla Balázs and others were, as yet, very personal in style, positing a sort of new musical Reform Era as their ideal. The turning point was to come in 1919 and its aftermath. Kodály became deputy director of the Academy of Music during the turbulent few months (March–July) of the Soviet Republic of Hungary and thus, for the first time, was in a position to think of implementing his educational ideas in practice, though the semi-Fascist Horthy régime that took over power in the autumn of 1919 was unimpressed and promptly slapped a ban on this reforming zeal that was to last several years.

In 1923 Kodály, along with Bartók and Dohnányi, was commissioned to compose a new piece to mark the 50th anniversary of the unification of Pest and Buda. That amounted to official rehabilitation and was also a sign that the political authorities had bowed, however reluctantly, to the dictates of the art world's own scale of values. The work that Kodály composed for this particular event was one of his masterpieces, the oratorio *Psalmus Hungaricus*. Through its biblical text, this setting of a sixteenth-century Hungarian paraphrase of Psalm 55 not only hinted at a national program rooted in the country's historical past but also carried an unequivocal message for Kodály's contemporaries, whilst the fact that King David's plaint, accusation and plea bore the inflections of Hungary's newly discovered rural music was a rallying call for the program of establishing a uniform musical culture.

Over the remaining decades of his career Kodály was also to accomplish his musical nation-building program in his own compositional output. The series of choral works for children's and adult voices on which he embarked in 1925 reveals that communal music-making and the up-and-coming generation, the children whom he wished to raise in a new world of music, had moved to the focus of his attention. Besides folk song and folk poetry, he sought to pass on the traditions of Hungarian poetry and church music, both Catholic and Protestant, to the ever growing number of choirs that were being formed, thereby underscoring both the Hungarian and the European as indissoluble twin objectives of his program. During the 1930s one school choir after another was set up to implement the Kodály principles, and these were soon bound together in a *Magyar Kórus* (Hungarian Choir) league, a movement with its own music publishing house and periodical.

Educating listeners as well as singers was also seen as important. As if to re-imagine the periods that were missing from Hungary's musical past due to its irregular development, Kodály created *Háry János* (1926), a scion of the népszínmű-type folk play now ennobled by being stocked with authentic folk songs, while *Székely fonó* (The Spinning-Room, 1932) was a one-act Singspiel that set Transylvanian folk songs on the theatrical stage. His orchestral compositions were memorials to nineteenth-century national dance music (*Galántai táncok*, Dances of Galánta, 1933) or brought the most characteristic melody of all ancient Hungarian folk songs into the concert hall in an art-music garb (*Felszállott a páva—Változatok magyar népdalra*, Variations on a Hungarian Folksong (Peacock), 1939).

The Stalinisation of the post-1948 era placed Kodály in a paradoxical situation. As a humanistic Catholic, he had a profound aversion to the Communist régime's dictatorial methods but, on the other hand, he had to admit that central control of cultural policy could greatly assist in implementing his ideas. New schools were established in which the cultivation of musical skills and instruction in solfege and Hungarian folk music received special emphasis, and even ordinary primary school education set an enhanced store on music lessons and choral singing. That enhanced role of music education has in fact survived the 1989 change in regime, so that learning music has remained one of the most readily acquired cultural appurtenances for those aspiring to middle-class status.

Hungarian Music since Bartók and Kodály: Neo-classicism, Modernism, Avant-garde, Post-modern

In looking back over Hungarian music of the twentieth century, several aspects need to be borne in mind. For all of the country's composers a primary issue was their relation to Hungarian tradition, though now, in the wake of Bartók and Kodály, that meant Hungarian folk music rather than the verbunkos. Linked to that was another ever-current question, throughout the whirlwind succession of stylistic developments that international music underwent in the course of the century, which was the stance to adopt towards foreign contemporary music: whether to adopt or reject, say, serialism, neo-classicism or later on, for that matter, chance music, repetitive (minimalist) music or neo-Romanticism. Tied in with both was specifically the relation to Bartók and Kodály, the country's two outstanding composers of the first half of the century, and the question of whether their examples were to be followed or rejected, indeed whether they could be continued at all. On top of that, they also had to contend with the new element of the attempts that successive totalitarian régimes made over four decades to bring the arts to heel under their political influence.

The oppressive burden of Bartók and Kodály fell most directly on the generation of musicians who had been born at the turn of the century. Bartók throughout his career refused to teach composition in any institutional setting, whereas Kodály, who was a professor of composition at the Academy of Music from 1908, accepted the teacher's role of preserving and building on tradition within the narrow circle of trained musicians before gradually assuming the role of, in effect, educator to the entire Hungarian nation. Most of the progressive young composers in Hungary of the 1920s received their schooling from him, learning the technique of arranging folk songs, to handle the modal harmonic world derived from folk songs, the distinctive stress patterns of the Hungarian language, the models of thinking in strict counterpoint supplied by Renaissance music, and most important of all, the link between national and classicising ideals.

In the 1930s it seemed that a new Hungarian school of composition was emerging whose members, for all their common features, displayed abundant divergences. One of the most open and receptive of these figures was Pál Kadosa (1903–83), whose chamber works and pedagogical pieces for piano, until the Nazi takeover in 1933, were published by the Schott publishing house in Germany. Ferenc Szabó (1902–69), who was also intensely interested in the literary and artistic avant-garde, had only a brief role in Hungarian musical life before he was forced to emigrate to Moscow in 1931 due to his involvement in the clandestine Communist movement. The first compositions of Sándor Veress (1907–92) were highly praised as the most promising of the whole decade, but his career in Hungary was to end when he emigrated to Switzerland in 1949. Even if none of these composers acquired a broad base of popularity, by the late 1930s it had at least become accepted by those interested in contemporary music that Hungary's contribution to modern music was worthily represented by a style that was traditional in form but displayed a plethora of pentatonic melodic lines and neotonal or modal harmonies. This was a folk-inflected classicising style in which any experimentation with new composing techniques was scarcely conceivable (Arnold Schoenberg's twelve-tone system, for example, elicited no reaction at all in Hungary).

The musical fruits of the brief political transitional period between 1945 and 1949 were determined by the sense of liberation that followed the end of the Second World War. The prevailing tone was of what has since come to be labelled the 'divertimento' style, the roots of which lay in the neo-classical sinfoniettas and concertinos of the 1930s, its harmonic optimism reflecting an important feature of the prevailing mood of those times, with the folk-based idiom allowing such stuff to be produced by the ream. After the 1948–49 Communist take-over of power, however, the blitheness of the divertimento style was deemed unfit to serve the socialist artistic ideals that were envisioned. Its place was taken by symphonies and imposing oratorios that were Romantic in conception and radiated revolutionary spirit, strings of choral works built on folk-song arrangements, mass songs designed to satisfy political desiderata, and cantatas dictated by the cult of personality's needs for festive events. Nevertheless, there were some composers, mainly of the older generation, who refused to submit their creative talents to such stylistic dictates. László Lajtha (1890–1963), for instance, continued to compose his French-tinged symphonies, but even his loyal pupil János Ferencsik had a hard job getting them a public performance.

The bleakest chapter of musical policy was the Bartók 'affair', kicked off around 1950 with the division of the composer's works into two categories in order to engineer a tricky process of partial stigmatisation. The anti-modernist cultural tsars had obvious difficulties tolerating such an uncompromisingly innovative artist as Bartók; at the same time, his name and significance, both from a musical and a national point of view, were just too considerable simply to anathematise his life work in its entirety. The only option was to split the Bartókian oeuvre in two: the works of what was then called the 'bourgeois-expressionist' Bartók (such as *The Miraculous Mandarin* and most of his chamber music) were proscribed for years, whereas the works that were directly based on folk music or reflected the 'folk classicism' of his late period were official blessed and widely propagated.

Composing underwent a profound change during the 1960s. This renewal was sparked in part by the reopening of possibilities for composers to re-orient themselves internationally, in part by the emergence of a new generation that had been born around 1930. They had several decades of catching up to do, since healthy ties between the musical life of Hungary and western Europe had effectively been sundered back in the mid-1930s; yet in under a decade the country's composers took advantage of the window of opportunity to undergo a crash-course in everything it had missed out on, mastering the Second Viennese School's twelve-note system even as it began to explore the new composing techniques it encountered in the works of Witold Lutosławski and Krzysztof Penderecki.

The opening moves were made by older composers. In his *Hat zenekari darab* (Six Pieces for Orchestra, 1959), Endre Szervánszky (1911–77) took a huge step forward: emulating Anton Webern, he bid farewell to his well-practised reflexes in Hungarian-style melody, symphonic structure and traditional instrumentation. András Szőllősy (1921) appeared in the late 1960s as a completely sovereign individual, unaligned to any trend, whose vigorous, impelling music stands out chiefly on account of its philosophical and even occasionally tragic tenor. An important sign of the advent of a new generation was marked by the premieres of two operas: *C'est la guerre* (1962) by Emil Petrovics (1930) and *Vérnász* (Blood Wedding, 1964) by Sándor Szokolay (1931). In these, a new dramaturgy and new musical resources were introduced that departed markedly from former tendencies of Hungarian music theatre. The composers did not aim to imitate traditional forms but sought to achieve an eclectic and effective mixture of divergent styles, combining folklorism, jazz and structural tendencies. For a long time, the young generation that emerged in this decade seemed a fairly consistent group, like the Polish School of the same period. Dodecaphony and serialism were the most conspicuous principles in the organising of musical structures, and most composers tended to use a kind of free atonality. In contrast to the 1950s, chamber music was cultivated more and more, serving as a favourite field for experimentation; political control was rapidly relaxed and the general atmosphere became much more permissive to those experiments. In the search for new methods of composing, Hungarian tradition became less important. Most composers of the new generation who began their careers in the '60s are still active, although they now show clearly distinct personal styles, which have been modified in various respects by internal changes and external influences.

For some younger composers of the 1960s a major task was to establish a balance between fresh influences from contemporary Western music and Hungarian tradition. Zsolt Durkó (1934–97) began his career mainly as an instrumental composer, although he later developed an individual style of vocal writing in his opera *Mózes* (Moses, 1977) and in his oratorios (e.g. *Halotti beszéd*, Burial Prayer, 1972). Durkó's most familiar idiom remained instrumental music, in which he crystallised a clear personal style characterised by small gestures and a high sensitivity to tonal colour. In the works of Sándor Balassa (1935) a strong melodic impulse and a lyric personality can be observed. Among his compositions, prominent vocal works, such as the oratorio *Requiem Kassák Lajosért* (Requiem for Lajos Kassák, 1969) or the opera *Az ajtón kívül* (Beyond the Threshold, 1977), show his interest in verbal expression. In the '60s and '70s Attila Bozay (1939–99) went further in experimenting than his contemporaries, though he tried to combine those experiments with traditional elements, as in his own playing on zither and recorder, which explored potentialities that went far beyond their use in folk music. From the late 1980s on Balassa and Bozay turned to neo-Romanticism, with an emphasis on nineteenth-century Hungarian music.

The most significant event of the '70s was the formation of the New Music Studio in 1970. This grouping co-opted itself as a genuine workshop, an alliance of composers and performers, to the service of a radically new paradigm in Hungarian music. Their chief model was John Cage, who had rethought the compositional process to admit chance operations that completely eliminated subjective creative intentions. The radicalism of the New Music

Studio's out-and-out avant-garde concept horrified many, with older colleagues and Marxist arts administrators seeing it as an attack and accusing the group of both abandoning the national tradition and serving 'pernicious' ideologies (e.g. Zen Buddhism). Part of the reason for the uproar may have been the apparent loss of steam on the part of the '60s generation of composers, and the consequent growing readiness of the younger, most receptive segment of the public to throng to the New Music Studio. This prompted greater vigilance on the part of the guardians of Hungary's cultural life, and indeed the Ministry of the Interior. A remote member of the New Music Studio was Péter Eötvös (1944), who had left Hungary back in 1966 to study at Cologne. Eötvös soon became one of the best-known international conductors of contemporary music, while in recent years he has won wide recognition throughout Europe as a composer in his own right, for instance, with his opera *Three Sisters* (1998). During the 1980s, in part as a result of speedier responses to contemporary developments outside Hungary but also undoubtedly in an attempt to win back the audiences who were deserting the concert halls, various neo-Romantic trends were embarked upon, which held an appeal for many 'thirtysomethings' as well as the young.

György Ligeti: a Curious Hungarian in Europe

György Ligeti (1923) has lived barely more than a decade in Hungary. Born in Transylvania, his homeland had by then been annexed to Romania as a result of the border changes imposed by the Treaty of Trianon. Ligeti grew up in Kolozsvár (Cluj), and this was where he started composing, considerably under the influence of his early encounter with the works of Bartók. After the Second World War he entered the Budapest Academy of Music, later going on to teach music theory there. This was the period when the Stalinist régime under local party leader Mátyás Rákosi mouthed

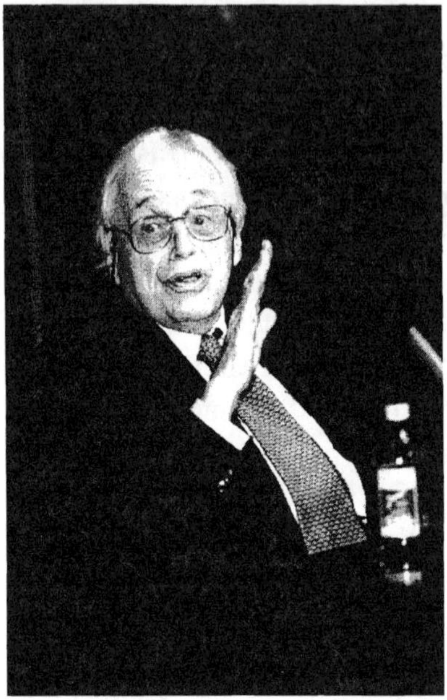

György Ligeti

the slogans of 'easy intelligibility' and 'socialist realism' as justifications for its rampantly hostile stance towards artistic freedom and experimentation in any shape or form. When the 1956 national uprising against Communism failed, Ligeti was among the 200,000 Hungarians who chose to flee the country.

Ligeti had shown his early promise to Hungarian audiences primarily through folksong-based choral works, but his truly important compositions, the series of piano pieces entitled *Musica ricercata* (1951) and the string quartet *Métamorphoses nocturnes* (1953–55), were not allowed to be performed in public, even though they show Ligeti striving more searchingly and intensively than any of his contemporaries to work out and take further the lessons that could be drawn from Bartók's art. Yet this was also the very period when Ligeti was nurturing the idea of a completely new kind of music that he did not yet feel quite ready to realise, a music which would be expressed in note clusters, rather than melody, harmony and rhythm, with its events consisting of the slow movement and transformation of clusters.

Once in Western Europe, Ligeti was quick to realise that he had several decades of missed musical developments to catch up on. Bartók was by then considered passé; the Darmstadt group of leading young composers—notably Luigi Nono, Pierre Boulez and Karlheinz Stockhausen—were taking the serialism of Anton Webern as their lodestar. Over the period 1957–59 Ligeti worked in

Westdeutscher Rundfunk's Studio for Electronic Music in Cologne, acquiring technical skills in the medium but meanwhile still preoccupied by his earlier musical goals. The result was *Atmosphères*, first performed in 1961, a hugely influential composition in which every single member of the orchestra plays an individual solo line, yet what emerges from this micropolyphonic 'chaos' is a body of sound of startlingly original and overwhelming power. What was considered amazingly novel about this way of treating the orchestral palette was that, at a stroke, it swept aside the exquisitely detailed, finicky scorings of the 1950s and replaced them with a monumental yet appealing and more readily comprehensible style.

Through this, Ligeti rapidly established his name during the 1960s, at least in Germany (and of course the use of passages from *Atmosphères* to give the feel of 'space music' in the soundtrack for Stanley Kubrick's 1968 science-fiction epic *2001: A Space Odyssey* had a big hand in promoting his work further afield). His next major works, however, seemed to mark a volte-face with the musical principles of *Atmosphères*. The paired speech pieces *Aventures* and *Nouvelles Aventures* (1963–65) set three voices and a small chamber orchestra on the concert platform, though it might be just as apposite to say on the stage, since the singers enact a series of diverse communicational situations in a phonetic language of the composer's invention. The protagonists run through a succession of situations, often switching abruptly, without transition, from love duet to quarrel, overbearing pride to pedantic sermonising, light-hearted chatter to bleak solitude—an absurd music theatre, and indeed a linking of irony with tragedy, that undoubtedly owed much to the plays of Eugène Ionesco. (As a political émigré, Ligeti could not get his music performed in Hungary until a breakthrough was achieved in 1972, when the Puppet Theatre in Budapest mounted a brilliant adaptation of *Aventures* in which the human characters were replaced by various objects: a hat, umbrella, coat, etc.)

In his next few compositions—*Lontano* (1967), *Chamber Concerto* for 13 instrumentalists (1970) and *San Francisco Polyphony* (1974)—Ligeti went further down the various new avenues that he had opened up. Some of these (e.g. *Continuum*, his 1968 piece for virtuoso harpsichord) quickly assumed an emblematic status, becoming frequently cited in music education, especially in Germany, as exemplary models of contemporary music. Having its world premiere in Stockholm in 1978, the opera *Le Grand Macabre*, the plot and text of which are freely adapted from a 1934 satirical Doomsday play by the Belgian writer Michel de Ghelderode, represents a further shift in style with its evocation, in a sort of post-modern collage, as a garbage heap of hopelessly moulded debris, of centuries old operatic clichés. With the *Horn Trio* (1982) and the ongoing series of *Piano Études*—6 in Book I (1985), 8 in Book II (1988–93) and a further three up to 2001—he seems to have returned to a more transparent, gentler style that is influenced as much by the percussion music of Central Africa as by the complex rhythmic devices of late medieval motets. Even these more recent works, however, have not retreated from the expression of his philosophically and indeed scientifically inspired inquiries into the very foundations of music-making.

Ligeti's musical thinking was being decisively shaped even during the 1960s by new mathematical findings in systems theory and catastrophe theory, well before those areas were widely known about. His compositions are made up of sequences of dynamic systems in which the musical process can be described as a transition from one system state to another. The allowable changes in the musical variables of pitch, volume and tempo within a given system are strictly limited; the composer can only determine the direction and speed of change of these attributes and co-ordinate the different, simultaneously proceeding movement types. There are two basic movement types: one a continuous transition of indistinctly defined, gently fluctuating sheets of sound, or 'clouds', to use Ligeti's preferred term; the other, abrupt shifts of sharply defined honeycombs of discontinuously articulated tone clusters, or 'clocks' in Ligeti's language. (These terms, which Ligeti used in the title of *Clocks and Clouds* (1973), are metaphors introduced by the philosopher Karl Popper in the mid-'60s to denote determinacy and indeterminacy in physical systems.) Ligeti's musical systems are always dynamic, the various internal changes to which they are subject being characterised by the different extents and speeds of their spread within the system: 1) *fluctuations* are purely localised interventions that have a colouring role; 2) *ramifications* slowly transform the entire system as they push ever further into the foreground; and 3) *catastrophes* alter the whole system in an instant. In effect, Ligeti sets the notion of process in the place that themes and forms occupy in earlier musics, and it is perhaps this that makes Ligeti's music relatively easy to understand as the listener can orient surprisingly well in this systematised world even at first hearing.

Ligeti's music does more than just illustrate mathematical theories, however; it is a totality of sounds, instruments and tonal relationships that certainly does not shy away from the traditional ingredients of music. For all the extraordinary precision that Ligeti put into their composition, their abstract structures are nevertheless embodied in a musical reality. Though the aforementioned processes undoubtedly lie behind the unity of Ligeti's music, it would be wrong to forget that these processes are manifested in musical forms. Indeed, these forms possess definite expressive characters that have nothing to do with the music's function as a process but deliberately embrace external musical features that yield just as readily to analysis as the underlying constructive methods of composition.

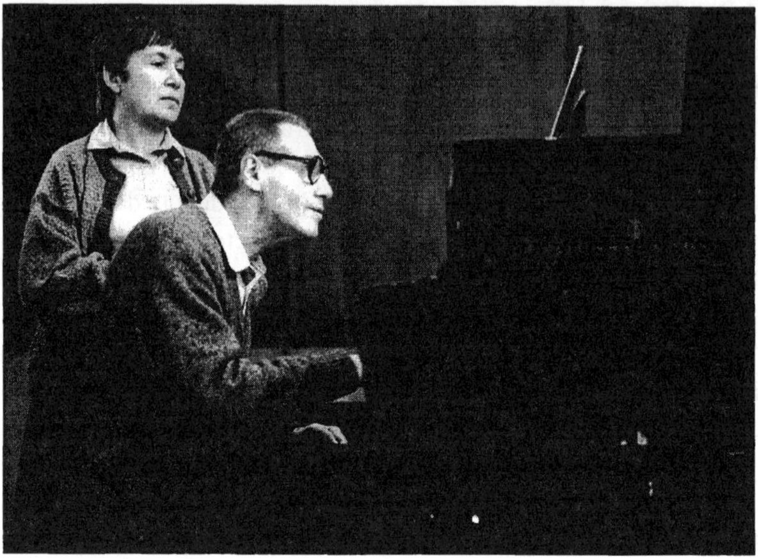

György Kurtág and Márta Kurtág

György Kurtág: Revival of the Personal Tone

Like Ligeti, György Kurtág (1926) arrived in Budapest from Transylvania in the autumn of 1945 in order to study at the Academy of Music. Graduating both in piano and composition, he was awarded a major state prize for the *Brácsaverseny* (Viola Concerto, 1954), his graduation piece, which received its first performance in 1955. This work too is a typical example of the direct influence that Bartók had on composers of that period, though Kurtág was soon to veer away in a completely different direction.

He spent a while during 1957–58 in Paris, not with any intention of defecting but to look around at what was happening. For him, as for most young composers then, the starting point was the guiding light of Anton Webern, whose works, since his money did not run to purchasing the published scores, he copied out by hand. Kurtág attended composition classes given by the likes of Darius

Milhaud and Olivier Messiaen at the Paris Conservatoire, but the profoundest impact on him was made by the conversations he had under psychoanalysis with Marianne Stein, an art psychologist. She gave him the advice: "As a composer, start off by connecting two notes." This minimal program was to open the way to Kurtág's later career. From then on, in his own words, his artistic credo was "to achieve a brevity in which every moment is essential and important. To discard everything superfluous, in other words, convey the greatest expression and content with the fewest notes." This is the key to understanding his endeavour to make the fullest possible use of the potential locked within the simplest music devices. At the centre of his oeuvre lies a paradoxical attempt to intensify the impact of a work to the greatest degree whilst keeping the composer's presence to the minimum.

Kurtág indeed composed no more than a handful of pieces during the five years after his return from Paris, and without exception those compositions consist of just a few very lapidary movements. Amounting to barely more than half an hour of music altogether, these were enough to secure Kurtág an exceptional position in Hungary's music life, bestowing on him a nimbus of mystery and respect. Over the next five years, between 1963 and 1968, he worked on *Bornemisza Péter mondásai* (The Sayings of Péter Bornemisza). Here Kurtág displays sentences taken from the sixteenth-century Lutheran preacher's book, *Ördögi kísértetek* (The Temptations of the Devil, 1576), through the filter of a kind of Freudian psychology. The work's four parts, entitled *Vallomás* (Confession), *Bűn* (Sin), *Halál* (Death) and *Tavasz* (Spring), appear to have subsumed Kurtág's entire world view at that time. It is in *The Sayings of Péter Bornemisza* that Kurtág introduced the topos of "*Flower we are*"—a simple sentence that encapsulates the defencelessness of the individual confronting his transience and also the beauty of existence that is worth preserving—that he was subsequently to return to many more times.

In 1973 Kurtág started on the first of the *Játékok* (Plays and Games) series of piano studies, which is still being added to right up to the present day and currently comprises more than 300 pieces. *Plays and Games* was originally conceived as an informal piano tutor, somewhat along the lines of Bartók's *Mikrokosmos*, with the aim of enabling young children who have just taken up the piano to approach the instrument without feeling any constraints. Rather than having to learn strictly prescribed rhythms, awkward fingerings or carefully designed articulations, the child gets training in the mastery of the whole keyboard, the joy of movement, and sensitivity to musical structures. At the same time, *Plays and Games* is a composition workshop-cum-diary, with frequent citations from the scores of predecessors or, for that matter, contemporaries as well as musical homage to dead friends. It has also become a personal vehicle inasmuch as the always-reticent composer only ever himself performs in public—and only ever when accompanied by his wife Márta—in concerts of selected pieces from *Plays and Games*. With their intense, intimate, almost ceremonial feeling, these concerts have had a substantial impact, over recent years, in building up Kurtág's international reception and the public's image of his personality.

During the 1970s and '80s Kurtág produced a series of vocal compositions in which he achieved renewed and effective expression of abiding human values through the texts that he set. *A boldogult R. V. Truszova üzenetei* (Messages of the Late R. V. Troussova), a series of 21 Russian-language songs for soprano voice accompanied by chamber ensemble, was completed in 1980. The little tesserae of the concise movements go to make up a mosaic that juxtaposes the happy and horrible, the memorably splendid and distressingly unforgettable moments of a former love. The world premiere of *Troussova* in Paris, in 1981, now drew an international audience under the ineluctable spell of Kurtág's music. In *Jelenetek egy regényből* (Scenes from a Novel, 1982)—at once a continuation and a counterpart to *Troussova*—the nimbus of cosmic loneliness, tormented sensuality and boundless disillusionment that surrounds the heroine of *Troussova* here shrinks to a closed world. The singer is accompanied by a 'Gypsy band' of fiddle, cimbalom and double bass, with the opulent colours of *Troussova* giving way to an alternation of dreamily lyrical, indistinctly floating and, more than once, grotesque musical images. *Kafka-Fragmente* (1986), Kurtág's next major composition forms a sort of 'Central European encyclopaedia', as Václav Havel aptly put it. Excerpts from Franz Kafka's diary and letters are performed by a soprano and violin only, thereby reducing the musical setting of the selected texts to a bare minimum of vocal and instrumental expression. The collision of opposites, the correspondences that chime together at a distance, the breathtaking excitement and irreverent playfulness carry the listener along step by step until, in the very last fragment, the impulse that is in all of Kurtág's music to search for beauty between tradition and novelty culminates in an apotheosis of melody and rhythm, voice and violin tone.

Right up to the end of the 1980s Kurtág's output still lacked an orchestral composition. Admittedly, most instruments of the orchestra have a part in the solistically orchestrated accompanying ensemble of *...quasi una fantasia...* (1988) written for piano and instrumental groups; however those instruments do not sit in a single bloc but at various points in the concert hall, not just on the platform but also above and behind the audience. Although the resulting spatial acoustic is not in itself anything novel in contemporary music, its sophistication and the magical way in which it is employed here are utterly individual. The same technique was used by Kurtág in his setting of the text of Samuel Beckett's last completed work in *What is the Word?*. Beckett's thoughts on the futility of words and expression accurately capture Kurtág's own decades-long struggle with his creative being. This is a summation of the idea of absence, which occupies a central place in the whole

Kurtág oeuvre: that despairing, unbearable, inhuman absence which makes composition at once impossible and yet absolutely necessary.

By the 1990s Kurtág had indisputably become one of the best-known figures in European contemporary music. In his courses on chamber music he takes his students through what they all sense to be the school of a great European performing tradition and profound analytical appreciation. The dominant ethical message of his teaching and composing alike is the painful actuality of today's obligation for strength of personal presence and for winning back human emotions.

Dispersed and Holding Together: Hungarian Performers in and outside Hungary

A majority of the internationally acclaimed Hungarian-born performers in the twentieth century received tuition from one of three great teachers—in violin from Jenő Hubay (1858–1937), in piano from Ernő Dohnányi (1877–1960), and in chamber music from Leó Weiner (1885–1960)—all three of whom played a role as composers in Hungary's musical life, though none would be described as one of the innovative composers in twentieth-century music. Hubay chiefly copied French models (Massenet) in stubbornly clinging on to late Romanticism and the last redoubts of nineteenth-century Hungarian folk-style music, whereas both Dohnányi, who started as a Brahms epigone above all, and Weiner, with his impressionist gifts, were influenced to varying degrees, from the 1920s onwards, by Bartók and Kodály's folkloristic agenda. Here, though, it is primarily their activity as teachers that comes under scrutiny, and the host of performers who passed through their hands to go on and establish world-wide reputations for themselves.

Jenő Hubay was the founder of the Hungarian school of violin-playing, but mention should first be made of two older violinists of Hungarian birth. First there is József Joachim (1831–1907), who held master classes for violinists in Hanover and Berlin over many years but wins a place in music history, in part, for restoring Bach's solo violin works to the concert repertoire but more particularly for his close collaboration with Brahms (he was the dedicatee of the latter's *Violin Concerto*). The second was Leopold Auer (1845–1930), a Joachim student himself and then, from 1868 to 1917, professor at the St. Petersburg Conservatory, and the original dedicatee of violin concertos by both Tchaikovsky and Glazunov. Following the Russian Revolution, Auer emigrated to the USA, teaching at the Juilliard School in New York and later at the Curtis Institute in Philadelphia. Virtually all members of the so-called Russian school who similarly took up residence in the US, including Mischa Elman, Nathan Milstein, Efrem Zimbalist and Jascha Heifetz, were Auer students, leaving their imprint on twentieth-century performance culture with a Romantic, intensely emotional style of matchless virtuosity.

Hubay too travelled around Europe before returning to Hungary in 1886. He too had studied with Joachim but in Paris and Brussels he had also come into contact with another of the major European traditions, that of the French school, whilst having Liszt himself as a frequent partner in sonata recitals gave him direct access to the very pinnacle of nineteenth-century standards of performance. The stream of outstandingly talented musicians who emerged from the Hubay school of violin-playing from the early years of the twentieth century onwards had an inspirational effect not only on audiences but also on many European composers. Sibelius, for example, dedicated his violin concerto to Ferenc Vecsey (1893–1935), whilst Bartók wrote both his Sonatas for Violin and Piano for Jelly d'Arányi (1893–1966), who was also the inspiration for Ravel's *Tzigane*. József Szigeti (1892–1973) was one of Bartók's favourite chamber music partners, and Zoltán Székely (1903–2001) also made a name as a performer of Bartók's works, giving the first performance of the *Violin Concerto* (1939), before becoming the leader of Hungarian String Quartet during the 1950s and '60s. Sándor Végh (1912–97), who was hugely respected as a transmitter of the musical traditions of Central Europe both as a chamber musician then, in the last twenty years of his life, as a conductor, was likewise a Hubay pupil.

David Popper (1843–1913) arrived in Budapest from Prague at the same time as Hubay was starting off his teaching career and performed a similar role for cellists. Although none of his immediate students could claim to have been a major international success, his successors include the likes of János Starker (1924), long resident in the United States, and Miklós Perényi (1948), who started his career as an infant prodigy.

That Ferenc Liszt should be seen as the true ancestor of the Hungarian piano school is more than just a symbolic affiliation but can be attested by a chain of teacher–pupil links, for the retinue of several dozen students who accompanied Liszt as he made his annual procession around the Weimar–Rome–Budapest triangle during his later years always included a number of Hungarians. Ernő Dohnányi, for example, who returned to Hungary in 1916, having spent a decade based in Berlin, tapped into the Liszt legacy directly from his teachers, Stefan Thomán and Eugen d'Albert, who had both been Liszt pupils. Dohnányi's renewed presence was of crucial importance, since he became a central figure in Hungarian musical life over the next three decades not only as a solo artist and conductor but also as a piano teacher whose pupils included Andor Földes (1913–92), Annie Fischer (1914–95), Géza Anda (1921–76), and György Cziffra (1921–94). A fresh clutch of outstanding pianists was to emerge from the Budapest Academy of Music under Pál Kadosa in the early 1970s, notably Zoltán Kocsis (1952), who in recent years has increasingly devoted himself to conducting, together with Dezső Ránki (1951) and András Schiff (1953).

Leó Weiner taught a course in chamber music at the Academy of Music for nearly four decades, between 1923 and 1960, his students including an impressive list of Hungary's up-and-coming string players, pianists, and conductors; indeed, very few musicians of any significance who trained in Hungary during the middle part of the century were not indebted to Weiner for the deepest insights that they received during their student years. Of singular importance is the series of distinguished string quartets—the Léner, Hungarian and Végh String Quartets in particular—who over many decades stood in a class of their own for chamber music lovers around the world.

A continuous chain of succession does not exist amongst Hungarian-born conductors in the way that it does for string players or pianists. The first truly major talent was Hans Richter (1843–1916), who from 1875 to 1898 was chief conductor of the Vienna Philharmonic Orchestra, certainly the most prestigious ensemble in the Austro–Hungarian Monarchy and, arguably, anywhere in Europe. As such, he gave the world premieres of many of Brahms' works, though equally he was one of the supreme Wagnerian conductors of his era, having worked as the master's assistant in Lucerne and been entrusted with directing the first complete *Ring* cycle at the Bayreuth Festival in 1876. Having established himself in London's musical life with the Richter Concerts that he initiated in the 1880s, he took up residence in England in 1897 to become director of Hallé Orchestra in Manchester (until 1911) and also conducted the first performances of many of Wagner's works at Covent Garden in London. Richter's visits to Hungary may have been infrequent but they involved more than guest appearances as he was instrumental in securing the first English engagements for several of the country's outstanding young performers, including Dohnányi (1902) and Bartók (1904).

From then on, the orchestras of Western Europe were to offer the route to world fame for Hungarian conductors. Some, indeed, such as Arthur Nikisch (1855–1922) or George Szell (1897–1970), were really only Hungarians in the sense that this was their birthplace, for they received most of their training in Vienna. Although Nikisch was director of the Budapest Opera House between 1893 and 1895, he continued his career as conductor of the Leipzig Gewandhaus (1895–1904) and the London Symphony Orchestra (1904–14). Szell worked in several opera houses in Germany until 1933 then in 1939 emigrated to the United States, where his name is indelibly associated with the Cleveland Orchestra (1946–70).

The major American orchestras owe a lot to Hungarian-born and trained conductors for their spectacular blossoming after World War II. Just about every prominent ensemble on the East Coast has had one Hungarian musician or another at its helm for a considerable spell to leave an enduring stamp on the orchestra's performing style. Fritz Reiner (1888–1963) is particularly associated with the Chicago Symphony Orchestra (1953–63), though prior to that he had already been chief conductor at Cincinnati (1922–29), New York (1929–31), Pittsburgh (1938–48) and at New York's Metropolitan Opera House (1948–53). Eugene (Jenő) Ormándy (1899–1985) first went to the United States in 1921 to further his career as a solo violinist but had to make do with rear-desk orchestral posts before switching to conducting, which brought an invitation to conduct the Minneapolis Symphony Orchestra for five seasons from 1931. He went on to share the conductorship of the Philadelphia Orchestra with Stokowski and then become its sole artistic director from 1938 to 1973. Antal Doráti (1906–88) had become resident in the USA in 1941, and after playing a big hand in reviving the Dallas Symphony Orchestra (1945–49), he succeeded Dimitri Mitropoulos at Minneapolis (1949–60), going on to direct the Washington National Symphony (1971–78) and the Detroit Symphony Orchestra (from 1977) and meanwhile making regular appearances as chief conductor for several European orchestras (Stockholm Philharmonic, London Symphonic). That was also true of his younger colleague, Sir Georg Solti (1912–97), who displayed his matchless sense of drama first at the State Opera in Munich then, from 1961 to 1971, as music director at Covent Garden, London. He brought the same dramatic expressiveness to his work in the concert hall, most notably during his long spell as Reiner's successor at the head of the Chicago Symphony Orchestra (1969–91). Diverse as they were in personality, there is a certain shared set of tastes that links these conductors, most obviously a respect for the values of European classical music and a love of precision of orchestral ensemble that is nowadays felt to be a typically American attribute. Hungarian music is greatly indebted to them, particularly for their efforts in promoting the music of Bartók and Kodály in the United States. It would be remiss, however, not to mention in this connection the work of János Ferencsik (1907–84) throughout this same post-war era as the leading domestic conductor in Hungary itself.

As with the conductors, the list of Hungarian-born singers who achieved major international acclaim is more a matter of sporadic successes than a continuous tradition. Among singers born around the turn of the century the basses Dezső Ernster (1898–1981) and Mihály Székely (1901–63), baritone Sándor Svéd (1906–79) and tenor Kálmán Pataky (1896–1964) established their names in the opera houses of Europe and the United States, whilst the main star of the international opera stage and recording studio today, particularly in the Wagnerian and Straussian repertoire, is Éva Marton (1943).

Note

1. Ferenc Liszt, *A czigányokról és a czigány zenéről Magyarországon* (About Gypsies and Gypsy Music in Hungary) (Pest, 1861), p. 279.

Gyöngyi Földesi-Szabó

SPORT INSPIRED BY THE IRON CURTAIN

"Unite Mind, Morals and Might"

Top-level sport divides society, having its ardent proponents, but also its detached observers and even fierce opponents. Given that, it may seem paradoxical that international sporting successes are nevertheless still widely recognised, followed by many who are not even committed fans. The reason for this, beyond any national pride, is probably that in sports it is so clear who is the winner and who the loser, who is first and who last.

There are few fields of human endeavour in which the international standard of individual and team achievements at the national level can be so clearly pinpointed as in sport. There are ways of roughly assessing the importance of scientific achievements through their long-term impacts on theory and practice, it is true, but comparisons of the merits of similar works of literature, fine art, film, photography, music, architecture, etc. are rendered extremely problematical by the lack of generally accepted precise, objective criteria. Here value-judgements usually have a consensual basis and often provoke heated debate within specialist circles or in the general public. In any rank-ordering of this sort—where it is at all possible and customary, as with film festivals, music prizes or architectural competitions, for example—subjective criteria inevitably play a role. Though that element cannot be completely eliminated from certain branches of sport either, the ranking of performances according to measurable criteria, laid down in explicit rules, is the very essence of sport.

Even in the ancient world and during the Middle Ages sports had a competitive side. What differentiates modern sports, as institutionalised in the nineteenth century, from their precursors, apart from such aspects as their secularisation, enforcement of principles of fair competition, and specialisation, is the drive to quantify results and to utilise them to establish records. It is no accident that many people regard the stop-watch as a symbol of the emergence of modern sport. Allen Guttmann asserts, in *From Ritual to Record*, that the reason why we do not know how fast runners performed in the track events of ancient Greece is probably not so much that the Greeks did not have accurate chronometers as that, on the contrary, they did not have accurate chronometers because they did not consider measuring how fast their champions ran as of any importance. They were not interested in the absolute performances in a given race, nor did they consider it important to know if the new 'Olympic' champion had thrown the discus further than the champion four years earlier, or the winner at a game held elsewhere that same year. What characterises today's sport, by contrast, is that every aspect of competitive performance is measured as exhaustively and accurately as possible, turned into statistics and placed on record. It is that recording of performances which makes it possible to compare and rank sportsmen and women who are separated from one another in space and time. And the measurement and recording of sporting performances in turn inevitably entail rationalisation and bureaucratisation.[1]

Hungary was one of the countries where modern sports spread and became popular early on. The first national sports clubs and federations were already established during the closing decade and a half of the nineteenth century: the Federation of Hungarian Sports Clubs in 1885, the Hungarian Rowing Federation in 1893, the Hungarian Cycling Federation in 1894, the Hungarian National Rifle Federation, and in 1897 the Hungarian Athletics Federation and the National Swimming Association. These were followed by national associations for football (soccer) (1901), tennis (1907), university and college sports (1907), pistol shooting (1907), chess

(1911) and skiing and alpinism (1913). From the outset, the standards of the national championships initiated during this period, and also recognised by these federations, signalled a desire to be internationally competitive.

Hungarian sports diplomats even assisted at the birth of international sports bodies and frameworks for competition, their sound grasp and preparedness doing much to give Hungarian sport a good name from the start. The one who rose to greatest prominence was Ferenc Kemény (1860–1944), headmaster of Eger's secondary modern school, who, on the recommendation of Hungarian pacifist circles and freemasons, was personally approached by Baron Pierre de Coubertin, the father of modern Olympianism, to join the movement to revive the ancient Olympic Games. Kemény was one of the dozen founding members of the International Olympic Committee, formed in 1894, and subsequently became one of de Coubertin's closest associates. Such was the esteem in which he was held that for a while, when idea of staging the first Olympiad at Athens (1896) looked to be in jeopardy, Budapest was seriously considered as an alternative.

Mental resources have been a mainspring, directly or indirectly, of Hungarian sporting success throughout the twentieth century. There are innumerable cases it has been the cognitive component of sport—the advanced qualifications, expertise and innovativeness of physical education instructors and trainers—that was the linchpin (and guarantee) for the discovery, unfolding and progress of competitive talent. For the most part, the coaches remained in the background, especially during the decades of hard-line socialism, when, in order to safeguard the secrets of Hungarian sporting achievements, they were given no opportunity to take their skills abroad.

Would-be PE instructors and trainers were considerably helped in the job of spotting and developing talent by the comprehensive, up-to-date training they themselves received. One product of the 1921 'Sports Act' (Law LIII), sponsored by Count Kunó Klebelsberg as the Minister of Education, was the establishment of a College of Physical Education (CPE) on Alkotás utca, Buda, in 1925, which offered college- and later (from 1975) university-level tuition to the best European and, in some fields, world standards. The motto that the College set for itself, and held itself to, was 'Unite mind, morals and might', and a good number of the key achievers and shapers of Hungary's sporting successes have had links (as undergraduates, doctoral candidates, teachers, coaches, etc.) with what might rightly be termed this 'intellectual bastion' of sport in the country.

Investment in sport has paid off handsomely in other areas too. Dr. Ferenc Mező (1885–1961), a provincial grammar-school Classics teacher who had played sport for fun as a student and was an enthusiast of the Olympic movement, won a gold medal at a literary 'Olympics' held in association with the 9th Olympiad at Amsterdam in 1928 for his book *Az olimpiai játékok története* (The History of the Olympic Games). He was unable to attend the event in person but still became a celebrated person overnight. The story goes that a telegraph was sent to him from Vienna addressed simply to 'the Olympian of the Mind, Nagykanizsa'. Mező's fame spread well beyond his town and even country, not just on account of that gold medal but also his subsequent activities as a sports writer and organiser. A fair number of the 44 books on sports subjects and close to 300 essays on sports history that he wrote were translated, with editions in 10 languages published in 14 countries. From 1948 until his death, he served as Hungary's representative on the IOC and still had the energy to deliver talks on Greek culture and sport in any forum, from university down to small village school.

Fulfilment of the CPE motto of uniting 'mind, morals and might' had a big part in Hungarian sport diplomacy's post-war international successes. Hungarian experts had held various offices in international sports bodies before World War II, but their functions in those bodies were given a good deal more weight and their numbers boosted substantially during the socialist era (1945–89). Domestic governing bodies in sport became very keen to nominate and secure the election of Hungarian candidates for such posts. Their aim was, firstly, to tilt decisions taken by a given international body in favour of the interests of Hungary and the Soviet bloc as a whole and, secondly, for the prestige such representation was deemed to bestow. For that reason, and somewhat against the government's stated policy, in the decades of soft dictatorship that followed the 1956 Revolution more emphasis was placed on talent and suitability for the job, rather than political commitment, as primary selection criteria for possible candidates. Often enough, keeping one's nose politically clean was accepted as good enough. It may be taken as a tribute to Hungary's sports diplomats that during the time when the world order was polarised into two opposing camps, it was they, out of the whole 'socialist bloc', who were the figures most welcome—as generally open-minded representatives of its 'happiest barrack'—to international sports bodies, as experts in various branches of sport, sports administrators, sports scientists, or sports reporters.

That stature was won by being able to approach sport from a theoretical as well as a purely practical perspective. The career of Dr. Ferenc Hepp (1909–80) provides an excellent example of such linkage of theory with practice. After wide-ranging studies in physical education, philosophy and psychology, he finally settled on basketball and had no mean career as a player whilst still pursuing scientific research, publishing significant works on the specifics of sports research and the detection of movement in sport. His *Sports Dictionary in Seven Languages* (1962) and *Olympic Sports Dictionary in Seven Languages* (1972) are still in print to this day. A high level of education, refinement and language skills have likewise been decisive factors in the careers of Drs. Pál Schmitt (1942) and Tamás Aján (1939) as sport diplomats, both members of the IOC and Schmitt a vice-president. He started as a fencer, winning Olympic gold medals twice with the Hungarian epée team, at Mexico City (1968) and Munich (1972). Like most Hungarian fencers at that time, he graduated from university whilst pursuing his sport then went on to obtain a doctorate at the College of Physical Education. Since the 1990s, his career as a sport diplomat has been combined with a regular diplomatic posting, first as the Republic of Hungary's ambassador to Spain (1993–97) and since 1999 as ambassador to Switzerland. Tamás Aján, unusually for a gymnast, graduated in aesthetics at the Budapest University before going on to gain a diploma and later a doctorate in physical education from the CPE. Since the late '60s, he has been a prominent figure in international weightlifting.

One point that emerges clearly from the following sections is that Hungary's sportswomen have performed with equally high distinction as their male peers over the past century. The reader might reasonably ask, then, why it is that no women's names are to be found in the list of prominent sports diplomats. The simple truth is that sport, 'macho' world that it is, has for decades singularly failed to include women, for no fault of their own, in its international governing counsels. That male dominance has begun to slip a bit in recent years, but only after a tough struggle. Hungary must be man enough to admit that, unlike some other sports-loving countries (e.g. Scandinavia, Canada and the USA), it has not been amongst the pioneers of equal opportunities in sport. Nor is that any wonder, because the division of power within Hungarian sports themselves is not fair either: leaving aside the testimony of day-to-day experience, there is hard research to demonstrate that in every area of sport Hungarian women are underrepresented in positions associated with any power and influence.

Despite that, there are some Hungarian women who have served in important positions in international sports bodies. Ilona

Novák (1925), a one-time world record-holder and 1952 Olympic gold medallist in the 4x 100 meter relay, was a member of the Board of the European Swimming League between 1974–94, whilst Dr. Judit Temes (1930), another member of the same relay team and also a European champion and world record-holder, served on the League's Technical Committee between 1976 and 1994. Sarolta Monspart (1944), a world champion in orienteering, excellent skier and marathon runner, despite having to overcome the handicap of severe disease, has been active in the International Orienteering Federation since 1982, and its vice-president since 1994. *A futás csodálatos világa* (The Marvellous World of Running, 1990) is a book that Monspart has co-authored with Dr. Márta Miltényi, a former national-team athlete and now a fine sports anatomist, whose textbooks have been primers for generations of students, and who in 1990–96 was on the Board of the International Marathon Medical Directors Association before becoming its deputy chair (1996–98) and the chair (1998–2000). Gyöngyi Szabó Földesi served as deputy chair of the Inter-national Committee for the Sociology of Sport between 1984 and 1991, then as a member of its Executive Committee (1991–94). Since 1998, she has been the vice-president for scientific affairs of Trim and Fitness International Sport for All Association, a world organisation for recreational sports.

Selection Difficulties

Some nations have written their names into the global history of sports through their victories in just one or two branches of sport—Norway in skiing, India in field hockey, the Far East in martial arts, the Middle East in wrestling, South America in football, to give a few examples. Hungary's sporting achievements have been much broader-based. The country has taken part in European and World Championships, as well as the Olympic Games, from the start, and very successfully at that. The state provided the funds needed to send a team to the first Olympiad in 1896, because, up to the end of the First World War, this active engagement and desire for success in sport was a rare opportunity to demonstrate the country's independence, for on the international sporting field it was treated as a sovereign state and not as a part of the Austro–Hungarian Empire. It achieved its first successes in swimming, athletics, fencing and wrestling.

The defeated nations of World War I were not invited to the 7th Olympiad at Antwerp in 1920 and a number of other international sports events. Hungary's desire to regain acceptance and prove its worth to the outside world thus became strong motivations to turn in outstanding sports performances, and that goes some way to explaining the tradition of success that was laid or reinforced during this period in fencing, wrestling, athletics, water polo, shooting, boxing and gymnastics.

As a result, Hungary had already won notice as a sporting nation in the eyes of the world during the first half century of modern organised sport. In the decades that followed the change in political regime in 1948–49, the country—in its 'popular democratic' then 'socialist' manifestation—it was again political considerations that prompted a quest for 'great power' status in sport. One of socialist Hungary's secrets of sporting success was that, as in the other post-war East European states, the intrinsic value of élite sport was vastly overrated; crack athletes were exploited for both domestic and foreign political purposes by bestowing diplomatic missions on them as people's ambassadors. Their international sporting successes were supposed initially to betoken the viability and later the superiority of, in turn, 'popular democracy' and 'socialist' society. A key reason for supporting them was that they helped Hungary emerge from its isolation behind the iron curtain, and in the era of 'two-bloc' ideological struggle they were intended as symbols of the country's commitment to peace. Sporting performances were regarded as a good vehicle for gaining international prestige—a curious form of compensation for the absence of any genuine social, political or economic achievements. Thus, most traditional sports were sponsored to the hilt, even some branded as reactionary, as long as they served the 'higher' goal of greater international acceptance and recognition. The compensatory function also endowed élite sports and athletes with a particular domestic political role, with competitive sport being treated as a surrogate for the national consciousness and identity that were suppressed in the name of internationalism and as an alternative channel for those who were opposed to party policy to get ahead at a time when careers were tied to compulsory party membership.

Such was the craving for international success that even during the depths of the Stalinist cult of personality, at a time when denunciation of certain fields and intellectual trends in the arts and sciences was widespread, the party leadership was willing to countenance even politically suspect branches of sport. During the sixties and seventies, when the guiding policy in science and cultural life was crystallised in the so-called 'Three Ts' approach, which labelled work as either officially 'touted', merely 'tolerated', or strictly 'taboo', in sport it was the officially 'touted' category that dominated. To be more specific, the main criteria for receiving enhanced state aid, whether a sport was old or new, was conspicuous success in the international arena. Sports administrators sought deliberately to boost sports in which fewer countries were active and so competition was less intense. The list of top sports that had enduring runs of success included not just those with no mean past record, such as football, kayaking, canoeing, weightlifting, table tennis, chess, and modern pentathlon, but also the likes of tennis, basketball, handball, volleyball, judo, and even ice-skating.

Hungarian sport has enjoyed many triumphs that, for all the difficulties of comparing them directly, are alike in being world-class in their own field. There is no space here to mention all the athletes individually, but one highly significant representative group ought to be particularly highlighted, and that is the Olympians of the post-war years. The choice falls on them because for Hungarian sport, even though not wanting for international recognition before then, this was its golden era, and it was the country's Olympic competitors who were the embodiments of the greater part of that success. The 42 medals (16 gold, 10 silver, and 16 bronze) won at the Helsinki Games in 1952 were a peak, but the eleven golds won at both the 1988 Olympiad in Seoul and the 1992 Games in Barcelona, and the 10 golds each in 1948 (London), 1964 (Tokyo) and 1968 (Mexico City) count as stunningly good performances by any standard. The sports surveyed below are those that the world has already acknowledged as areas of distinctive Hungarian achievement, and the sportsmen and women are those who, over and above the glory they have gained for their nation, have distinguished themselves by their exceptional personal qualities and integrity. Both for practical reasons and on principle, those sportsmen and women of Hungarian birth or descent who achieved honours competing as citizens of foreign countries, rather than in Hungarian colours—gratifying though the reflected glory may be—have not been included here.

From the 'Human Dolphin' to Ágnes Kovács

Alfréd Hajós (1878–1955), nicknamed the 'human dolphin', was Hungary's first Olympic champion. On a single April day at Athens in 1896, Hajós, whose father had drowned in the Danube after being caught up in a whirlpool, won two events, the 100 meters and 1,200 meters freestyle, in the freezing water of the Saronic Gulf. "I did not feel the water's appalling, numbing chilliness, all that hovered before my eyes was the finishing line, which I wanted to reach with every last ounce of energy of my mind and body," he wrote in his memoirs.[2] So cold was the water that many of the competitors in the 1,200-meter event abandoned the race, but Hajós took such a huge lead that victory was assured and the Hungarian flag was hoisted even before he reached the finishing line. He was so numbed on completing the course that he no longer had the strength to get out of the water, so he was hauled out by Greek sailors, whose admiration for his feat was such that they celebrated him almost as if he had been their own countryman. The Hungarian flag, incidentally, had been brought by Hajós himself, having had it thrust in his hand at the railway station in Budapest, just before his departure, by a patriotic fan who feared that the Greeks might not have a red, white and green tricolour.

An all-rounder, Hajós also excelled in other sports, including gymnastics, athletics and football; in the latter he was member of the very first national squad in 1902. But he also had major achievements in his subsequent professional career as an architect, being the designer for the indoor National Competitive Swimming Pool on Margaret Island, Budapest (1930), the Újpest football stadium, the Aranybika Hotel in Debrecen, and the Bank Palace in Szabadka (Subotica, Croatia). At the 'intellectual Olympics' held in association with the Paris Games in 1924, he and Dezső Laubner jointly won a silver medal for architecture with their plan for a sports stadium, though no gold medal was awarded. A better 'first' champion than Alfréd Hajós could not be found if one tried: his two swimming golds, and the silver medal for architecture set no ordinary standards for posterity.

A worthy successor to Hajós in the pool, if not outside, was his close contemporary, Zoltán Halmay (1881–1956). Another all-round sportsman, he was an excellent ice skater and ice hockey player, a national champion in speed-skating, and also earned a place in the national football team. In swimming he won not only 14 Hungarian national champion's medals but British, Austrian and German championships too with his 'Hungarian style', the torso held almost still and relying on powerful arm strokes. A triple world record-holder, he collected two Olympic gold medals at

St. Louis (1904), a total of four silvers at Paris (1900) and London (1908), and the one bronze (Paris), making him a swimmer whose achievements gained universal respect. It was grossly unfair that when the Hungarian team failed to win the 4 x 200 meters relay on his account, that single failure was a bigger talking-point in Hungary for years afterwards than of all his victories put together.

The following decades were not unproductive. Admittedly, at the Amsterdam Olympiad in 1928, in a year when Johnny Weismüller was king of the swimming pool, István Bárány was only able to grab second place behind the American swimming prodigy in the 100 meters freestyle. At the 1936 Berlin Olympics, though, Ferenc Csík managed to win over the same distance against three Japanese and two American rivals. A World and European record-holder, Csík later also served his country as a military physician, being killed on active duty in a bomb attack just a few weeks before World War II ended.

Finding its place again in the aftermath of those hostilities, Hungarian swimming enjoyed a fresh era of success during the early fifties. Four of the country's 16 gold medals at the 15th Olympiad in Helsinki were brought home by swimmers: Valéria Gyenge won the 400 meters freestyle, Éva Székely the 200 meters breaststroke, Katalin Szőke the 100 meters freestyle, and the squad of Ilona Novák, Judit Temes, Éva Novák and Szőke (with Mária Littomeritzky) the 4 x 100 meters freestyle relay. The 400 meters freestyle and 200 meters breaststroke victories were in fact double triumphs, because Novák was second-placed in both events. This was an exceptionally talented women's squad, setting world records and also reigning as European champions, and there is little doubt they would have been world champions too, had those events been held for swimming at the time. Quite apart from their sporting prowess, the team was highly educated, with its ranks including a later physician, a pharmacist, a coach, a PE instructor, and a photographer.

Though all have been held in high regard ever since, Éva Székely (1927) in particular won the hearts of sports lovers, earning herself the nickname 'Madame Butterfly'. The title of her first book, *Sírni csak a győztesnek szabad* (Crying's Only for Winners, 1992), became a catchword, and the willpower, perseverance and honesty that it attests to an exemplary model. As a Jewish girl, she had to live through the black years of state-sponsored anti-Semitism when the Arrow Cross Party, Hungary's home-grown Nazis, came to power in October 1944, but she gained the will to survive and fight from an inner conviction that one day the Hungarian national anthem would be played at the Olympics in her honour. The Socialist exaltation of sport, whatever its contradictions, gave an opportunity to fulfil that decade-long wish, but then illness intervened. Even then she did not give up. Securing a fourth place at the first post-war Olympiad, in London (1948), represented failure in her own eyes, but she still had enough determination to give it another go. Four years later, at the Helsinki Games, Éva at last took first place on the winner's stand and heard the Hungarian anthem played for her. The true significance of that moment is best described in her words: "The way I see it, sport, the Olympic Games, saved my life. Having overcome all those many, many difficulties, when I finally took my place on top of the winner's stand, I knew that this gold medal would be a determining factor throughout the rest of my life... In this uncertain, ever-changing world, in my life that Olympic gold is like a fixed star in the firmament. Its light might change according to the circumstances, but its place will always remain the same and it has given me miraculous strength during the periods of struggle in life. Whenever I felt I could take no more, or wanted to give up the hopeless fight, time and time again that is what I clung on to."[3]

During the 1950s Székely was branded a 'class-alien' because her father had run a wholesale business, and although, thanks to her sporting achievements, that did not have the often dire consequences that were visited on less prominent athletes who carried the same stigma, it was frequently impressed on her that this privileged position might be withdrawn at any time. Despite that, she was able to maintain her moral integrity, only allowing her actions to be guided by her own principles and not selfish interest. After her Olympic victory, she was called upon to read out on the radio a statement which contained the then obligatory wording: "My thanks to Mátyás Rákosi, the great leader of the Hungarian Communist Party," and she had the guts to refuse to do so.

Éva Székely had every cause to hang on to that Olympic gold. For all the 36 Hungarian national records and even world records that she set, the 33 individual national championships and even a World Collegiate Championship that she won, the follow-up Olympic golds that were her own measure of achievement were not to be: history and politics interfered once again. The 1956 Melbourne Olympiad was staged just weeks after the Hungarian Revolution. Worried for her baby and family back home, as well as her country, 'Madame Butterfly' was unable to give swimming her full concentration. In the final of the 200 metres breaststroke, "the wall would not come", or rather it came a touch too late: she finished 'only' second. In 1960 Éva did not get the chance to participate at the Rome Olympics. In the spring of 1957, she and her husband, Dezső Gyarmati, himself an outstanding water polo player, with their little daughter, stayed on in the USA for a year without official permission. On their return home, being now deemed unreliable in the eyes of the Ministry of the Interior officials who issued exit visas for foreign trips, retaliation took the usual form for those times of not allowing the couple, top athletes or not, to travel together to the West. As the water polo team looked the more promising bet, it was Dezső Gyarmati who was allowed out.

The story does not end there, as the couple raised their daughter to be a swimmer too, cheering her on from the poolside during races, hand in hand. To Andrea Gyarmati (1954) it seemed entirely natural that she would be a competitive swimmer, and her mother became her coach. Andrea was already competing at the age of nine, won a European champion's medal as a teenager and later as an adult. She set a world as well as European and national records. In 1968, she took fifth place at the Mexico City Olympics, then in 1972 won silver and bronze at Munich and a bronze at the first World Swimming Championships in Belgrade. For years she was voted 'Sportswoman of the Year'—a sporting career of unbroken success by most people's standards. Traditions, ambitions and expectations within the Gyarmati–Székely family, however, which by then included Andrea's husband, kayaker Mihály Hesz, a holder of Olympic gold (Mexico City) and silver (Tokyo), were different. Andrea's Olympic second and the third placings in 1972, when there had been realistic chances of a gold, were a major blow in their eyes. That frustration was communicated so vividly that, for a while, half the country felt they too had been let down. A younger generation is more grateful: paediatrician Andrea Gyarmati is ranked by them as an outstanding athlete, who could win crying and lose with a sporting smile. In the end, Andrea herself cannot have been too disillusioned by sport, for otherwise it would be hard to explain how her son, Máté Hesz, has become a top sportsman too as one of the up-and-coming talents in Hungarian water polo.

The great swimming successes of the fifties were all founded on top-class coaching. Éva Székely, for one, always spoke about her master, Imre Sárosi, in the sincerest terms of esteem. For a long time, the coaches somehow remained in the shadows, until the eighties ushered in a new era of the great coaching 'magicians'—a bit along the lines of what happens in the theatre world when directors take the actors' place in the spotlight.

Tamás Széchy (1931), the legendarily successful 'director' now referred to as the 'pope' of Hungarian swimming, originally began in athletics, only making the switch when the country's sports administrators set up the Central Sports School in the mid-sixties as a new type of institution to bring on the next generation. Széchy became one of its architects. Though Soviet Russian and East German models were looked at, CSS offered a more independent and much more humane program. Széchy was initially the head coach there, but in time, owing to a matchless string of successes, he moved on to act as head coach for top-notch sports clubs and the national swimming team then captain of the National Swimming Federation. He has worked out special training routines that have attracted interest from other coaches around the world. Though the details have remained closely guarded professional secrets, one thing is for sure: during a period when the competition for sporting glory became really tight and, as is now common knowledge, East Germany's coaches and, indeed, sports physicians certainly resorted to drugs for enhancement of performances, and the Chinese demonstrably likewise, whilst several other nations were strongly suspected to have done so, Széchy's methods for getting his protégés to the top were clean.

Széchy brought on several generations of swimmers, from the children of the fifties to those of the seventies, but one thing they have in common is that they all set world or European records, became Olympic or World champions, but also had the education to follow careers outside sport. The first of those big successes was that of Sándor Wladár (1963), now a veterinarian, who won the 200 meters backstroke at the 1980 Moscow Olympiad, when he was just 17, and later went on to gain a silver medal at the World and gold at the European Championships, thus proving that his Olympic victory could not simply be dismissed as due to the boycott by the USA, West Germany and other Western states.

The most recent Olympic champions to have come out of the Széchy school were Attila Czene (1974) and Norbert Rózsa (1972). Rózsa had already taken a silver in 1992 at Barcelona, in both the 100 and 200 meters breaststroke, golds at the World Championships in 100 meters (1990 and 1994) and 200 meters (1994) as well as gold, silver and bronze at the European Championships, but the peak of his career was his Olympic gold in 200 meters breaststroke at Atlanta in 1996. In both his events he also won the US National Championships and broke a number of records. Australia granted him a permanent residence permit, as an 'outstanding talent', but he has since returned to live in Hungary. Czene likewise became an Olympic champion in 1996, in his case in the 200 meters individual medley, adding that to a collection of silver and bronze medals from World and European Championships and the Barcelona Olympiad. By the time of the 27th Olympiad in Sydney, Czene was a university undergraduate but still picked up points for taking fourth place.

Spanning the period between Wladár and Rózsa and Czene are a string of other giant figures in swimming who are similarly testimony to Széchy's magic. András Hargitay (1956) enjoyed huge popularity as a swimmer then a sports manager (and more recently as a veterinarian), particularly through his World Championship gold medals in 1973 (400 meters individual medley) and 1975 (200 and 400 meters individual medley), but he also collected several golds from the European Championships and a bronze at the 1972 Olympics, setting a number of world and European records and, indeed, national records for events such as 800 and 1,500 meters freestyle that he normally did not race. Hargitay's exact contemporary, Zoltán Verrasztó (1956), now a surgeon, added to Széchy's string of successes principally through a World Championship title he took in 200 meters backstroke (1975), but a European gold and Olympic silver and bronze medals are hardly less lustrous. He too set world and European records and around two dozen national ones as well. Verrasztó's younger sister, Gabriella, likewise a Széchy protégé, should not be overlooked either. She swam to a number of national records in the latter half of the seventies, and was several times voted the woman 'Swimmer of the Year'.

József Szabó—'Joe', as he was popularly known—delivered everything a coach could expect of his charge: in the 200 meters breaststroke he was successively World (1986), European (1987) and Olympic (1988) champion as well as seventeen times Hungarian champion. He ended his competitive career early but did not turn his back on the sport as he now runs his own swimming school.

One factor that may have helped prompt Szabó's early retirement was that his own achievements were cast somewhat in the shade by the phenomenal success of Tamás Darnyi (1967), another Széchy prodigy. Darnyi's exceptional abilities came to light at an early age, gaining him admission into the Széchy school. His sports career was later threatened when his doctors advised him to stop swimming because of an eye problem, but he kept on consulting specialists until he found one willing to give the green light. To outside observers it must have seemed that for him swimming took priority over his eyesight. Fortunately, he can still see today, despite the incredible amount he must have swum during the period when he captured four Olympic golds (200 and 400 meters individual medley in both 1988 and 1992), four World Championship golds (the same distances in 1986 and 1991), eight European Championship golds, and fifty-seven national titles. With world and European records also to his credit, Darnyi is Hungary's most successful male swimmer of all times. American David Wharton, the former world record holder, tried to regain his title from Darnyi several times at different world events, and fellow Americans Eric Namesnic and Gregory Burgess later also made fruitless attempts to wrest the crown in the 400 and 200 m individual medley, respectively. Overblown sobriquets are not exactly uncommon in sports, but the style 'king' or 'emperor' is not lightly bandied about; nevertheless 'King Thomas' is what Darnyi was called in the pool. Nowadays a swimming school carries his name.

A royal epithet was likewise handed out in the eighties to Krisztina Egerszegi (1974), the greatest protégé of another master

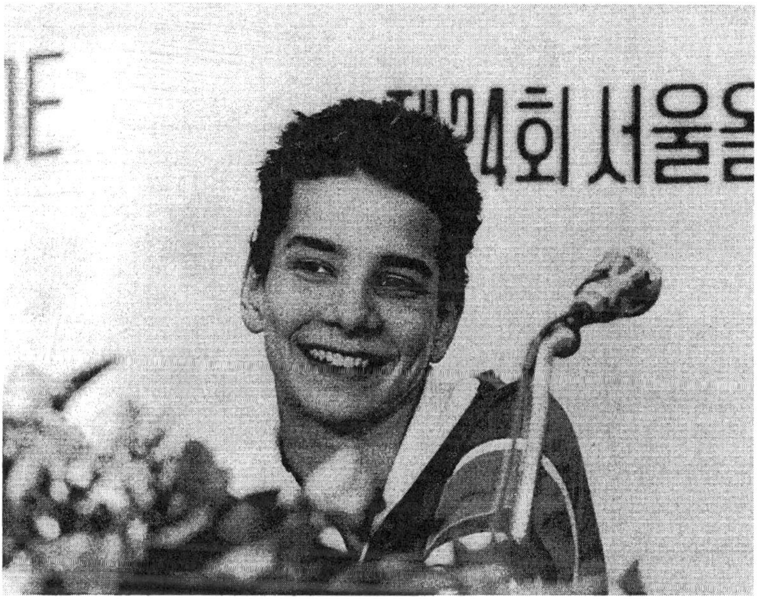

Krisztina Egerszegi

coach, László Kiss (1940). Yet even when she became a star 'Queen Krisztina', as Hungarian newspaper headlines dubbed her, kept the other nickname of the nation's 'Little Mouse' that was bestowed on her when she unexpectedly won gold at the Seoul Olympics at the age of only 14. It was not simply her results that made her a favourite, not just back home but with swimming fans around the world, for even as the most successful Hungarian sportswoman, with a fantastic international sports career, she remained a straightforward, smiling young lady. Egerszegi's main events were the 100 and 200 meters backstroke, but Kiss was shrewd enough to push her into doing the individual medley too, which alone yielded a respectable haul of medals. She won altogether five Olympic golds (200 meters backstroke in 1988, 100 and 200 meters backstroke and 400 meters individual medley in 1992, and 200 meters backstroke in 1996)—the only woman swimmer ever to do so—as well as two World Championship golds (for 100 and 200 meters backstroke), nine European Championship golds, and eighty-two Hungarian titles. Only superlatives fit her: the most medals, the most championships, the best Hungarian woman swimmer, perennial 'Sportswoman of the Year' in her own country and, in 1992, even the sports journalists' 'European Sportswoman of the Year', with Mónika Szeles coming in second to her (those positions were reversed in the voting for 'World Sportswoman of the Year'). She ranks for swimming's 'Hall of Fame' not only in Hungary but also in the USA. If one were to calculate the sheer distance covered in training and races over her career, that total would also most likely be up with the very highest, for already at the age of 14 she was swimming that many kilometres a day. Egerszegi's talents extended beyond sports as from the late eighties, in a Hungarian market that was still nominally 'socialist' but already changing, she spearheaded a successful campaign to promote a new pharmaceutical product. Ever since then she has been a media star, still 'Queen Krisztina' and the 'Little Mouse' in her own country.

Over and above her superb physical and psychological endowments and the prodigious amount of training work she put in, Krisztina Egerszegi also, crucially, had the superbly tailored competition pacing techniques of her coach, László Kiss, to thank for her success. That same approach also gave a solid point of support for another talent who exploded into women's swimming by taking an Olympic bronze for 200 meters breaststroke in 1996, at the age of 15 years. For a while, Ágnes Kovács's (1981) medal was considered to be a fluke, but her subsequent career showed that it was no such thing. The peak, for herself and her coach alike, was the gold she won at the Sydney Olympiad. Even to outsiders it was perfectly obvious that László Kiss had worked out the race tactics (and Ágnes followed them) down to a split-second. Leaving the finishing spurt in a 200-meter final to the last 10–15 meters is a highly risky business, but if someone is capable of carrying it off, that can give a huge advantage over more defensively swimming rivals.

Still in Deep Water: Water Polo, Kayaking and Canoeing

After the Hungarian water polo team's victories in the 1952 and 1956 Olympics, Dezső Gyarmati (1927) failed to win a third gold in 1960 but did so four years later, at the Tokyo Games. That gold in Tokyo was the fifth the country had won in the history of the Olympics, and it was followed by a sixth at Montreal in 1976. Nor has there been any shortage of other successes for water polo over the past seven decades, what with medal placings at Olympic and World Championships, European golds and other placings. The series started in 1928 with a silver at the Amsterdam Olympiad, then in 1932, at Los Angeles, the team was led to victory by the enthusiasm, unselfishness and, not least, innovativeness of the legendary Béla Komjádi (1892–1933) ('Uncle Komi', as he was known to his countrymen), whose name is now perpetuated by the Komjádi National Competition Pool, on the Buda embankment facing the Hajós Pool on Margaret Island. At the Berlin Olympics in 1936, captain László Beleznay was able to convince his team that they could emulate their predecessors and win the final.

The coach and manager of the 'golden team' at Helsinki and Melbourne was Béla Rajki (1909–2000), an internationally respected expert whose books on water polo and competitive swimming were translated in English, German, Russian, Estonian, Polish, Bulgarian and Japanese. Under him, the team won, in addition to those two Olympic golds, the European Championships and an Olympic bronze place. He was a true manager and superlative coach; able to devise new and appropriate tactics for every match they played. Hungarian polo players are not generally known for brute force, more for inventiveness and humour, but they came through one of the roughest matches in the history of the game, with blood literally flowing in the water, when they beat the USSR squad at the Melbourne Olympics, just a few weeks after the Soviet Union had ruthlessly suppressed the Hungarian Revolution.

As most players served in the national squad for years, new generations of replacements were brought on seamlessly. A number of the members of Károly Laky's (1912–2000) golden team in Tokyo carried on to be part of the success of the subsequent decade. The international prestige of Hungarian polo players was further en-hanced during the seventies, under Dezső Gyarmati's captaincy, when they won the first World Championship (1973) and took the silver medal at the second (1975), became European champions (1974 and 1978) and then Olympic champion (in 1976), rounding the decade off with a World Cup victory and a bronze at the Moscow Olympics. All that was done in spectacular, elegant style and in a milieu that expected academic study as well as training, with squad members qualifying as doctors, lawyers, coaches and managers. Hungarian water polo has kept its place in the top international tier during the two decades since then. That means that outstanding players and coaches are much sought after by foreign club teams, and now that the way is free there has been a substantial haemorrhaging of talent, making it ever harder to forge a truly effective, incisively winning national squad. For all that, a great team has once more emerged with the squad that won at the 2000 Olympics in Toronto.

In recent decades a steady stream of gold medals has also been secured in kayaking and canoeing. Comparatively short though the history of these sports may be in Hungary, it has already racked up an impressively long list of outstanding achievements. Before World War II, it was rowing that had been popular as a water sport; canoeing and kayaking were only taken up seriously during the fifties, when Hungary's sports administrators saw the potential in promoting a sport that awarded plenty of medals but in which competition was relatively thin. Youngsters took to it from the beginning and, once the basic facilities had been provided, one success followed another in short order. Here a brief sketch will be given of a few of the milestones and exceptional figures in these sports.

The first person to gain international recognition was István Grenache (1926–71), a fanatical advocate of the sports within Hungary and umpire at numerous world competitions who won election onto the governing board of the International Canoeing Federation. The calculated nurturing of the sports yielded its first fruits already in 1954, when Hungarian kayakers and canoeists brought home six gold medals from the World Championships in Magon, France. The real breakthrough in public acceptance, however, only came when the duo of János Urányi (1924–64) and László Fábián (1936) took an Olympic gold at the Melbourne Games in 1956. Journalists dubbed them the 'odd couple', not in any derogatory sense but because of the age gap of 12 years between the two, Urányi having been one of the very few who started canoeing before it became a mass sport whereas Fábián was one of the new recruits. The combination of experienced old hand and fresh young blood meshed superbly. Despite being chalk and cheese as characters, they reached the top because they were able to set those differences to one side and work together for victory with grit and determination, keeping an eye out for and adjusting to one another. Both continued to work for their sport after retiring from competition: Fábián as a coach, Urányi as manager of a workshop that built competition boats.

Boatmaking and repairs also became the hobby of another outstanding canoeist, János Parti (1932–99), a champion at the World (1956) and Olympic (1960) as well as European (1957, 1959, 1961) levels. Though a qualified English translator, gaining a diploma at the Foreign Language College, before later gradating in law from Budapest University, he devoted his energies to his sport. His Olympic gold medal earned him the special regard of fans and colleagues within the sport on two scores: firstly, for having had the mental toughness to pick himself up, after coming in second at successive Olympics, in 1952 and 1956, to have a third try at going for gold; and secondly, despite the fact that his original successes had come in the long-distance races, when the 10,000 meters event was taken off the Olympic program for 1956 he still managed to make the switch to shorter distances. He subsequently served his sport as a manager and captain of the national team and as a technical innovator, earning no little acclaim outside the country in both spheres. Under his stewardship, Hungarian kayaking and canoeing climbed to world stature, with the country's paddlers gaining ten Olympic gold medals between 1968 and 2000, four of them at the last Games.

Given the variety of events that kayaking and canoeing offer, everyone can find a discipline that suits himself or herself, so that the unhealthy rivalries that are seen within national squads in many other sports are less common here. That is possibly why Hungary's kayakers and canoeists still act more like a big extended family, even in these individualistic times, providing a background of emotional (and financial) security that many find absolutely indispensable in the search for world fame.

The women of the kayak and canoe 'tribe' have not disgraced the family either, having equally been a force to be reckoned with internationally for decades, with a welter of world, Olympic and other titles to their credit. The undisputed No. 1 amongst them, with some two dozen world and Olympic medals to her name, almost half of them gold and split equally between those for individual events and those for pairs and fours, is Rita Kőbán (1965). Her main events were the kayak individual and fours over 500 meters, but she won silver at the World Championships even over 5,000 meters. The greatest triumph of her career, after a gold in the fours won at the Barcelona Games, together with Kinga Czigány (1972), Éva Dónusz (1967) and Erika Mészáros (1966), was first place in the individual 500 meters at Atlanta in 1996. Her opponents around the world would 'swot up' on Rita, studying her kayak, her technique and her tactics. Whilst such aspects can be 'pinched' from others, Rita Kőbán has that something extra which cannot be learnt:

she is not only clever, hard-working and talented but also obliging by nature, devoted to her sport, her fans and her country. After her Atlanta victory, she said in an interview: "...I swear by everything I hold dear that over the final meters I was no longer conscious, and all that took me across the finishing line was the knowledge, in every fibre of my body, that a whole country was expecting this medal of me and I couldn't let that country down."[4] It is precisely on that account that she was so disappointed to take only the silver in the 500 meters kayak fours at the 2000 Olympics in Sydney, despite giving it her all. Silver does not have quite the same sparkle next to a haul of golds.

In kayaking and canoeing no official track is kept of race timings because conditions vary so greatly between courses. That is a matter of great regret to statistics buffs, and particularly to commentators and journalists, as it means they cannot resort to the easy tool of dramatising fights against the clock. However, that eschewal of record-chasing is all to the good of the sport, giving it added allure over and above the fact that in this urbanised world training and competition take place in nature, on water, in the open air and sunshine, so it stands a better chance of avoiding modern-day sport's twin plagues of drug doping to enhance performance and ruthless competition to make a living, and thus of preserving a human face.

Fencing—the "Most Hungarian" of Sports

In a break at an international scientific conference on sports, I once asked a fair sprinkling of non-Hungarian colleagues which sport they associate most closely with Hungarians. Almost without exception, the answer was sabre fencing. They even proffered a reason why they supposed this was so: the Magyars were a warrior nation, they said, and so had to be able to handle swords to fight for sheer survival; fencing was then kept alive over a protracted period, indeed raised to a badge of high social prestige, by the anachronistic institution of duelling. Though that is an oversimplification, it is true that sabre fencing as we know it, in the form of the modern sport, did spread out from Hungary, after adaptation to the Hungarian 'temperament' and incorporating characteristically Hungarian elements. It is also a fact that Hungarian sabre fencers, having learnt the tricks from an Italian, Santelli Italo, were unmatched for decades, winning medals at all twenty-seven Olympiads to date, and gold at twenty-five of them.

That series of victories, unparalleled of its kind, was initiated by a feat that in itself is worthy of inclusion in the *Guinness Book of Records*: at the London Olympics in 1908, five of the six finalists fighting for the three individual medals were Hungarians. First place was taken by Dr. Jenő Fuchs (1882–1955), whose fencing style fascinated no less than King Edward VII of England, earning him an invitation to a private soirée at the court to demonstrate his skills. Both Fuchs and the team replicated that dominance at the Stockholm Olympics four years later and would almost certainly have pulled off a treble had World War I not intervened.

Between the two world wars, the Olympic crown in sabre fencing was passed from one Hungarian to the next, going to Dr. Sándor Posta in 1924, Ödön Tersztyánszky in 1928, György Piller in 1932, and Endre Kabos in 1936. Except for a 'mere' silver in 1924, the sabre team took gold at all the Olympiads from 1928 to 1936. One member of two of those gold-winning Hungarian teams, Attila Petschauer (1904–43), tragically found that even fencing as a native Hungarian in Hungarian colours was not enough to save him from the fate of so many of his Jewish compatriots during World War II: he died at the Ukrainian front after deplorable degradations in a forced labour battalion—a shocking story that is commemorated, in fictional form, in the 1999 film *Sunshine* by Oscar-winning Hungarian film director István Szabó.

That same period also saw the emergence of Hungarian foil fencing amongst women as well as men. The exceptionally talented Ilona Elek (1907–88)—she was made an honorary member in perpetuity of the International Fencing Federation—won her first individual Olympic gold for foil in 1936. Her next Olympic gold was no ordinary repeat performance but simply remarkable, because she won it twelve years later, in 1948, after the enforced interval imposed by the Second World War, in 1948 and even had the stamina to capture an individual silver medal a further four years on, at Helsinki, at the age of 45! Her record at World Championships covers an even longer period, for between 1937 and 1955 she was five times a member of the winning foil team, whilst in the individual competition she won gold, silver and bronze—the latter at the age of 49. Ilona Elek was thus a dominant international figure in women's fencing for the best part of three decades—a feat that remains unequalled in women's foil fencing and matched in men's fencing only by Hungary's sabre fencers.

The trio of Aladár Gerevich (1910–91), Rudolf Kárpáti (1920–99) and Pál Kovács (1912–95), as a team, collected altogether close to two dozen Olympic medals for sabre, among them 19 gold, and a similar number of golds at World Championships. In the Olympic individual competitions they took turns, Gerevich winning in 1948, Kovács in 1952, Kárpáti in 1952 and 1956. That trio can rightfully be seen as co-equal architects of the Olympic team victories, though several other colleagues were likewise consistently in gold-winning teams: Tibor Berczelly and László Rajcsányi three times (1936, 1948 and 1952), and Bertalan Papp twice (1948 and 1952). Gerevich and Kovács themselves had been members of the winning team already in Berlin (1936) and Gerevich even four years earlier, in Los Angeles. Thus their amazing Olympic fencing careers too, like that of Ilona Elek, include a gap of 12 years, and by the time they stood with the Hungarian team on the top tier of the podium for the last time, at Rome in 1960, Kovács was 48 and Gerevich 50.

They and Kárpáti are the all-time outstanding figures of Hungarian and international men's fencing, and also the most successful Hungarians in any branch of sport. Their Olympian successors have included Tibor Pézsa (individual sabre, 1964), the sabre team of Imre Bujdosó, László Csongrádi, Imre Gedővári, György Nébald and Bence Szabó (in 1988) and Bence Szabó (individual sabre, 1992), and although that list is far from complete, it should not omit Béla Bay, who was active as a coach and team manager for nearly half a century and widely respected for his work as a teacher with young fencers.

Turning again to the other disciplines which have contributed to Hungary's abiding supremacy in fencing, there are a fair number of foil fencers, besides Ilona Elek, whose achievements have also been noteworthy—the ladies first and foremost on their overall performance. In the individual foil, World Championship titles were taken by Lídia Dömölky (1936) in 1955 and Ildikó Rejtő (1937) in 1963, and Olympic titles by Rejtő in 1964 and Ildikó Tordasi (1951) in 1976, with the team taking gold at all four competitions. Rejtő provides a further example of how a long career is possible in women's fencing too where the skill is exceptional, for even into her forties she was still a member of a bronze medal-winning Olympic team in Montreal, 1976. On retirement, she put a coaching diploma to good use as a technical advisor, whilst Tordasi was a qualified referee, and Dömölky became a well-known journalist. Quite apart from fencing prowess, many of Hungary's women fencers were much admired for their good looks, with Lídia Dömölky, for example, still stylishly attractive long after she had retired, and Gertrúd Stefanek (1959), a member of victorious World and European Championship teams during the 1980s, also winner of a sportswomen's beauty contest.

In foil Hungary's men have had the problem that the French and Italians have striven to stamp the same domination on the event as the Hungarians have enjoyed in the sabre: whilst in the front rank, they did not gain first placings so often. That has meant that the World Championship gold medals won by József Gyuricza in 1955 and Mihály Fülöp in the individual foil and the team event (both in 1957), and even Jenő Kamuti's (1937) Olympic silvers (1968 and 1972), are all the more treasured. Kamuti's extraordinary sportsmanship was also highly appreciated, twice earning him a Hungarian Fair Play prize as well as an award from UNESCO's Fair Play Committee. Whilst pursuing his medical career, he also held office as general secretary of the International Fencing Federation, president of the European Fencing Federation, and chairman of the International Fair Play Committee.

The extraordinary career of Pál Szekeres (1964) is also very much bound up with Hungarian fencing. A highly talented foil fencer, he won bronze individual medals at the World Championships (1987) and the Olympics (1988) as well as both gold and silver in the team event at the Universiade (1985, 1989). A severe car accident then confined him to a wheelchair but, far from ending his sports career, he fought back from the trauma to take a gold medal in foil at the Barcelona Paralympics in 1992. During the nineties, he was chairman of his club's Sportsmen's Committee and a board member of the Fencers' Lobby Group; since 1999, he has been a deputy state-secretary in the Ministry of Sport and Youth Affairs, representing the interests of the handicapped, sporting and non-sporting alike.

To turn finally to the epée, the leading figure amongst the men has undoubtedly been Győző Kulcsár (1940), a holder of three World Championship golds (for the individual event in 1970 and 1971, and the team event in 1978) and also of four Olympic golds—in the individual epée in 1968, and as a member of the Hungarian teams that won in 1964, 1968 and 1972. On two of those occasions the team also included Zoltán Nemere, Pál Schmitt and Csaba Fenyvesi, the latter of who also gained an individual gold in 1972. Amongst the women epéeists, Gyöngyi Szalay and Ildikó

Mincza stand out for their excellent performances during the 1990s, primarily at World and European championships, between them collecting a dozen first places in team events and generally being highly placed in individual events. That has been capped by the individual gold won at the 27th Olympiad in Sydney by Tímea Nagy, a modest, personable thirty-year-old with several professional qualifications to her name.

In recent years, it has to be said, the strong influence that the media, television in particular, have had through the enhanced attention that they give to more spectacular sports, is now placing the viability of fencing in question. Whatever the future might hold for its international position, though, the achievements of Hungarian fencing are assured of a place in the history of world sport.

Gymnastics—a Sport with a Great History

At the 2000 Sydney Olympics, 30-year-old Szilveszter Csollány (1970) took the gold for the rings with a meticulous, faultless performance, while Zoltán Supola (1970), at the same age, was placed sixth in the pommel horse. Women gymnasts did not figure amongst the top finalists at these Games, though in the past, especially before the mid-seventies, they could generally be relied on to finish well; indeed, they had a particularly illustrious era in the fifties. In truth, gymnastics is no longer a sport for women so much as for teenage girls; it is now inconceivable for a grown woman to produce the sort of virtuosity that the men do, because they lose any chance of making the national squad long before they reach the age of 30. In the glory days of the sport, though, that was certainly feasible. Ágnes Keleti (1921), for instance, was 35 when she won golds in three individual events and—along with Erzsébet Köteles, who was also in her thirties, and Margit Korondi, Olga Tass, Andrea Bodó, and Alíz Kertész, who were all in their twenties—one in the team hand apparatus at the Melbourne Olympics.

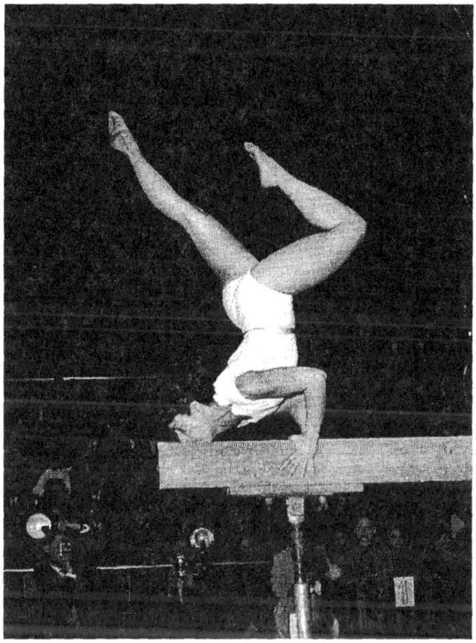

Ágnes Keleti

Ágnes Keleti had a natural talent, with a superb physical flair that was set off by her sense of rhythm and a feel for music that came through from her cello playing. Her career was almost ended by Hungary's rampant anti-Semitism in the years up to and during World War II, and she was later dogged by injuries, but her undaunted persistence did finally bear fruit. In total she won ten Olympic medals, five of them gold, as well as mounting the victory podium several times at World Championships. In purely numerical terms, she was the most successful Hungarian sportswoman there has been to date, but for her fans that was not the only consideration: her performances simply enchanted Hungarian and international spectators alike, her free exercises being regularly hailed by prolonged applause even at world competitions. Some of that applause also went to Margit Korondi for her eight Olympic medals, including two golds. Fragile as a butterfly, she looked effortless when executing even the most difficult elements of exercises on her favourite apparatus, the asymmetric bars, her disarming smile making her a ready favourite with the public.

This fabulous team broke up after the 1956 Revolution, but several of them remained in the sport as coaches, managers, or teachers. Keleti found a new home in Israel, but Köteles, Kertész and Olga Tass stayed in Hungary. Thanks to them and also to the coaching efforts of Éva Romák, Sára Berczik, Klára Förstner and Mrs. Jenő Nagy ('Aunt Vali' as she was known), the temporary vacuum left by the retirement of that great generation was at least partly filled by new talent. Olympic medal-winners Katalin Makrai (1964), Anikó Ducza (1964), the 1972 national team, and Márta Egervári (1976) might have bridged the way to future better fortune, but as it turned out the only glimpses of gold were afforded by Henrietta Ónodi (1974), a protégé of the Unyatyinszkis, an extraordinarily dedicated husband-and-wife coaching pair. Henrietta, a diligent and meticulous gymnast, was just fifteen when she became European champion in vaulting and still in her teens when she took gold in that discipline at the World championships and the Olympics in 1992, as well as silver in the floor exercises on both occasions. She then retired from the sport to attend college in the US and although she participated again at the Atlanta Games and did moderately well; it was certainly not the big comeback she may have been hoping for. That in no way diminishes Ónodi's earlier achievements, but it clearly shows that in women's gymnastics today even the most talented 22-year-old is 'over the hill'.

Male gymnasts do not face that threat, yet the passage of time has left its mark on their sport too. István Pelle (1907–86), the first Hungarian gymnast to become champion at both World (1930) and Olympic (1932) level, was technically brilliant and yet he performed with casual ease at no mean level, assembling his exercises from elements that he subsequently put to good use as a globetrotting acrobat. Ferenc Pataki (1917–88) likewise won his gold at the London Olympics by incorporating a series of acrobatic leaps in his free exercise yet maintaining his elegant style. Nor was he limited to that discipline as he also gained medals in international competition for the pommel horse, horizontal bar and parallel bars as well as in the free exercise. He, too, later moved into the circus world as an instructor at the State Institute for Acrobatics.

A quarter of a century later, innovation was again the key to the invincibility in international competition of Zoltán Magyar (1953), 'uncrowned king' of the pommel horse. Working with his fantastic coach, László Vígh (1941), he came up with some absolutely original elements—the 'Magyar walk', the 'spindle' and the 'leaping walk'—that helped him to win golds at World Championships (1974, 1978, 1979) and the Olympics (1976, 1980). Already in those days, achieving that sort of success took sustained and punishing effort: Zoltán Magyar himself started at the age of 12 and trained for nine years before winning his first World Championship, but the stylishness of what he did was still there to see, time and time again.

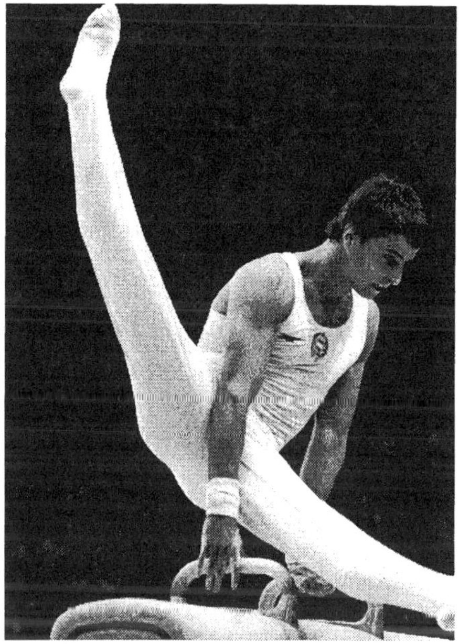

Zoltán Magyar

Bygone Achievements: Track and Field

If there is one branch of sport that every Hungarian child, from times of old, must have had a taste of, then it is athletics. As part of physical education, jumping, throwing and running have probably been in school curriculums from the very earliest days. Athletics also has a long history as a competitive sport, with Hungary being amongst the early pioneers in its development on continental Europe. An institutional framework (national federation, sports clubs) had been set up before the close of the nineteenth century, and for more than fifty years athletics counted as of the country's most popular sports. For all that successes have been disappointingly limited, both in time and in the spread of events, there have been occasional moments when the world paid attention to the performances of Hungarian athletes.

For her own era, Ibolya Csák (1915) was an excellent athlete—only a national champion in long jump but taking the high jump as far as gold at the Olympics (1936) and also the European championships (1938), setting a European record in doing so which was not broken nationally for nearly a quarter of a century. Restarting her career in the late forties, after the interruption of the war, the versatile Olga Gyarmati added to the growing reputation of Hungary's women athletes by winning gold in the long jump at the London Olympiad.

The disciplines in which Hungary won major international acclaim, however, were running and throwing. A world-class running school developed around coach Mihály Iglói (1908–98), with his interval training schedules tailored to individual medium- and long-distance runners. Under his tutelage, Sándor Iharos (1930–96), István Rózsavölgyi (1929) and László Tábori (1931) improved fifteen world records in individual and relay events. Iharos's feat was of similar calibre to that of Paavo Nurmi, the Finnish running prodigy of the 1924 Games, in that in 1955 he set world records at all seven distances between 1,500 and 10,000 meters, making him 'Athlete of the Year' abroad as well as back home. Tábori was a specialist at 1,500 meters, 1 mile and 5,000 meters. In the 1,500 meters he set a European and world record. Emigrating to the USA after 1956, he went on there to win at the same distance at the US Nationals. Rózsavölgyi ran distances between 800 and 3,000 meters. His 1,500 meters world record remained on the books for fifteen years. The most spectacular success that the three runners notched up, however, was the world record that they, together with Ferenc Meixner, set as the Honvéd Sports Club team in a 4 x 1,500 meters relay in 1955. Such was the popularity of athletics at the time in Hungary—in part because of them—that tens of thousands were present to cheer them on during such record-breaking attempts. A similar crowd was on hand in Budapest to celebrate long-distance runner József Kovács (1926–87) when he became the first athlete to beat Emil Zatopek, the 'flying Czech', and when Hungary's other European champion runners, Sándor Rozsnyói (3,000 meters), Lajos Szentgáli (800 meters), the 'greased-lightning' relay squad of Zarándi, Varasdi, Csányi and Goldoványi (4x 100 meters) or long jumper Ödön Földessy competed.

The first major Hungarian mark to be made in throwing events internationally was that of Rudolf Bauer (1879–1932), who took gold in the discus at the Paris Olympiad in 1900 and subsequently switched to rowing (he was also a passionate hunter). A concerted string of success came only after World War II. Imre Németh (1917–89), who later became director of the Népstadion in Budapest, brought home an Olympic gold in the hammer throw from London. In short order, József Csermák (1932–2001), who as a novice hammer thrower had received detailed instructions on technique from Németh by mail, put that advice to good use because by the time of the 1952 Olympiad the 35 year old master had to be content with 'only' a bronze whilst 20-year-old

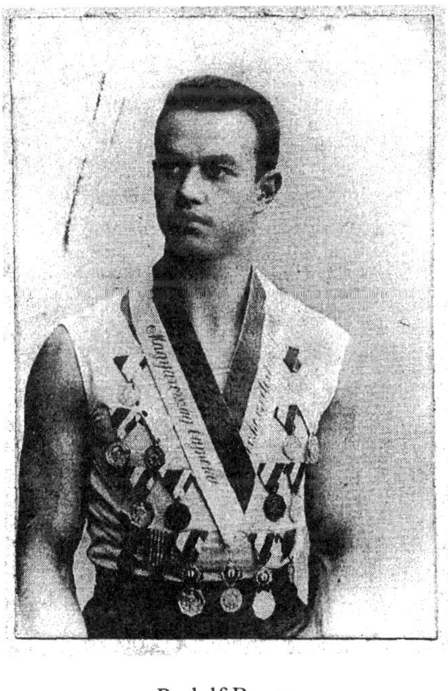

Rudolf Bauer

Csermák took the winner's stand, having set a world record by throwing the hammer for the first time ever beyond 60 meters. Csermák in turn became both exemplar and coach to Gyula Zsivótzky (1937). The latter considered Olympic silvers in 1960 and 1964 to be just half-successes, despite the fact that he had undergone a series of major operations in the meantime, spending two hundred days in hospital and losing some twenty kilos in body-weight. Many people thought Zsivótzky was crazy to pick up the hammer again after so close a brush with death from his disease, but being able to train again boosted his physical and mental condition. With unending diligence, willpower and determination, he fought back to his old top form, became the first in the world to throw the hammer over 70 meters, and achieved his ambition of winning an Olympic gold at the 1968 Mexico City Games.

During Zsivótzky's time the throwing events were hives of activity in Hungary. At Mexico City two fellow countrymen were also placed in the hammer (Lázár Lovász with the bronze medal, Sándor Eckschmiedt in fifth place), whilst javelin thrower Gergely Kulcsár won medals at three successive Games (1960, 1964 and 1968). The women throwers were also up there with them, Márta Antal winning an Olympic silver for javelin in 1964, then Jolán Kontsek a bronze in the discus, and Angéla Németh, at 22, a gold in the javelin in 1968. Angéla Németh was almost lost to field athletics as she also excelled in basketball and had difficulty deciding between the two sports. Miklós Németh, the javelin thrower who won gold at the Montreal Olympics with a world record throw of almost 95 meters, is not related to Angéla Németh but is the son of hammer thrower Imre Németh.

Many of the best throwers later passed on their skills to the up-and-coming generation, working as coaches. Amongst them was József Várszegi, winner of a bronze in javelin at the 1948 Olympics, who became head and later the first rector of the Department of Athletics at the College of Physical Education, and Jenő Koltai, who himself used to be a top javelin thrower. Koltai gained an international name not only through protégés who went on to become Olympic champions, but also with his theoretical work. As a recognised expert on the throwing events, he was regularly invited to hold international presentations. Hungarian throwing coaches were welcome all over the world, and by the eighties several even got official permission to work abroad—a loss which probably contributed to the break that was seen in the chain of success. It picked up again only two decades later, when hammer thrower Balázs Kiss (1972) won at the 1996 Olympics, though he was studying in the US at the time. A few top competitors and coaches did stay in the old country, however, including hammer thrower Tibor Gécsek (1964), who under the guidance of Pál Németh took a gold medal at the 1999 World Championship.

Hungary has a good reputation for staging sporting events as well as turning out great athletes. The long record of hospitality and demonstrated quality of organisational expertise that the country offers makes international federations happy to award it the rights to host their competitions. One of the most lavish events of recent years, a spectacular festival of sport and culture, was the World Athletic Championship held at Budapest in 1998.

Football, the Most Popular Sport in Hungary

It is customary to divide Hungary's sporting life, with a touch of malice, into two: football and other sports. That may not be justified on the basis of long-term results, but it certainly is in terms of popularity. In the heyday of Hungarian football the evidence that it had more fans than any other sort was simply a matter of everyday experience, but the opinion polls and scientific surveys of today still yield the same message: the present decrepit state of football in Hungary seems, if anything, to have enhanced rather than diminished the value of long-bygone results. It is highly likely that most mortals the world over, especially in the era before television became a mass medium, formed their first impression of Hungarian

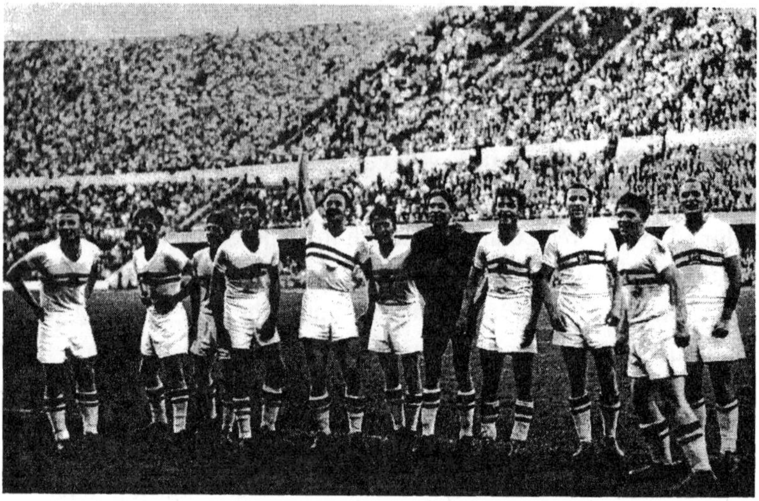

The Golden Team

sport, and even of the country itself, through its football and its football players. If there is one globally familiar Hungarian name, it is that of Ferenc Puskás (1927), but for Hungarians travelling abroad, whatever continent they may be visiting, even in the unlikeliest of situations and places, it is not uncommon, even today, for them to come across people who will cite off the top of their head the names of that now-legendary forward line: Puskás, Kocsis, Hidegkúti, Czibor, and Budai. And plenty can put names to the rest of the line-up: Grosics, Buzánszky, Lóránt, Lantos, Bozsik, and Zakariás, though fewer are aware that Kárpáti, Börzsei, Kovács, Szojka, Sándor, Palotás and Tóth M. also played on the team several times.

From its early days onwards, Hungarian football had players who were known about beyond the country's borders. In the 1910s the names of Gáspár Borbás, Sándor Bródy and Imre Schlosser would get a smile of recognition abroad, whilst in the twenties the Fogl brothers, fullbacks József and Károly, were famed across

Europe, and for many cognoscenti in the countries where he worked as a coach, from Italy to Peru, as well as in his homeland, the 'goal king' György Orth was the most versatile, intelligent player ever to come out of Hungary.

The national team had its Olympic successes after the momentous win in 1952, bringing back gold from Tokyo (1964) and Mexico City (1968) and silver from Munich (1972), but there is no question that the true glory days were the period when the 'Golden Team' reigned supreme during the early fifties. Of course, it has become a habit to call winning World Championship and Olympic teams in every sport 'golden', but whenever a Hungarian uses the term 'the Golden Team' without a qualifier, everyone knows that it is a reference to this football squad. And anyway the squad did not get the sobriquet particularly for its victory at the 1952 Olympics, much more for the brilliant style of its game over a period of several years. Indeed, if results in world competition were the sole determinant, the crushing disappointment for the country of being on the losing side in the 1954 World Cup final at Zurich would entitle them, at best, to be known as 'the Silver Team'. But in a historical context there was the incontestable feat of Hungary's 6:3 win over England at Wembley Stadium in November 1953—the first time in 90 years that an English team had ever been beaten on home ground. That still stands vividly in many English (and not just Hungarian) memories as one of the matches of the century. For the soccer world that was only rubbed in further by the Hungarian team's 7:1 victory in the return match, at Budapest's Népstadion, in May 1954.

The reasons why the Golden Team was so widely cherished by the Hungarian general public went beyond mere statistics, however. All its players, captained by centre-forward Ferenc Puskás—fondly known as 'Öcsi', or 'Kiddo'—were uncommonly talented, and these talents were readily exploited by the country's political leaders to appease a public that had been starved of success—even literally starved. The players were cosseted with all they could want, including privileges unimaginable today, to allow them to concentrate on playing football, but they had enough talent and genuine love of the game not to let that go to their heads. The football skills they displayed were so fiendish, amazing Europe and half the world, because even as stars under the spotlight they did not regard training as an unpleasant duty or hard work but as fun, an agreeable pastime, and they were more than happy to play in their own free time, outside the official practice sessions. That is where their instincts for switching into such dazzling passing movements, for shooting and heading such magnificent goals, were polished; that was why watching their clever, seemingly effortless game gave such pleasure to football fans, Hungarian and non-Hungarian alike. For most Hungarians, though, it was much more than that: in a grim world that offered little cause for genuine communal rejoicing, it is fair to say that the Golden Team gave literally hundreds of thousands of people high days to celebrate, restored a measure of faith and confidence to an unsure nation that was fearful or despaired of ever having faith again.

It would be unrealistic to ask any team to emulate, let alone surpass, this legendary team's feats on the football field. Though the national side has enjoyed memorable victories since then—beating the Brazilians at a World Cup in 1966, for instance—it has also tasted bitter defeats. The game has had its ups and downs in Hungary over the years (currently it is in one of the slumps), but it has always managed to produce a crop of above-average players, not a few of whom have gone on to flourish in foreign leagues. Even though the national team of recent decades has not shaped up too well in international competition, some individual players have made their mark. In 1967, Flórián Albert (1941), the leading footballer of the sixties, capped 75 times for his country, won the prestigious Golden Ball award as Europe's 'Player of the Year', previous recipients of which had included England's Stanley Matthews, Spain's di Stefano, the USSR's Yashin, and Portugal's Eusebio.

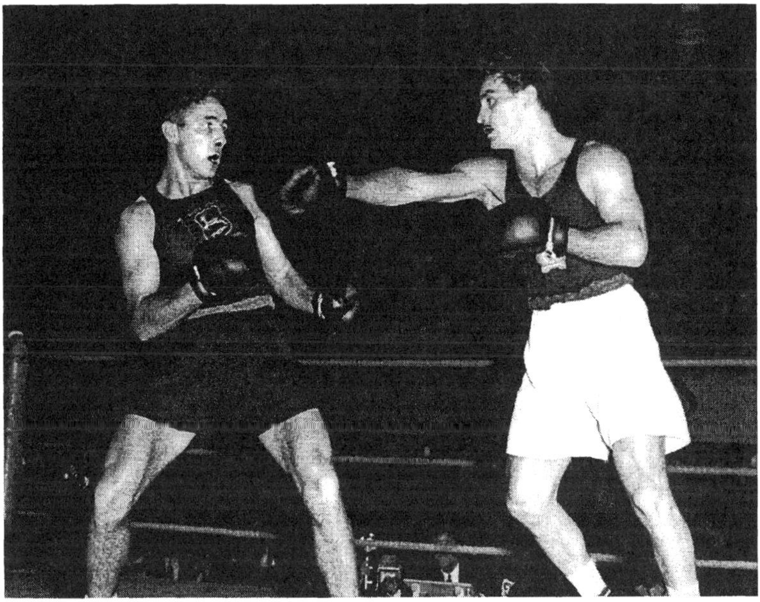

Laci Papp

In the Ring: 'Laci' Papp, 'Koko' and the Rest

Boxing already reached a high standard in Hungary between the two world wars, a measure of that being evidenced in gold medals won at two Olympiads (Antal Kocsis in 1928, István Énekes in 1932) and a fair sprinkle of placings at European championships, but also on the administrative front, with the secretary-general of the International Amateur Boxing Association from 1933 to 1945 being a Hungarian. After World War II, one of the older generation of boxers, Imre Harangi, secured gold at the 1948 Olympics, but by then a lot of fresh talent was also emerging, and another gold was taken by the almost unknown Tibor Csík. Even that, however, was nothing compared to the stir that László Papp (1926) caused on bursting onto the scene at the same Games.

In Hungary, as elsewhere, boxing strongly divides opinions, some being unapologetic supporters, others fiercely opposed to it, and many having very mixed feelings. Those attitudes tend to be more positive, at least temporarily, if the topic is raised in the context of Hungarian achievements in the sport, most people losing sight of their objections as soon as the name of the country's most distinguished, modest and sportsmanlike boxer is uttered. 'Laci' Papp's fantastic career in the ring, as both an amateur and a professional middleweight, and later as a trainer, commands respect even from those who are ordinarily no friends of boxing. "My own preference is for graceful sports, like gymnastics and eurhythmics, but because of Laci Papp one has to pay some attention to boxing," the celebrated Hungarian actress Klári Tolnay has been quoted as saying.[5] Béla Abody, a music critic, has noted, "After '45, the joys of living were more strident, both in culture and sport; a person could get to see great performances. There was a resurrection—both in opera and in sport."[6] As part of that 'resurrection', the special atmosphere that boxing matches generate attracted huge crowds: club championships were always sell-outs, with fans queuing up at dawn for tickets and touts being paid incredible sums by those who had missed out.

Laci Papp's innate talent had shown itself early in life—albeit in football, which he still loved to play even after he became famous as a boxer—and he was a star in the ring from the very beginning. Never a crude 'slugger', he used his astonishingly delicate hands, or rather fists, to box cleverly, with a highly refined technique. Under the eyes of his coach, Zsigmond Ádler (1901–82), his international career soared ever higher, from one Olympiad to the next. As the only boxer (until 1980) to have taken the winner's stand at three consecutive Olympics—1948, 1952 and 1956—fans admired him and took him to their bosom. At Melbourne in 1956, the spectators on finals night—as a way of expressing profound solidarity with those who had fought for Hungary's freedom—cheered Papp to the rafters when he won his third title.

To an outsider, Papp's subsequent professional career, between 1957 and 1964, seems to have been equally successful, when he eventually retired, after 29 fights, as undefeated European champion (1962). He personally was disappointed, however, because he was offered the chance to fight for the world middleweight title, but was denied permission to do so by Hungary's political leaders. He consoled himself with the feats of his pupils. Although Hungary's next Olympic boxing champion, Gyula Török (in 1960), was a protégé of former champion Árpád Énekes, by the seventies and throughout the eighties Papp—still held in universally high regard—was the team captain, as well as coach and manager to a number of well-known boxers. Papp demanded the same high standards that he had set, and that era was associated with a steady trickle of European and Olympic championship medals. At the Munich Games, for instance, the team's medal haul included György Gedó's gold and two silver medals. In 1989 the World Boxing Council awarded László Papp a belt as all-time best amateur and professional middleweight, but even as the recipient of such distinctions he has stayed the straightforward, generous and considerate sportsman that he was at the start of his career.

After Papp's retirement, boxing in Hungary went through a fallow period until a new world star emerged in the mid-nineties. István 'Koko' Kovács's (1970) exceptional talent at featherweight, his adroit and intelligent boxing skills, were likewise manifest early in his amateur career. He was already European and World champion (both 1991) by the age of 21, and he went on to take gold at the Atlanta Olympiad in 1996 and regain the World Championship the next year. On turning professional, he won the WBC World title in 1999, the European Boxing Union title in 2000, and the World Boxing Organisation title in 2001. His extraordinary popularity in Hungary was undimmed even when he lost the latter title a few months later—temporarily, his fans hope.

A Sense of Mission: Balczó and Other Modern Pentathletes

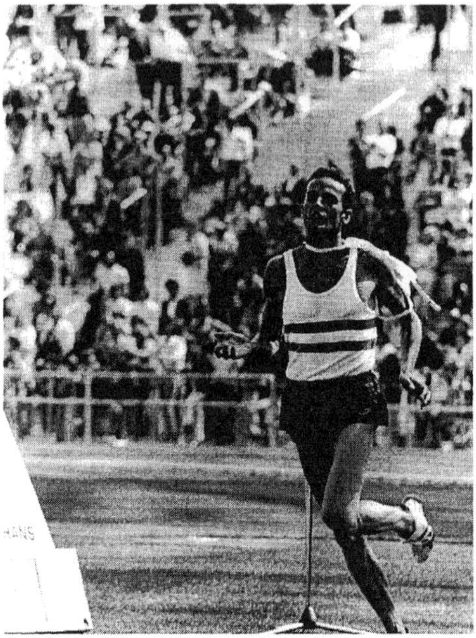
András Balczó

In the mid-seventies, tens of thousands of Hungarians went to the cinema to see Ferenc Kósa's captivating *The Mission*—sports fans largely on account of the film's main protagonist, modern pentathlete András Balczó (1938), intellectuals who had no great interest in sport primarily for its deeper-lying hidden message. Balczó's name became a symbol for heroism, determination and fighting to the utmost, both within and outside sport.

By the time Balczó appeared on the scene, Hungary had already gained a high reputation in the modern pentathlon through the deeds of Aladár Kovácsi, István Szondy (who moved to West Germany in 1956), and Gábor Benedek (who for political reasons was forbidden to compete after 1956) in winning the team gold at the 1952 Olympics and a series of World Championships in the early fifties. Balczó's fellow team members at the Olympics and other world events were likewise great figures in their own right: Ferenc Németh—the individual champion at the Rome Olympiad, Ferenc Török—the same at Tokyo, and István Móna and Imre Nagy alongside them. Yet Balczó stood out even in this amazingly talented team with his quite exceptional abilities and incredible willpower. His sister, the ceramic artist Edit Balczó, commented, "...it runs in the family that whatever we turn our hands to, we do it obsessively."[7] We may infer that András Balczó outstripped the other members of his family in this. His talent was evident early on, propelling him to the top echelons of the sport when barely out of his teens (he was already in the victorious 1960 Olympic squad), but it took him several years more before he could deliver his very best in the individual event. He did not become individual World champion until 1963, but that was just the start of a unique string of victories in which he won the title every year up to 1969. Most memorable of all was the championship held at Budapest in 1969, when the entire crowd of around 30,000 people rooted for him in the decisive final discipline of running. True, in those days a lot more people attended sport events than do today, but even then Balczó was the sole reason why quite so many had turned out that day. It was this knowledge, which gave him strength to go on and, at 34, win an individual Olympic gold at Munich. Over much of his career he had no coaching guidance but instead prepared himself for competitions with a training system that was unique in its own right. Many have attempted to adopt his methods, but they have trouble supplying the willpower it requires, yet the key to Balczó's success, over and above superb physical fitness, is ambition, endeavour and, as was evident to anyone who followed his career, unimaginable determination.

Balczó's team-mates at the Munich Olympics, Pál Bakó and Zsigmond Villányi, were already distinctly younger than him. They and subsequent generations have been equally well schooled and continue to figure regularly in the top international bracket of their sport. The team of Tamás Kancsal, Tibor Maracskó, and Szvetiszláv Sasics won the World Championship in 1975 and also brought back medals from the 1976 and 1980 Olympics, but for more golds the country had to wait until 1988, when János Martinek took the individual event and the team of himself, László Fábián and Attila Mizsér was also victorious. Golds have continued to come regularly in individual, team and relay events at World and European championships throughout the eighties and nineties. Hungary in fact took a prominent role in working out the rules for the relay event and getting it established in an effort to ensure the long-term viability of the sport by making it more spectacular and thus more attractive for television audiences.

Modern pentathlon is usually associated with men, because the women's events have only quite recently been included in world competitions. Despite that short history, Hungary's female pentathletes deserve at least a mention as they have already captured World Championship golds in both the individual and the team events and look set to follow the traditions of their male colleagues.

Chess—A Clever Way to Bring up Geniuses

The large number of outstanding sportswomen who get a mention in this chapter tends to hide the documented fact that women have less chance of making it to the top in competitive sports than their male counterparts. That inequality may have shrunk, bit by bit, over the course of the twentieth century, but even today there are fewer branches of women's sports and fewer events, and it is still rare—due to obvious differences in physical build—for women to be allowed to compete with men. Yet women have had to, and still have to, fight hard to gain equal treatment even in areas such as chess where gender differences in physical abilities have no bearing on the outcome.

For decades, organised chess in Hungary was a male preserve. The Pest Chess Circle, established in 1839 and presided over for three decades by the composer Ferenc Erkel, had a team exclusively of men. Nor were any woman chess players to be found in the second half of that century, even though by then four Hungarian men (Ignác Kolish, Izidor Kunzberg, Miksa Weiss and János Lőwenthal) were world-renowned players; indeed, that still held into the 1920s and 1930s, when Hungary won gold at the first international team 'Olympics' in 1927 and several silvers thereafter. Oddly enough, Czech-born Vera Menchik (later Mrs. Stevenson of England), the first women's World chess champion and the invariable winner of that title up to the Second World War, had originally been coached by Géza Maróczy, the first board in the Hungarian Olympic team—and during that period she was the only woman to be invited to the men's world championships. A breakthrough has only occurred over the last fifty years, though for a good while it was still the men who maintained the advantage. At various times, László Szabó and Pál Benkő, then subsequently Lajos Portisch, András Adorján and Zoltán Ribli, all played their way through to FIDE Candidates' Tournaments for the world championship; indeed, Portisch was a candidate several times and for decades ranked among the world's top players. Gedeon Barcza, an international grandmaster was a widely acclaimed coach and chess writer.

Among the women, Mária Ivánka and Zsuzsa Verőczi reached international grandmaster rank in 1978, Éva Karakas in 1982, and Ildikó Mádl (1969) in 1984, with Verőczi subsequently going on to become a women's World Championship candidate. With the notable exception of Ildikó Mádl, though, who won a men's international title in 1992, the women distinguished themselves predominantly in women's competitions. With Mádl, however, a new era was opening in chess, kick-started by the three extraordinary daughters of a brilliant (chess) teacher, their father László Polgár (1946). The clever and amiable Polgár girls, Zsuzsa (1969), Zsófia (1974) and Judit (1976) were not born child prodigies: it was their father's deliberate, pre-planned educational strategy which raised them as chess-playing phenomena, and doing so, what is more, in a socio-political milieu that was intolerant, or at best very grudging, in its attitudes to anyone seeking to rise above the average. In the preface to Polgár's book on the family, András Mezei describes that climate as follows: "Here [in Hungary] generations have been taught not to think for themselves. Not to doubt, not to argue... This brilliantly average upbringing could tolerate only docile genius, minds it had already tamed. Anyone who withstood that 'test' was allowed to become a 'great subject'. Here generations were taught to accommodate to non-action, and these are the generations with which we are drifting around today in the desert of quasi-reformism."[8]

Polgár went to considerable pains to extract his daughters from this 'stultifying terror of mediocrity'. He sought to prove that every child has innate promise; provided one has faith and confidence in it, a consistent, unflagging approach to education is capable of wonders. He did not back down even when he faced incredulity and injustice, or when, instead of support, major obstacles were put in his way, as with the official refusal, for many years, to issue passports for his daughters to travel to the West. Subsequent events, however, have vindicated his stance to a remarkable degree, as even those one-time opponents must be obliged to admit.

The three Polgár girls between them hold some forty *Guinness Book* world records in chess. To mention just a few from their earlier years: Zsuzsa was unbeatable in the schoolchildren's class in Hungary by the time she was just four years old and 12 when she won every board to take first place in the unofficial World Championships for girls of under 16; Zsófia was Hungarian schools' champion at five, and undefeated in a couple of adult international tournaments at only eleven; whereas Judit was still just seven when she won, without dropping a point, at her first international youth (boys') tournament and ten when she took the title in the World Championships for girls under 14. Their subsequent careers were equally unprecedented, all three earning international master and grandmaster rankings at very young ages and going on, together with Ildikó Mádl, to win the women's team competitions at Olympic and World Championships—all because

László Polgár had accustomed them to competing with males from the very outset.

His argument is that it is a mistake to make gender distinctions in chess on the grounds that it is a sport; chess is a science, an art, a game and, above all, an intellectual pursuit as much as it is a sport, and in intellectual pursuits it is not customary to award separate prizes for men and women. Just as there are no separate Nobel Prizes for men and women, it seems reasonable to suppose that intrinsic capabilities in chess are not linked to gender. Polgár was considered an oddball for suggesting this, yet the results that Zsuzsa, Zsófia and Judit achieved in holding their own in competition with men substantiate their father's postulate. They all met the international masters' standard for men whilst still young girls, and as teenagers they went on to defeat strings of male grandmasters. Whilst the precise details of those performances are naturally significant, they are nevertheless secondary, merely illustrative of the central point, which is that by their now celebrated example the Polgár family have furnished proof that (chess) genius can be deliberately nurtured. Furthermore, by furnishing that proof in the field of chess they have removed the habitual trump card in the arguments of all those who maintain there is intrinsic intellectual difference between men and women.

Of all aspects of sport in Hungary, it is the Polgár girls' achievements which have gained the greatest coverage in the foreign press, but on the subject of exceptional chess talents, mention should also be made of Péter Lékó (1979), who was likewise still very young when he met the standards required for grading as an international master (at 13 years) and grandmaster (at 14). Now in his twenties, Lékó's FIDE rating puts him amongst the top ten chess players in the world, and he is thought to have a good chance of challenging the world champion.

The Afterlife of Success

Sic transit gloria mundi, the old Latin saw has it—So passes away the glory of the world. Naturally, the achievements of some top Hungarian sportsmen and women have proved short-lived, but transience is not typical of the cases that have been touched upon in this chapter. The influence they have may be observed individually over the long as well as the short term. On examining careers after retirement from competition, it is found that success in sport is followed, more often than not, by success—often conspicuous—in one's profession.

The example of Alfréd Hajós, Hungary's very first Olympic champion and later a distinguished architect, proves to have been just at the head of a long series of instances where individuals have made as much of a mark in their 'civilian' lives as they did in their sports career. To take a number of examples from the last decade or two alone, swimmer Valéria Gyenge is now a professionally acknowledged photographer, fencer Rudolf Kárpáti—a musicologist, swimmer Éva Székely—a writer, fencer Jenő Kamuti—a surgeon, István Préda—deputy rector at a medical school, fencer Lídia Dömölky—a journalist, modern pentathlete Ferenc Török—a lawyer and member of parliament, swimmer Judit Temes—a doctor and lawyer, swimmer András Hargitay—a veterinarian, swimmer Andrea Gyarmati—a pediatrician, javelin thrower Angéla Németh—a university lecturer and head of department, javelin thrower Miklós Németh—an innovator and businessman, and gymnast Zsuzsa Csisztu—a TV journalist. Many other former top sportsmen and women likewise have higher educational qualifications, which is all the more noteworthy because Hungary's universities and colleges, unlike those in the USA, have not traditionally provided a platform for competitive sports. That was especially true in the Socialist era, when the powers-that-be not only did not encourage but actually did not tolerate students organising in any form that was not under firm party control, despite which a higher-than-average proportion of leading sportsmen chose to study at and graduate from institutions of higher education.

Top competitors are often accused of being 'muscle-heads', concentrating on developing their physical abilities at the expense of their minds. Sociological surveys that were carried out on cohorts of Hungarian Olympic champions from the periods 1948–76 and 1980–96, however, showed exactly the opposite: they had a significantly higher than average level of education on retiring from sports, as compared with non-sportsmen of the same age. In the fifties, sixties and seventies, when about 2–5% of the Hungarian adult population had university-level degrees, half of the male and 42% of the female Olympic champions had graduated from university or college. The percentages were lower for those reaching their sporting peak in the eighties and nineties but still three times higher than for the comparative sample from the general population.[9]

It was made clear to those coming into many sports (fencing, swimming, water polo, modern pentathlon, canoeing, etc.) that studying and gaining qualifications were just as important as high levels of sporting performance. Admittedly, many top athletes took longer to complete their university courses, and graduated later, than fellow students who did no sports. However, the best of them made a conscious effort to keep their intellectual as well as their physical abilities in good shape for the time after they retired from sports, since for them studying was a pleasure rather than a problem. One particular achievement that stands out amongst many fine examples is that of Dr. Judit Temes, who qualified as a physician when in her twenties but in later life, when she was 69, took a law degree and now, in her seventies, puts her irrepressible energy into working on patients' rights.

Many of the qualifications that top athletes obtain are diplomas in coaching, PE instruction, or sport management. The willingness of former stars to put their experience back into their sports has played a big part in breeding continued success. To pick out a few notable examples where sports have profited from having the knowledge and expertise of such stars at their disposal as head coaches or in advisory capacities over many years, there are István Jónyer in table tennis, Mihály Iglói in athletics, Csaba Hegedűs in wrestling, Ferenc Török and Imre Nagy in modern pentathlon, László Papp in boxing, László Hammerl in shooting, Dezső Bordán, Klára Förstner and Alíz Kertész in gymnastics, Dezső Lemhényi and Dezső Gyarmati in water polo, Béla Bay and Tamás Kovács in fencing, or Éva Székely in swimming.

Substantial numbers, on retiring from active sport, go on to managerial posts in jobs that have no relation to sport. Having been leading figures in their sports, for shorter or longer periods of time, former champions are often loath to work as employees under a boss and find it quite natural that they should likewise seek to get to the top in their new profession. The positive character traits of assertiveness, creativity, steadfastness of purpose, resolution, psychological toughness, etc. that are indispensable to sporting success, and in part developed through participation in competitive sports, also predispose them to leadership roles. Of course, it is also true that the prestige won in sport can often be turned to great advantage in later careers outside sport. Financially, they are much less wealthy on retiring from sport than their counterparts in more industrialised countries, but they have been able to use their connections to much better effect in starting their new careers. That was particularly true in the Socialist era, given the privileges that top sportsmen enjoyed.

The fact that former top sportsmen and women also hold their own in the fields of study and the workplace amply bears out a comment made by Albert Szent-Györgyi, one of Hungary's Nobel Prize winners, in a speech that touched on his own observations of the early blossoming of modern sport when working as a research biochemist at Cambridge University, England, during the 1920s: "Sport is not just a training for the body but also the most powerful and noblest means of training the mind... it is not a matter of the body being subdued through the mind but of the mind being opened up through the body."[10] That influence can be assessed directly in the attitudes towards sport of the progeny of top sportsmen. Around fifty Hungarian Olympic champions are known whose own children in turn went on to become members of an Olympic sports squad.

The best known of these sports dynasties are the Bogen–Gerevich family, three generations of which have won Olympic medals, and the Székely–Gyarmati–Hesz family, whose medal collection matches that of the Gereviches and whose third generation has already set off on the track to the top. Such collections of Olympic medals within a single family, whether through direct or collateral line of descent, a married couple, parents and children, or other ties of relationship, is not uncommon. The fencer Pál Kovács, for instance, won six Olympic golds and one bronze, whilst his son, Tamás, won two bronzes, also in fencing; the wrestler Gyula Bóbis's Olympic gold is complemented by his daughter Ildikó's three silvers and one bronze; and in water polo Tamás Kásás and Barnabás Steinmetz both outdid their fathers by taking Olympic golds, as compared with the silver of Zoltán Kásás and János Steinmetz's bronze. Extraordinary as it may be, there is also more than one case in Hungarian sports of father and son taking the winner's stand at the Olympics in the same sport (though not necessarily the same event): Imre Németh in the hammer (in 1948) and Miklós Németh in javelin (in 1976), then water polo players István Szívós Sr. (in 1952 and 1960) and István Szívós Jr. (in 1976).

Yet the impact of sporting achievements reaches far beyond the immediate family. Indirectly it is discernible in the attitudes of a country's entire population towards sport and their patterns of participation in sports. Although many top sportsmen and women have urged the merits of regular physical training—the orienteer

Sarolta Monspart, for example, has strongly championed jogging—the Hungarian general public would win no medals for its level of sporting activity. At the same time, if one discounts recently spread football hooliganism, competitive sports in Hungary have nurtured a noticeably sports-loving and sporting fan base. One manifestation of this is the alacrity with which Budapest and other cities (e.g. Szeged, Szombathely, Miskolc, Nyíregyháza, and Tiszaújváros) undertake to organise World and European championships, and also the serious consideration that has been given more than once to mounting a bid for Budapest to host the Olympic Games.

In 1896, as was mentioned earlier, the IOC lined up Budapest as a reserve venue, whilst the first time the city actively bid for the honour—for the 1920 Olympiad—it was so well received that Budapest beat off Antwerp by 21 votes to 7 in the first round of voting. Due to the intervention of the First World War, those rights were subsequently reassigned to Antwerp; Hungary, like the other defeated nations of that war, was not even invited. Any thought that Hungary might host the Games in 1940 or 1944 was, of course, dashed by the Second World War. In principle, the best chance to bring them to Hungary was the 17th Olympiad in 1960. Budapest was well prepared, with a government guarantee, and was in contention to the very end: its bid was the last to be discarded after Tokyo, Mexico City, Brussels and Detroit, before Rome emerged as the winner. From among the losers then, of course, Tokyo and Mexico City have since staged the event.

The idea that Budapest should try again by entering a bid for the 2012 Olympiad was raised at the dawn of the new Millennium, in 2001. Whether this will remain a pipe-dream or has a realistic hope will depend on two basic sets of factors: firstly, on how well other applicants are prepared and enjoy international support; and secondly, on how well placed Budapest (and the country as a whole) is to accept the financial burden, the standard of its organisational skills, and on the ability of the Hungarian general public to cope with foreign languages. Thorough studies are currently being prepared to gauge these aspects. Modest as its chances may be, there can be no doubt that the idea will get the backing of a sports-loving nation brought up on sports success.

Notes

1. Allen Guttmann, *From Ritual to Record. The Nature of Modern Sport* (New York, 1978), pp. 15–55.
2. Walter Umminger, *A sport krónikája* (Chronicle of Sport) (Budapest, 1990), p. 172.
3. Éva Székely, *Sírni csak a győztesnek szabad.—Jöttem, láttam... vesztettem?—Megúsztam* (Only the Winner May Cry.—I've Come, Seen... Lost?—I Have Survived) (Budapest, 1992), p. 109.
4. Gyöngyi Földesi-Szabó, *Félamatőrök, félprofik* (Half Amateurs, Half Professionals) (Budapest, 1999), p. 105.
5. Zsuzsa Hegedűs, *Papp László hatvan éves* (László Papp at Sixty) (Budapest, 1986), p. 22.
6. András Kő, *Papp Laci* (Budapest, 1973), p. 20.
7. Sándor Dávid, *Balczó* (Budapest, 1973), p. 180.
8. László Polgár, *Nevelj zsenit* (Bring up a Genius) (Budapest, 1989), p. 5.
9. See Gyöngyi Földesi-Szabó, *Magyar olimpikonok önmagukról és a sportról* (Hungarian Olympic Athletes on Themselves and on Sport) (Budapest, 1984) and *Félamatőrök, félprofik* (Half Amateurs, Half Professionals) (Budapest, 1999).
10. Albert Szent-Györgyi, "Az iskolai ifjúság testnevelése" (Physical Education of School Children), *Testnevelés* 3 (1930), pp. 928–933.

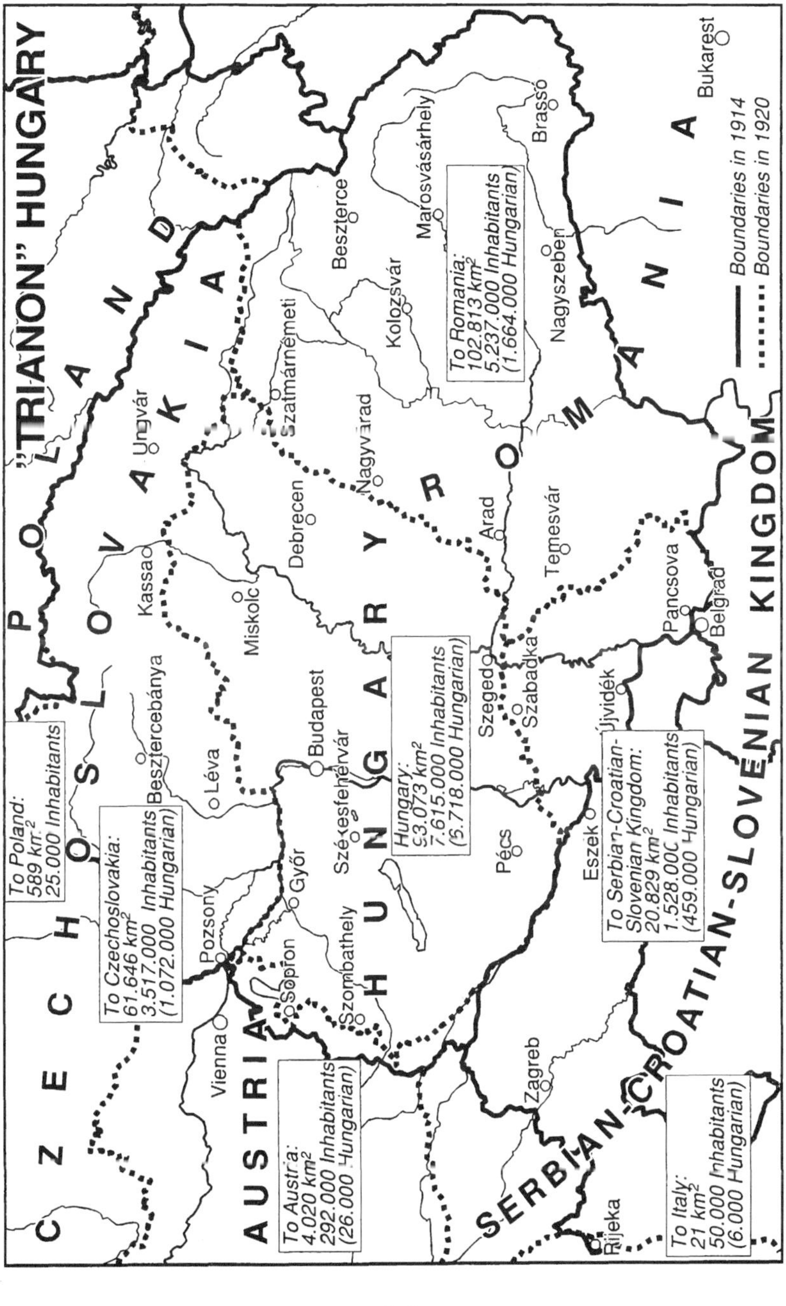

"TRIANON" HUNGARY
P O L A N D
C Z E C H O S L O V A K I A
A U S T R I A
H U N G A R Y
R O M A N I A
SERBIAN-CROATIAN-SLOVENIAN KINGDOM
To Poland:
589 km²
25.000 Inhabitants
To Czechoslovakia:
61.646 km²
3.517.000 Inhabitants
(1.072.000 Hungarian)
To Austria:
4.020 km²
292.000 Inhabitants
(26.000 Hungarian)
Hungary:
93.073 km²
7.615.000 Inhabitants
(6.718.000 Hungarian)
To Romania:
102.813 km²
5.237.000 Inhabitants
(1.664.000 Hungarian)
To Serbian-Croatian-Slovenian Kingdom:
20.829 km²
1.528.000 Inhabitants
(459.000 Hungarian)
To Italy:
21 km²
50.000 Inhabitants
(6.000 Hungarian)
Vienna
Pozsony
Sopron
Szombathely
Győr
Székesfehérvár
Budapest
Léva
Besztercebánya
Kassa
Miskolc
Ungvár
Debrecen
Nagyvárad
Szatmárnémeti
Beszterce
Kolozsvár
Marosvásárhely
Nagyszeben
Brassó
Bukarest
Arad
Temesvár
Szeged
Szabadka
Pancsova
Belgrád
Újvidék
Eszék
Pécs
Zagreb
Rijeka
Boundaries in 1914
Boundaries in 1920

BIOGRAPHIES*

Ádler, Zsigmond (1901–1982)

Boxer, coach. Hungarian champion at flyweight both individually (1925) and as club team member (1928). Club coach from 1927, national team coach (1945–62). A master of physical condition, technical and tactical training, who cultivated a 'fencing' style of boxing. His protégés included three times Olympic champion László Papp. Has an international junior competition named after him.

Ady, Endre (1877–1919)

Poet. Having published the insignificant volume *Versek* (Poems, 1899), from 1900 until his death he earned his living mainly as a journalist. His second collection, *Új versek* (New Poems, 1916), struck Hungarian literary life like a bombshell, however. Revolutionary in form, language and content, these "new songs for new times", as A. himself put it in the opening poem, the public was faced with an abrupt transition for which the work of previous poets had not prepared it. Though even the most conservative reader could not cavil with A.'s strict rhyming schemes, his frequent repetition of whole lines was found irritating, while his unconventional, robust language, with its unusual choices of adjectives, genuinely shocked many, the outrage being completed by the general tone of the poems. A. was an angry young man whose preceding encounter with worldly Paris had provoked him to unleash a storm of excoriatory abuse against Hungary's backwardness. A.'s understanding of his country and the social and political

* Compiled by György Kovács and Zsuzsa Szentgyörgyi (natural sciences and engineering), Ferenc Vámossy (architecture), Edit Dömötör, Zoltán Imre and Orsolya Kelemen (literature), Gergely Bikácsy (film), László Beke, Erzsébet Tatai and Zsuzsa Varga (visual arts), Péter Halász (music), Gyöngyi Szabó Földesi and Rita Raffai (sport).

evils which beset it continued to inspire him, over the next 12 years, to seek new expression—in a further ten volumes of poetry, as well as short stories and countless articles—for his sense of pain, anger and revulsion, which attained apocalyptic dimensions with the sufferings inflicted by World War I. A. was concerned, above all, with eternal problems, and his love poems, mostly inspired by a remarkable woman whom he called Leda, are striking in their originality and their mystical approach to physical love. His religious poems, which seemed blasphemous to many, reveal a search for the hidden God "behind all things, to whom all the bells toll and on whose left, I, alas, sit."

Aján, Tamás (1939)

Sports administrator. Degrees in aesthetics (1964) and physical education (1965); doctorate in latter (1983). Deputy Chairman of National Office of Physical Education and Sport (1983–89). Vice-president (1986–89) and secretary-general (1989–) of the Hungarian Olympic Committee; chairman of the Hungarian Olympic Academy (1989–). Member of International Olympic Committee's Sport for All Committee (1985–); board member of the Association of Summer Olympic International Federations (1989–); member of IOC since 2000. Deputy chairman (1970–76), general secretary (1976–2000), then chairman (2000–) of the International Weightlifting Federation. Member (1978–) then vice-president (1986–) of the General Association of International Sport Federations. Recipient of Sports Prize of the Republic of Hungary (1992) and the IOC's Olympic Order (1995).

Albert, Flórián (1941)

Footballer. Coach's certificate (1988). Member of the Olympic bronze winning football team (1960). Top scorer in the Hungarian League Division 1 (1961, 1965). Several times selected for World XI team. Hungarian Footballer of the Year (1966, 1967); European Footballer of the Year ('Golden Ball'

Award, 1967). Playing career with Ferencváros Athletic Club (FTC), then its coach and manager (1974–87), director (1987–90) and chairman (1990–) of the club's football section. Elected to FTC's 'Hall of Fame' (1974); FTC Golden Diploma (1984); Hungarian Hall of Fame (1992).

Alpár (Stöckl)**, Ignác** (1855–1928)

Late-historicist architect. Beginning his career as a mason's apprentice, he qualified as architect at the Berlin Bauakademie (1877) then worked as Hauszmann's assistant at the Technical University (1882–88). His best-known and enduringly popular work is *Vajdahunyad Castle* (1896), the assemblage of replicas of historical Hungarian buildings in Budapest's City Park, originally conceived as a temporary structure for the Millennial Exhibition but rebuilt in more permanent form (1902–07) to fulfil its present function as an Agricultural Museum. His other important works, all in the Pest inner-city area, were also marked by free adaptations and often highly eclectic combinations of styles, with a preference for neo-Renaissance and Baroque: the *Stock Exchange Building* (1899–1905), since 1975 the headquarters of Hungarian Television, and the *Austro–Hungarian* (since 1924: National) *Bank* (1903–05), both in Szabadság tér; the *Bank Palace in Kristóf tér* (1913–15), off Váci utca, built for the First Pest Savings Bank and now occupied by the Budapest Bank; the *Hungarian General Credit Bank at No. 2–4 József Nádor tér* (1909–13), which since 1945 has housed the Ministry of Finance; and the *Pest Hungarian Commercial Bank at No. 3–4 Roosevelt tér* (1907, with Zsigmond Quittner), which since 1945 has housed the Ministry of the Interior. Sheer stamina alone made him a leading figure in the architectural profession, as a member of the capital's Metropolitan Board of Public Works (1900–23), president of the Steindl Guild (from 1903), member of the National Committee for the Preservation of Historic Monuments (from 1910), president of the Budapest Architects'

Craft Society and of the National Association of Applied Arts (both from 1920), not to speak of being a member of the judging panel for many competitions.

Altorjai, Sándor (1933–1979)

Painter. Having studied at the Pharmaceutical Faculty of the Medical University, Budapest (1951–57), A. enrolled as a student at the Academy of Fine Arts, Budapest, whilst working as a pharmacist. After study trips to Moscow and Leningrad in 1960 (Matisse influence) and Prague in 1962 (Kokoschka influence), he became an advocate of the figurative post-surrealist approach of 'surnaturalism'. In 1965 A. moved to Szigliget, on Lake Balaton, to cultivate his own vineyard, paint, and also participate in several happenings, actions and film projects. A major work, the huge collage diptych *Süllyedjek felfelé (Integető kép)* (Let Me Sink Upwards (Waving Picture)), was rejected by the selection panel for the Studio of Young Artists' 1967 exhibition. The neo-Dadaist private mythology of the world of 'Uncle Gyagya' portrayed in the semi-figurative, drip-painted pictures displayed at his sole one-man show, at the Budapest Exhibition Hall in 1971, provoked official furore. Elements of pop art-type collage and assemblage were increasingly to be identified in A.'s subsequent oeuvre, with details from earlier works sometimes being reused, mixed with lyrical pastel details and (later) spray-over-painting. *Painting for the Blind* (a hot plate of asbestos slates in frames complete with an iron grate) was rejected from a showing at a 1976 exhibition. After several spells of hospitalisation had culminated with a diagnosis of laryngeal cancer in 1979, A. finished his semi-completed works then took his own life.

Anda, Géza (1921–1976)

Pianist. A pupil of Ernő Dohnányi at the Budapest Academy of Music, he moved to Zurich in 1942. He was noted particularly for his interpretations of the works of Mozart and Bartók.

Apáthy, István (1863–1922)
Neuroanatomist. After qualifying in medicine at Budapest, he began researching on nervous tissues in Naples. His theory of neuronal continuity in the transmission of the nerve impulse throughout the nervous system did not survive the test of time, but his research techniques were of enduring value. A. was appointed a lecturer of zoology at Kolozsvár (Cluj) University in 1890, but also taught histology and evolutionism to medical students, and ventured into the social sciences, founding the Hungarian Association for Social Sciences. Having opted to take a political role in the Károlyi government (1918–19), he was sentenced to death by the Romanian authorities who subsequently overtook Transylvania, but French intervention secured his release and he was amongst the staff who transferred with the university in its move to Szeged. His major works are: *Die Mikrotechnik der tierischen Morphologie* (1896); *Das leitende Element des Nervensystems* (1901); *A társadalom élettudományi alapjai* (The Physiological Bases of Society, 1911).

Arany, János (1817–1882)
Poet. After spending a short time in a troupe of strolling players, A. became first an assistant teacher then a village notary. After initial success with a mediocre satirical epic, *Elveszett alkotmány* (The Lost Constitution, 1846) in a competition, A. made his real entry on the literary scene in 1847 with his popular epic *Toldi*, which won him another prize from the same Kisfaludy Society, which earned the admiration and friendship of Petőfi. After the defeat of the Revolution in 1849, A. was again obliged to return to school-teaching, re-emerging into public life in 1858, when he was elected a member of the Hungarian Academy and later (1860) director of the Kisfaludy Society. From 1860 he edited the literary periodical *Szépirodalmi Figyelő* (Literary Observer) and then *Koszorú* (Wreath) until elected first secretary of the Academy in 1865, and its

secretary-general in 1870. *Toldi* was meanwhile expanded to a trilogy with *Toldi estéje* (Toldi's Eve, 1847–48) and *Toldi szerelme* (Toldi's Love, 1879), whilst other important narrative poems include the ironical, autobiographical *Bolond Istók* (Crazy Stevie, 1850), the mock-heroic *A nagyidai cigányok* (The Gypsies of Nagyida, 1852), and *Buda halála* (The Death of Buda, 1863), the first part of another planned epic trilogy. A.'s ballads, e.g. *V. László* (László V., 1853), *Szondi két apródja* (Szondi's Two Pageboys, 1856) and *A walesi bárdok* (The Bards of Wales, 1857), mostly concerned with some terrible crime and its punishment, are unparalleled in virtuosity in Hungarian literature. The poems of his two great lyrical periods are fraught with melancholy. The earlier poems, written in the 1850s, are overshadowed by the loss of Petőfi and A.'s despair for the Hungarian nation and himself, whilst the cycle *Őszikék* (Autumn Bouquet, 1877–80) is his beautiful swan song. A. was also a great translator of Shakespeare (*A Midsummer Night's Dream*, 1864; *Hamlet*, 1867; *King John*, 1867) and of Aristophanes.

Árkay, Aladár (1868–1932)

Architect, painter and industrial designer. From 1886 he pursued parallel studies at Budapest Technical University and under painters Bertalan Székely, Károly Lotz and Károly Kernstok, then at the Faculty of Architecture of the Paris Académie des Beaux-Arts. After a spell in the office of Viennese architects Fellner and Helmer and later with Alajos Hauszmann, he started working with his father-in-law, the architect Mór Kallina. Amongst their Historicist-inspired joint projects were the *Ministry of Defence Building* (1895–97) which stood between Dísz tér and Szent György tér on Buda's Castle Hill; the *Buda Vigadó* (Assembly Rooms) *at No. 8 Corvin tér* (1900; rebuilt 1923), on the Danube embankment at the foot of Castle Hill; the *St. Gellért Monument*

(1904) overlooking the Buda end of the Elizabeth Bridge. On setting up an independent practice, Á. broke with Historicism, designing many family houses in a strongly individual take on the Hungarian Art Nouveau style, e.g. the *Babockay Villa on Andrássy út* (1905), at the corner with Heroes' Square (remodelled by Lajos Kozma in the thirties, and since 1945 the Yugoslav Embassy). Between 1903 and 1913 Á. built a *Villa Community* for judges and public prosecutors *in Buda's Twelfth district* at the corner of Istenhegyi út and Kékgolyó utca. His main works, however, were churches: the *Calvinist Church in Városligeti fasor* (1913) in Pest's seventh district; the initial plan for a *Roman Catholic Church in Buda's Városmajor Park* (1919), which was abandoned before implementation; and the *Church in Gyárváros* in Győr (1928). The *Parish Church of the Holy Heart of Jesus* (1932–33) that Á. and his son, Bertalan, jointly redesigned for the Városmajor site shows a startling assimilation of modernist trends, its ferroconcrete structure early even by international standards.

Auer, Leopold [Lipót] (1845–1930)

Violinist. Studied in Vienna and Hanover, where he was a pupil of József Joachim. From 1868 to 1917 he taught at the Imperial Conservatory, and from 1874 was solo violinist to the Russian tsar. He lived in the US from 1918. Among his students at St. Petersburg were Mischa Elman, Jascha Heifetz, Nathan Milstein, and Efrem Zimbalist.

Babits, Mihály (1883–1941)

Poet, novelist and essayist. B. read classics and Hungarian literature at the University of Budapest then initially went into teaching. His first volume of poetry, *Levelek Iris koszorújából* (Leaves from Iris's Wreath, 1909), appeared in 1909. Having resigned from his teaching post during World War I because of his pacifism, he thereafter devoted all his energies to literature, with his first novel, *A gólyakalifa* (The Nightmare/The Stork

Caliph, 1916) appearing during this period. Along with Ady, Móricz and Kosztolányi, B. was at the centre of the group of leading writers and poets who turned *Nyugat* into Hungary's single most important literary magazine. A shy, reticent man, B. found himself charged with responsibilities he had not sought, including, in 1929, curatorship of the important Baumgarten Prize, awarded annually to a writer "who had remained faithful to his principles." An intellectual imbued with the spirit of antiquity, whose style is remarkable for its compression, B.'s poems can be abstruse. Self-centred and withdrawn in his early period, he later emerged from his ivory tower to deal with the pains and problems of his fellows. In the novel *Halálfiai* (The Sons of Death, 1927) he gave a poignant picture of the disintegration of Hungary's middle classes. A prolific essayist, with the scholarly *Az európai irodalom története* (A History of European Literature, 1934) to his credit, he was also a brilliant translator of, among other works, plays by Sophocles and Shakespeare, Dante's *Divine Comedy*, medieval Latin hymns, and works by Goethe.

Bak, Imre (1939)

Painter. An important representative of the generation of geometric abstract artists who emerged during the 1960s. With works like *Sávok I–II* (Stripes I–II, 1968), B. succeeded in marrying hard-edge and minimalist approaches to Hungarian avant-garde traditions. Most of his works in the early 1970s, such as *Arckép* (Portrait, 1972), were conceptualist, whence he went on to the use of form as a visual sign-constructing language, as in *Nap-ember-arc* (Sun-man-face, 1975–76). Around 1980 these geometries assumed more complexity; during the ensuing period his art could be variously associated with the trans-avant-garde, post-geometry and eclectic new painting (e.g. *Hommage à Giorgio de Chirico*, 1986). By the 1990s, structures comprising architectural components were the principal motifs. *E. Hopper Budaörsön* (E. Hopper in Budaörs, 1997) exemplifies the monumental unity achieved with a 'landscape' of geometrical forms and perspective elements.

Bakó, Pál (1946)

Modern pentathlete. Coach's certificate (1980). World Championship gold in individual (1970) and silver in team (1969) events; Olympic silver in the team event (1972). Coach and technical manager since 1978.

Balassa, Sándor (1935)

Composer. A student of Endre Szervánszky at the Budapest Academy of Music, from 1965 to 1980 he was a recording producer for Hungarian Radio. As measure of his acclaim, the Koussevitzky Foundation and the Boston Symphony Orchestra have commissioned works from him. His best-known compositions are *Requiem Kassák Lajosért* (Requiem for Lajos Kassák, 1969) and the opera *Az ajtón kívül* (Beyond the Threshold, 1977).

Balázs, Béla (1884–1949)

Film aesthete, screenwriter, director. Having had little interest in cinema before leaving Hungary in 1919, B. established his name in the cinema field with early seminal works on film aesthetics, published in Berlin: *Der sichtbare Mensch, oder die Kultur des Films* (1924), *Das Geist des Films*,1930. In Germany he worked as a screenwriter (e.g. for Reinhardt, Piscator and Alexander Korda, as well as for Pabst on *Die Dreigroschenoper*, 1931); with Leni Riefenstahl, he co-directed *Das blaue Licht* (The Blue Light, 1932). After emigrating to the USSR on Hitler's rise to power, B. taught at the Moscow film school. On returning to Hungary after the war, he was screenwriter and artistic adviser on Radványi's *Valahol Európában* (Somewhere in Europe, 1948) and a founder of the Budapest School of Film Art.

Balczó, András (1938)

Modern pentathlete. Coach's certificates (1968, 1974). The world's most successful modern pentathlete, he was Olympic champion in the individual (1972) and team (1960, 1968) events, and won 10 gold, 7 silver and 2 bronze medals at World Championships. Hungarian Modern Pentathlete of the Year (1965–69, 1972) and Sportsman of the Year (1966, 1969, 1972). Now a businessman, he was awarded a Hungarian Heritage Prize (1996).

Bánki, Donát (1859–1922)

Mechanical engineer. Despite completing the engineering course at Budapest Technical University in 1880, he only graduated officially 13 years later, having immediately joined the Ganz Works, where he worked for 18 years, meanwhile reaching the position of chief engineer. He was the designer and leader of many significant projects (e.g. rotary plough, railway switches, turbines, etc.) but is usually remembered for his work on internal combustion engines, including the first description of a petrol carburettor (with János Csonka in 1893). He continued to innovate after becoming head of department at the Technical University in 1899; a hydraulic turbine that he patented in 1916 is still considered to be modern. A member of the Hungarian Academy of Sciences, his major published work was *Energiaátalakulások folyadékban* (Energy Transformations in Liquids, 1916).

Barabás, Miklós (1810–1898)

Painter. The first Hungarian to make a living from painting, B. played a key role in fostering a national culture and artistic life in Pest during the Reform Era. He is celebrated principally for the more than five thousand portraits and miniatures that he painted, with virtually everyone of any standing in public life sitting for him (e.g. Sándor Petőfi, Mihály Vörösmarty, Ferenc Liszt, Ferenc Deák, István Széchenyi). He was elected a corresponding member of the Hungarian Academy of Sciences as early as 1836, despite his rather skimpy formal training, having attended the Akademie in Vienna for just a brief period as a pupil of Johannes Ender, amongst others. He picked up the technique of lithography during a stay in Kolozsvár, in his native Transylvania, and constantly returned to it throughout his career, using it, for instance, to prepare his *Group Portrait of the 1848 Government*. He learnt the technique of water-colour painting from the English painter W. Leighton Leitch, whom he met during a study tour of Italy in 1834–35, employing it in works such as *A Lánchid alapkövetétele* (Laying the Foundation Stone of the Chain Bridge, 1842). The best known of his occasional genre paintings are *Vásárra induló oláh család* (Romanian Family Going to Market, 1843) and *A művész három leánya* (The Artist's Three Young Daughters, 1849). He founded the Fine Arts Society of Hungary in 1859. His autobiography included a comprehensive catalogue of his oeuvre.

Bárány, Róbert (1876–1936)

Clinical physiologist. Born into a Hungarian family but spending most of his life in Vienna, B. qualified at a medical school there then undertook postgraduate studies in Germany before returning to clinical practice in Vienna, devoting himself to studying inner ear function and its innervation. His discovery that vertigo can be induced by washing the external eardrum with a cold fluid (the 'caloric response') led him to show that this and also involuntary eye movements (nystagmus) were linked to disturbances of the vestibular organ of balance; further elaborations of the method aided in the location of certain brain tumours and focal abscesses. His theories were questioned, though, and there were disputes with Endre Hőgyes on the issue of priority in regard to certain findings. Enlisting in the army in 1914, B. soon fell into Russian captivity. In the meantime he was awarded the 1914 Nobel Prize for Medicine, but he only received it a year later, when an exchange of prisoners of war got him to Sweden. As his former Vienna clinic did not want his services, he secured a teaching position at Uppsala University. His major published works were *Physiologie und Pathologie des Bogengang-Apparates beim Menschen* (1907) and *Funktionelle Prüfung des Vestibular-Apparates* (1911).

Barcsay, Jenő (1900–1988)

Painter, graphic artist. Occupying a peculiar place in the Szentendre school of painting, B. drew on Lajos Vajda in his handing on of traditions of icon painting and in his artistic treatment of Szentendre (and occasionally provincial Baroque) motifs whilst at times reducing what he had to say to a 'pictorial architecture' of Kassák-style rigor. His work after 1945 began with groupings of abstract female figures and was carried on in mosaics depicting monumental female forms. His source of inspiration was the architecture of Szentendre through a creative simplification of which he produced lyrical abstract paintings that have a close kinship to constructivism. The device of a painter's easel (a recurrent subject from 1957) links in with these motifs. During the 1960s B. painted many non-figurative, geometric canvases, the severe, highly pared-down black-and-white substrata of which he was to exhibit publicly at a one-man show put on at Budapest's Műcsarnok (Exhibition Hall) a few years before his death. His work as a teacher was extremely important: his *Anatomy for Artists* (1953) and two other manuals have been widely translated and used as primers.

Barcza, Gedeon (1911–1986)

Chess player. Reached international grandmaster grade in 1954. Winner of bronze at Chess Olympics (1956), and silver (1970) and bronze (1961, 1965) at European championships. Outstanding coach and author of chess books.

Baross, László (1865–1938)

Plant breeder. The best known of Hungary's 'wheat improvers'. After graduating from the Mosonmagyaróvár Agricultural Academy B. spent most of his life after 1895 at Bánkút, on the Bükk Plateau just west of Miskolc. The spring and winter strains of 'Bánkúti' wheat that he introduced commercially in the 1930s, based on crosses of the frost-resistant Canadian 'Manitoba' variety with domestic wheats, were superbly adopted to the local soils and climatic conditions and were the most widely sown wheat strains in Hungary for around quarter of a century. The Bánkúti 1201 variety won a Gold Medal for 'Best Winter Wheat' at the World Wheat Exhibition, Regina, Canada, in 1933.

Bartók, Béla (1881–1945)

Composer. Studied and taught at the Budapest Academy of Music, where he and Kodály undertook research into Hungarian folksong. In 1940 he emigrated to the USA. His music blends elements of East European folk music with dissonant harmonies into an astringent and often percussive style. He composed among others *Mikrokozmosz* (Mikrokosmos, 1926–39). His stage works include the opera *A Kékszakállú herceg vára* (Duke Bluebeard's Castle, 1911) and the ballet *A csodálatos mandarin* (The Miraculous Mandarin, 1919).

Bay, Béla (1907–1999)

Fencer, master coach. Qualified lawyer (1937); coach's and master coach's (1961) certificate. Head coach and captain of the national team (1968–76). A titular professor at the College of Physical Education, he revived training of master fencers; as the College's first head coach (1961–69), he developed the Hungarian epée school, preparing several Olympic and World champions. Awards include the Star with Wreath of the People's Republic of Hungary (1987), and the International Olympic Committee's Olympic Order (1997).

Bay, Zoltán (1900–1992)

Physicist. After obtaining a doctorate in 1926 at Budapest University for a dissertation on the theory of magneto-optic phenomena in dispersed media, B. spent four years in Berlin, familiarising himself with the theory of relativity and quantum physics. After returning to Hungary in 1930 as professor of physics at Szeged University, in 1936 he accepted a post as director of the research laboratory at United Incandescent Lamp Co. (Egyesült Izzó), and worked on early radar location systems. Alongside that position, 1938–48 he also taught nuclear physics at the Technical University in Budapest. In early 1946, he was the second person (barely one month after American J. DeWitt) to demonstrate that radar signals could be bounced off the Moon's surface. He went to the US in 1948, first teaching at the University of Washington, then from 1955 to 1972 headed the Department of Nuclear Physics at the US National Bureau of Standards, designing and constructing new instruments in order to carry out his experimental work in nuclear physics. He perfected the sensitivity of electron multipliers (being probably the first to use signal repetition and integration) to the point that they could register individual particles. A major achievement was his reform of the definition of the standard meter in terms of the distance travelled by a specified wavelength of light in a vacuum within a given time frame. His books include *Reflection of Microwaves from the Moon* (1946), *The Speed of Light and the New Meter* (1974), and *Life is Stronger* (1991).

Beke, Manó (1862–1946)

Mathematician. After graduating from Budapest University, he worked for a period as a secondary school teacher before his appointment as professor at the university in 1900; he was also elected a corresponding member of the Hungarian Academy of Sciences in 1914. When stripped of both titles under the Horthy régime that assumed power with the fall of the Soviet Republic in 1919, B. worked at a publishing house until rehabilitated in 1945. He tackled problems of algebra, geometry and physics, but was most widely recognised for his work on linear differential equations. A great populariser of science and also one of the leading modernisers of mathematics education in Hungary, B.'s most important published works were *Determinánsok* (Determinants, 1915) and *Differenciál- és integrálszámítás* (Differential Calculus and Integration, 1916).

Békésy, Georg von [György] (1899–1972)

Biophysicist, acoustician. B. graduated at Bern, Switzerland, but obtained his doctorate in physics from Budapest University and went on to work for the Hungarian Post's Research Institute, where he began to study the physical basis of hearing within the cochlea. Teaching himself the necessary medical background and building his own instruments, he carried out highly sophisticated and unusual experiments to test his theoretical insights, opening up many new approaches in physiological acoustics. Though he continued this work after leaving Hungary in 1946 for the USA, first at Harvard then at the University of Honolulu, the Nobel Prize for Medicine that B. received in 1961 rested to an important degree on the pre-war results. His published works include *Experiments in Hearing* (1960) and *Sensory Inhibition* (1967).

Benedek, Gábor (1927)

Modern pentathlete. College of Commerce graduate (1949); coach's certificate (1965). Won gold medals in the Olympic team event (1952) and World Championship individual (1953) and team (1954) events. Banned for life from competition for political reasons after 1956. Coached Hungary's gold-winning Olympic modern pentathlon team (1960–69), then coach for pentathlon fencing at Warendorf, West Germany (1969–90).

Benedikt, Ottó (1897–1975)

Electrical engineer. Having fled Hungary with the fall of the Hungarian Soviet Republic in 1919, he qualified in Vienna, where he also designed his first innovation, a single-phase commutator motor with excellent characteristics, which earned him an invitation to act as a consultant on its manufacture in the Soviet Union. He stayed on there until 1955, working particularly on asynchronous induction motors for use in diesel-electric tramcars. His major invention, however, was the *autodyn* amplifier machine, considered to be cutting-edge technology until the thyristor came into use. His theoretical research on the calculation of the saturated magnetic fields was of considerable practical significance, his nomographic methods being widely used by engineers in the Socialist bloc countries. On his return to Budapest, B. was made a professor, then rector at the Technical University; he was also director of the Automation Research Institute (SZTAKI) from 1964, and a member of the Academy of Sciences. His books include *Különleges villamosgépek* (Special Electric Machines, 1957), *Autodyne: a New Electrical Machine* (1960), and *Beiträge zur Weiterentwicklung der Gleichstrommaschine* (1975).

Beöthy, Étienne [Béöthy (also Beothy), István] (1897–1961)

Sculptor. Commenced studies at the Architectural Faculty of the Technical University, Budapest, at the end of World War I, focusing on problems of proportionality—in particular, the 'golden sequence' of Fibonacci numbers (...0, 1, 1, 2, 3, 5, 8, 13, 21, 34, 55, 144...). His thesis, completed in 1919, was later published as *La Série d'Or* (1939). His first works were models for cubist-abstract monuments, but he also produced traditional busts. From 1925 B. lived in Paris. His first exhibition, at the Galerie Sacre du Printemps in 1928, was of figurative statues in which he explored the 'golden sequence'. He went on to produce abstract-organic manifestations of this sequence, became deputy chairman of the Abstraction–Création society (1932–36), and in 1938 organised a collective show of Hungarian and French abstract artists at Budapest's Tamás Gallery. After leading a Hungarian resistance group in Paris during World War II, in 1946 he was a founding member (later deputy chairman) of the Salon des Réalités Nouvelles, again drawing in abstract painters from Hungary. In his late oeuvre B. developed an interest in issues of colour. He had a joint show with close friend Auguste Herbin at the Galerie Denise René in 1946, and the Galerie Maeght mounted a one-man retrospective in 1948. B. worked with Fernand Léger and Le Corbusier to launch the periodical *Formes et Vie* in 1951. His wife, Anna Beöthy-Steiner, was also a notable abstract artist.

Berczeller, Rezső [Rudolf] (1912–1992)

Sculptor. After studies at the Budapest School of Design (1927–31) and Academy of Fine Arts (1931–32), where János Vaszary was his teacher, B. began painting but turned to sculpture, not least under the influence of Endre Szöllősi. His early small sculptures and portraits were developed, during the 1950s and '60s, into characteristic interpretations of human masses (demonstrators, fighters, dancers, etc.) as constructions, machinery or complex monsters, with models being found in ancient civilisations and early Western art (Egyptian sculpture, pre-Columbian and Indian cultures, Grünewald's paintings and sculptures, medieval winged altarpieces). Wood was the chosen material for his first large-scale works, e.g. *Dózsa* (1972; since damaged) and *Karóbahúzott* (Impaled, 1968–1969; now at Middelheim Museum, Germany), and his huge bridge construction for the Nagyatád Symposium of Wood Sculpture. He also experimented with non-traditional materials such as the zinc netting used during the 1980s in the shocking *Nyomorékok* (Cripples) series. B.'s consistent concern with bodily suffering (hangings, birth, physical defects), cataclysms, catastrophes and apocalyptic visions is typified by his (unrealised) wish to see his group of 32 hanging angels installed in Budapest either in the former chapel of Kiscell Museum or under the dome of the National Gallery.

Berczelly, Tibor (1912–1990)

Fencer. Attended Ludovika Military Academy (1929–32); coach's (1940) and master coach's (1964) certificates. Won Olympic gold medals in sabre team (1936, 1948, 1952) and bronze in individual (1952) events. Of all Hungarian sabre fencers, he accumulated the highest career tally of individual wins.

Berczik, Sára (1906–1999)

Choreographer, coach. Master coach's certificate (1963). Introduced rhythmic gymnastics in Hungary and coached in that discipline for the Hungarian national gymnastics team (1953–72). Head coach and choreographer of the national rhythmic gymnastics team (1963–78). Teacher of movement at the College of Theatre and Film (1978–83).

Berzsenyi, Dániel (1776–1836)

Poet. A retiring gentleman farmer for most of his life, it was the discovery of his poems by a mutual friend that led Kazinczy to publishing them. His early poems were mainly influenced by Horace and Gessner, with odes such as *A magyarokhoz* (To the Hungarians, 1807), inspired directly by the Napoleonic wars and concerns about Hungary's situation, and the elegy *A közelítő tél* (The Approaching Winter, 1804–08). His friendship with Kazinczy played an important role in his life and poetry between 1809 and 1817, but when B.'s first volume of poetry was published in 1813, younger neologists, notably Kölcsey, criticised him for provincialism and his laboured, jerky style. As a result, B. fell almost totally silent as a poet, instead choosing to pursue theoretical studies in order to formulate a worthy reply, which, when it eventually emerged as *Észrevételek Kölcsey recensiójára* (Remarks on Kölcsey's Review, 1825), offered cogent arguments for the Romantic poetic attitude and images.

Bessenyei, György (1747?–1811)

Novelist, playwright, philosopher, poet. Acquaintance with the ideas of the English and French Enlightenment whilst serving in Maria Theresa's Hungarian Bodyguard in Vienna from 1765 pushed B. into his activities as a novelist, playwright, philosopher, poet. Publication of his *Ágis tragédiája* (The Tragedy of Ágis) in 1772 is traditionally taken to mark the start of Hungarian Enlightenment, with B. taking the lead in organising a group of writers amongst his fellow guardsmen (e.g. Lőrinc Orczy, Ábrahám Barcsay). This and *Hunyadi László tragédiája* (The Tragedy of László Hunyadi, 1772) and *Buda tragédiája* (The Tragedy of Buda, 1773) can be read as veiled comments on the unease of Hungary's nobility with Austrian rule. *A filozófus* (The Philosopher, 1777) is the most significant of several later comedies. *Magyarság* (The Hungarians, 1778), *Törvények útja* (The Path of Laws, 1778), whilst *Egy magyar társaság iránt való jámbor szándék* (A Humble Proposal for the Establishment of a Hungarian Learned Society, 1781) was the first to float this idea. *A holmi* (A Miscellany, 1779) showed that B. was also deeply influenced by Spinozist ideas on the identity of God and nature, whilst *A lélekrül* (About the Soul, 17) by the concept of the soul being a slave to the body. In the novel *Tarimenes utazása* (The Journey of Tarimenes, 1804), unpublished for more than a century, B. depicts a utopian society.

Bihari, János (1764–1827)

Composer, violinist. He founded a Gypsy band in Pest in 1801. His improvised works are known through the notated records left by schooled contemporaries.

Birkás, Ákos (1941)

Painter. Made his debut in the mid-1960s with a series of portraits based on spontaneity of gesture, typified by *Önarckép* (Self-Portrait, 1969). By the mid '70s this program had gradually assumed the hyperrealistic qualities of *Kerítés* (Fence, 1974), soon turning to photographic and filmed interpretations of images as objects, as in *Kép és fal* (Painting and Wall, 1974). From the early 1980s he went back to painting, focusing on the virtuoso spontaneity of highly subjective gestural painting while maintaining his own principles regarding picture structure, with landscapes like *Hegyvidék* (Hilly Country, 1981) painted in strident colours with firm brush strokes. By the end of the decade he became preoccupied with an extremely reduced head motif and a much darker palette, as in *Fej 63* (Head 63, 1989).

Bláthy, Ottó Titusz (1860–1939)

Mechanical engineer. Joining the Ganz Works in 1883, after qualifying in Vienna, B. was the one of the team of himself, Déri, and Zipernowsky who in 1885 patented the first alternating-current transformer to remain loyal to the firm, eventually becoming chief engineer and a director. It was B.'s idea to use a closed iron core in the transformer design, which made AC transmission of electricity a practical proposition, and to measure consumption he also designed the first watt-hourmeter (1889), utilising a small induction motor, which is still the basis for household electricity meters around the world. Most of B.'s subsequent work related to power generation, including generators, turbines, voltage regulators, and an automatic hydraulic turbine regulator which permitted a fuel-burning power station and a hydro-electric station to feed a grid network simultaneously. He was the first to design magnetic circuits for electric machines with the help of magnetisation curves.

Bodó, Andrea (1934)

Gymnast, sports administrator. Degrees in physical education and psychology from University of California, Berkeley, USA. Member of the Olympic hand apparatus team, winning bronze (1952) then gold (1956) medals, and a silver in the combined team event (1952). A US resident since 1956, she teaches at California State University, San Francisco. Office holder in the American Gymnastics Federation and the International Gymnastics Federation since 1984. Technical director at the Los Angeles Olympics and technical deputy chairperson at the Barcelona Olympics in rhythmic gymnastics events.

Bolyai, Farkas (1775–1856)

Mathematician. During his studies in Göttingen he was a fellow student of Gauss. After returning home, he taught at the Reformed Church (Calvinist) College in Marosvásárhely (Tîrgu Mureş, Romania) for 47 years. His two-volume *Tentamen... in Elementa Mathesios Purae* (1832–33) contains several important mathematical results, amongst them one now known as the Bolyai–Gerwien equidecomposability theorem (any two polygons of identical area can be cut up into congruent pairs of polygons), and the three related theorems.

Bolyai, János (1802–1860)

Mathematician. One of the founders of non-Euclidean geometry, he is revered as the greatest of all Hungarian mathematicians to date. After studying under his father Farkas at Marosvásárhely College, he attended military academy in Vienna and started his career as a sub-lieutenant posted to Temesvár (Timişoara, Romania), where he worked out his non-Euclidean theory, published as an appendix to the first volume of his father's *Tentamen...* (1832). It gained no attention until the year of his death, when German R. Baltzer published a book noting the *Appendix*'s importance.

Bozay, Attila (1939–1999)

Composer. A student of Ferenc Farkas at the Budapest Academy of Music, he himself became a teacher there in 1979. His compositions show a marked partiality for the zither and the recorder. The most important works are *Pezzo sinfonico, No. 1* (1967) and the opera *Csongor és Tünde* (Csongor and Tünde, 1985).

Bozsik, József (1925–1978)

Footballer, sport manager. Member of the gold-winning Olympic team (1952), the team that beat England in the 1953 'match of the century' at Wembley Stadium, and of the 1954 World Cup silver-winning team. Chairman of his club's (Budapesti Honvéd) football section (1962–64).

Brassaï [Halász, Gyula Jr.] (1899–1984)

Photographer. Resident in France from 1924, B. owed much to André Kertész for his early knowledge, though Eugène Atget's shots of Paris (1930) were probably equally influential for his first published collection of photographs, *Paris de nuit* (1932). The technical novelty, lyricism and social sensitivity of these pictures captivated contemporaries and rapidly propelled B. to fame. It was more than forty years before the photographs that he also took of drug-users, prostitutes, and other underworld figures were published as *Le Paris secret des années 30* (1976). B.'s interest in his later years turned to photographing graffiti (*Graffiti*, 1960), whilst friendship with Picasso resulted in two books: *Les Sculptures de Picasso* (1949) and *Conversations avec Picasso* (1964; in Eng. as *Picasso and Company*, 1966).

Breuer, Marcel (1902–1981)

Architect, designer. Born at Pécs, he studied with Walter Gropius at the Weimar Bauhaus (1920–24), then when the school transferred to Dessau was appointed head of the joinery and cabinet workshop (1925–28). There followed spells of work in Berlin (1928–33), Zurich (1934–36) and London (1936–37) before he emigrated to the United States, where he became professor of design at Harvard University (1937–46) and also worked for a while in a partnership with Gropius (1937–40). His early works included the first tubular chairs (1925), the first aluminium chairs (1933), and two *Apartment Buildings in Dodertal*, near Zurich (1934, with Alfred and Emil Roth). His early American work, all with Gropius, includes the *Breuer House in Lincoln, Mass.* (1939); a *Housing Project in New Kensington, Penn.* (1941–43); *Buildings for Harvard University* (1945–50); the second and third *Breuer Houses in New Canaan, Conn.* (1947 and 1951); *Exhibition Building in the park of the Museum of Modern Arts, New York* (1949), the *Sarah Lawrence Art Centre in Bronxville, New York* (1952); and the *UNESCO Headquarters in Paris* (1953–58, with P. L. Nervi and B. Zerfuss). These led to prestigious commissions all over the USA.

Bródy, Imre (1891–1944)

Physicist. Finished his studies in Budapest. Having gained his doctorate with a thesis on quantum chemistry and after several years of studying abroad, he joined the United Incandescent Lamp C.'s research laboratory. It was B.'s discovery (1934) that a krypton-filled light bulb had superior properties to any previous design; he then worked with Egon Orowan and Michael Polanyi to develop a practical manufacturing process and particularly to overcome the problem of mass-producing the hitherto prohibitively expensive krypton gas. The lamp was introduced commercially in 1937.

Bródy, Sándor (1863–1924)

Writer, journalist. B. was considered by fellow young writers to be leader of the anti-academic movement around the turn of the century. With Jókai and Zola as his twin literary idols, he was fascinated by the naturalistic features that are also present in Romanticism: the depiction of extremes, exaggerations, and destructive passions. His typical themes are the problems of urban life, illustrated by drifting, exploited and self-surrendering figures. B.'s literary significance is more measurable in his works for the stage than in his prose pieces, his three plays—*A dada* (The Nurse, 1902), *A tanítónő* (The Schoolmistress, 1908), *A medikus* (The Physician, 1911)—inaugurating a new naturalistic, and more progressive, period in Hungarian drama. His fictional works included the novels *Hófehérke* (Snow-White, 1894) and *A nap lovagja* (The Knight of the Sun, 1902), and two short-story cycles, *Nyomor* (Poverty, 1884) and *Rembrandt* (1925).

Budai, László (1928–1983)

Footballer. Many national caps. Member of the gold-winning Olympic team (1952), the team that beat England in the 1953 'match of the century' at Wembley Stadium, and of the 1954 World Cup silver-winning team.

Bujdosó, Imre (1959)

Fencer. Has certificates in coaching (1982) and sports management (1991). Member of gold-winning Olympic (1988) and World Championship (1982, 1991) sabre teams. Awarded the Golden Cross of Merit of the Republic of Hungary (1992).

Buzánszky, Jenő (1925)

Footballer, coach, sports administrator. A qualified railway officer (1960); coach's certificate (1964). Member of the gold-winning Olympic team (1952) and the team that beat England in the 1953 'match of the century' at Wembley Stadium. Club coach and technical director (1972–85). Deputy chairman of municipal and later county football federations (1985–93), then chairman of latter (1993–). A recipient of the International Olympic Committee's Olympic Order (1995) and the Ferenc Csík Prize (2001).

Capa, Robert [Friedmann, Endre] (1913–1954)

Photographer. One of the all-time great press photographers, C. was a self-educated man who achieved world fame largely through the photographs he took during the Spanish Civil War and, as a reporter for *Life* and *Collier's* magazines as well as the American Army, during the D-Day landings in World War II. The aesthetic essence of his war photographs was their synthesis of chance and classic style; he also had a particular talent for blending sharp-eyed, psychological observation with a moral view. C. became an American citizen in 1946. The joy of restarting life re-emerged in C.'s post-war photographs. His portraits, many of celebrities who were his friends (Ingrid Bergman, Hemingway, Picasso, Steinbeck, Matisse, Gary Cooper etc.) unite the qualities of carefully composed pictures and spontaneous snapshots. In 1947 he was a co-founder, with H. Cartier-Bresson, D. Seymour and G. Rodger, of the photographic agency Magnum, and their top representative until his death. He made three trips to Israel between 1948 and 1950. The photographs taken in Hungary during the years of the communist take-over (1947 and 1949) are interesting documents, largely overlooked in the international history of photography. After a spell based in Paris (1950–53), he was killed by a mine whilst covering the war in Indochina for *Life*.

Csáth, Géza (1887–1919)

Writer, music critic and psychiatrist. Cs.'s early short stories are characterised by their naturalistic, symbolic features and their break with the typically anecdotal tradition of previous Hungarian prose-writing. The young Cs. was influenced both by a Darwinist biological approach and a Symbolist stance towards the mystery of life. In his best-known short story collections, *A varázsló kertje* (The Magician's Garden, 1908) and *Délutáni álom* (Afternoon Nap, 1911), the protagonists are typically driven by repressed drives and thoughts, with not only adults but also seemingly innocent children capable of displaying murderous instincts and sadistic inclinations. For Cs., human life has to be seen as a whole, embracing fantasy as well as mere physical existence, which is why his characters often seek the totality of childhood experience and the world of dreams—presented, in the best examples, with a masterly handling of time and objectivity of language. Of eight works for stage (several fragmentary), the puppet play *Hamvazószerda* (Ash Wednesday, 1911) can be regarded as forerunner of absurdist drama. Cs. was also one of the first critics to discern the importance of Béla Bartók and Zoltán Kodály, and in 1912 published the personal journal of one of his female patients as *Egy elmebeteg naplója* (Diary of a Mental Patient).

Csermák, Antal (ca. 1774–1822)

Composer. The best schooled of the cultivators of the *verbunkos* style, he is historically important for taking the first steps in creating a nationally flavoured and programmatic chamber music. His main work is *Az intézett veszedelem avagy a Haza szeretete* (The Peril Addressed, or Love of One's Country, 1809).

Csermák, József (1932–2001)

Hammer thrower, coach. Qualified train driver. Olympic hammer champion with world record throw (1952); first to throw the hammer over 60 meters. Throwing coach of the Hungarian Athletics Federation (1967–70), producing several Hungarian champions.

Csernus, Tibor (1927)

Painter. A master of sur-naturalism and hyper-realism, he was a student of Aurél Bernáth at the Academy of Fine Arts. His early work, in landscapes such as *Sashegyi látkép* (View of Sas Hill, early 1950s), portraiture and other figurative compositions, including *Ferenc Juhász* (1954) and *A három lektor* (Three Publisher's Readers, 1955), was recognised with the award of a Munkácsy Prize in 1953. During a study trip to Paris in 1958, however, he picked up new techniques of scoring and texturing paint surfaces to give his realist pictures a surrealist 'feel', leading him to produce 'sur-naturalist' masterpieces like *Lehel téri piac* (Lehel Square Market, 1962). Cs. settled permanently in Paris in 1964, working initially on applying a Caravaggesque chiaroscuro to hyper-realism and later applying lessons drawn from other great masters to scenes from the Bible and modern everyday life, as in a recent series of paintings based on Hogarth's *The Harlots' Progress*. His most important exhibitions have been mounted in Paris, New York, Chicago and Budapest.

Csete, György (1937)

Organic architect. After graduating from Budapest Technical University (1961), he was appointed architect to Mineralimpex, the Hungarian Trading Company for Oils and Mining Products (1961–69). He then moved to the Pécs regional design office (1970–77), amongst whose young architects he organised the 'Group Pécs' atelier in 1973. However, the motifs adopted in one of the group's earliest projects, the embellishment of dwellings on a housing estate for workers at the Paks Nuclear Power Station (1973) provoked official disapproval—the so-called 'tulip debate'—which resulted in their disbandment. Cs.'s key work in this period was a *Water-pumping Station (the Spring House) at Orfű* (1974), just outside Pécs, the striking fructiform design of which was to become a symbol for one of the main branches of organic architecture. His only other major project for many years was the *St. Elizabeth Church at Halásztelek* (1976, with Jenő Dulánszky), on the southern outskirts of Pest. Cs.'s manner of reconciling technically innovative approaches with ancient architectural forms and the advocacy of ecological principles is exemplified by such masterpieces of more recent years as the *Museum* and ecumenical *'Church of the Forest' for the National Historical Memorial Park at Ópusztaszer* (1992–96), north of Szeged. The *Chapel of Reconciliation at Beremend* (1995–97), near the Serbian border, with its two slim towers, stubby central core of petaloid concrete shells and lily-shaped crosses in front, powerfully evoke the natural budding process.

Csík, Tibor (1927–1976)

Boxer. Twice Hungarian champion at bantamweight (1946, 1948), then Olympic champion (1948); took gold at the World University Championship (1949). Left Hungary in 1956, dying in Australia.

Csokonai Vitéz, Mihály (1773–1805)

Poet. A student at the Calvinist Debrecen College for 15 years, he was expelled in 1795 for nominal infractions of college rules, or rather because of his sympathy for the Jacobins. His first literary works are the animal fable *Bagoly és kócsag* (The Owl and the Heron, 1790) and the Homeric 'travesty' *Béka-egérharc* (Batrachomyomachia, or the Battle of the Frogs and Mice, 1791), both express his disappointment in the resistance of the nobility, but his satirical poems also attack the citizens. Rococo style and popular elements appeared in his work by the early 1790s. The unfinished satirical play *Méla Tempefői avagy Az is bolond, ki poétává lesz Magyarországon* (Musing Tempefői, or Whoever Becomes a Poet in Hungary is a Fool, 1793–95) criticises the deplorable state of education and civilisation in the country, and depicts the personal sorrows of the poet. *Konstancinápoly* (Constantinople, 1793) is an anti-religious poem deeply influenced by Voltaire, whilst the yearning for an uncorrupted earlier age of *Az estve* (The Evening, 1793) is more Rousseauesque. A further comic epic, *Dorottya vagyis A dámák diadala a fársángon* (Dorottya, or The Triumph of the Ladies at the Carnival, 1799), criticised the imitation of foreign fashions, its prologue emphasising the priority of content over form. The love poetry of the *Lilla* cycle, possibly composed before 1795 but not published until after his death, broke new ground in Hungarian. The *Anakreoni dalok* (Anacreontic Songs) of Cs.'s final years manifest the inspiration of the Greek poet, whilst his last great philosophical poem, *Halotti versek* (Funeral Verses, 1804), is a meditation on the immortality of the soul.

Csollány, Szilveszter (1970)

Gymnast. Coach's certificate (1992). Olympic silver (1996) then gold (2000), and golds in World Cup (1998) and European Championships (1998), all on the rings; European gold on the pommel horse. Coach in the USA (1995–98). Awarded Officer's Cross of the Order of Merit of the Republic of Hungary (2000).

Csongrádi, László (1959)

Fencer. Coach's certificate (1991). Member of gold-winning Olympic (1988) and silver-winning World Championship (1990) sabre teams.

Csonka, János (1852–1939)

Mechanical engineer. With only four years of elementary schooling, Cs. picked up his skills (including fluency in several foreign languages) whilst working as an apprentice in various machine workshops around Europe. At the age of 25 was appointed to run the workshop at Budapest's Technical University. A prolific inventor of motor engines, he was asked by András Mechwart to help put the Ganz Works their foreign-made motors into operation in 1887. His most significant work was the carburettor that he invented in 1893, together with Donát Bánki, which already had a spud, a butterfly strangler valve, or 'choke', and a damped air feeder. Cs. appears to have been the only Hungarian ever to whom the Chamber of Engineers ever awarded the right to call himself a qualified mechanical engineer without receiving a formal university training.

Csontváry Kosztka, Tivadar (1853–1919)

Painter. The most idiosyncratic of Hungary's *fin-de-siècle* artists, his work, though rooted in Romanticism, displays a visionary world of highly symbolic subjects, painted in pure, bright colours with odd perspectives and spatial effects. From 1894 he studied painting in Munich, Karlsruhe, Düsseldorf and Paris, and made visits to Palestine and Egypt. The expressionist *A Panaszfal bejáratánál Jeruzsálemben* (At the Entrance to the Wailing Wall in Jerusalem), with its multitude of figures, and the big landscapes *A nagytarpataki völgy a Magas-Tátrában* (Tarpatak Valley in the High Tatras) and *Görög színház Taorminában* (Ruins of the Greek Amphitheatre at Taormina) all stem from the period 1904–05, while *Baalbek*, his masterpiece, was completed in 1906. A trip to the Lebanon resulted in two significant works, *Magányos cédrus* (Solitary Cedar, 1907) and *Zarándoklás a cédrusokhoz* (Pilgrimage to the Cedars in Lebanon, 1907), which were followed by *Mária kútja Názáretben* (Mary's Well in Nazareth, 1908). Though Cs. made further trips to the Middle East, he stopped painting in 1910 and was already of failing mental health when, during World War I he resumed work on a massive cartoon drawing for *A magyarok bejövetele* (Arrival of the Magyars), also commencing an autobiography and writing pamphlets. Though showings of his work were put on in 1905, 1908 and 1910, recognition of his true significance only came when his collected works went on display in 1930, later cemented by several successful exhibitions abroad.

Curtiz, Michael [Kertész, Mihály] (1888–1962)

Director. Originally an actor, a study trip in 1912 to the Nordisk studio in Denmark set him off as a director of close to fifty silent films back in Hungary, ranging from crime thrillers like *Rablélek* (Captive Soul, 1913), through an adaptation of József Katona's stage tragedy *Bánk bán* (1914) to the anti-war *Jön az öcsém!* (My Brother's Coming Home!, 1919). Moving to Austria in 1919, he directed several high-budget epics (e.g. *Sodom and Gomorrah*, 1922). In 1926 he took up residence in the USA, working for Warner Bros. for several decades as a prolific director in many film genres, including the biblical epic *Noah's Ark* (1929), the earliest film musicals (*The Jazz Singer*, 1927, starring Al Jolson), 'films noirs', horror movies (*20,000 Years in Sing Sing*, 1933; *Mystery of the Wax Museum*, 1933), but above all swashbuckling adventure spectaculars (*Captain Blood*, 1935; *Robin Hood*, 1938; *The Sea Wolf*, 1941; *The Sea Hawk*, 1940), in which Errol Flynn typically starred as a hero intrinsically committed to order and legitimate power but forced to become a rebel in order to fight injustice. Most enduring, however, is the psychological melodrama *Casablanca* (1942), in which Humphrey Bogart's acting accurately reflects the film's anguished split between idealism and cynicism. Even his late films (*Huckleberry Finn*, 1960; *Francis of Assisi*, 1961; *The Comancheros*, 1961) are characterised by a variety of subjects and story-lines.

Czene, Attila (1974)

Swimmer. Coach's certificate (1998). In 200 metres individual medley, he took bronze (1992) then gold (1996) at the Olympics. Runs the Czene–Savanya swimming school.

Czibor, Zoltán (1929–1997)

Footballer. Member of the gold-winning Olympic team (1952) and the team that beat England in the 1953 'match of the century' at Wembley Stadium. Top scorer in the Hungarian League Division 1 (1955). A long-time resident in Spain (1958–91), he was also a famed goal scorer in Spanish club colours. Has since returned to Hungary. Recipient of the International Olympic Committee's Olympic Order (1995).

Cziffra, György (1921–1994)

Pianist. An infant prodigy, he was a pupil of Dohnányi at the Budapest Academy of Music from 1930. His career was interrupted by spells as a POW during World War II then as a political prisoner in Rákosist Hungary (1950–53). He moved to France in 1956, where he was hugely successful, particularly with interpretations of Liszt and Chopin in the tradition of 19th-century virtuoso pianists.

Czigány, Kinga (1972)

Kayaker. In kayak fours over 500 meters, a winner of gold at the Olympics (1992), and silver and several bronzes at World Championships. Woman Kayaker of the Year (1993). A television personality and journalist in Hungary, she now lives in Canada (1999–). Awarded the Small Cross of the Order of Merit of the Republic of Hungary (1992).

Darnyi, Tamás (1967)

Swimmer. Coach's certificate (1994). In the individual medley, he holds Olympic golds over 200 meters (1988, 1992) and 400 meters (1988, 1992), and World Championship golds over 200 meters (1986, 1990) and 400 meters (1986, 1990), as well as being multiple European champion and world record holder. European Sportsman of the Year (1991); inducted into the Hungarian Hall of Fame (1991). Runs the Darnyi swimming school. Awarded Middle Cross of the Order of Merit of the Republic of Hungary (1992) and the International Olympic Committee's Olympic Order (1996).

Déri, Miksa (1854–1938)

Mechanical engineer. Although beginning his studies in Budapest, he qualified as a water construction engineer in Vienna. He continued to study electro-technology, and then joined the Ganz Electrical Works in 1882, where he was part of the team (with Zipernowsky and Bláthy) which developed the AC transformer (1885), doing most of the experimental work and the international promotion as the firm's representative in Vienna (and later its director). He also worked with Zipernowsky on a self-starting, compound-wound synchronous generator and himself designed a compensated DC motor and single-phase commutator motors.

Derkovits, Gyula (1894–1934)

Painter. Rather to follow his father's trade as a carpenter, he volunteered for army service, was wounded and contracted tuberculosis. He settled down in Budapest, learning graphic art at a night school from Károly Kernstok, amongst others. He joined the Communist Party of Hungary in 1918; during the Soviet Republic was sent to the artists' colony of Nyergesújfalu, on the Danube between Komárom and Esztergom. After the collapse of the Commune he was again reduced to making his living from manual labour, but then made a fresh attempt to break through as an artist during a spell in Vienna from 1923–26, where he won a successful showing at the Galerie Welhburg. In 1927 D. had a one-man show at the Ernst Museum, Budapest. His style evolved from cubism towards compositions in which the monumental scale is expressionistically heightened by dramatic lines. His series of 12 woodcuts entitled *1514* (1929) counts as the most signal body of graphic art of the period. D. moved increasingly away from urban landscapes to more explicitly social subjects, as in *Téglahordók* (Hod Carriers, 1932), *Zsákolók* (Sackers, 1932), *Szövőmunkások* (Textile Workers, 1933), *Homokszállítók* (Sand Deliverers, 1934), and the like. His marriage marked the return of double portraits and intimate scenes of family life as subjects of his canvases, with marvellous gradations of reds and silvers dominating in his masterfully restrained palette. The most outstanding of his works include *Szőlőevők* (Grape Eaters, 1929), *Végzés* (Execution, 1930), *Téli viharban* (Winter Storm, 1931), *Nemzedékek* (Generations, 1932), *Hídépítők* (Bridge Builders, 1932), and *Híd télen* (Bridge in Winter, 1933). D. was posthumously awarded a Kossuth Prize in 1948.

Detre, László (1906–1974)

Astronomer. After studies in Budapest and Berlin, he joined the staff at the Budapest Observatory (Institute of Astronomy), becoming its director in 1943 until his death. Unable to continue his original work on astronomy statistics due to the inadequate data provided by the modest available instrumentation, he switched to the study of variable stars. Observation of long-term periodic changes in these bodies yielded valuable results, and this area has remained one of the most important areas of research in Hungarian astronomy to the present day. In 1961 D. founded the *Information Bulletin on Variable Stars*, which ever since has been edited in Hungary. He was member of the Hungarian Academy of Sciences.

Dévényi, Sándor (1948)

Organic architect. On graduating from Budapest Technical University (1973) he worked briefly for IPARTERV (1973–75), the state planning office for industrial construction. He was thereafter with various planning offices in Pécs and meanwhile a beneficiary of the Young Architect's Circle (FÉK) postgraduate training scheme of the Architects' Union (1980–82). Of particular note amongst the earlier products of his long-term involvement (1975–98) with projects relating to the regeneration of Pécs city centre were the *Wedding-Party House* (1983–84), which is reminiscent of Gaudí's style, and his *House Struck by Lightning* (1978–84), which displays striking post-modern affinities in the way it preserves the Eclectic-style remains of a war-ruined house whilst turning the interior spaces inside-out towards the street. He has since developed his own highly individual brand of organic architecture, with a strong sense for decoration and an original formal approach as manifested by: the '*Bulls-head' House at No. 3 Kazinczy utca* (1981–86); the Zsolnay Tea House (1981), in which a model of Pécs Cathedral is displayed; the *Roman Courtyard* (1991), incorporating Roman archaeological relics and finds; the *Munkácsy Courtyard* (1992); the *Bagolyvár Restaurant* (1992); and the *Hungária Offices* (1993). He also worked with István Kistelegdi on the plans for Kálvária

Tunnel (1981–85), on the northern approach to Pécs. Outside Pécs his projects have included an dynamically-roofed, stone-walled *Villa in the Mecsek Hills* (1986); *Spa Hotels at Harkány* and *Csongrád* (both 1990) and *at Szigetvár* (1991); and the *Farm at Makár* (1992–98). In 1991–96 he participated in the Lágymányos Bridge Competition for a new Danube bridge design for Budapest. D. participated in the Hungarian Living Architecture exhibition (1985), was a founding member of the Károly Kós Association (1989), and exhibited at the ground-breaking 1991 International Architectural Bienniale at Venice. Awarded the Ybl Prize in 1988, he is also a recipient of the 'Pro Architectura' Prize (1994) and Kossuth Prize (1999) and his work has three times won the 'House of the Year' Prize (1986, 1987, 1990).

Dohnányi, Ernő (1877–1960)

Pianist, composer. Following studies in Budapest, in 1898 his spectacularly successful pianist career started in Europe and the US. He had two spells as a professor at the Budapest Academy of Music (1916–19 and 1928–41) and was also the long-time chief conductor of the Budapest Philharmonic Orchestra (1921–44). After 1945 he lived in the US. The emotional profundity and yet seeming effortlessness of his technique made him a legendary interpreter of Beethoven and Liszt. His compositions early on were in the late Romantic tradition but later took on neo-classical characteristics.

Dónusz, Éva (1967)

Kayaker. Olympic champion in kayak fours over 500 meters (1992). Runs a fitness club. Awarded the Small Cross of the Order of Merit of the Republic of Hungary (1992).

Doráti, Antal (1906–1988)

Conductor, composer. He was a pupil of both Kodály and Weiner at the Budapest Academy of Music. After 1928 he conducted in Germany. He moved to the US in 1945, moving on to appointments as conductor to the Dallas (1945–49), Minneapolis (1949–60), Washington National (1971–78) and Detroit (1977) Symphony orchestras.

Dömölky, Lídia (1936)

Fencer. Coach's certificate (1978). An individual gold in foil at the World Championships (1955), as well as team golds both there (1955, 1959, 1967) and at the Olympics (1964), with two Olympic team silvers (1960, 1968). Coach in Hungarian sports schools then at Harvard University. Now a journalist and specialist writer.

Dudich, Endre (1895–1971)

Zoologist. Having studied in Budapest and Szeged, he was professor of zoology at Budapest University from 1933, as well as setting up and directing the Institute of Zoological Taxonomy. Active in many areas of his subject, D.'s most enduring contributions were to the morphology and classification of arthropods, though he also did pioneering work in cave biology and population ecology. A member of the Hungarian Academy of Sciences, his main published works were *Barlang-biológia és problémái* (Cave Biology and its Problems, 1959), *Állatrendszertan* (Zoological Taxonomy, 1969), and *Rendszeres állattan. Ízeltlábúak* (Systematic Zoology. Arthropoda, 1972).

Durkó, Zsolt (1934–1997)

Composer. A pupil of Ferenc Farkas at the Budapest Academy of Music, he went on to study with Goffredo Petrassi in Rome (1961–63). In his works he strove to unify modern techniques with the traditions of Hungarian music. His best-known compositions are *Una rapsodia ungherese* (1965), *Halotti beszéd* (Burial Prayer, 1972), and *Mózes* (Moses, 1977).

Egerszegi, Krisztina (1974)

Swimmer. Olympic champion over 100 meters (1992) and 200 meters (1988, 1992, 1996) in backstroke, and over 400 meters in individual medley (1992). Multiple world record holder.

Top in European and 2nd in World Sportswoman of the Year (1992); Hungarian Sportswoman of the Year (2000). Inducted into the Hungarian Hall of Fame (1991). Awarded Middle Cross of the Order of Merit of the Republic of Hungary (1992), the International Olympic Committee's Olympic Order (1997 and 2001), and a Life Fair Play Prize (2001).

Egry, József (1883–1951)

Painter. E.'s uniquely atmospheric works do not neatly fit into any of the 'isms' that succeeded impressionism. Having studied in Munich and Paris, he entered the Budapest Academy of Fine Arts. The early paintings, *Menhely előtt* (Before a Night Asylum, 1907) and *Kikötőmunkások* (Stevedores, 1912), apart from their references to his own hard life, show influences of Millet and Meunier. A one-man show was held in the House of Artists as early as 1909. After a fairly long study trip to Belgium, he found a true outlet for his pantheistic leanings in the elementary experience of nature around Lake Balaton, where he settled permanently, except for a few short trips to Budapest or abroad, first at Keszthely in 1918 then, in the '20s, at Badacsonytomaj. The light and horizon of paintings like *Szivárvány* (Rainbow, c. 1926), *Visszhang* (Echo, 1936), and *Napfelkelte* (Sunrise, 1940) express a universal human desire for freedom, whilst the animals and figures of pictures like *Hazatérő nyáj* (Flock Returning Home, 1932), *Delelés* (Siesta, 1937) and *Halász* (Fisherman, 1937–38) bestow an intimate voice to his love of nature. He also painted a number of self-portraits and a few pictures on religious subjects. E.'s work can be seen at the National Gallery, Budapest, and at Kaposvár as well as the Egry Museum, Badacsony.

Elek, Ilona (1907–1988)

Fencer, sports administrator. Graduate of the Budapest Academy of Music. She was Hungary's first woman Olympic champion, taking individual golds (1936, 1948), as well as silver (1952) in the foil; she also won individual (1951) and team

(1937, 1952, 1953, 1954, 1955) golds at World Championships, making her the most successful woman fencer in the history of fencing. An honorary member in perpetuity of the International Fencing Federation, she was awarded the International Olympic Committee's Olympic Order (1983).

Énekes, Árpád (1921–1992)

Boxing coach. Certificate as PE instructor (1946), coach (1946) and master coach (1964). Coached several Olympic and European boxing champions and other medal winners. Author and publisher of textbooks on boxing.

Eötvös, József (1813–1871)

Writer, statesman. E. devoted his energies to the creation of modern Hungarian literature and the establishment of a modern democratic Hungary, especially through educational reform. E. was deeply influenced by his first-hand studies of social conditions in England and France in 1836–37. Proclaiming literature's social mission, notably the alleviation of poverty, his first novel, *A karthausi* (The Carthusian, 1839–41), a tale of a poor girl's seduction by an aristocrat set in Paris of the day, was a masked criticism of conditions in Hungary. *A falu jegyzője* (The Village Notary, 1845) is a bitter satire of old Hungary, whilst his great historical novel *Magyarország 1514-ben* (Hungary in 1514, 1847) is a picture of one of the great sixteenth-century peasant rebellions. In subsequent essays E. advocated modernisation of the penal code and prison reform, whilst in *Szegénység Irlandban* (Poverty in Ireland, 1837–40) he pointed out that Western Europe had its social problems too. E. became Minister of Education in the revolutionary government of 1848 and prepared a plan to raise educational standards. In his political writings he wished to base relationship between Austria and Hungary on the principles of 1848, and the Compromise of 1867 was partly his work. He played a distinguished part in the reorganisation of the Hungarian Academy and became

Minister of Education again after 1867, pushing through modernisation of the educational system.

Eötvös, Loránd (1848–1919)

Physicist. A graduate from Heidelberg, he became first a lecturer (1872) then professor (1878) and eventually rector of Budapest University (which has borne his name since the war). His initial research interest was in capillarity, carrying out much work on measuring the surface tension of liquids and its relationship to molecular weight and the effect of heat. E.'s most enduring work, however, were the major advances he made in the measurement and theoretical understanding of gravitational fields with his self-designed double-armed torsion balance. Use of the balance allowed him to determine the gravitational constant to an accuracy of one in 200 million (improving on previous measurements by three orders of magnitude), and also had widespread practical use in geodetics and geophysics for detecting local variations in the gravitational field. He founded the Hungarian Mathematical and Physical Society (1891) and was also briefly (1894–95) the country's minister of education.

Eötvös, Péter (1944)

Composer, conductor. Studied under Ferenc Szabó at the Budapest Academy of Music before moving to Cologne from 1966, where he became closely associated with Karlheinz Stockhausen's ensemble. From 1979 he led the Paris-based Ensemble InterContemporain; since 1994 he has been chief conductor of the Radio Netherlands Chamber Orchestra. As a composer, he was originally allied to the New Music Studio circle in Budapest; more recently he has gained widespread success in Europe with his visionary, psychologically slanted compositions. His best-known works are *A szél szekvenciái* (Sequences of Wind, 1975), *Psychokosmos* (1993) and *Three Sisters* (1998).

Erdély, Miklós (1928–1986)

Architect, writer, visual artist, film director. The most versatile of the avant-garde figures who emerged in the latter half of the 1960s, E.'s theoretical bent led him to a concept of art that focused on the relationship between science and the arts and on privileging new expressive approaches in a wide range of art forms. Much of his output was thus designed to demonstrate principles derived from his theoretical concerns. Having published poems and short stories in the early '60s, he turned to conceptual art and a regular involvement in producing films, notably *Partita* (1974), *Álommásolatok* (Dream Copies, 1977), *Verzió* (Version, 1979–81), *Vonatút* (Train Trip, 1981), and *Tavaszi kivégzés* (Spring Execution, 1985). During the late '70s and early '80s E. made carbon-copy graphics and asphalt paintings, as well as installations and environments such as *A kalcedoni zsinat emlékére* (In Memory of the Council of Chalcedon, 1980), which incorporated a multiplicity of materials (e.g. bitumen, matzos, glass), some bearing clearly symbolic significance. From 1984, when he completed an installation entitled *Hadititok* (Military Secret), E.'s works were displayed at various exhibitions and he lectured extensively.

Erdős, Pál (1913–1996)

Mathematician. Internationally acknowledged as one of the most outstanding figures in his discipline, he was the first Hungarian to receive its highest distinction, the Wolf Prize (1983), for his contributions particularly to combinatorics and number theory. Already noted at school as a brilliant solver of mathematical problems, he was a pupil of Lipót Fejér at Budapest University but also deeply impressed by Dénes Kőnig's lectures on graph theory at the Technical University. A paper on number theory that he jointly wrote with László Kalmár whilst still a student won him instant recognition, and on completing his doctorate in 1934, he was invited to

Manchester. Based mainly in the US from 1938, he published over 1,500 papers, covering a wide range of topics, mostly in co-authorship with others. He was member of eight academies and was awarded the title of honorary doctor at six universities.

Erkel, Ferenc (1810–1893)

Composer. Musical director of the Budapest National Theatre (1838–74). He was director of the Hungarian National Academy of Music (1875–88) and one of the first striving to establish a national opera. His first opera, *Bátori Mária* (1840), was followed by *Hunyadi László* (1844) and *Bánk bán* (1844–52).

Esterházy, Péter (1950)

Novelist. By common consent, the leading (and most popular) writer of the generation whose members began their careers in the 1970s. E. made his mark with *Termelési regény* (Novel of Production, 1979), a parody of socialist realist dictates, which are subverted through pointed use of literary quotations, notably the nineteenth-century Hungarian anecdotal tradition, and the absurdity of a modern-day Johann Peter Eckermann annotating the life of his 'Master'. *Bevezetés a szépirodalomba* (An Introduction to Literature, 1986), which brings together seven separately published works—*Függő* (Depending, 1981), *Daisy* and *Ágnes* (both 1982), *Fuharosok* (The Transporters, 1983), *Kis Magyar Pornográfia* (Little Hungarian Pornography, 1984) and *A szív segédigéi* (Helping Verbs of the Heart, 1985)—with a further 17 texts, all of markedly different character, into an huge inter-textual structure, succeeds in being both completely new and radical yet at the same time rooted in a multitude of earlier traditions. More recent works include *Tizenhét hattyúk* (Seventeen Swans, 1987), published under the pseudonym Lili Csokonai, which is the purported diary of a young, uneducated but clearly very modern woman couched in seventeenth-century Hungarian; *Hrabal könyve*

(The Book of Hrabal, 1990), a memoir of E.'s encounter with the old Czech writer; *Hahn-Hahn grófnő pillantása* (The Glance of Countess Hahn-Hahn, 1992), nominally a Danube travelogue; and *Egy nő* (She Loves Me, 1995), an enigmatic set of texts describing the relations of the narrator(s) to a female figure(s). In E.'s most recent novel, *Harmonia Caelestis* (2000), the narrator employs constant changes of characters, perspectives, and narratives to explore his family's history, and especially the significance of the father figure, within a structure of two seemingly very discrepant parts.

Fábián, László (1936)

Kayaker, coach. Coach's certificate (1978). Olympic champion in kayak doubles over 10,000 meters (1956) as well as World Championship golds in both doubles (1958, 1963, 1966) and fours (1963) and seven times European champion. Head coach, head of the kayak section at Újpesti Dózsa (1969–90) and ÚTE (1991–), with several world champion students.

Fábián, László (1963)

Modern pentathlete. Coach's certificate (1987); degree in economics (1995). Won a team gold at the Olympics (1988), golds at World Championships in individual (1989), team (1987, 1989), and relay (1991, 1993) events and at European Championships in individual (1993), team (1991, 1993), and relay (1991, 1993). Several times Hungarian Modern Pentathlete of the Year, and Sportsman of the Year (1989). Vice-chairman of the Hungarian Modern Pentathlon Federation.

Fábri, Zoltán (1917–1994)

Director. After training as an artist, F. worked in theatre as a stage designer and (after 1945) director before turning to films. After *Életjel* (Fourteen Lives in Danger, 1954), *Körhinta* (Merry-Go-Round, 1955), his third feature film, and the subsequent *Hannibál tanár úr* (Mr. Hannibal, Teacher,

1956), with their expressionist style and dramatic handling of topical subjects, were widely acclaimed abroad. International attention was not regained until *Két félidő a pokolban* (Two Halves In Hell, 1961), which examined the moral implications of war. F.'s many film adaptations of literary works included *Édes Anna* (Anna Édes, 1958), *Húsz óra* (Twenty Hours, 1965), *Pál utcai fiúk* (The Boys of Paul Street, 1967) and *Tóték* (The Tóth Family, 1969). *Twenty Hours*, winner of the Grand Prix at the Moscow Film Festival, played a key role in Hungarian cinema by overcoming the official barrier to more candid treatments of previously taboo subjects (in this case the days preceding the 1956 revolution). *141 perc a befejezetlen mondatból* (The Unfinished Sentence in 141 Minutes, 1974), a two-part adaptation of Tibor Déry's novel, used narrative association to highlight the subjective power of human memories. After *Magyarok* (Hungarians, 1977), F.'s last major achievement was *Az ötödik pecsét* (The Fifth Seal, 1975), set in 1944 and reflecting his uncompromising attitude to moral problems.

Farkasdy, Zoltán (1923–1989)

Architect. Whilst training at Budapest Technical University in 1944, F. was one of a group of sixty students who avoided army service by getting themselves evacuated via Halle to Denmark, where he worked until 1946, including a spell in the office of Jørn Utzon (later architect of the Sydney Opera House). This left a clear modern North European imprint on his own subsequent professional approach. 1949 he moved to MATI, the state high-rise design office (1949–54), also opening a private office in 1952. His most important projects in this period were: the *High School for Adult Education* (1952–53) in Buda's Zugliget (twelfth) district, now the University of Applied Arts, and one of the few products from the era of 'socialist realist' dogma to have enduring value. During a subsequent move to KÖZTI, the state planning office for public buildings, he partnered István Janáky on the work for the new *Miskolc Technical University for Heavy Industries* (1951–52) and himself headed the team building the *Dunaújváros Hospital* (1956–62). Other work at this time included the *Apartment Houses at No. 32 Úri utca and No. 15 Fortuna utca* (1957–59) and at *No. 8 Dísz tér* (1963–66), all on Buda's Castle Hill. F. meanwhile changed office again to become the leading architect at TTI (1961–71), the state standard design office, where his major projects included: the three 18-storey, rubble-walled *High-Rise Apartment Blocks in Kelenföld* (1967–69); the *Hotel Olympia on Eötvös út* (1968–72), on Buda's Széchenyi Hill; and the *Children's Hospital at Miskolc* (1967–1975). From 1971 he taught in the Department of Public Building at Budapest Technical University, becoming its head and a full professor in 1981. He was a founding member of the Union of Hungarian Architects (MÉSZ, 1951) and later one of its leaders until his demonstrative resignation in 1985. Twice awarded the Ybl Prize (1963, 1973), he also received the Kossuth Prize in 1988.

Fehér, László (1953)

Painter. In early works such as *Brigádnapló* (Work-Brigade Journal, 1975) F. adopted a sociologically slanted photo-realist approach that was reminiscent of amateur black-and-white snapshots, but by the end of the '70s he was openly avowing his Jewish identity, as in works like *Diaszpóra* (Diaspore, 1982). After a period of painting highly stylised figurative landscapes such as *Vízesés* (Waterfall, 1986), he developed the cartoon-like outline figure that was to become a hallmark of his photograph-based depictions of defenceless solitude, which have gained international recognition. *Domus Aurea I* (1987), a product of a stay in Rome, marks the appearance of colour in his work, and after 1988 the figure and architectural element of each picture were painted on a uniform surface of yellow, pink, silver or black (during successive periods), though the colour contrasts have mellowed since the late '90s. From 1990 onwards F. has occasionally produced installation-like adaptations in which the outline figure appears as an iron-plate silhouette. Since the mid '90s he has also turned to human portraiture, with works like *Balatoni fotózás* (Lake Balaton Photography, 1998) evidencing a re-emergence of his photo-realist beginnings.

Fejér, Lipót (1880–1959)

Mathematician. Taught by Manó Beke, Gyula Kőnig and József Kürschák in Budapest, F. was only twenty when he made his most significant individual discovery: a new method for summing Fourier series and the related Fejér theorem. Supported by Loránd Eötvös, he began teaching at Budapest University in 1911, where he remained until his death, supervising whole generations of scientists and teachers. Even before the First World War, he had built up a strong school of functional analysis, several of whose leading members subsequently moved abroad and won international recognition there. He himself continued to produce significant work on constructive functions and interpolation theory. A member of the Hungarian Academy of Sciences, he was also elected to the Polish and Bavarian academies.

Fejős, Pál [Fejos, Paul] (1898–1963)

Director. Originally studied medicine, but having turned to film in the turmoil of post-World War I Hungary, F. had to abandon shooting his first film in 1923, leaving the country for America, where he embarked on medical research at the Rockefeller Institute, New York. The short film *The Last Moment* (1928) brought him to the attention of Universal Film Studios, for whom he then directed *Lonesome* (1928). A move to Paris in the early 1930s to remake Hollywood films for the local market brought him back to Hungary to direct *Tavaszi zápor* (Marie, une légende hongroise, 1932) and *Itél a Balaton* (The Waters Decide, 1932). He left Hungary for good to work next in Vienna (e.g. *Sonnenstrahl*, 1933) then Denmark. In the mid-'30s he gave up feature films to join various archaeological and anthropological expeditions to South America and the Far East as a documentary film-maker.

Fekete, Mihály (1886–1957)

Mathematician. After graduating and completing his doctorate in Budapest, F. worked for a short while with Manó Beke then, under Lipót Fejér, did research into the theory of complex functions, especially on set tools. After the 1919 fall of Hungary's Soviet Republic, however, he was removed from his post and maintained himself with work as an actuary, though in 1928 he also private tutored the young John von Neumann. After World War II, on the recommendation of E. Landau, he was invited to teach at the Hebrew University of Jerusalem, where he soon became director of the Einstein Mathematics Institute, and subsequently dean and later rector of the university. Though F. did work of lasting value in several fields of mathematics, his major achievement was to introduce and apply the concept of the transfinite diameter, for instance, to measuring certain polynomials.

Fényi, Gyula (1845–1927)

Astronomer. A teacher at Kalocsa gymnasium, he became director of the school's new observatory for 28 years from 1885 onwards. The three decades of regular observations of solar activity from a single locality provided a useful series of data on flares and their relation to sunspot activity. As a meteorologist, F. worked on storm forecasts and the stratification of atmospheric pressure and its causes, particularly the links with solar flares. He was a corresponding member of the Hungarian Academy of Sciences.

Fenyvesi, Csaba (1943)
Fencer. Qualified physician (1968). Olympic golds in individual (1972) and team (1968, 1972) epée, as well as World Championship golds (1970, 1971, 1977) and several silver and bronze medals in team event. Hungarian Fencer of the Year (1972). As a cancer researcher since 1974, he is a member of the United Virginia Cancer Institute (1980–) and now medical director of the Jenner Foundation Cancer Institute (1994–).

Ferencsik, János (1907–1984)
Conductor. Conducted at the Budapest Opera House from his debut in 1930. From 1953 he was chief conductor of both the Budapest State Opera House and the Hungarian State Orchestra, though continuing to appear regularly in the Vienna Staatsoper and with other European orchestras. As a noted musical interpreter of the Viennese Classical and, most particularly, the Hungarian repertoire, he was the most prominent conductor to have lived and worked in Hungary in the twentieth century.

Ferenczy, Károly (1862–1917)
Painter. Peerless representative of the Nagybánya approach in the nobility, moral conviction and profundity of the example he set. A supreme individual, who picked up his craft from studying the works of Bastien-Lepage and was not much influenced by anyone else subsequently. After a protracted period of working on his own, he attended the Académie Julian in Paris, before settling in Szentendre, thereby launching it as an artistic colony. In search of new impulses, he moved with his family to Munich, where as well as intimate family portraits such as *Béni vörös ruhában* (Beni in Red Clothes, 1894), he painted splendid *plein-air* pictures like *A neuwittelsbachi kertben* (In the Garden at Neuwittelsbach, 1893). There he met István Réti and Simon Hollósy, later moving with them to Nagybánya in Transylvania. After several Secessionist works,

such as *Nápolyi emlék* (Reminiscence of Naples, 1896), the first canvas to be completely inspired by the new locality was *A hegyi beszéd* (Sermon on the Mount, 1896). Earlier equestrian studies were ultimately to bear fruit in Biblical compositions like *Három királyok* (The Three Magi, 1898). *Józsefet eladják testvérei* (Joseph Sold by His Brothers, 1900) and *Ábrahám áldozata* (The Sacrifice of Abraham, 1901) were unsurpassed in their power, representing the Nagybánya spirit to the core. This deeply religious tendency culminated in various drafts of *Keresztlevétel* (The Deposition from the Cross, 1914; subsequently cut up by the artist) and *Pietà*. His second Nagybánya period was characterised by magnificent *plein-air* pictures like *Márciusi est* (Evening in March, 1902) and *Napos délelőtt* (Sunny Morning, 1905). Studding his *oeuvre* throughout are marvellous portraits and female nudes.

Feszl, Frigyes (1821–1884)

Neo-Romanesque architect. Following studies at the Munich Academy of Fine Arts (1838–41), he travelled widely in France, northern Italy and Germany. Failing to gain admission into the master-builders' guild, F. sought a career as an independent architect. His early works were the *Oswald Apartment House at No. 22 Nádor utca*, Pest (1846–51) and the *Capuchin Parish Church and Monastery in Fő utca*, Buda (1852). The design he entered in the competition for the *Central Synagogue in Dohány utca,* Pest (1854) was passed over in favour of one by the Viennese architect Ludwig Förster, but the lavish oriental interior decorations of the huge galleried 'nave' and the cupolaed twin towers flanking the main entrance were F.'s work (1858). His masterpiece was the Rundbogenstil *Vigadó* (1859–65), built on the foundation walls of Pollack's destroyed Redoute building. The main façade, with its two massive lateral projections plus their fluted corner towers and the five round-arched windows of the recessed central section echoed in the arcades at ground level, was for a long time the most distinctive element on the Danube embankment. With its more restrained interior spaces and lavishly ornamented staircase, it is one of the outstanding buildings of the period. Badly damaged during World War II, the Vigadó's side wings were demolished in 1960 but the remainder was restored, with particular attention to improving the Large Hall's formerly notoriously poor acoustics (completed 1980).

Finta, József (1935)

Architect, graphic artist. Following graduation from Budapest Technical University (1985), he entered the third cycle of the Union of Hungarian Architects' Master School postgraduate training scheme (1959–62), meanwhile joining LAKÓTERV, the state planning office for housing, later becoming its chief architect until 1995. His highly proficient yet strong individual design approach, based on functional analysis of needs and a willingness to employ advanced technology where this is advantageous, have made him Hungary's most popular and sought-after architect. The many commissions for prominent public buildings and hotels on which he has worked have powerfully shaped the Budapest cityscape in particular over the last three decades. Amongst these are: the *Pécskő Department Store in Salgótarján* (1966–68); the former *Duna–Intercontinental* (now Marriott) *Hotel* (1966–69), on the Pest embankment of the Danube; the *Novotel Hotel on Alkotás utca* (1979–81), in Buda's twelfth district; the *Hotel Forum in Apáczai Csere János utca* (1978–81), near the Pest end of the Chain Bridge; the *Budapest Congress Center* (1983–84) besides the Hotel Novotel; a *Family House on Gellért Hill* (1984); in Váci utca, Pest's main inner-city shopping precinct, the *Taverna Hotel* (1982–85) and the *International Trade Center* (1982–84) opposite; the *Hotel Liget on Dózsa György út* (1985–90), overlooking Pest's City Park; the *County Library in Salgótarján* (*c*. 1986); the *Representative Office Block in Erzsébet tér* (1988), and in the same Pest inner-city square, the *Grand Hotel Corvinus Kempinsky* (1988–92, with Antal Puhl); and the *International Bank Centre in Szabadság tér* (1991–95). Notable amongst many competition successes during this period was that in the 1974 international competition for the town centre of Perugia (Italy). Projects undertaken by his private office since 1995 include: the new *Central Police Headquarters in Deák tér* (1994–98); the *Informatics Building for the Technical University and Eötvös Loránd University* (1995–97), on the Buda embankment south of Petőfi Bridge; and the *Westend City Center* (1998–2000). Known also as a graphic artist and theorist, F. has published *Építőkockák–épületkockák* (Building blocks–Block buildings, 1970) and *Tervek, gondok, gondolatok* (Plans, concerns, thoughts, 1979). A full member of the Hungarian Academy of Sciences since 1994, he was also one of the founders of the Hungarian Academy of Arts. Twice a recipient of the Ybl Prize (1965, 1972), he has also been awarded the State Prize (1970), Budapest's *Pro Urbe* Prize (1983), and the Kossuth Prize (1996).

Fischer, Annie (1914–1995)

Pianist. She was a pupil of Dohnányi at the Budapest Academy of Music. She won the first Liszt Piano Competition in 1933. She regularly toured the world, gaining particular acclaim for her interpretations of Beethoven, Schubert and Schumann.

Földes, Andor (1913–1992)

Pianist. A pupil of Dohnányi (1932) at the Budapest Academy of Music, he performed widely in Europe from 1930 then settled in New York (1940–48) before finally moving to Switzerland. Known chiefly as an interpreter of the Viennese classics and of Bartók's works.

Fülep, Lajos (1885–1970)
Philosopher, art historian, writer. F.'s studies embraced periods spent in Italy and England. He translated and wrote the foreword to the Hungarian version of Nietzsche's *The Birth of Tragedy* and, with György Lukács, co-edited the journal *Szellem* (Mind) during its year of existence in 1911. Having defended a doctoral thesis on philosophy, art history and Italian literature, he went into teaching, first at various secondary schools, later at Budapest University. With writings on a wide range of topics, from theatre reviews to essays on art history, literature, folklore and music, F. was one of the first Hungarian critics to acclaim the significance of Ady, Cézanne, Rippl-Rónai, and Bartók.

Fülöp, Mihály (1936)
Fencer. Certificates as coach (1972) and master coach (1977). At World Championships won golds in both individual and team foil events (1957) and a team bronze both there (1959) and at the Olympics (1956). Fencing master to the Hungarian modern pentathlon national team (1966–72), with several Olympic and world champions among his students.

Füst, Milán (1888–1967)
Poet, novelist, and aesthetician. Making a name for himself initially as one of the poets around the literary magazine *Nyugat*, F.'s first volume, *Változtatnod nem lehet* (Nothing Canst Thou Change, 1914), is archaicly stylised, its Secessionist tone and long lines marking an end to traditional poetics. His dramas—notably, *Boldogtalanok* (Unhappy Ones, 1914), *Catullus* (1927) and *IV. Henrik király* (Henry IV, 1931)—are naturalistic, with protagonists who are locked in inner turmoil; unable to resolve their dilemmas, they are destroyed by the shame of realising the impossibility of complete life. In his early novels F. drew on the tradition of nineteenth-century psychological novels, but none rises to the level of *A feleségem története* (The Story of My Wife, 1935–42). In this,

now F.'s best-known and most-translated work, F. employs a mixture of the sublime, the ridiculous and grotesque self-irony to tell a narrative about Dutch sea captain Störr's futile attempts to stifle his growing jealousy for his wife—a tale which, ultimately, is about the insoluble loneliness, homelessness and estrangement of human existence. After World War II he taught aesthetics at the University of Budapest, writing on moral philosophy and psychiatry.

Gabor, Dennis [Gábor, Dénes] (1900–1979)

Physicist. Began his studies at Budapest Technical University but graduated in 1924 from Berlin as an electrical engineer, and continued to regard himself primarily as that. In Germany, he started to research into transient phenomena, oscillography, then electron optics. On moving to Imperial College, London, with Hitler's rise to power, he turned to information theory and carried on his detailed studies of the focusing of electron beams to improve electron microscope resolution. This led him to discover holography—a phenomenon that was not fully exploited until after lasers were developed, but even so having useful applications. He received the Nobel Prize for Physics in 1971. Towards the end of his life, G. was increasingly preoccupied by basic questions about human society.

Ganz, Ábrahám (1814–1867)

Ironmaster. Swiss-born, G. trained to be a carpenter but also learnt foundry work. Arriving at Pest in 1841 after a lengthy study tour in Europe, he worked for several employers before establishing his own workshop in 1844. After 9 years spent acquiring skill in chill-casting technology, previously used only in the USA, his own process for manufacturing hard-wearing railway wheels (patented in 1853) became the foundation for his company's rapid growth, becoming a supplier to some 60 manufacturers of rolling stock around the world by the time he died. For a long time he was the company's sole director, handling most of its business and commercial relations, as well as overseeing its design, development and manufacturing activities.

Gárdonyi, Géza (1863–1922)

Novelist. Lacking a proper formal education and coming from a rural provincial background, G. may be considered a prime example of the oversensitive outsider. He had a gift for achieving at least a temporary reconciliation between conflicting stylistic means and theoretical problems, dealing with modern topics of death, dreams, the fate of the artist, power and loneliness, whilst showing an interest in all manner of spiritual and political movements, including Spiritism, Occultism, and Oriental mysticism. His early works were stories based on village life, culminating in *Az én falum* (My Village, 1898), before he turned to the historical topics of *Egri csillagok* (The Stars of Eger, 1899), *A láthatatlan ember* (The Invisible Man, 1902), Az öreg tekintetes (The Old Sir, 1905), and *Isten rab jai* (Prisoners of God, 1908). In the eyes of his contemporaries and later generations alike, this gave his *oeuvre* a markedly inconsistent quality.

Gécsek, Tibor (1964)

Hammer thrower. First Hungarian to throw over 80 meters. Fourth place at the Olympics (1992); gold at European Championships (1998). Hungarian Sportsman of the Year (1988, 1992, 1993).

Gedó, György (1949)

Boxer. Olympic champion in light flyweight (1972). Hungarian Boxer of the Year (1968, 1972). Head coach at Vasas (1980–82, 1994–95).

Gedővári, Imre (1951)

Fencer. Qualified engineer (1972) and lawyer (1987). Olympic gold in sabre team (1988) and bronzes in individual and team events; World Championship gold (1978, 1981, 1982), silver

(1977) and bronze (1974, 1977) in team; several World Cup golds. Hungarian Fencer of the Year (1978, 1984). Secretary-general of the Hungarian Fencing Federation (1989–91); Chairman of sports club UTE (1991–97). Awarded the Ferenc Kemény Prize (1992).

Gerevich, Aladár (1910–1991)

Fencer. Degree in economics; master coach's certificate. The most successful Hungarian Olympic sportsman of all times, with golds in the individual sabre (1948) and six times in the team event (1948–60), two individual (1951, 1955) and seven team golds at World Championships, one individual (1935) and four team golds at European Championships. Inducted into the Hungarian Hall of Fame (1991). Recipient of the International Olympic Committee's Olympic Order (1988 and 1991) and an International Fair Play Prize (1989).

Gerlóczy, Gedeon (1895–1975)

Modernist architect. Qualifying in Munich, he was leading architect on a wide range of projects, including: *Villa at No. 21 Napraforgó utca* (1931), in Buda's Pasarét (Second) district; the *Magdolna Trauma Hospital and Central Clinic on Fiumei út* (1939, with Nándor Körmendy), in Pest's Józsefváros (eighth) district, which set standards in contemporary European hospital design; the *Madách Chamber Theater at No. 6 Madách tér* (1941), just off Pest's Károly körút; the *Apartment House and Shopping Gallery on Párizsi utca* (1942–44), at the intersection with Petőfi Sándor utca, one of the capital's most modern residential buildings at the time; the *National Sports Hall* (1941–43, with Gyula Rimanóczy and others), now alongside the later Népstadion. He taught for a few years at the Academy of Fine Arts (1944–48) but was then sidelined. Throughout the fifties he worked for IPARTERV, the state design office for industrial projects. He was part of the team that designed the *Tower Block for the Semmelweiss*

University of Medical Sciences in Nagyvárad tér, completed during the seventies. G. was responsible for saving the Tivadar Csontváry Kosztka estate after the painter's death in 1919. He wrote *Építészet, városépítészet* (Architecture and Urban Planning, 1949).

Gothár, Péter (1947)

Director. Started his career with Hungarian Television before becoming a theatre director (Kaposvár Theatre from 1979, then József Katona Theatre, Budapest), with a parallel career in films. His first feature film, *Ajándék ez a nap* (Priceless Day, 1979), was awarded a prize at the Venice Film Festival, whilst *Megáll az idő* (Time Stands Still, 1981), a look at life in Hungary from the 1956 revolution on into the sixties, was a major international success, picking up awards at Cannes, New York and Tokyo. Most of *Volt egyszer egy Amerika* (Just Like America, 1987) was shot in New York, whilst the more recent *A részleg* (The Outpost, 1994) and *Haggyállógva Vászka* (Let-me-hang Vaska, 1995) were widely shown at international film festivals.

Gothard, Jenő (1857–1909)

Engineer, astronomer. Whilst qualifying as an engineer at Vienna Technical College, he also studied astronomy. On returning to his own estate at Herény, near Szombathely, with assistance from Konkoly-Thege he set up his own observatory. His pioneering work on astronomical photography and spectrography, particularly of star clusters, comets and gaseous nebulae brought him international acclaim and played a big part in gaining acceptance of photography's value to astronomy. His major published works were *Az újabb kori csillagászat módszerei és megfigyelés módjai* (The Methods and Observational Techniques of Modern Astronomy, 1886), *A fotográfia* (Photography, 1890), and *Spektrálfotográfiai tanulmányok* (Spectrographic Studies, 1891). A corresponding member of the Hungarian Academy of Sciences and a member of Britain's Royal Astronomical Society, as an engineer G. in 1881 also constructed Hungary's first long-distance (178 km) telephone line, and in 1895 designed and ran the country's first hydroelectric power station (at Ikervár on the River Rába below Sárvár).

Grosics, Gyula (1926)

Footballer, coach, sport manager. Certificate as coach (1960) and sports organiser (1965). Member of the gold-winning Olympic team (1952), the team that beat England in the 1953 'match of the century' at Wembley Stadium, and of the 1954 World Cup silver-winning team. National inspector and coach to Kuwaiti team (1963–68). Sports club Budapesti Volán chairman (1969–86), then honorary chairman (1991–92). Awarded the Middle Cross of the Order of Merit of the Republic of Hungary (1992), the International Olympic Committee's Olympic Order (1995), the Hungarian Heritage Prize (1996), and the Ferenc Csík Prize (2001).

Gulácsy, Lajos (1882–1932)

Painter. After studies at the Budapest School of Design, he visited Rome in 1902 and again in 1904–05, with interludes at the Florence *Accademia* (1903) and as stage and costume designer for some of the Thália Society's ground-breaking theatrical productions in Budapest (1905). Though he made a trip to Paris (1906), it was to Italy that he kept returning, identifying closely with Italian culture, not just in the subjects of his paintings, but also, for example, in having himself photographed in Renaissance costume. His paintings at this stage, typified by *A varázsló kertje* (The Magician's Garden, c. 1904), were in a symbolist–Art Nouveau style somewhat in the manner of the English Pre-Raphaelites. G.'s first major show, jointly with Ödön Márffy, was put on by the Uránia Art Gallery, Budapest, in 1907, which led to his being awarded the Franz Joseph Prize in 1908. *A mulatt férfi és a szoborfehér nő* (The Half-Caste and the Woman White as a Statue, 1910–13), *Az ópiumszívó álma* (The Opium Smoker's Dream, 1913–16) and *Rózsalovag* (Rosenkavalier, 1914) display the continuing hold of antique and nostalgic moods and Rococo, symbolic, exotic and esoteric subjects. Mental problems of increasing severity emerged during a year spent in Venice (1914–15), prompting G. to withdraw more and more into his own fantasy world of Na'Conxypan, which was the subject not only of paintings but also poems and stories. That surrealistic dream world was displayed in 1918, when G. exhibited his works jointly with the activists, despite the chasm that separated him from the constructivism favoured by Lajos Kassák's *Ma* group. Much of the last fifteen years of G.'s life, including the period when a major retrospective of his work was mounted in 1922, was spent as in mental hospitals.

Gulyás, Zoltán (1930–2000)

Architect. After graduating from Budapest Technical School (1952) he worked with István Janáky at KÖZTI, the state planning office for public buildings, before being sent on a consignment to North Korea, where he planned the *Medical University and Bacteriological Institute for Pyongyang* (1954–55). On returning home, he entered the second cycle of the Union of Hungarian Architects' Master School postgraduate training scheme (1955–58). During a long association with IPARTERV (1957–86), the state planning office for industrial construction, he aimed to implement the clarity and elegance of proportions of his models, Gunnar Asplund and Alvar Aalto, in his own work. Amongst his early projects in Hungary were an *Apartment Block on Fehérvári út* (1957–58), in Buda's Kelenföld (eleventh) district; an *Apartment Block on Madách út* (1961–62); a *Shopping Precinct at Tihany* (1962), on Lake Balaton; and the highly influential *Office Building for the Chemolimpex Trading Co. at No. 4 Bécsi utca* (1962–64), in Pest's inner-city area (reorganised in 1997 as the great hall for the National Savings Bank (OTP)). Through the agency of London-based Ernő Goldfinger, G. joined the team designing a building for the London Commercial Marine Ministry (1964–65). His projects on returning home included: a *Café and Confectionery Shop in Hévíz* (1966), his own *Summer House at Ábrahámhegy* (1967), on the northern shore of Lake Balaton; the *Headquarters Building for the Medicor Works at Lehel tér* (1973–74; rebuilt 1996), in Pest's Terézváros (Sixth) district; a *House on Tárnok utca* (1974), on Buda's Castle Hill; and a *School at Révfülöp* (1980–81), also by Lake Balaton. G. lectured at Budapest Technical University from 1960 to 1976, and from 1969 to 1979 was also a tutor for the Young Architect's Circle (FÉK) postgraduate training scheme. He was awarded the Ybl Prize in 1962.

Gyarmathy, Tihamér (1915)

Painter. A key figure of post-war abstractionism in Hungary, Gy. was inspired during the 1940s by Ernő Kállai's 'New World Vision' of a unity of macro- and microcosmic forces that could not be rendered by the traditional artistic means of representing space. A few digressions aside, this imaginary space and its events—exemplified by *Elmozduló formák* (Shifting Forms, 1959)—were the sole topic of Gy.'s subsequent work. Gy. retreated from society in the late '40s, initially composing small-scale abstract-surrealistic paintings and making Moholy-Nagy-style photograms. From the late '50s onwards, he gradually developed a language of geometrical elements of lapidary simplicity which became the most important ingredient of his mature style over the next decade, though his studies of organic natural systems threw up unsettlingly surrealist forms which not only betoken a metaphysical anxiety but impart an intertwining counterpoint of magnetic forces to a structured space of colour grids.

Gyarmati, Andrea (1954)

Swimmer. Qualified physician (1978). Olympic silver in 100 meters backstroke (1972), bronze in 100 meters butterfly (1972); European champion in 200 meters backstroke and 100 meters butterfly (1970). World record holder for 100 meters butterfly (1972). Hungarian Woman Swimmer of the Year (1968–73) and Sportswoman of the Year (1970, 1971, 1972). Was coached by her mother, Olympic champion Éva Székely.

Gyarmati, Dezső (1927)

Water polo player, coach, sports administrator. Degree in art history (1956); certificates as coach (1966) and master coach (1973). Olympic golds as member of the water polo team (1952, 1956, 1964). Set up the Swimming and Water Polo section of the Central Sports School (1963). Coach to Hungarian club teams (1966–69); head coach and captain of the Colombian national team (1970–71) then the Hungarian national team (1973–80, 1985–88), the latter winning golds at the Olympics (1976) and World (1973) as well as European Championships (1974, 1977). Board member of the Hungarian Water Polo Federation (1991–), its vice-chairman (1992–96) and co-chairman (1996–98). Member of Parliament (1990–94). Chairman of the Hungarian Hall of Fame (1994–). Recipient of the International Olympic Committee's Olympic Order (1992).

Gyarmati, Olga (1924)

Athlete. Olympic champion in long jump (1948). Points winner at international competitions in hurdle, sprints, and in pentathlon. Lived in the UK from 1957, later moving to the US.

Gyenge, Valéria (1933)

Swimmer. Studied at the New York School of Photography (1974). Olympic champion in 400-metre freestyle (1952). Multiple world and European record holder. Has lived in Canada since 1956. Once a swimming coach in Toronto (1963–65), she later became a professional photographer.

Gyulai, Pál (1826–1909)

Literary historian, critic, novelist, poet. Studied in Kolozsvár (Cluj, Romania) before moving to Pest after the War of Independence. Having visited Berlin, Paris and Munich as a tutor in 1855–56, he returned to Hungary to become a leading figure in a literary group that supported the Liberal Party under Ferenc Deák, which negotiated the Compromise with Austria. Gy. became editor of the influential literary magazine *Budapesti Szemle* (Budapest Review) in 1873, a university professor in 1876, and director of the Kisfaludy Society in 1879. His novel *Egy régi udvarház utolsó gazdája* (The Last Master of an Old Manor-house, 1867) was among the first to suggest that the nobility could no longer maintain their traditional way of life. A verse novel, *Romhányi* (1869–72), which shows the influence of Pushkin, remained fragmentary. As a critic Gy. recognised the significance of Petőfi early on, but he did not note his revolutionary quality. He fought for the aesthetic triumph of the so-called *népnemzeti* ("National Classicist") trend, supporting the works of János Arany and Zsigmond Kemény, for instance, but passionately rejecting the likes of Mór Jókai and Lajos Tolnai. His major works of literary history are *Petőfi Sándor és lírai költészetünk* (Sándor Petőfi and Our Lyrical Poetry, 1854), *Katona József és Bánk bánja* (József Katona and His Ban Bánk, 1860), and *Vörösmarty életrajza* (A Biography of Vörösmarty, 1866).

Gyuricza, József (1934)

Fencer. Certificate as PE instructor (1968), coach (1970) and master coach (1977). Won World Championship golds for foil in individual (1955) and team (1957) events and a team bronze at the Olympics (1956). Was also a member of the national team in sabre and in epée. Professional military officer (1958–78), reaching the rank of lieutenant colonel (1979–), as well as fencing coach (1967–85).

Haar, Alfréd (1885–1933)
Mathematician. He won first prize at the 1903 Eötvös Competition for secondary-school pupils before starting his studies at Budapest. He obtained his doctorate under David Hilbert at Göttingen in 1909 then lectured for a while at Zurich before joining the staff at the University of Kolozsvár in 1912. With the university's relocation following the loss of Transylvania to Romania after World War I, H. moved with it to become one of the founders of the 'Szeged school'. As well as a distinguished expertise in integral theory (the creator of the Haar measure and integral in 1933), he contributed significantly to the analysis of orthogonal functions and the calculus of variations. Member of the Hungarian Academy of Sciences.

Hajas, Tibor (1946–1980)
Performance artist, filmmaker, poet. Whilst studying English at Budapest University, H. was arrested in 1966 for participating in a street demonstration. Though later granted an amnesty and released from prison, he did not resume his course but trained as bookbinder and, from 1970, worked for a co-operative firm making wood and paper products. Though he considered himself to be a poet, contacts with various groups of avant-garde artists spurred him to produce his first political-conceptual art works, *Hús és vinil* (Meat and Vinyl, 1969) and *Minden csak a saját másolata* (Everything is Just a Copy of Itself, 1969), which also used photographs. Other works, such as *Levél barátomnak Párizsba* (Letter to My Friend in Paris, 1975) and *Élő comics* (Living Comics, 1975), can be regarded as photographic documentations of actions. His oeuvre in the visual arts includes poems, actions, performances and body-art actions, which are perpetuated in the form of photographic tableaux, a few pages of graphics, a film and three videos. Later labelling some of the projects 'pre-performance', H. went on to devise 11 performances (1978–80), creating what

still remain the most prominent Hungarian examples of a 'hard-edge' genre in which the performer's enactments of suffering, sacrifice, self-revelation, etc. are formally framed by fixing key moments on the viewer's retina or memory by the use of camera flashlights (with the photographs themselves as an ancillary means).

Hammerl, László (1942)

Sport shooting. Qualified physician (1968); certificates as coach (1982) and master coach (1985). Has won Olympic gold (1964), silver (1968) and bronze (1964) medals. Captained the Hungarian national shooting team (1977–96) and more recently that of New Zealand (1996–99).

Hantaï, Simon (1922)

Painter. Influenced in his early career by Csontváry, post-impressionism, Bonnard and Matisse, H. took up residence in Paris in 1948. After a decisive shift to surrealism in 1950, following contacts with André Breton, he inhabited his canvases with fantastic creatures and biomorphic formations. These gradually ceded place to a more gestural language of signs, leading to 'totally' gestural compositions like *Sexe-Prime. Hommage à Jean-Pierre Brisset* (1956), the white cross on dark background of *Peinture—Souvenir de l'avenir* (1956), and calligraphic works like *Écriture rose* (1958–59). In 1960 he started to use the 'pliage' method of painting a pre-crumpled canvas that is subsequently stretched after the paint has dried. The gradual elimination of colour, starting with the series *Les Meuns* (1966–68), resulted in monochrome pictures; the *Tabulas Lilas* canvases, completed in 1982, were produced with the use of white alone. In 1994 H. started to cut up his own older canvases, re-using certain details in new works through a reproduction technique.

Haraszty, István (1934)

Sculptor. The pioneer and most significant exponent of kinetic art and mobiles in Hungary, he was self-taught, having trained as a machine fitter then, from 1953, as a decorative ironworker, when he also began to paint. He started to exhibit regularly from 1964, unsettling critics by exemplifying an ideal Socialist career as worker-turned-artistic genius whilst creating amusing, ironic emblems of the régime's anomalies, as in *Egyenlet* (Equation, 1969); *Hathúszas gyors* (The 6:20 Express, 1970); *Madárkalitka* (Birdcage, 1971); *Acélmosoly* (Steel Smile, 1972), a pun on the name of the country's cultural tsar, György Aczél, which is a strong-box that dispenses machine-generated laughs instead of cash; *Központi irányítás* (Central Control, 1973), etc. The kinetic machine *Fügemagozó* (Fig-Pipper, 1970) is seen as his outstanding creation. Employing Plexiglass and other plastics as well as metal, H.'s transformation of his former cramped home into a matchless *Gesamtkunstwerk* is an on-going project.

Hargitay, András (1956)

Swimmer. Qualified veterinarian (1982); coach's certificate (1982). In individual medley, he won bronze at the Olympics over 400 meters (1972) and gold at World Championships over 400 meters (1973, 1975) and 200 meters (1975), as well as several European championships. Set world record in 400 meters individual medley (1974). Hungarian Swimmer of the Year (1969–76, 1978) and Sportsman of the Year (1975). Captain of the Hungarian Swimming Federation (1985–88).

Harsányi, John [János] (1920–2000)

Mathematician and economist. Born in Hungary, his first degree was in pharmacy (1942) but he was immediately drafted into a forced labour battalion for the next three years. After the war he enrolled in the Arts Faculty at Budapest and gained a doctorate in philosophy. He emigrated to Austria in 1950, obtaining a degree in economics and teaching the subject there for a while before finally settling in the USA in 1961, where he taught at the Berkeley Business School until retirement. In 1994, H. was the joint recipient, with J. Nash and R. Selten, of the Nobel Prize for Economics (the first Hungarian to be so honoured), for his pioneering work on game theory in the area of equilibrium analysis within the theory of non-co-operative games in various economic activities (e.g. designing auctions).

Hatvany, József (1926–1987)

Computer engineer. Born into the wealthy Hatvany-Deutsch family, he received most of his education in France and in England, qualifying as an engineer, despite strong interests in physics and the humanities. As the head of the mechanical engineering team at the Automation Research Institute (SZTAKI), Budapest, after 1964 he became an internationally recognised expert on mechanical automation and computer-aided design and manufacture. With his original ideas and persuasive ability, he was one of the founders of information technology in Hungary. He was elected a foreign associate of the U.S. National Academy of Engineering and received an honorary doctorate from the Institut National Politechnique de Toulouse.

Hauszmann, Alajos (1847–1926)

Late-Historicist architect. Starting as a mason's apprentice, he spent two years studying at the Joseph Polytechnic in Buda and graduated from the Berlin Bauakademie (1868). H. gained a teaching post at Budapest Technical University in 1872 and meanwhile built up a private office. In contrast to Lechner's 'individual' language of form, H.'s early neo-Renaissance designs gave way later to a neo-Baroque influence, though towards the end of his career he too adopted a muted Secessionist style. His most important works are: the *County Hall and Theatre at Szombathely* (1879–80), the *State*

Secondary Modern School in Markó utca, near the Parliament building (1882); *St. Stephen's General Hospital* (1876–80), in the Ferencváros (ninth) district of Pest, and the *Elizabeth Hospital* on Alkotás utca (1882) in Buda's twelfth district (now housing the Sports Hospital); the *Palace and Apartment house at Nos. 13–15 Teréz körút* (1884), modelled on the Palazzo Strozzi in Florence; the *Royal Law Courts*, occupying the entire block bounded by Alkotmány–Nagy Ignác–Markó–Bihari János utca (1888–91); the *Commercial School and Technological Exhibition* on the József körút, at the intersection with Népszínház utca (1889); the *New York Palace* on the Erzsébet körút (1891–95), built for an American insurance company incorporating, with the 'Hungária' coffee-house on the ground floor; the *Supreme Court* in Kossuth tér (1893–96), which after 1945 first housed the National Gallery and now the Ethnographic Museum; and the *Government Building in Fiume* (1893–96). After the death of Miklós Ybl in 1891, H. took over and completed the construction of the *Royal Palace in Buda* (1892–1903), modifying Ybl's plans for the West wing (which now houses the National Széchenyi Library), elongating the Danube façade up to Szent György tér by adding the wing that now houses the Contemporary Art Museum (Ludwig Collection) and part of the National Gallery collection, and placing an almost Art Nouveau-style false dome on what is now the central block. H.'s last important commission was the 220-meter long *Main Building of Budapest Technical University* (1903–09), the Danube façade of which now forms an important element of the cityscape. H. was a member of the Hungarian Academy of Sciences and received a baronetcy from Emperor Francis Joseph.

Heller, László (1907–1980)

Mechanical engineer. Having qualified in Zurich, H. was a major figure in building up the Hungarian power generation industry, particularly through his contribution to the Heller–Forgó system, utilising a special aluminium heat exchanger (designed by László Forgó), which allowed the steam from turbines to be cooled by air instead of water. This system was widely used internationally, especially in areas where water assets were scarce. In his theoretical work, he made particular allowance for entropy in specific design jobs. He was a professor at Budapest Technical University and a member of the Hungarian Academy of Sciences.

Herman, Ottó (1835–1914)

Zoologist, ethnographer. One of the great self-educated scientists, H. originally trained as a machine fitter in Vienna then, after a spell fighting in the Polish insurrection of 1863 and as a photographer, he was employed by the Transylvanian Museum Association at Kolozsvár (Cluj, Romania) as a workshop assistant, developing an existing interest in birds and spiders. After being engaged by the Natural History Society to study Hungary's spiders, he moved to Budapest in 1876. An active supporter of the principles of Hungary's 1848 revolution and, by then, friend of Kossuth, he was a parliamentary representative for the Independence Party from 1879. He extended interests into the ethnography of fishing and shepherd life, ornithology, archaeology, and the preservation of viniculture, publishing more than a thousand articles and 14 books, the most important of which are *Magyarország pókfaunája* (The Spider Fauna of Hungary, 1876–79), *A magyar halászat könyve* (A Handbook of Fishing in Hungary, 1887–88), *A madarak káráról és hasznáról* (On the Damage and Use of Birds, 1901), and *A magyar pásztorok nyelvkincse* (The Word Stock of Hungarian Shepherds, 1914). He set up and directed the Hungarian Ornithology Center and in 1894 founded the ornithological journal *Aquila*.

Hesz, Mihály (1943)
Kayaker. Qualified dentist. In single kayak won silver (1964) then gold (1968) over 1,000 meters at the Olympics, as well as gold at European Championships over 10,000 meters (1966), and gold at World Championships in the 4 x 500 kayak relay (1971). Has lived in (West) Germany since 1980.

Hevesy, George von [György] (1885–1966)
Chemist. Although he started his university studies in Budapest, H. graduated in Germany and gained his doctorate at Zurich. Returning to Budapest in 1913, he was appointed professor in 1918. After the fall of the Hungarian Soviet Republic, he was forced to leave the country and chose to go to Copenhagen. He discovered the principle of isotope tracing in 1913, and isolated the element hafnium in 1920. He spent eight years at Freiburg University, developing the application of radiotracing *in vivo*, initially in animals, before returning to Denmark in 1934, with intervals in Stockholm during the German occupation, to study activation analysis. The remainder of his working life was devoted to studying metabolic processes with radiotracing—work that essentially founded the theory and practice of nuclear medicine. H. was awarded the Nobel Prize for Chemistry in 1943. His published volumes include *Lehrbuch der Radioaktivität* (1923), *Radioactive Indicators* (1948), and *A Scientific Career: Collected Papers* (1962).

Hidegkúti, Nándor (1922–2002)
Footballer, coach. Member of the gold-winning Olympic team (1952), the team that beat England in the 1953 'match of the century' at Wembley Stadium (scoring three of Hungary's six goals), and of the 1954 World Cup silver-winning team. He was coach to Italian club side Fiorentine when they won European Cup Winners' Cup (1962), then held coaching posts in Cairo (1973–81) and Dubai (1983–85); consultant in El-Ihihad (1996–97). He now lives in Alexandria. Awarded the

International Olympic Committee's Olympic Order (1995) and the Ferenc Csík Prize (2001).

Hollósy, Simon (1857–1918)

Painter. Founder and guiding light of the artists' colony at Nagybánya (Baia Mare, Romania). After training at the Royal Hungarian School of Design in Budapest, he went to the Munich Academy of Art, where he initially fell under the influence of genre painting, as in *Jóbor* (Good Wine, 1884). His work runs from the naturalism of *Tengerihántás* (Corn Husking, 1885) through to realism, but glimpses of an impressionistic approach and even a certain expressionism are seen in paintings on patriotic subjects, such as *Ország bajai* (The Country's Woes, 1893) or the unfinished *Rákóczi induló* (Rákóczi March). It was as a teacher in his own private school that H., with a group of pupils, first visited Nagybánya in 1896. He mastered *plein-air* painting there but in 1902 left the colony, thereafter taking his Munich pupils on summer school trips chiefly to Técső (Tyachev, Ukraine). The outstanding work of this period is his *Önarckép* (Self-Portrait, 1916). His charisma and modern pedagogical principles had a potent influence on his pupils (e.g. István Réti, János Torma, Gyula Rudnay).

Hőgyes, Endre (1847–1906)

Neurophysiologist, bacteriologist. A major figure in several fields of experimental medicine, H. was professor of pathology and pharmacy at Kolozsvár (Cluj, Romania) and then (from 1883) Budapest. His achievements included establishing that the renal cortex and medulla have separate blood supplies; demonstrating the role of the convoluted tubules in urine formation; and establishing a connection between involuntary eye movements and the semicircular canals of the inner ear (even though he had published this work in Hungarian well before Bárány, he lost the priority debate). On his appointment to Budapest, he undertook to test Pasteur's anti-rabies vaccine and in the process made various improvements, including attenuation of the virus, which were later adopted by the rest of the world. He was made director of Hungary's Pasteur Institute on its foundation in 1890 and was a member of the Hungarian Academy of Sciences. H.'s main published works were *Az asszociált szemmozgások idegmechanizmusáról* (On the Nervous Mechanism of Involuntary Eye Movements, 1880–85) and *Die experimentelle Basis der antirabischen Schutzimpfungen von Pasteur* (1889).

Hubay, Jenő (1858–1937)

Violinist, composer. He studied with Joachim in Berlin (1873–76) then became professor at the Brussels Conservatoire (1882–86). He returned to Budapest to take up a teaching post, which he held until he died. As founder of the Hungarian school of violin playing, his pupils included Ferenc Vecsey, József Szigeti and Zoltán Székely. His compositions are mostly in the French late-Romantic style.

Huszár, Vilmos (1884–1960)

Painter. After studies at the College of Applied Arts, H. moved to the Nagybánya artists' colony and later to Munich. During a trip to the Netherlands in 1906 he came under the sway of Kees van Dongen, though in somewhat later works such as *Walrech műkereskedő portréja* (Portrait of the Art-dealer Walrech, 1916) influences from Les Nabis are also identifiable. Joining the newly formed De Stijl group, he cultivated a totally planar, neo-plasticist style in a *Geometric Composition* series (1917–18), though its application in what he termed 'paintings of humans'—*Táncospár* (Dancing Couple) and *Kettős figura* (Double Figure, 1917)—was, perhaps, more characteristic. *Kártyások* (Card Players, c. 1920) shows a preference for allowing the dynamics of colour and brushwork, rather than the dividing stripes typical of the constructivists, to articulate a canvas. In his mechanical figures and designs for the stage a streak of Dadaism was evident, and H. also became involved in poster and model design. Seuphor organised a joint exhibition of his and Lajos Kassák's works in Paris during the late 1920s, but H. later decided to go his own way, returning to portraits and still lifes.

Huszka, Jenő (1875–1960)

Operetta composer. One of the founders of Hungarian-language operetta through his score for *Bob herceg* (Prince Vagabond, 1902). He successfully grafted a Hungarian tone onto international light-music fashions of the day in later operettas like *Lili bárónő* (Baroness Lili, 1919) and *Mária főhadnagy* (Lieutenant Mary, 1942).

Hutÿra, Ferenc (1860–1934)

Veterinary surgeon. After studying medicine in Budapest, he went on to qualify as a veterinary surgeon. Starting his career as a lecturer at the Institute of Pathology in the Veterinary School, at the age of 28 he secured the chair in internal medicine there. After securing major reforms in veterinary training, including an overhaul of the syllabus, he worked on the massive *Állatorvosi belgyógyászat* (Veterinary Internal Medicine, 1894–96), which, particularly in the revised edition (1904), with József Marek as co-author, was the standard work on the subject for many decades. His research work in the field of infectious diseases, especially in demonstrating the viral causation of swine fever (hog cholera) and developing a hyperimmune serum to treat the disease, gained him international acclaim. He was a member of the Hungarian Academy of Sciences, and, amongst many foreign distinctions, an honorary doctor of the Vienna and Berlin veterinary colleges, and chairman of the International Animal Health Office in 1929.

Iglói, Mihály (1908–1998)

Coach. Certificates as PE instructor (1934) and master coach (1955). A middle-distance runner in the national team, he was best known as a running coach, particularly for his innovative interval training methods, schooling many Hungarian and international world record-holders and Olympic champions. Emigrated to Brazil (1956–58), before residing in the USA (1958–70) and Greece (1970–98).

Iharos, Sándor (1930–1996)

Athlete. Coach's certificate (1962). A medium- and long-distance runner, he set world records over seven distances from 1,500 to 10,000 meters and also a member of 4x 1,500 meters and the 4 x 1 mile relay squads. World Athlete of the Year (1955). Head coach (1963–70). Recipient of Helms Foundation Memorial Plaque (1955).

Illyés, Gyula (1902–1983)

Poet, novelist, essayist, and playwright. Already recognised as an outstanding poet by the late 1920s, I. became a leading figure in the emerging group of Populist writers with poems such as *Három öreg* (Three Old Men, 1931), *Ifjúság* (Youth, 1932) and *Hősökről beszélek* (I Speak of Heroes, 1933), with he and Lajos Nagy representing them at the Soviet Writer's Congress in 1934. During this period he also wrote a biography of the poet Sándor Petőfi (*Petőfi*, 1936) and the major prose work *Puszták népe* (The People of the Puszta, 1936), which employs a mixture of autobiography, fiction, sociography, diary and memoir to analyse a social phenomenon with rational clarity—an approach also applied to the post-war novels *Hunok Párizsban* (The Huns in Paris, 1946); *Ebéd a kastélyban* (Dinner at the Castle, 1962); *Beatrice apródjai* (Beatrice's Page Boys, 1979). With the approach of war, I. became editor of *Nyugat*'s successor, *Magyar Csillag* (Hungarian Star, 1941–44). With the imposition of Communist dictatorship, and particularly in the immediate wake of the 1956 revolution, he largely withdrew from public life. Poems of this and his later period are typified by *Bartók* (1955), *Mozgó világ* (Moving World, 1962) and *Koszorú* (A Wreath, 1970). I.'s work is characterised first and foremost by its objectivity, clarity, straightforward sense of order and a clear connection to social problems, though any apparent naiveté is leavened by mild irony, the lyric carefully blended with narrative. I. also wrote numerous historical plays, chronicling tragic or heroic events of various periods, notably *A tű foka* (The Eye of the Needle, 1944), *Fáklyaláng* (Torchlight, 1952) and *Kegyenc* (The Favourite, 1963), as well as folk comedies or satires ridiculing various aspect of contemporary life.

Istvánffy, Benedek (1733–1778)

Composer. From 1766 until his death he was Kapellmeister at Győr cathedral. His extant sacred music makes him the most important Hungarian-born composer of the eighteenth century.

Izsó, Miklós (1831–1875)

Sculptor. The pre-eminent exponent of Hungary's 'national Romanticism', it was as a fugitive second-lieutenant following the defeat of the country's 1848–49 War of Independence that I. worked for six years as a stone-dresser's apprentice in Rimaszombat (Rimavská Sobota, Slovakia). When the classicist sculptor István Ferenczy, also living there, asked for an assistant to finish off busts of the poets Kazinczy and Kölcsey, the master recommended I. Having received much devoted attention from Ferenczy during his three-year apprenticeship, I. continued his studies at the Munich Academy of Art. His best-known masterpieces are the marble *Búsuló juhász* (Grieving Shepherd, 1861–62) and the *Csokonai Monument* in Debrecen (1867–71). His terracotta maquettes of dancing peasants are unrivalled in managing to portray movement and capture a national flavour whilst also masterfully incorporating influences from antiquity and personal experiences. He gave the title *Gypsy Laocoon* to a group of figures modelled from Gypsy musicians. Due to his early death from tuberculosis, his designs for a *Monument to András Dugonics* in Szeged

(1874) and the famous *Statue of Petőfi* in Budapest (1871–72) were executed by Adolf Huszár.

Jacobi, Viktor (1883–1921)

Operetta composer. A pupil of Hans Koessler at the Budapest Academy of Music, he gained success mainly with English-style operettas. He took up residence in the US in 1915. He is best known for *Leányvásár* (Girls' Fair, 1911) and *Sybill* (1914).

Jancsó, Miklós (1921)

Director. Strongly influenced by his native Transylvania, J. was early on attracted towards ethnography and folk culture, working for a long time as director of newsreels, shorts and documentaries. After *Így jöttem* (My Way Home, 1964), the Cannes Film Festival attention won by his third feature film, *Szegénylegények* (The Round-Up, 1965), turned him overnight into a focal figure in European cinema, consolidated by a string of equally controversial and much-acclaimed subsequent works, including *Csillagosok, katonák* (The Red and the White, 1967) and *Csend és kiáltás* (Silence and Cry, 1967). Typically these take turning points in Hungarian history as their setting, with the protagonists having to choose between loyalty, betrayal or lying in the face of an overwhelming, Kafkaesque oppressive power. The films are not based on identifiable individual faces so much as ritual, choreographed movements of crowds or groups of people, with authority figures in black uniform, fighters or victims in white. Their ostensibly simple plots are bafflingly elusive to pin down, with more importance accorded to conveying the subject matter through purely cinematic means, notably the long and complicated single takes that are J.'s trademark. Following major controversies in Hungary over *Fényes szelek* (The Confrontation, 1969) and *Égi bárány* (Agnus Dei, 1972), J. spent six years in Italy, where his films included *La Pacifista* (The Pacifist, 1970) and *Vizi privati, pubbliche virtu* (Private Vices,

Public Pleasures, 1975), before returning home again to shoot *Magyar rapszódia* (Hungarian Rhapsody, 1978). J.'s output over the next decade and a half included notable parables—*Szörnyek évadja* (Season of Monsters, 1986), *Jézus Krisztus horoszkópja* (Jesus Christ's Horoscope, 1988), *Isten hátrafelé megy* (God Walks Backwards, 1990) and *Kék-Duna keringő* (Blue Danube Waltz, 1991)—but these have been eclipsed by the startling shift in tone and treatment of the recent slapstick trilogy of *Nekem lámpást adott kezembe az Úr—Pesten* (The Lord's Lantern in Budapest, 1998), *Az anyád! A szúnyogok...* (Damn You! The Mosquitoes..., 1999) and *Utolsó vacsora az Arabs szürkénél* (Last Supper at the Arabian Gray, 2000), which cast a satirical look at the travails of haves and have-nots alike in post-Communist Hungary.

Jánossy, György (1923–1998)

Architect. Like Zoltán Farkasdy, one of the group of architecture students who escaped to Denmark in 1944, where he worked in the office of Ole Falkentorp. Returning to Hungary, from 1949 he worked at KÖZTI, the state planning office for public buildings, partnering István Janáky on the initial designs for the *Miskolc Technical University for Heavy Industries* (1951–52). The first to be promoted from the 'Master School' postgraduate training courses organised from 1953–60 by the Union of Hungarian Architects, his main works included: the *Cultural Center for Hódmezővásárhely* (1949–50); a set of *Student Residential Blocks for Miskolc Technical University* (1952); the *Water Tower for the University of Agricultural Sciences, Gödöllő* (1952–54, with István Czebe); the *House at No. 38 Úri utca* (1957–63), on Buda's Castle Hill; and the 450-bed *Hospital and Polyclinic at Kazincbarcika* (1960–65, with József Hrecska); the *Lidó Sport Hotel and Boathouse* (1963, with Zoltán Walkó), just beyond the Old Buda Roman Settlement of Aquincum; the *Karancs*

Hotel in Salgótarján (1960–64, with Hrecska) and the town's *Main Square* (1964–68, with György Szrogh, Géza Magyar and József Finta). From 1963 J. taught in the Department of Architecture and later the Department of Interior Design at the Academy of Applied Arts. For many years J. was also a tutor for the Young Architect's Circle (FÉK) after the scheme of postgraduate training that was resurrected by the Architects' Union in 1970. Chief architect of a private practice until he retired in 1983, his projects during included the *House in Gyorskocsi utca* just below Castle Hill (1966–68) and the *Diplomat's House at No. 5 Szentháromság tér* on Castle Hill (1980–83, with László Laczkovich). In the field of public ornamental art, J. designed the *Martyrs' Monument for Hősök tere in the Pestszenterzsébet* (twentieth) district of Pest (1965, with sculpture by Tibor Vilt) and the *Liberation Monument for Pécs* (1975, with sculpture by Agamemnon Makrisz). Twice awarded the Ybl Prize (1965, 1982), he also received the State Prize (1985) and was a member of the Hungarian Academy of Arts.

Jávorka, Sándor (1883–1961)

Botanist. After graduating from Budapest University, J. entered the service at the Botanical Collection of the National Museum for the remainder of his life, including two spells as its director. His sixty years of work as an exploratory botanist won him wide acclaim, leading him to collect a vast number of samples and identify 120 new species. His *magnum opus* was the 1,400-page *Flora Hungarica* (1924–25), the product of two decades of work, which is a critical description of some 4,000 species, which is supplemented by the invaluable picture atlas *Iconographia florae partis austro-orientalis Europae centralis* (1975), co-produced with the botanical illustrator Vera Csapodi. J. was a member of the Hungarian Academy of Sciences.

Jedlik, Ányos István (1800–1895)

Physicist. A Benedictine monk, who taught in his order's institutions until 1839, when he became professor of physics and mechanics at the Budapest University. There he tackled many areas of physics, particularly those where his superb practical skills could shine. He was widely acclaimed, for instance, as a maker of the finest-resolution optical gratings for astronomical telescopes. However, his major work was as an inventor of electromagnetic machines. In 1861, some six years before Siemens and Wheatstone, he constructed the first single-phase (i.e. direct-current) dynamo, working on the principle of self-induction, but—as with his many other inventions, including an early condenser cascade (1873), electric motors and a high-voltage power source—failed to apply for a patent. His main publications were *Compendium Hydrostatice* (1847), *Súlyos testek természettana* (The Nature of Heavy Bodies, 1850), and *Fénytan* (Optics, 1851).

Jendrassik, György (1898–1954)

Mechanical engineer. Graduating from Budapest Technical University, he immediately joined the Ganz Works, later becoming its managing director. He made his mark early on by producing internal combustion engines, including diesel motors, for track inspection railcars and ships, which enjoyed extraordinary demand during the 1930s, going on in 1938 to present a design for the world's first low-output gas turbine with an independent combustion chamber. Under the constraints of Hungary's post-war politics, he chose to leave the country in 1947, choosing to settle in Britain.

Joachim, József (1831–1907)

Violinist. Studied in Vienna (1839–43) and Leipzig (1843–47). He lived and taught in Hanover (1853–65), then in Berlin until his death. He was an intimate friend of Brahms, collaborating with him on the composition of the *Violin Concerto*, of which he gave the first performance.

Jókai, Mór (1825–1904)

Novelist. Even today one of the most widely read authors in Hungary, J. played an active part in the events leading to the outbreak of the Hungarian revolution in 1848, though not an extremist by nature. During the repression that followed its defeat he had to hide and only escaped arrest through the cleverness of his wife, the famous actress Róza Laborfalvi. A prolific writer (and probably translated more widely than any other Hungarian writer), J.'s early work, from the first novel, *Hétköznapok* (Weekdays, 1845), onwards, shows a strong influence of French Romanticism, his gentle humour and optimism leading him to draw a fairly rosy picture of both past and present. The many historical novels, such as *Erdély aranykora* ('Midst the Wild Carpathians, 1852) and *Török világ Magyarországon* (The Slaves of the Padishah, 1853), have little claim to accuracy, but their evocations of the past are compelling. Even in novels dealing with contemporary Hungarian society or with the Reform Era leading up to the Revolution, e.g. *Egy magyar nábob* (A Hungarian Nabob, 1853–54), *Kárpáthy Zoltán* (1854), and *Az új földesúr* (The New Landlord, 1862), J. managed to gloss over harsh realities. Of the works written after the Compromise, *Fekete gyémántok* (Black Diamonds, 1870) and *Az arany ember* (Timar's Two Worlds, or The Man With the Golden Touch, 1873), in which J. tackles some of the problems of a mercantile society, are outstanding.

Jordán, Károly (1871–1959)

Mathematician. With studies at universities in Paris, Zurich, Manchester, Geneva and Budapest behind him, he became a professor at Budapest Technical University, was interested primarily in mathematical statistics and probability calculus, and their potential practical applications. One of those areas was in earthquake prediction, which led to his appointment as head of the Hungarian Seismological Institute for seven years. The other main area was meteorology, in which he was the first to apply techniques of statistical correlation. His major published works were: *Matematikai statisztika* (Mathematical Statistics, 1927), *Véges differenciák módszere* (Method of Finite Differences, 1947). J. was elected a member of the Hungarian Academy of Sciences in 1947.

Jovánovics, György (1939)

Sculptor. Of all the artists who started their career during the late '60s, J. has perhaps had the most ambiguous relationship with up-to-date trends. His early masterpiece, *Részlet a Nagy Gillesből* (Detail from the Great Gilles, 1967), is a strange mixture of history, memory and myth. In 1974 he completed a series of conceptual photographs entitled *Liza Wiathruck: HOLOS GRAPHOS*. Virtually all his sculptural works are executed in plaster, with the prominent exceptions of *Trójai papírgyár* (Trojan Paper Mill), a monumental concrete statue in Seoul's Olympic Park, and the Budapest *Memorial to the Martyrs of the '56 Uprising* (1989–92). A retrospective exhibition of J.'s collected works at the 1995 Venice Bienniale was crowned with his paraphrase of Giorgione's *Tempesta*, a painting in the Venice Accademia collection. J. is a recipient of the Kossuth Prize and currently teaches in the Department of Sculpture at the University of Fine Arts, Budapest.

József, Attila (1905–1937)

Poet. J. published his first book of poems, *Szépség koldusa* (Beggar of Beauty), in 1922. Having entered Szeged University in 1924, he was expelled the next year over the controversy aroused by publication of his poem *Tiszta szívvel* (With a Clear Heart, 1924). He spent the next few years in Vienna and more particularly Paris, taking on Surrealist and Expressionist influences, but was also inspired by the Hungarian idiom, most notably in the forms and rhythms of Hungarian folk songs. J. reveals his full maturity as a poet in the volume *Külvárosi éj* (Night in the Slums, 1932), yet critics failed to comprehend him. Two further collections appeared during his lifetime: *Medvetánc* (Bear Dance, 1934) and *Nagyon fáj* (It Hurts a Lot, 1936). After 1934, his mental condition deteriorated, and with it his capacity for overcoming the conditions of his chronic poverty and the increasingly chaotic atmosphere of his time, his poetry became ever more brilliant. His final poems written during the three years preceding his suicide are intensely subjective and painfully revealing soliloquies of his suffering.

Juhász, Ferenc (1928)

Poet. Before publishing his first volumes, *Apám* (My Father, 1949) and *Tékozló ország* (The Prodigal Country, 1954), J. studied art history at Budapest University. Although his early poems cultivated a simple lyrical voice, he increasingly turned to the narrative, epic dimension of poetry, developing an intensely personal language and a private universe inhabited by monads, mythical figures, strange beasts and other figures. This is perhaps best exemplified by *A szarvassá változott fiú kiáltozása a titkok kapujából* (The Boy Changed into a Stag Clamours at the Gate of Secrets, 1955), which describes the confrontation of an ancient world with the complexities of the modern era, with man caught between the two, wrestling with the cosmic secrets of life and death. *A szent Tűzözön regéi* (Legends of the Holy Flood of Fire, 1969) revived an ancient style of narrative myth with clarity and feeling, with mankind's self-destructive flirtation with incompletely mastered forces providing the theme. Subsequent volumes include *A virágok hatalma* (The Force of Flowers, 1956), *A tenyészet országa* (The Land of Profusion, 1957), *A halottak királya* (The King of the Dead, 1971), and *A Föld alatti liliom* (The Underground Lily, 1991).

Jurcsik, Károly (1928)

Architect. After graduating from Budapest Technical University (1952), he stayed on as an assistant lecturer under Károly Weichinger in the Department of Public Building until

1965. His first major works were a Swimming Hall and Spa for Kecel (1961), between Kalocsa and Kiskunhalas. Towards the end of this period he had the opportunity to work in the planning department of the then London County Council (1963–64). Moving to LAKÓTERV (1965–79), the state planning office for housing, he worked on the *Cultural Centre for Orgovány* (1968), a village south-east of Kecskemét; a *Store in Budafok* (1969), Buda's twenty-second district; and *Szekszárd Town Centre* (1968–72). Between 1969 and 1973 he developed a family of plans for nursery schools, utilising a permutable 'panel skeleton system' of small modules, which were adopted on dozens of housing projects in Hungary. In 1970 he restored a peasant house in Vérteskozma near Tatabánya. In 1979 he moved to Győr, where his projects included *Panel-Construction Family Houses on Páva utca* (1982); a *School for Health Professionals* (1984), and the development of a family of plans for family houses in Győr County (1987). Projects undertaken in his private practice have included: a *House for an Art Collector* in the Buda hills (1979); an *Apartment House on Aradi Vértanúk útja, Pécs* (1982), near Dóm tér; *House in Győr's inner city* (1991), awarded the 'House of the Year' Prize in 1992; and the *School in the Woods at Sopron* (1993–95). J. taught at the Budapest Academy of Applied Arts from 1966 to 1971, and was a tutor for the Young Architect's Circle (FÉK) scheme of postgraduate training. Twice awarded the Ybl Prize (1964, 1972), he has also been a recipient of the State Prize (1980) and the Kossuth Prize (1993).

Kabos, Gyula (1887–1941)

Actor. Establishing himself as a comic dancer and actor in cabaret theatres during World War I, K.'s stage roles were mainly the sort of comic or tragicomic lower middle-class figures of his later film career, rarely dramatic figures (e.g. Chekhov's *The Cherry Orchard* or Dostoievsky's *Crime and*

Punishment). Success playing the put-upon little man of *Hyppolit, a lakáj* (Hippolyt, the Lackey, 1931) made K. the most sought-after comic film actor of the 1930s through his ability to redeem even mediocre, formulaic films with his skill in satirically subverting the realities of a society in which the upper classes still had a firm grip on authority. With few exceptions—Béla Gaál's *Meseautó* (Wonder-Car, 1934) and István Székely's *Lovagias ügy* (An Affair of Honour, 1937)—the nearly 40 films in which he appeared are eminently forgettable. Hungary's anti-Jewish laws in 1939 forced K. to leave the country for America, where he died in New York.

Kacsoh, Pongrác (1873–1923)

Composer. After studying at Kolozsvár, from 1912 he was music director for the Budapest Music School. His most enduring success was the outstanding piece of Hungarian operetta *János vitéz* (John the Valiant, 1904). Despite his own conservative musical style, he was an enthusiastic advocate of Bartók and Kodály.

Kadosa, Pál (1903–1983)

Composer, pianist. A private pupil of Kodály, he then taught piano at the Academy of Music from 1945 until his death. As a pianist, he was a prominent exponent of the contemporary repertoire; as a composer he was mainly influenced by Bartók and Paul Hindemith. His pupils included Zoltán Kocsis, Dezső Ránki and András Schiff.

Kálmán, Emmerich [Imre] (1882–1953)

Operetta composer. Studied at the Budapest Academy of Music, contemporaneously with Bartók and Kodály. After the success of his first operetta *Tatárjás* (The Gay Hussars, 1908; revised as *Herbstmanöver* for Vienna, 1909) he moved to Vienna; in 1940 he emigrated to the US. His most important works are *Die Csárdásfürstin* (1915) and *Gräfin Mariza* (1924).

Kalmár, László (1905–1976)

Mathematician. Winning the Eötvös Competition for secondary school pupils in 1922, he obtained his first degree and doctorate from Budapest University before moving to Szeged, where he won a chair. With a wide range of interests, he did successful research on mathematical logic. He contributed significantly to the development of the decidability problem and its relation to the Church and Gödel theorems. He became a pioneer of computer science in Hungary after 1955, constructing a highly innovative programmable logic device that served to introduce many students to information technology. The Szeged University Computer Centre bears his name today. K.'s achievements were also recognised by awards of the Kőnig and Kossuth Prizes, and he was a member of the Hungarian Academy of Sciences.

Kamuti, Jenő (1937)

Fencer, sports administrator. Qualified physician (1963). Four times an individual winner at the World Student Games, he went on to gain silver medals in individual foil at the Olympics (1968, 1972) and World Championships (1967), as well as a gold (1957) and four silvers at the latter in the team event. Chairman of the Hungarian Fencing Federation (1989–); secretary-general of the International Fencing Federation (1992–96); chairman of the European Fencing Federation (1996–); chairman of the International Fair Play Committee (2000–). Awarded UNESCO Fair Play Prize (1977) and the International Olympic Committee's Olympic Order (1995).

Karinthy, Frigyes (1887–1938)

Novelist, critique, humorist and translator. Without writing an acknowledged major work, K. nevertheless exerted considerable influence on his own and later generations. He earned particular success with *Így írtok ti* (Here's How You Write,

1912), a collection of telling caricatures of contemporary writers. K. developed the persona of an ordinary man fumbling, with humour as his sole weapon, to come to terms with the vicissitudes of modern urban life. *Tanár úr kérem* (Please, Sir!, 1916) is a sharp evocation of the typical characters and basic experiences of school life, whilst the science fictions *Utazás Faremidóba* (A Journey to Faremido, 1916) and *Capillária* (1921)—modern-day extensions of *Gulliver's Travels*—typify his fascination with science and technology. *Krisztus és Barabás* (Christ and Barabbas, 1918), by contrast, was K.'s deeply felt protest against the stupidity and immorality of wartime slaughter—a theme later picked up again in two volumes of poems, *Nem mondhatom el senkinek* (I Cannot Tell Anyone, 1930) and *Üzenet a palackban* (Message in a Bottle, 1938), in which he warns against society's alienation and dehumanisation. K.'s last novel, *Utazás a koponyám körül* (A Journey Round My Skull, 1937), is the story of his own brain tumour and convalescence from surgery, whilst *Mennyei riport* (Celestial Report, 1937) is a meditation on life after death.

Kármán, Theodore von [Tódor] (1881–1963)

Aeronautical engineer. After graduating from the Budapest Technical University, he stayed on to work with Bánki before spending six years in Göttingen learning about aerodynamics. During World War I, he and colleagues developed the first military helicopter (though it never flew). Having played an active role during the Hungarian Soviet Republic, he was compelled to leave the country in 1919, gaining a post at Aachen for over a decade until, with Hitler's rise to power, he emigrated to the USA to head the aerodynamics laboratory in Pasadena. There he made pioneering contributions to the early supersonic flight and missile programs. His theory of boundary surface is still an accepted approach in the design of aircraft today, and his results in inelastic deformation and in crystal lattice structures are still valid. His published works include *Collected Works* (1956), *From Low-Speed Aerodynamics to Astronautics* (1963) and *The Wind and Beyond* (1967).

Kárpáti, Rudolf (1920–1999)

Fencer, sports administrator. Graduate of the Budapest Academy of Music (1942); coach's certificate (1965). Won golds at the Olympics in individual sabre (1956, 1960) and the team event (1948, 1952, 1956, 1960), as well as at World Championships in the individual (1954, 1959) and five times in the team event. Hungarian Sportsman of the Year (1959, 1960). Board member of the Hungarian Fencing Federation (1961–91). Career army officer and musicologist. Chairman of the Hungarian Hall of Fame (1991–94). Awarded the Olympic Order of Merit (1990).

Kassák, Lajos (1887–1967)

Painter, graphic designer, poet and writer. From the 1910s to the end of the '20s and again after 1945, K. was a key figure in the Hungarian avant-garde movement, working in many visual art genres, from typography to painted wood relief, and idolised by young artists for his purity of thought. K.'s early poems were regularly published in the daily newspaper *Független Magyarország* (Independent Hungary) before he launched his own periodical, *A Tett* (The Deed), in 1915. When this was banned the next year, he promptly started up *Ma* (Today), the editorial office of which was also used to display the works of artists K. had discovered. Forced to emigrate in 1919 after the downfall of the Hungarian Soviet Republic, in which he had played a publicly prominent role, K. continued to publish *Ma* in Vienna until his return to Hungary in 1926. During the '20s K.'s intellectual stance shifted from a strongly Dada-influenced approach to a pragmatic constructivism with his formulation of a concept of *Bildarchitektur* ('image architecture') that brought him into the mainstream of avant-garde concerns. He became the editor successively of the periodicals *Dokumentum* (1926–27) and *Munka* (Work, 1928–39). Between 1946 and 1954 he once again took up graphic art, though after 1949 was yet again forced to retreat from the public eye. The works of his final years were more lyrical abstractions.

Katona, József (1791–1830)

Playwright. Having become interested in the theatre whilst qualifying as a lawyer in Pest, it was only after securing a post in his native town of Kecskemét, that K. was able to pursue his enthusiasm, initially by translating works by popular German authors like Iffland and Kotzebue and dramatising romances such as *A Luca széke karácsony éjszakáján* (Luca's Chair on Christmas Eve, 1812). His first original play is *Ziska a calice, a táboriták vezére* (Ziska, the Hussite Leader, 1813) but his most important work by far is *Bánk bán* (Ban Bánk), the story of a real early thirteenth-century person which raises questions of fighting against foreign tyrants and the problem of serfs. Written in 1815 for a competition run by the periodical *Erdélyi Múzeum* but, following its rejection by the jury in early 1818, K. published it privately in 1820, however it gained its status as the Hungarian national drama largely because it was the source of the libretto for Ferenc Erkel's hugely popular opera of that name, completed in 1852 but itself not staged until 1861. K.'s also wrote a significant theoretical work on drama: *Mi az oka, hogy Magyarországon a játékszíni költőmesterség lábra nem tud kapni* (The Reason Why Dramatic Poetry Can Find No Footing in Hungary, 1821).

Kazinczy, Ferenc (1759–1831)

Poet, writer, translator. K. was the leading figure of the Hungarian 'neologist' language-reform movement in the early nineteenth century. Educated at the Calvinist Sárospatak College, he was an enthusiastic advocate of emperor Joseph

II's modernising efforts and in 1788, with Dávid Baróti Szabó and János Batsányi, published *Magyar Museum*, the first prominent periodical of the Hungarian Enlightenment, before taking sole editorship of *Orpheus* (1789–92). This was when his sole prose fiction, the sentimental epistolary novel *Bácsmegyei öszveszedett levelei* (The Collected Letters of Bácsmegyei, 1789), was published. Imprisoned for 2,387 days (1794–1801) in various notorious Austro–Hungarian gaols (Buda, Spielberg, Obrovic, Kufstein, Munkács) for participation in the Hungarian Jacobin movement—the subject of *Fogságom naplója* (The Diary of My Captivity, 1831). After his release he settled at Széphalom in N.E. Hungary, to set about reviving the country's literary life and promote language reform. K. translated many works by Gessner, Holbach, Rousseau, Voltaire, Shakespeare, etc., but these and his own volume of epigrams, *Tövisek és virágok* (Thorns and Flowers, 1811), are now of historic interest only. The prison 'diary' and the similarly autobiographical *Pályám emlékezete* (Reminiscences of My Career, 1828) are his most enduring testaments.

Keleti, Ágnes (1921)

Gymnast, coach. PE instructor's certificate (1954). Won individual golds at the Olympics for floor exercises (1952, 1956), asymmetric bars (1956) and beam (1956) as well as with the hand apparatus team (1956), and at World Championships for the asymmetric bars (individual) and hand apparatus team (1954). Since 1957 she has lived in Israel, establishing the sport there as head of department at Wingate College (1957–80) and head coach to the national gymnastics team (1958–80). International judge (1988–). Inducted into the Jewish Hall of Fame (1980), the American Jewish Hall of Fame (1985), and the Hungarian Hall of Fame (1992).

Kemény, Zsigmond (1814–1875)

Novelist, journalist, politician. A baron, descended from a prince of Transylvania, K. lost many of the advantages of his birth through his father's early death, but his private means and title gave him easy access to political life. This he approached through journalism, first in his native Transylvania then, from 1847, on the staff of *Pesti Hírlap* (Pest News). In 1855 K. became editor of *Pesti Napló* (Pest Diary) and, as an active politician, helped prepare the way for the 1867 Compromise between Hungary and the Habsburgs. K.'s reputation nowadays, however, is principally as an author of primarily historical novels. *Gyulai Pál* (Pál Gyulai, 1847), *Özvegy és leánya* (A Widow and Her Daughter, 1855); *A rajongók* (The Fanatics, 1858); *Zord idő* (Bleak Times, 1862). These have been praised by critics but less appreciated by the public due to their stylistic difficulties, their emphasis on description rather than action, and their gloomy fatalism: entangled in personal and political conflicts, K.'s protagonists move inexorably toward destruction.

Kepes, György (1906–2001)

Painter, photographer, theoretician. Though a pupil of István Csók at the Budapest Academy of Fine Arts, by 1928 he was already turning towards the avant-garde of Lajos Kassák's Munka (Work) Group. His first exhibition was as a member of the New Association of Artists (KUT) Group in 1930, in which year he was invited by Moholy-Nagy to the Bauhaus in Berlin, where he took up film and photography. From 1937 K. taught at the New Bauhaus in Chicago, pursuing experiments with light and colour, then from 1945 in the Department of Architecture at the Massachusetts Institute of Technology, producing works of kinetic art. Through his commissions for murals, noted for the boldness of their technological innovation (e.g. a light wall for Milan), he came into contact with

Walter Gropius, Alvar Aalto and Marcel Breuer. In 1967 he founded and directed the Centre for Advanced Visual Studies at MIT, which brought scientists and artists together in developing ecologically slanted projects as responses to challenges of the present day.

Kertész, Alíz (1935)

Gymnast, coach. Certificates as PE instructor (1960), coach (1967) and master coach (1974). Won golds as team member in both hand apparatus and composite events at World Championships (1954) and Olympics (1956). Coach (1961–81). Captain of the women's national gymnastics team (1980–81). International judge.

Kertész, André [Andor] (1894–1985)

Photographer. Taking up photography at the age of 18, K.'s early pictures were of the rural landscapes and village life around Szigetbecse, at the southern end of Csepel Island below Budapest, but after the outbreak of World War I he soon showed an accomplished eye for war reportage as well more abstract approaches, like the well-known *Underwater Swimmer* (Esztergom, 1917). *Blind Musician* (Abony, 1921) and other pictures of beggars and cripples are evidence of his critical approach to social issues. On moving to Paris (1925–34), K. worked as a freelance reporter for *UHU* magazine, *La Sourire*, the *Berliner* and *Frankfurter Illustrierte*, the London *Times*, etc. and contributed to art journals such as *Vu*, *Art et Médecine*, and *L'Art vivant*, mixing with many prominent members of the Parisian artistic community, including fellow Hungarian émigrés such as Brassaï (whom he inspired to take up photography), István Beöthy, Robert Capa, József Csáky, and Lajos Tihanyi. Many of these friends featured in the well-known portraits and shots of bistro and nightclub life of those years: *Satiric Dancer* (1926), *Piet Mondrian in his Studio* (1926), *Meudon* (1928), *You and I* (1931). K. developed a unique genre of picture report, which he later referred to as 'literary reportage', several examples of which (e.g. *Fallen Horse*, 1927) were clear forerunners of the 'sequences' of the '60s. During a 1936 trip to New York, at the invitation by Keystone View Co., K. decided to settle there, becoming an American citizen in 1944, though he eventually became disillusioned with New York life and its tenets of photography. The innovative *Distortion* series (1937) of nudes (reflections in distorting mirrors) attracted considerable attention but some pictures had to be cropped. Towards the end of his life, K.'s artistic activity was limited to pictures taken from the window of his flat in Washington Square and polaroids of his flat and his wife's favourite objects.

Kertész, Imre (1929)

Writer. In 1944, he was deported to Auschwitz and then to Buchenwald, where he was liberated in 1945. In Hungary, he worked for the journal *Világosság* (Light) until his dismissal in 1951. From then on, he has been an independent writer and translator. Among his translations are German authors such as Nietzsche, Hofmannsthal, Schnitzler, Freud, Roth, Wittgenstein and Canetti, each of whom influenced his own writing. Working for almost 15 years on his most significant novel, in 1975 *Sorstalanság* (Fateless) was published but remained largely without resonance in Hungary. The novel, based on K's autobiographical experience, is the story of a Holocaust survival from a child's point of view. The idea that there is no experience beyond the Holocaust is central to the novel and to all of K's later writings. *A kudarc* (Fiasco, 1988) is regarded as the second volume in a trilogy that began with *Sorstalanság* and of which the third volume is *Kaddis a meg nem született gyermekért* (Kaddish for a Child not Born, 1990). In 1992 *Gályanapló* (Galley Diary), a diary in fictional form, covering the years 1961–1991 was published. His lectures and essays have been collected in *A holocaust mint kultúra* (The Holocaust as Culture, 1993), *A gondolatnyi csend, amíg a kivégzőosztag újratölt* (Moments of silence while the execution squad reloads, 1998), and *A száműzött nyelv* (The exiled language, 2001). In 2002 Kertész was awarded the Nobel Prize for Literature.

Kiss, Balázs (1972)

Hammer-thrower. Studied at the University of Southern California (1992–97). Olympic champion (1996); golds at World Student Games (1995, 1997). Has trained in the USA since 1993; twice winner at US Nationals.

Kiss, László (1940)

Coach, sports administrator. Certificates as coach (1972) and master coach (1983). Swam in the national team (1958–60). Head coach of the swimming division of Budapesti Spartacus (1967–2000); swimming coach to the national modern pentathlon team (1969–83); head coach to the women's national swimming team (1993–2000) and individual coach to Olympic champion Krisztina Egerszegi; captain of the woman national team (1999–). Hungarian Coach of the Year (1991–93, 1996–97, 1999). Vice-chairman of the Hungarian Swimming Federation (1993–). Awarded the Officers' Cross of the Republic of Hungary (1992).

Kocsis, Sándor (1929–1979)

Footballer. Member of the gold-winning Olympic team (1952) and the team that beat England in the 1953 'match of the century' at Wembley Stadium; as a member of the 1954 World Cup silver-winning team was top scorer (11 goals) in the tournament. Three times top scorer in the Hungarian Championship Division 1 (1951, 1952, 1954). Leaving Hungary in 1956, he was a successful club player in Switzerland and Spain. Took his own life due to acute leg disease.

Kocsis, Zoltán (1952)

Pianist. A pupil of Pál Kadosa at the Budapest Academy of Music, he had already embarked on an international solo career by 1971. He recorded the complete piano works of Debussy and Bartók. With Iván Fischer, he was co-founder of the Budapest Festival Orchestra in 1983; since 1997 he has been chief conductor with the National Philharmonic Orchestra.

Kodály Zoltán, (1882–1967)

Composer. He studied at the Budapest Academy of Music and studied arts at the University of Budapest. He undertook research into Hungarian folksong with Bartók. He travelled widely then taught theory at the Academy in 1907, and took over the composition classes from Koessler in 1908. From that time he was closely involved with the music curriculum in Hungarian schools. After the Second World War he travelled to France, England, the USA, and USSR to conduct his own works. His main compositions are *Psalmus Hungaricus* (1923); other works include the opera *Háry János* (1926), *Galántai táncok* (Dances of Galánta, 1933).

Koestler, Arthur (1905–1983)

Writer, journalist. Born in Budapest, Koestler left Hungary in 1919. After studying electrical engineering and psychology at Vienna, he spent a short time at a kibbutz in Palestine before becoming a foreign correspondent for German papers in Paris and elsewhere. Having joined the Communist Party in 1931, during the Spanish Civil War K. was captured by Franco's troops in 1937 as a correspondent for a British newspaper; after his release had been secured from there, he was detained in France, escaping in 1940 to make his way to Britain. His novels *Darkness at Noon* (1940), *Arrival and Departure* (1943) and *Thieves in the Night* (1946) reflect these personal experiences, also explored in the volume *The Yogi and the*

Commissar and Other Essays (1945). K. was thereafter preoccupied by aspects of psychology and the mind, notably in his trilogy on human creativity: *The Sleepwalkers* (1959), *The Act of Creation* (1964), and *The Ghost in the Machine* (1967).

Komjáthy, Jenő (1858–1895)

Poet. Though he produced just a single volume of poems, *A homályból* (From the Dark, 1895), K. was a significant representative of the *fin-de-siècle* search for new lyrical modes. Inspired by a mix of Spinoza's pantheism, Nietzsche and Gnostic philosophy, his poetry is strongly spiritual, sometimes even mystical, albeit flawed both artistically and philosophically. The recurring notion is one of redemption, the attainment of freedom and bliss, through love. His work is closely linked to the Romantic tradition in Hungary, showing a strong national disposition, but at the same time he turns towards the inner world of the individual and the universal problems of existence. As such, it can be seen as a forerunner of later Symbolist poetry in the country.

Kondor, Béla (1931–1972)

Painter, graphic artist, poet and also musician. K. began his training in the painters' section at the Budapest Academy of Fine Arts but graduated as a graphic artist. He early on evolved a distinctive style, notably in etchings like the *Dózsa* series (1956), though also using wood-, lino- and papercut, lithography, monotype and other graphic techniques, painting on canvas, wood and walls, and also taking up photography in his final years. He made his living mostly from commissions for graphic works, with the Petőfi Literary Museum a willing patron. Intellectual in his approach, with a philosophically demanding bent, K. was his own man, bringing a high degree of freedom to his penetrating interpretations of the classical tradition, from Giotto and Dürer through to Goya and Picasso, always seeking answers to the questions of his own times. He expressed scepticism towards technological progress in apocalyptic visions based on Christian subjects and iconography, as in *Apokalipszis bábokkal és masinákkal* (Apocalypse with Dolls and Machines, 1956), *A műtücsök felbocsátása* (Launching of the Artificial Cricket, 1958), *Atomágyú felállítása* (Mounting the Atomic Cannon, 1959), *A géprepülés géniusza* (The Genius of Mechanical Flight, 1964), etc. Another focal point of his enquiries was the ambivalent nature of man's loftily soaring ambitions, probed through his use of the emblematic motifs of an insect, as in *Darázskirály* (Wasp King, 1963), bird, angel or aircraft. The struggle between Good and Evil is sometimes given direct expression, as in *Párharc* (Duel, 1967), *Szent Antal megkísértése* (Temptation of St Anthony, 1966), etc. He was also much preoccupied by man's relation to Christ—e.g. in *Krisztus a kereszten* (Christ on the Cross, 1960), *Szerb Krisztus* (Serbian Christ, 1962), and *Pléh Krisztus* (Tin Christ, 1964)—and to angels, as in the engraving *Négy angyal* (Four Angels, 1957), *Angyal és a festő* (Angel and the Painter, 1961), and *Angyal a város fölött* (Angel above the City, 1972). Most of his designs for murals, such as *A szentek bevonulása Jeruzsálembe* (Entry of the Saints into Jerusalem, 1971–72)—a composition submitted to the jury for the Budapest Chamber of Commerce—were never executed. One on the subject of Dominican nuns, for a shrine to St Margaret on Margaret Island, Budapest, was a rare exception.

Konkoly-Thege, Miklós (1842–1916)

Astronomer. Following studies at Pest (with Jedlik) and Berlin, in 1871 he set up his own observatory, then the only one in the country, on his estate at Ógyalla (Hurbanova, Slovakia), which in 1899 was donated to the state, becoming the Hungarian Royal Astrophysical Observatory. In 1872 he commenced a long series of observations of sunspots and solar flares, and planetary surface features (e.g. the 'Great Red Spot of Venus'), regularly publishing his findings until his death. As well as being a pioneer of astronomical spectroscopy and photography, he made and re-designed many instruments, including a solar telescope (1877) and a heliograph (1884), fully documented in manuals: *Praktische Anleitung zur Anstellung astronimischer Beobachtungen mit besonderer Rücksicht auf die Astrophysik* (1883), *Anleitung zur Himmels-photographie* (1887), and *Handbuch für Spectroscopie im Cabinet und am Fernrohr für Anfänger auf dem Gebiete der Spectralanalyse* (1890). Between 1890 and 1911 he was the director of the Hungarian Meteorological Institute, initiating construction of its present building.

Konrád, György (1933)

Novelist, essayist. K. established his name with the novel *A látogató* (The Case Worker, 1969), based on his own chilling experiences as a sociologist and social worker from 1959 onwards. In this he expresses fears not so much over the inexpressibility of the world but rather the impossibility of changing it; language is considered a possible salvation, as it at least offers the possibility of fresh perspectives on the world. K. became an international *cause célèbre* after his brief arrest in 1974 over a long essay *Az értelmiség útja az osztályhatalomhoz* (The Intellectuals on the Road to Class Power), co-written with Iván Szelényi, which attempted to clarify the role of intellectuals in a socialist society. A smuggled manuscript was first published in Paris (1978) then in translation before eventually appearing in Hungary itself in 1989. K.'s second novel, *A városalapító* (The City Builder, 1977), presents the reflections of an architect who, through the role he and his family have played in designing an unnamed East European city, has contributed to the undeniable inhumanities of a society born from a clash of left-wing idealism and cynical pragmatism. Later novels, which include *A cinkos* (The Loser,

1983), *A kerti mulatság* (A Feast in the Garden, 1987) and *Kőóra* (Clock of Stone, 1994), place more emphasis on quasi-autobiographical themes and narrative structures, their objective narrative voices are distillates of the quandaries of the modern East European intelligentsia. These identities, attitudes, ideological and ontological strategies are also the dominant subjects of his essay volumes, e.g. *Az autonómia kísértése* (The Temptation of Autonomy, 1980), *Antipolitika* (Antipolitics, 1984), *Európa köldökén* (In the Navel of Europe, 1990), and *Az újjászületés melankóliája* (The Melancholy of Rebirth, 1996).

Korda, Sir Alexander [Sándor] (1893–1956)

Director, producer. Began as a journalist and film reviewer in Budapest before moving into film direction in 1914. After purchasing the Corvin Film Studio, he directed or produced 25 films, including *Gólyakalifa* (The Stork Caliph/ The Nightmare, 1917), based on a psychological novel by Mihály Babits, and the 3-hour *Aranyember* (Man of Gold, 1918; directed by Ödön Uher), the first Hungarian full-length silent feature film, based on Mór Jókai's novel. Forced to emigrate in 1920, K. made films in a string of countries (four in Austria, six in Germany, nine in Hollywood between 1926 and 1930, two in France in 1931) before finally taking up residence in England and forming London Film Productions in 1932. There *The Private Life of Henry VIII* (1933) and *The Private Life of Don Juan* (1934) proved big box-office hits and established a new genre of historical portraiture, combining realism and romanticism, that came to define the British film industry. This enabled K. to turn his Denham Studio into the most modern facility in Europe, producing a steady stream of memorable films featuring world stars, including *Catherine the Great* (1934; dir. Paul Czinner), *The Ghost Goes West* (1936; dir. René Clair), *Elephant Boy* (1937; dir. Robert Flaherty and Zoltan Korda), *The Thief of Baghdad* (1940; dir. Michael

Powell), *The Jungle Books* (1941; dir. Zoltan Korda), and *The Life and Death of Colonel Blimp* (1943; dir. Michael Powell and Emeric Pressburger). Of the six films that K. himself directed during this period, the most important were *Rembrandt* (1936), *That Hamilton Woman* (1941), and *An Ideal Husband* (1948). After 1947 K. produced a further 45 films—amongst others, *Anna Karenina* (1948; dir. Julien Duvivier), *The Third Man* (1945; dir. Carol Reed) and several directed by David Lean—the last of which was *Richard III* (1955), directed by Laurence Olivier.

Korniss, Dezső (1908–1984)

Painter. A pivotal figure in the history of Hungarian twentieth-century visual arts, with connections both to avant-garde reform trends and the Szentendre school. Early archaising cubo-surrealistic compositions, such as *Galambász* (Pigeon Breeder, 1927), were based on constructivist principles. From 1935, K. sought motifs in folk art and architecture, incorporating them with abstract forms in townscapes like *Szentendrei motívum I–IV* (Szentendre Motif I–IV, 1935). After the war this world of symbolic folkloric motifs was expanded to include insect-like forms and fabulous creatures in vividly coloured figurative yet constructivist-based pictures such as *Kántálók* (Chanters, 1946) and *Tücsöklakodalom* (Crickets' Wedding, 1948) as well as more expressive pictures like *Ellentét* (Contrast, 1947). During a retreat into privacy after 1950 K. made small surrealist paintings and collages, from which he re-emerged after 1956 to produce large enamel-painted calligraphies. He then changed medium yet again to work with György Kovásznai on a series of eight animation films (1963–68), only to revisit the folk-art repertoire in hard-edge treatments of the traditional motifs sewn on the felt cloaks worn by herdsmen (*Szűrmotívum*, 1968) and eventually the abstract geometries of the *Horizontal–Vertical* series of his late period, from 1975 onwards.

Korondi, Margit (1932)

Gymnast. At the Olympics she won individual gold for the asymmetric bars (1952) and a team gold for hand apparatus (1956), team silvers for the composite event (1952, 1956), as well as individual bronzes for beam and floor exercise, and a team bronze in hand apparatus (1952). She has lived in the USA since 1956, managing a gymnastics school.

Kós, Károly (1883–1977)

Architect, painter, graphic artist and writer. On graduating from Budapest Technical University (1907) he became the leader and exhibition organiser of the *Fiatalok* ('Young Ones') group from the university whilst working in the office of Móric Pogány and Géza Maróti. His most important works were the *Parish Church at Zebegény* (1908–09, with Béla Jánszky) and the *Deer House* and *Aviary* in Pest's *Zoological Garden* (1909–10, with Dezső Zrumeczky), which shows affinities to Finnish Art Nouveau; a *School on Városmajor utca* (1910–11, with Dénes Györgyi), in Buda's twelfth district; the *Central Square* (now Kós Károly tér) *for the Wekerle Housing Estate* (1912–13) in the Kispest (nineteenth) district of Pest. In Transylvania he built a summer house and studio for himself at Stana, near Kolozsvár (Cluj, 1911); the *Székely National Museum at Sepsiszentgyörgy* (now Sfîntu Gheorghe; 1911–12); and the *Calvinist Church on Monostor út, Kolozsvár* (Cluj, 1912–13). Though appointed a teacher at the Academy of Applied Arts in Budapest in 1918, he abandoned his architectural career in the capital to return to his birthplace, by then annexed to Romania, and in 1921 set up the Transylvanian Folk Party, in 1924 the Erdélyi Szépmíves Céh (Transylvanian Crafts Guild), and in 1926 the Helikon Literary Society. Working as a writer, he hand-printed and bound his own books. After World War II he was one of the founders of the Hungarian National Alliance, then became deputy head at the School of Agronomy. His books include

A régi Kalotaszeg (The Kalotaszeg (Calata) District of Old, 1911), *A lakóház művészete* (Art of the Dwelling, 1930), *A székely nép építészete* (Architecture of the Székely People, 1944), *Falusi építészet* (Village Architecture, 1945), and *Hármaskönyv* (Tripartitum, 1969).

Kosma, Joseph [Kozma, József] (1905–1969)

Composer. Following his studies in Budapest, he moved to Berlin in 1929, then Paris in 1933, where he became a noted composer of film scores, e.g. for Jean Renoir's *La grande illusion* (1936) and Marcel Carné's *Les enfants du paradis* (1945). His chansons, with texts by Jacques Prévert, were performed by the likes of Edith Piaf, Juliette Gréco and Yves Montand.

Kosztolányi, Dezső (1885–1936)

Journalist, poet, writer. On finishing his studies at Budapest University, K. originally intended to teach but instead became a journalist, joining the circle of the literary magazine *Nyugat* and publishing his first volume of poetry in 1907. He won wide recognition in 1910 with the publication of a cycle of poems *A szegény kisgyermek panaszai* (Lament of a Poor Little Child), a small boy's subtle and moving impressions of his surroundings. K.'s sensitivity to experience and mastery of expression enabled him to write poetry of great lucidity and deceptive simplicity. A sympathetic observer of human frailty, with a gentle humour and a penchant for the macabre, he was an accomplished short story writer. K. also wrote several novels, of which *Édes Anna* (Anna Édes, 1926), the story of an exploited servant girl, is perhaps the best. An outstanding verse translator, K. paid much attention, especially in his later years, to the preservation of the purity of the Hungarian language. He showed little interest in what others considered the major problems of the age: for him, a poet was a 'soldier of castles in Spain', a guardian of beauty, civilisation and humanity.

Kovács, Ágnes (1981)

Swimmer. In 200 meters breaststroke she won Olympic bronze at the age of 15 (1996) then golds at the World Championships (1998) and Olympics (2000); several times European champion. Inducted into the Swimmers' Hall of Fame (2000). Awarded the Officers' Cross of the Republic of Hungary (2000).

Kovács, István (1970)

Boxer. As an amateur, won golds at flyweight at Olympics (1996) and World Championships (1997). Hungarian Boxer of the Year (1990, 1991) and Sportsman of the Year (1991). As a professional since 1997, won title belts from the World Boxing Council (1999), the European Boxing Union (2000), and the World Boxing Organisation (2001). Awarded the Bronze Cross of Merit of the Republic of Hungary (1992).

Kovács, József (1926–1987)

Athlete, coach. As a long-distance runner, he had notable wins in Budapest's Népstadion over 5,000 meters against Vladimir Kuts (1953) and over 10,000 meters against Emil Zatopek (1954). Won silvers at the Olympics (1956) and European Championships (1955). Was an athletics club coach.

Kovács, Pál (1912–1995)

Fencer, sports administrator. Graduated from Ludovika Military Academy (1935). Won golds at the Olympics for individual sabre (1952) and five times for the team event, as well as at the World Championship for the individual (1937, 1953) and seven times for the team event. Chairman of the Hungarian Fencing Federation (1963–68); member of the International Olympic Committee (1963–68); member (1968–80) then vice-chairman (1980–88) of the International Fencing Federation Executive Committee. Inducted into the Hungarian Hall of Fame (1991). Recipient of France's Sports Medal of Merit (1976), the Middle Cross of the Order of Merit of the

Republic of Hungary (1992), and the International Olympic Committee's Olympic Order (1993).

Kovácsi, Aladár (1932)

Modern pentathlete. Qualified physician. Won a team gold at the Olympics (1952), whilst at World Championships gained individual bronze (1955) as well as a team gold (1955) and silver (1958).

Kozma, Lajos (1884–1948)

Architect, industrial designer, graphic artist and book illustrator. On graduating from Budapest Technical University (1906) he became a member of the *Fiatalok* ('Young Ones'). At this stage he was of note particularly for his 'Baroque' Art Nouveau graphic work, incorporating many folk motifs, amongst which were his designs for a fantasy house he called the '*Lapis Refugii*', his album *Utolsó ábrándok* (Last Reveries, 1908), illustrations for volumes of Endre Ady's poems, and many woodcuts. He returned from a spell in Paris (1909–10) to run Béla Lajta's office, providing the interior decorations (ruined in 1945) for the Rózsavölgyi Music Shop in Lajta's *Office Block at No. 5 Martinelli* (now Szervita) *tér* (1911–12). In 1913 K. founded the Budapest Workshop and became a teacher at the School of Industrial Design. Due to his appointment to the Directorate for Arts and to a professorship at Budapest Technical University during the Hungarian Soviet Republic in 1919, he was never to receive any government-sponsored commission. The *Kner Villa in Gyoma* (*c.* 1925), in south-eastern Hungary, still reflects K.'s 'Baroque' style, but by 1930 his work was in the modernist mainstream. Larger projects included: the *Hungarian Fashion Salon* at Rákóczi út in Pest (1930, destroyed in 1944); the *Glass House Workshop, Office and Warehouse at No. 29 Vadász utca* (1934), in Pest's Lipótváros (Fifth) district; the *Apartment House at No. 39* (before 1934) and the three-storey *Villa at No. 67 Bimbó út*

(1935), on Buda's Rózsadomb; the *'Átrium' Apartment Building and Cinema* (1936) on the Margit körút in Buda; and the *Apartment House in the Régiposta utca* (1937) in Pest's inner-city area. Smaller dwellings, all in Buda's Pasarét (second) district unless otherwise indicated, included: the *Villa at No. 21 Orsó utca* (*c.* 1930); the *Villa at No. 5* and the *Semi-detached Villas at No. 6–8 Napraforgó utca* (1930); the *Villa at the Ruszti–Hankóczy Jenő utca* intersection (1932); the *Klinger Villa at No. 10 Hermann Ottó utca* (1934); the *House at No. 10 Csopaki utca* (*c.* 1935); *Summer Houses on Lupa Island* in the Danube (1935); the three-storey *Family House at No. 19 Berkenye utca* (1936) on the Rózsadomb; and a two-storey *Villa on Somlói út* (1938) on Gellért Hill. K. was also noted for simple, clear, fabric-covered furniture. In 1946 he was made director of the Academy of Applied Arts. His one published book was *Das neue Haus* (Zurich, 1941).

Kőbán, Rita (1965)

Kayaker. Sports manager's certificate (1997). Won Olympic golds in kayak fours (1992) and single kayaks (1996) as well as a World Championship gold in kayak fours (1986), all over 500 meters. Woman Kayaker of the Year (1984, 1987, 1991–92). Board member of the International Olympic Committee (2001–). Recipient of the Officers' Cross of the Order of Merit of the Republic of Hungary (1992) and the Hungarian Heritage Prize (1999).

Kölcsey, Ferenc (1790–1838)

Poet, critic, orator. After completing his law studies, K.'s interest was drawn to the literary movement started by Kazinczy, whom he admired and supported vigorously in the discussions aroused by the language reforms. With a small poetic *oeuvre* of his own, K. is remembered chiefly as the author of *Himnusz* (Hymn, 1823), which a generation later became the text of the Hungarian national anthem, and a handful of other poems, including *Vanitatum vanitas* (1823), *Zrínyi dala* (The Song of Zrínyi, 1830) and *Zrínyi második éneke* (Zrínyi's Second Song, 1838). However, his moral rectitude and devotion to his country are also reflected in his often severe but masterly literary criticism (e.g. of Csokonai and Berzsenyi) and his funeral orations and parliamentary speeches. Between 1832 and 1834, as a representative in the Hungarian Diet, he was a dignified exponent of the Liberal ideas that were then gaining support in the country. In his critical essays such as *Nemzeti hagyományok* (National Traditions, 1826) he laid the foundation of systematic literary criticism and aesthetics in Hungary.

Kőnig, Dénes (1884–1944)

Mathematician. Son of Gyula Kőnig, he won the 1902 Eötvös mathematics competition for schools before embarking on his studies in Budapest, though he graduated from Göttingen, and also obtained his doctorate there in 1907. He immediately joined the staff at Budapest Technical University, where he remained until his death. He wrote a significant monograph on descriptive topology as early as 1918, but his most significant work was his 1936 monograph introducing graph theory, on which the so-called 'Hungarian method' of linear programming is based *A véges és végtelen gráfok elmélete, Theorie der endlichen und unendlichen Graphen.*

Kőnig, Gyula (1849–1913)

Mathematician. Following studies at Vienna, Heidelberg and Berlin, he gained his doctorate from Heidelberg before appointment to a chair at Budapest University (1871) then at the Technical University (1874). The leading figure in Hungarian mathematics around the turn of the century, he made major contributions to the theory of sets (e.g. the Kőnig inequality among the numerous infinite sets), analysis, and number theory (the Kőnig–Rados theorem). His lectures and joint seminars with József Kürschák were a fabled training-ground for scientists. His main books are: *Analízis* (Analysis,

1887) and *Neue Grundlagen der Logik, Arithmetik und Mengenlehre* (1914). K. took an active public role as one of the organisers (and from 1891 a vice-chairman) of the Mathematical and Physical Society, as well as a director and section head of the Hungarian Academy of Sciences. He promoted improved standards in his subject by writing excellent textbooks for secondary and higher education.

Köteles, Erzsébet (1924)

Gymnast. Coach's certificates (1964, 1969). As a member of the hand apparatus team, she won bronze (1952) and gold (1954) at World Championships and gold at the Olympics (1956); in the team for the composite events, three Olympic silvers (1948, 1952, 1956). Coached a county gymnastics team (1950–52) and later the Hungarian national figure-skating team (1960–62). International judge in gymnastics (1954–), and rhythmic gymnastics (1962–).

Krúdy, Gyula (1878–1933)

Writer and journalist. K.'s first short story was published when he was just fifteen, and within two years he was already the principal contributor of a provincial weekly and getting articles published regularly in national dailies. Early influences included Jókai, Mikszáth and Reviczky amongst domestic writers, and Dickens, Maupassant, Zola and Turgenev amongst foreigners. He liked the tone-painting and character drawing of impressionism. Sindbad, the protagonist of three short-story cycles—*Szindbád ifjúsága* (Sindbad's Youth, 1911), *Szindbád utazásai* (The Travels of Sindbad, 1912) and *Szindbád megtérése* (The Conversion of Sindbad, 1925)—is a lover of life's pleasures, women above all, and tries to recapture the memories of his past. He is a mythical figure without precedent in Hungarian literature, but typical of Krúdy—one who invites his readers to embark on an adventure into a fantasy world. K.'s prolific output included more than 60 novels and approximately 3,000 short stories and other prose pieces.

Perhaps the most popular of his novels is *Vörös postakocsi* (The Crimson Coach, 1914), a nostalgic, even elegiac farewell to the old world, yet not without delicate touches of the grotesque and of irony. His volumes of short stories include *Üres a fészek* (The Nest Is Empty, 1897), *Őszi utazások a vörös postakocsin* (Autumn Journey in the Crimson Coach, 1917), and *Utolsó szivar az Arabs szürkénél* (A Last Cigar by the Arab Grey, 1928).

Kulcsár, Gergely (1934)

Javelin thrower, coach. Certificates as PE instructor (1957), coach (1963) and master coach (1977). In javelin won Olympic silver (1964) and bronzes (1960, 1968) and was first to throw over 80 meters in competition. Selected for European team (1967). Head coach for throwing events to the Hungarian national athletics team (1972–80), producing several Olympic and Hungarian champions, then to the Kuwaiti national team (1980–89).

Kulcsár, Győző (1940)

Fencer. Qualified engineer (1976); master coach certificate (1985). Won golds at the Olympics for individual epée (1968), and for the team event both at Olympics (1964, 1968, 1972) and World Championships (1970, 1971, 1978), as well as several World Championship individual silver and bronze medals. Hungarian Fencer of the Year (1968, 1971, 1974). Secretary-general of the Hungarian Fencing Federation (1979–80), captain of the national team (1980–88). Now a coach in Italy (1988–).

Kurtág, György (1926)

Composer. Studied at the Budapest Academy of Music (1946–55) then attended composition classes of Milhaud and Messiaen at the Paris Conservatoire. He worked as tutor at the Bartók music school in Budapest (1958–63) and became coach of the National Philharmonia (1960–68). Beginning as a disciple of Bartók and Kodály, he later was influenced by Webernian serialism.

Kürschák, József (1864–1933)

Mathematician. A student of Gyula Kőnig at Budapest Technical University, after graduating in 1886 he taught at secondary schools in Rozsnyó (Roznava, Slovakia), Debrecen and Budapest before obtaining his doctorate from Budapest University in 1891. Appointed a lecturer at the Budapest Technical University the next year, he was awarded a chair there in 1904. His wide-ranging researches, published in over a hundred articles and several books, included a simplification of Hilbert's proof of the Waring conjecture, extensions of Gyula Vályi's work on the theory of differential equations and that of Jenő Hunyady in linear algebra. His primary achievement, however, was to lay the foundation for the theory of evaluation to permit completion of the number fields. He involved himself particularly in Hungary's annual schools mathematics competition (this has borne his name since 1949) and in 1929 published all the problems of the competitions organised to date, later translated as *Hungarian Problem Book* (1963).

Kürti, Miklós (1908–1998)

Physicist. Hungarian born and schooled, he was obliged by the restrictions on numbers of Jewish students imposed in Hungary from the early 1920s onwards to obtain his university education in France and Germany before finally settling in England, where he eventually became a professor at Oxford. His main field of research was cryogenics, the physics of attaining very low temperatures employing adiabatic magnetisation, in which he achieved some unique results during the late 1950s.

Kvassay, Jenő (1850–1919)

Agricultural engineer. After qualifying as an engineer in Pest, he went on to study agriculture at Mosonmagyaróvár and hydraulic engineering in Paris. The idea of setting up an Institute of Agricultural Engineers in Hungary was inspired by a study tour of Europe. Through its dual concern with soil amelioration and water management, the Institute, which he directed, brought more than 3 million hectares (7.5 million acres) of land up to the standards required for modern farming and reclaimed nearly 1 million hectares (2.5 million acres) more by 1918. He helped draft legislation on water use. His most important publication was *Mezőgazdasági vízműtan* (Agricultural Water Engineering, 1882).

Lajta (Leitersdorfer)**, Béla** (1873–1920)

Architect. Coming from a wealthy family, after graduating from Budapest Technical University (1895) he toured England, France, Germany, Italy, Spain and Russia over the next four years. At the start of the new century, he worked with Ödön Lechner on the designs for the *Tomb of Sándor Schmidl* (1903) in the Jewish New Cemetery on Kozma utca, in Pest's Kőbánya (tenth) district, and also on a competition-winning design for an *Administration Building for the Post Office in Pozsony*. By then he was running his own office, employing a small group of trainee architects, including Béla Málnai, Lajos Kozma and Gyula Tálos, who worked particularly on planning the ornamentation and interior details of designs such as the *Jewish Institute for the Blind* on Mexikói út (1905–08) and the *Jewish Charity Home at No. 57 Amerikai út* (1909–11; now the National Institute for Neurosurgical Science), both in Pest's fourteenth district and characterised by high-pitched roofs and a stylistic debt to the English Arts and Crafts movement. There followed several projects, which are seen as pioneering examples of early modernism in Hungarian architecture. These include the *Parisiana Night-Club* (1908–09), later the home for the János Arany Theatre (this was restored in 1996 by Tamás Kőnig and Péter Wagner) and the Messel-influenced monumental brick architecture of the *Vas utca Commercial School* (1909–12), off Rákóczi út. The *Office Block at No. 5 Martinelli* (now Szervita) *tér* (1911–12) in

Pest's inner city area, now housing the *Rózsavölgyi Music Shop*, is exceptionally interesting for its honesty of structure and use of new materials: its first three floors are simple glass façades framed by columns, whilst its upper four apartment floors are clad in white ceramic tiles broken only by stark horizontal coloured bands. Other works included the *Erzsébetváros Savings Bank at No. 18 Rákóczi út* (1911–13; now housing the Gas Board) and many more funeral monuments in the Kozma Street Cemetery.

Lajtha, László (1890–1963)

Composer, ethnomusicologist. He studied in Budapest and Paris. From 1919 until 1949 he was a teacher at the National Music School, Budapest. He had a considerable compositional oeuvre, including nine symphonies and ten string quartets, which show the twin influences of Hungarian folk music and French impressionism.

Lánczos, Cornelius [Kornél] (1893–1974)

Mathematician, physicist. At Budapest University L. studied physics under Eötvös and mathematics under Fejér, gaining a doctorate in theoretical physics in 1921. He then did research at several European universities, working with Einstein in Berlin for two years, before moving to the USA between 1931 and 1954 and finally professor at the Institute for Advanced Studies, Dublin, Ireland, until he retired. His mathematical work on numerical analysis and the Fourier series was of enduring value, whilst in physics he made significant contributions to quantum mechanics and electrodynamics and the theory of relativity as well as writing an illuminating book on how the concept of space developed historically (*Development of the Concept of Geometric Space*, 1976). He received the Chauvenet Prize in 1960. The Irish Academy elected him a member. Other books by him include *A. Einstein and the Cosmic World Order* (1965).

Lavotta, János (1764–1820)

Composer. Between 1792 and 1804 he was music director at theatres in Pest, Miskolc and Kolozsvár. In his compositions he sought to blend elements of the *verbunkos* style with those of Classical chamber music.

Lechner, Ödön (1845–1914)

Art Nouveau architect. A fellow student at the Joseph Polytechnic and the Berlin Bauakademie with Alajos Hauszmann and Gyula Pártos, he initially went into partnership with them (1866–69). Trips to Italy (1868) and the chance to explore the Loire valley during a spell in the Paris office of Clement Parent (1875–78) drew him under the influence of the French Renaissance. The *Town Hall at Szeged* (1883–84), which L. designed with Pártos, is neo-Baroque in style, largely in deference to the older building that it replaced, but the *'Drechsler' Apartment House on Andrássy út* (1883), opposite the Opera House in Pest shows a clear French Renaissance influence. The five-storey steel-framed *Thonet House at No. 11/a Váci út* (1889, with Pártos) was the first building on which ceramics from the Zsolnay works were used as a cladding material. Persian and Indian ceramics seen at the Victoria and Albert Museum during a trip to London strongly influenced L.'s first major work, the *Town Hall at Kecskemét* (designed 1890, built 1893–95, with Pártos), in which the formal wealth characteristic of L.'s mature style is evidenced alongside the French and English architectural influences. In continued studies of the Oriental arts, L. became preoccupied with Hungarian folk art and its Asian roots—an interest exemplified by his masterpiece, the *Museum and School of Applied Arts* on Üllői út (designed 1891, built 1893–96). Beside L.'s typically original handling of space and mass, the decorative language and the use of Zsolnay's pyrogranite tiles on façades and roof ally him to the rich seam of fantasy that was being explored by his close contemporary Antoní Gaudí. Other important works from this period are *St. László Parish Church* in Pest's Kőbánya (tenth) district (1895–97), the *Institute of Geology* (1896–99), now next to the Népstadion, and the *Postal Savings Bank* in Váci utca (1899–1901, with Sándor Baumgarten).

Lehár, Franz (Ferenc) (1870–1948)

Operetta composer. He studied at the Prague Conservatory. In his operettas, his aim was to blend the Vienna style with various international musical trends. His most enduring works have been *Die lustige Witwe* (1905), and *Land des Lächelns* (1930).

Lékó, Péter (1979)

Chess player. Reached international master grade in 1992 and grandmaster in 1994, the youngest ever to achieve the latter. Won bronze (1989, 1990) then silver (1991, 1993) at Junior World Championships; European Junior (1992) and World Youth (1994) champion. Winner of several grandmaster tournaments (1995–99); several wins over the world champions in the WFG doubles. Currently 5th in the FIDE world rankings, with a score of 2,748 (2001).

Lenhossék, Mihály (1863–1937)

Neuroanatomist. Though qualifying as a doctor, he was not attracted by clinical work and after being appointed an unsalaried lecturer at Budapest University in 1888 he spent four years teaching and researching in Basel. From 1900 he held the chair in anatomy at Budapest. His work covered the whole field of histology, but L. is most remembered for the support he gave to the view that individual neurones are not physically interconnected, which led to a major dispute with Apáthy, who held that they are contiguous. His textbooks include *Tanulmányok az idegrendszer és az érzékszervek hisztológiájáról* (Studies on the Histology of the Nervous System and the Sensory Organs, 1893), *Az idegrendszer finomabb szerkezete* (1893; *Der feinere Bau des Nervensystems*, 1895), *Sejt és szövetek* (Cell and the Tissues, 1918), and *Az ember anatómiája* (Human Anatomy, 1922).

Ligeti, György (1923)

Composer. Studied at the Cluj Conservatory then at the Budapest Academy of Music. He taught music theory at the Academy until 1956, when he left the country. Worked at Cologne then settled in Austria. Taught in Darmstadt (1959–1972). Lived in Berlin and Vienna between 1969–73 then was professor of composition at the Hamburg Music Academy (1973–89). His orchestral writing uses precisely calculated textures and dense (but not thick) scoring, deriving ultimately from Webern.

Liszt, Ferenc [Franz] (1811–1886)

Composer, pianist. Went to Vienna in 1821, having lessons from Salieri and Czerny. Played in Paris in 1823 and in London in 1824. Returned to England in 1825 and 1826 then lived in Paris between 1823–35, becoming friend of Berlioz and Chopin. He returned to Vienna in 1838 and to London in 1840 and 1841. Until 1847 he toured widely. In 1848 he became Kapellmeister at the Weimar court, staying until 1859. From 1869 he divided his time between Rome, Weimar, and Budapest.

Lóránt, Gyula (1923–1981)

Footballer, coach. Member of the gold-winning Olympic team (1952), the team that beat England in the 1953 'match of the century' at Wembley Stadium, and the 1954 World Cup silver-winning team. After 1960 coached first division club teams in Romania, West Germany, and Greece. Died during a championship match of his Greek team.

Lukács, György (1885–1971)

Marxist philosopher, writer, literary critic. He influenced mainstream European left-wing thinking particularly post World War II. His contributions include the formulation of a Marxist system of aesthetics that opposed the political control of artists and defended humanism, and for an elaboration of Marx's theory of alienation of the industrial society. His essay collection *Geschichte und Klassenbewußtsein* (History and Class Consciousness, 1923), is a unique re-evaluation of Marx, arguing that in translating theory into revolutionary practice, proletarian self-awareness transforms, and does not merely reflect, the historical situation. On being denounced for deviating from Marxist–Leninist orthodoxy, L. soon repudiated this thesis; his later works adhered more strictly to the accepted Soviet view. Having lived mainly in Berlin and later Moscow since 1919, L. returned to Hungary in 1945, becoming a parliamentary representative and professor of aesthetics and cultural philosophy at Budapest University. L.'s brief spell as minister of culture during the Hungarian uprising of 1956, led to his arrest und deportation to Romania, but unlike other leading figures he was allowed to return to Budapest in 1957, thereafter devoting himself to his magnum opus, the two volume *Die Eigenart des Aesthetischen* (The Specificity of the Aesthetic, 1963), and works on ethics and ontology. L. wrote more than 30 books and hundreds of essays and lectures, including important philosophical works on Hegel and existentialism, and literary essays on realism, Thomas Mann and the historical novel.

Madách, Imre (1823–1864)

Poet, playwright. Author of the verse drama *Az ember tragédiája* (The Tragedy of Man, 1861), which won fame in Hungary and, in translation, wider afield for its lofty philosophical thought and its effectiveness on the stage. A nobleman by birth, M. spent his life in farming and public service, writing poetry and several dramas as a pastime, e.g. *Férfi és nő* (Man and Woman, 1843) and *Mózes* (Moses, 1861). However, it is the 15-act *Tragedy of Man*, covering mankind's history from the Creation to the Ice Age caused by the cooling of the sun, on which M.'s place in Hungarian literature rests. The central figures are Adam (in some scenes assuming the personality of a well-known figure, e.g. Miltiades, Tancred, Kepler, Danton), Eve and Lucifer, the latter of whom, in a spirit of negation and spite, wishes to demonstrate the failure of God's work and the futility of man's ameliorative endeavours. Lucifer, however, is finally vanquished through his inability to take into account God's infinite mercy, on which Adam pins his hope.

Madarász, Viktor (1830–1917)

Painter. The most significant of Hungary's Romantic historical artists, M. fought in the 1848–49 War of Independence before taking up painting, taught by Gusztáv Pósa, whilst studying law at Pécs. After a period at the Viennese Academy (1853–54) then as a student of Waldmüller, in 1856 he was the first Hungarian artist to make his way to Paris, where early works included the large-scale *Zrínyi Péter a fogságban* (Péter Zrínyi in Captivity, 1857). *Hunyadi László siratása* (The Lamentation for László Hunyadi, 1859) was awarded the French state's Gold Medal when shown at the Salon and propelled him into the Parisian intellectual élite. M.'s most mature masterpiece was *Zrínyi és Frangepán a bécsújhelyi börtönben* (Zrínyi and Frangepán in Wiener Neustadt Prison, 1864). He returned to Hungary for a painting competition in 1870 and although beaten by Benczúr in the historical category, he went on to gain wide popularity with canvases such as *Petőfi halála* (Death of Petőfi, 1875) and *Ónódi országgyűlés* (The Diet at Ónód, 1879). M.'s portraits are also of enduring value.

Mádl, Ildikó (1969)

Chess player. Coach's certificate (1990). Reached both women's (1986) and men's (1992) international grandmaster grades. As a member of Hungarian teams, she has won both silver (1986, 1994) and gold (1988, 1990) medals at the Chess Olympics.

Magyar, Zoltán (1953)

Gymnast. Qualified veterinarian (1985); coach's certificate (1979). On the pommel horse was European (1973, 1975, 1977), World (1974, 1978, 1979) and Olympic (1976, 1980) champion. He and his coach, László Vígh, devised several innovative moves. Hungarian Gymnast of the Year (1973–80) and Sportsman of the Year (1974, 1978, 1980). Member of the Hungarian Olympic Committee, and its vice-chairman (1986–89). Awarded the Hungarian Heritage Prize (1997).

Makovecz, Imre (1935)

Organic architect. After graduating from Budapest Technical University (1959), he initially worked for SZÖVTERV (1962–71), the state planning office for co-operative enterprises, where his main projects were the *Borostyán Motel and Inn at Sárospatak* (a converted monastery, 1965); a *Wine Bar at Balatonszepezd* (1966); the *Gulyás Tavern in Kőbánya* (1967), near the HUNGEXPO site in Pest's tenth district; and the *Csákányos Tavern in Tatabánya* (1967, burnt down in 1979). At the end of the decade M. started running a private master school, a 'subversive' activity in the eyes of the political authorities at that time, though he was also a tutor in the Architects' Union's official Young Architect's Circle (FÉK) postgraduate training scheme (1971–82). Moving to VÁTI (1971–77), the state urban planning office, he designed the *Sárospatak Cultural Centre* (exterior built 1974–77) and the *Mortuary at Farkasrét Cemetery* (1975) in Buda. Leaving under duress, he became architect to the Pilis Park Forestry

Commission (1977–84), where one of his works was the *Dobogókő Ski-lift Station* (1980), in the hills overlooking the Danube Bend north of Buda. M. established the Makona co-operative for his private practice in 1984; regrouped in 1989 as the Károly Kós Association. M.'s subsequent major projects up to the change in régime were the *Szigetvár Cultural Centre* (1985–88), the *Lutheran Church at Siófok* (1986–93), the *Catholic Church at Paks* (1987–91), and the *Árpád Gimnázium at Sárospatak* (1989–93). These fully display M.'s ideas about replacing mechanical principles of spatial organisation with the greater communal and artistic values that can emerge from a more organic, metabolic approach, utilising a language that emphasises natural materials and continuity with tradition. These and other approaches of the Hungarian organic architecture movement only received major international exposure when M.'s work was featured at the 1991 International Architectural Bienniale at Venice and in the *Hungarian Pavilion for the Seville World Exhibition* (1992) designed by M. His most recent works include: *Indoor Swimming Bath for Eger* (1996–2000); *Cultural Centre for Makó* (1998); and the *Catholic Church at Százhalombatta* (1998). In 1992 M. founded the Hungarian Academy of Arts, of which he is president. Awarded the Ybl Prize in 1969, he also received a Kossuth Prize (1990).

Manninger, Rezső (1890–1970)

Veterinary scientist. Dissuaded from pursuing chemistry after a meeting with Ferenc Hutÿra, M. opted to study veterinary science. Success in competitions whilst still a student landed himself a post at the Institute of Epidemiology as soon as he had qualified, where he quickly developed a vaccine against fowl cholera (1918). Over his career he did significant work on the epidemiology of such diseases as brucellosis, swine fever, rabies, and transmissible gastro-enteritis of pigs as well as the development of vaccines against foot and mouth disease, tuberculosis, swine fever, swine erysipelas, and anthrax. The National Animal Health Station that he set up (1928) and directed, along similar lines and aiming to the same standards as the provisions for public health, was one of the first of its kind in Europe. He was a permanent delegate to and elected officer of the international health organisation between 1933 and 1963. The 7th (German-language) edition of Hutÿra and Marek's classic *Veterinary Internal Medicine* (1938) included his own appendix on infectious diseases, whilst its successor, *Internal Medicine for Veterinary Surgeons and Veterinary Students*, which M. co-authored with János Mócsy, was first published in 1943. A member of the Hungarian Academy of Sciences at a relatively young age, he also won many distinctions from foreign academic bodies.

Márai, Sándor (1900–1989)

Poet, novelist. Mainly remembered as a prolific novelist and diarist, he achieved considerable success with a string of some 40 novels published from the late 1920s onwards, including *A zendülők* (The Mutineers, 1930), *Csutora* (1930), *Vendégjáték Bolzanóban* (Guest Performance in Bolzano, 1940). The autobiographical *Egy polgár vallomásai* (Confessions of a Citizen, 1934) uses his own experience of middle-class family life in the small but prosperous Upper Hungarian city of Kassa (now Košice, Slovakia) to explore the mentality and cultural identity of this milieu within an epic tableau. *Szindbád hazamegy* (Sindbad Goes Home, 1940) is nowadays considered a prime example of an intertextual dialogue, drawing on the modern Hungarian epic style established by Krúdy to interrogate the tension between autonomy and cultural dependence in Krúdy's and his own language. Having published an initial wartime instalment of his *Napló* (Diary, 1948), he was forced to leave Hungary and thereafter lived mainly in the United

States, publishing five further volumes of the *Napló. Föld! Föld!* (1991, transl. as Memoir of Hungary, 1944–1948) is a separate reworking by M. of his own experience of the modern crises and changing cultural values of the Europe's middle classes in the face of the totalitarian threats of Nazism and Bolshevism. Whilst insisting on the existence of central values, he attempts to come to terms with the world as he finds it in a dialogue between contemporary world literature and Hungary's modern epic tradition which propagates a liberal view of engagement with strangeness and otherness, whilst asserting the individual's right to preserve integrity in relation to different cultures.

Marek, József (1868–1952)

Veterinary scientist. Having qualified at Budapest Veterinary School in 1892, he initially practised as a country vet but, not abandoning his scientific ambitions, began publishing papers in professional journals. He secured a post heading the animal health laboratory in Budapest (1895), which opened the way to being appointed an assistant lecturer at the Veterinary School. M. quickly earned attention with his superb diagnostic skills and as a designer of several valuable instruments, but he is best remembered for first describing Marek's disease (1907), a tumour of young poultry flocks, and discovering its viral etiology. His subsequent development of filicin (extracted from the male shield fern) as an effective treatment for liver flukein ruminants was of major practical importance during the interwar period. Apart from co-authoring the 2nd edition of Hutÿra's *Veterinary Internal Medicine* (1904), his own chief publication was *Klinikai diagnosztika* (Clinical Diagnostics, 1902). He was a member of the Hungarian Academy of Sciences and many foreign academies.

Markó, Károly Snr (1791–1869)

Painter. The first major (albeit belated) exponent of Classical landscape painting in Hungary, and founder of a school of disciples. Having trained at academies in Pest and Vienna, he started painting purely topographical views, his early masterpiece being an oil of *Visegrád* (1826–30). He had prominent and wealthy patrons, including Archbishop Pyrker of Eger and the Grand Duke of Tuscany, and it was a Viennese banker (Geymüller) who made possible a trip to Italy, where he settled for good, though still maintaining close connections with his homeland. M. aspired to surpass the masters of the late Renaissance and Baroque classicism. Using studies from nature, he composed his pictures artificially, giving their Biblical subjects an archaising character. From the period 1832–38, spent in Rome, the most significant paintings are *Róma látképe* (Panorama of Rome), *Asszonyok a kútnál* (Women by the Well), and *Szüret* (Vintage), whilst the most characteristic work from the ensuing Pisan period is *Olasz táj aratókkal* (Italian Landscape with Reapers). In 1853 he made a return visit to Pest with *Pusztai tájkép* (Landscape of the Great Plain). The pictures of woods that M. painted from nature towards the end of his life took him close to the concept of a '*paysage intime.*' His pupils included Antal Ligeti, Károly Telepy, József Molnár, and Mihály Kovács.

Martinek, János (1965)

Modern pentathlete. Coach's certificate (1994). Won gold at the Olympics in both individual and team events (1988), at World Championships in the team (1987, 1989) and relay (1989, 1994) events, and in the team events at the World Cup (1993). Hungarian Modern Pentathlete of the Year (1988).

Marton, Éva (1943)

Soprano. Starting her career as a member of the Budapest Opera House company (1968–72), she moved to Frankfurt

(1972–77) and then Hamburg. She made her debut at the Vienna Staatsoper in 1973 and at the Metropolitan Opera, New York, in 1976. She has mostly sung the dramatic soprano roles of Wagner and Richard Strauss.

Martyn, Ferenc (1899–1986)

Painter, graphic artist, sculptor. Spending most of his early career in Paris (1926–40), M. was initially an engraver and aquarellist (e.g. *Parisian Portfolio*). After transiently allying with the surrealists (*Judgement of a City*, 1928–30), he was attracted to the abstract school and by 1939 had developed a distinctive approach to reconciling a planar formulation of space and motion with abstracted, sublimated representation of elements of reality, coupled with purity and harmony of colour. Returning to Hungary in 1940, M. continued to paint large-scale coloured abstract compositions which incorporate references to man and nature, including symbols of a desire for freedom (bird, boat, fight) and folkloric or mythological elements. He voiced his protest against war in the series of drawings *A fasizmus szörnyetegei* (The Monsters of Fascism). M. refused to relinquish his style after the war, even during worst excesses of socialist realism dogma in the 1950s and '60s, which denied him public showings for the bulk of his output until 1969, though more abstract graphics, such as *Vonalas szerkezet* (Lined Structure), became known after 1965. M. illustrated several literary works (Cervantes: *Don Quixote*, G. Flaubert: *Madame Bovary*, J. Joyce: *Ulysses,* Sándor Petőfi: *The Apostle*), whilst his *Napló* (Diary) is a series of drawings documenting his life.

Mathiász, János (1838–1921)

Viniculturalist. Having amassed some 1,600 samples of vinestocks within just a few years of purchasing a vineyard in the Rosalia Hills near Kassa (Košice, Slovakia), M. moved to Mád in the Tokaj region to run a 100-hectare (250-acre) vineyard, however the havoc wrought by phylloxera eventually induced him to move to the pest-free sandy soils of the Great Plain area between the Danube and Tisza. There he created some 3,500 hybrid strains of grapes, both for wine-making and dessert purposes, providing detailed description of some 180 of them, of which 67 are still cultivated, 12 internationally. Among the latter is 'Szőlőskertek királynője' (Queen of the Vineyards), the last variety that he introduced (1916).

Mattis Teutsch, János (1884–1960)

Painter, sculptor. A Transylvanian Saxon by birth, he trained first for the timber industry then as an industrial designer. As a young man he made visits to both Munich and Paris, where he was greatly influenced by Rodin, Toulouse-Lautrec and Matisse. Early linocuts published by Lajos Kassák in the periodical *Ma* (Today) still showed an imprint of Symbolism and Art Nouveau, but his increasingly abstract later work had more affinity to that of the *Blauer Reiter* group. M. T.'s book *Kunstideologie* (1931) was published in Potsdam. In his late period he experimented with somewhat schematic Art Deco-style compositions. His sculptures of painted wood are three-dimensional transpositions of pictorial forms and abstract human figures.

Maurer, Dóra (1937)

Graphic artist, painter. A versatile yet systematic experimental artist, M. started her career with surrealist-inspired graphic work in the early '60s, soon extended by the application of collage, montage and *pliage* techniques. By the end of the decade, however, she had branched into conceptual works such as *Nagyon magasról leejtett lemez* (Plate Dropped from a Very High Place, 1970). She also began producing series of objects by the systematic permutation of diverse materials, as in *Mennyiségtáblák 3* (Arithmetic Tables 3, 1972), ultimately deriving abstract geometrical painted and relief versions, such as *Mennyiségtábla 6* (1972). Meanwhile M. was also engaged in producing films, in new ways of looking at the use of texts, and in techniques of preserving imprinted traces, as in *V. május 1-jei felvonulása mesterséges talajon* (V.'s May Day Parade on Artificial Soil, 1971). In the mid '70s she embarked on applying geometric transformations to generate further series, as in *Kvázi-kép véletlen adatok szerint* (Quasi-Picture According to Random Data, 1977). This 'quasi-picture' approach was further developed during the 1980s by placing colour bands in spatial grids in environments where changing lighting effects create the illusion of virtual spaces (e.g. *Space Painting*, Buchberg am Kamp, Austria, 1982–83). By the mid '90s a less rigidly systematic approach to her compositions was evident.

Mechwart, András (1834–1907)

Mechanical engineer. Swiss-born, he was already a qualified engineer when he began working for the Ganz Works, initially reorganising and upgrading the company's own processes and equipment for the assembly of steam locomotives. After being made general manager of the firm, on its founder's death (1867), he made the decision to acquire the patent rights on a flour roller-mill design and, by applying the firm's own proven chill-cast technology, turned it into a huge commercial success. He followed that by setting up a new electrical engineering division (1878), which, through the pioneering work of Zipernowsky, Déri and Bláthy on the transmission and application of alternating-current electricity seeded an entire new industry.

Medgyaszay (Benkő)**, István** (1877–1959)

Pioneer architect in reinforced concrete, graphic artist, writer. Having graduated from the School for Industrial Construction (1896), he enrolled in the masterclass of Otto Wagner at Vienna's Technical University in 1900 and, after a spell in France, completed his studies at Budapest Technical

University (1904). Over the next two years he undertook field trips to survey folk architecture, collecting material for Dezső Malonyay's *A magyar nép művészete* (Hungarian Folk Art, 5 vols., 1907–22), and contributed to the decorations of Géza Maróti's *Hungarian Pavilion at the Milan World Exhibition* (1906). His main works were a *Villa and Double Studio in Gödöllő* (1904–06), the *Theatre for Veszprém* (1907–09), reconstruction of the *Sopron Theatre* (1908–09), and the *Parish Church at Rárosmulyad* (Mul'a, Slovakia; 1908–10), the cupola of which is an early stressed-skin structure. Two major buildings that went up in 1914, were the *Apartment House in Dorottya utca*, off Vörösmarty tér in Pest's inner city, and the *State High School for Caransebeş* (Romania). During the war he designed the pavilions for a *War Exhibit at Lemberg* (Lvov, Ukraine; 1916) and the *Pavilions and Art Gallery for a War Exhibit on Margaret Island*, Budapest (1918). During the post-war economic slump he won only small commissions; in 1925 he became an unsalaried lecturer at Budapest Technical University. Major projects included a *Housing Estate on Budaörsi út* (1925) in Buda's eleventh district; the *Theatre for Nagykanizsa* (1926); the *Sport Hotel at Mátraháza* (1927); an *Apartment Block on Németvölgyi út* (1928), also in the eleventh district; the *Baár–Madas Reformed Boarding College for Girls* (1928–29) on Lórántffy Zsuzsanna utca close to the Városmajor Park; and the *TÉBE Bank and Apartment House on Hild tér* (1938–39), in Pest's inner-city area. A growing interest in the Magyar race's origins in the East, and thus in Oriental art more generally, led to a design for the Persian–Turanian Museum in Bombay (1929; remained unrealised) then a study-trip to India in 1931. After 1945 he obtained no commissions; although employed in a state planning office for a short period during 1948, he died as an impoverished, neglected figure.

Medgyessy, Ferenc (1881–1958)

Sculptor. Going to Paris as a qualified physician, he continued studies first as a painter then as a sculptor. Fascinated by ancient Egyptian, Greek and Etruscan sculpture and the unfinished works of Michelangelo, he fashioned works that, besides the compactness of their stocky forms, show passages of self-abandoned, lifelike expression, the most telling example of which is *Súroló asszony* (Woman Scrubbing, 1913). His artistic aims have been linked by some with those of Maillol and Despiau. Having made a successful debut, he served as an army doctor at the front in World War I but still continued to pursue his art, developing his sketches at a later date, as with *Kis lovas* (Small Rider, 1915 and 1922), *Szoptató anya* (Mother Suckling Her Child, 1918), and *Kövér gondolkodó* (Fat Woman Meditating). In 1919 M. taught in the proletarian Apprentice Workshop for Fine Arts; he was also a founding member of the New Association of Artists (KUT) Group in 1924 and of the Ady Society. *Régészet, Tudomány, Művészet, Néprajz* (Archeology, Science, Art, Ethnography, 1930), a group of four allegorical figures created for the Déri Museum in Debrecen, won a Grand Prix at the 1937 Paris World Fair. The thirties was also the period when he completed sculptures such as the statue of *András Jelky* at Baja, *Magvető* (Sower), *Hortobágyi bika* (Bull of Hortobágy) and *Turáni lovas* (Turanian Rider). The inspiration of antiquity frequently recurred, as in the relief *Furulyázó faun* (Faun Playing the Pipes) and *Táncoló nő* (Dancing Woman, 1954). M. regenerated Hungarian sculpture by bringing to it a new, purely sculptural perspective, and that is equally the source of his universality.

Mednyánszky, László, Baron (1852–1919)

Painter. Studied at the Munich Academy of Art then at the École des Beaux-Arts in Paris, also spending several months at Barbizon, which was a powerful stimulus for his landscape painting. After a prolonged tour of Italy, he worked in Vienna—a period that yielded many of his well-known autumnal and winter landscapes. Meanwhile he made frequent visits home, exhibiting in both cities. From 1889 until 1892 he was mainly Paris-based and had a one-man show there in 1897. Carpathian mountain ranges, marshlands on the Great Plain, and suburban scenes alike figured in his landscapes. During World War I he sketched on virtually every front line, the shattering experience of which drew him towards a human portraiture of great dramatic force and social sensitivity. Although the bulk of his *oeuvre* is held by the Hungarian National Gallery, there are still many pictures in private hands and in the Bratislava Gallery. A large commemorative exhibition of his work was put on at the Exhibition Hall, Budapest, in 1952.

Mészáros, Erika (1966)

Kayaker. PE instructor's certificate (1988). Won an Olympic gold in kayak fours over 500 meters (1992). Hungarian Woman Kayaker of the Year (1986, 1989, 1990). Awarded the Small Cross of the Order of Merit of the Republic of Hungary (1992).

Mikszáth, Kálmán (1847–1910)

Novelist. Regarded by contemporaries and succeeding generations alike as the outstanding Hungarian writer of the turn of the century, K. initially studied law but soon took up journalism, first at Szeged, later in Budapest. In 1887, already a famous writer, he was elected as a parliamentary representative. M. scored his first success with two volumes of short stories, *Tót atyafiak* (Slovak Kinsmen, 1881) and *A jó palócok* (The Good People of Palocz, 1882) before publishing his first novel, *Beszterce ostroma* (The Siege of Beszterce, 1884), the story of a Quixotic Hungarian aristocrat. M. was primarily a writer of everyday life, which he described with understanding and sympathy, although he did not hesitate to pillory the shortcomings of society with sharp-witted satire, as in *A Noszty fiú esete Tóth Marival* (The Noszty Boy's Affair with Mari Tóth, 1908), the story of a frivolous young noble who tries to make a fortune by seducing a rich middle-class girl. *A fekete város* (The Black City, 1910) is his final and also one of his finest historical novels.

Mincza, Ildikó (1969)

Fencer. Qualified lawyer. Won gold as a member of the foil team (1991) then in individual (1997, 2001) and team (2001) epée events at European Championships, as well as gold at the World Cup and bronze at the World Championships (1999) in individual epée. Recipient of the Silver Cross of Merit of the Republic of Hungary (2000).

Mizsér, Attila (1961)

Modern pentathlete. Graduate of the College of Foreign Trade (1994); coach's certificate (1987). At the Olympics won gold in the team event (1988) and individual silver (1992); at World Championships took golds in individual (1985), team (1987, 1989) and relay (1989, 1993) events. Awarded the Gold Cross of Merit of the Republic of Hungary (1992) and an International Fair Play Prize (1994).

Moholy-Nagy, László (1895–1946)

Painter, sculptor, photographer, writer. Having originally studied law, M.-N. took up drawing during a period of internment as a prisoner of war in Russia during World War I. Returning to Budapest he became involved with the avant-garde circle around Lajos Kassák. A versatile experimental artist, M.-N.'s most productive creative period was associated with German art of the 1920s. After a brief stay in Vienna, in 1920 he moved to Berlin, where he familiarised himself with Dadaism and Russian constructivism and made 'industrial landscapes' under these influences. From 1922, M.-N. had regular shows at the Der Sturm gallery, and in that year he and Kassák published the influential anthology *Buch neuer Künstler*. M.-N. taught at the Weimar Bauhaus from 1923 and was also involved, on Walter Gropius' invitation, in editing the school's publications. This period was also marked by his most harmonious geometric-abstract paintings, photographs, photograms and 'photoplastics' as well as by publication of a screenplay (*Sketch for a Film: Dynamics of the Big City*), along with the books *Malerei. Fotographie. Film* (1927) and *Von Material zu Architektur* (1929). He presented his *Light–Space Modulator* in 1930, a kinetic mechanism also used in the making of the abstract film *Lichtspiel schwarz-weiss-grau.* After interludes in Amsterdam (1934–35) then London, M.-N. moved to America in 1937 to become director of the New Bauhaus in Chicago, leaving in 1939 to found the Institute of Design, which gained university status in 1944. M.-N. was still working on the book *Vision in Motion* (1947) when the died.

Molnár, Farkas (1897–1945)

International Modern architect. He spent several years at the Bauhaus (1921–25), where he attracted attention with his 'Red Cube House' project (1922), before graduating from Budapest Technical University. His first executed project was a *House on Delej utca* (1929, with Pál Ligeti), in Pest's Józsefváros (eighth) district. As a CIAM delegate he formed a Hungarian section of the Comité International pour la Réalization des Problèmes d'Architecture Contemporain (CIRPAC) and actively promoted the social principles of modern architecture. Exhibitions at Budapest's Tamás Gallery in March and September 1932 led to M. receiving a suspended sentence for 'anti-state agitation'. As a result, he had access only to small private commissions, the most notable of which were: *Houses on Kavics utca and Cserje utca* (1931), both on Buda's Rózsadomb; the *Schwarz House in Szeged* (1932); the *Dálnoki–Kovács Villa on Lejtő út* (1932) in Buda's twelfth district, which won the first prize at the 1933 Milan Trienniale; a *Summer House in Felsőgöd* (1933), north of Pest; the *Villa on Darányi* (now Hankóczy) *utca* (1933) and the *Building on Lotz Károly utca* (1934), both in Buda's Pasarét district; and an *Apartment House on Toldy Ferenc utca* (1933), below Castle Hill. Under political pressure, M. dissolved the CIRPAC group and was accepted back into the Chamber of Engineers. Commissions in this period, now including some from the state, were: *Apartment Block on Köztársaság* (then Tisza Kálmán) *tér* (1934, with eight partners), in the Józsefváros district; a *Staff Pavilion for the Pestújhely Hospital* (1935, with József Fischer); a *Villa on Csévi utca* (1935, also with Fischer), an *Apartment Building on Pasaréti út* (1936), and a *Villa on Mese utca* (1937), all in the Pasarét district. M. undertook a study-trip to Palestine to research his innovative design for the *Holy Land Church* for the Franciscan Order in Buda's Hűvösvölgy (second) district (1940–49); however, construction work was interrupted by the war and Molnár's death, and finally abandoned when confiscated by the Rákosi régime.

Molnár, Ferenc (1878–1952)

Playwright and novelist. M. achieved international acclaim for plays about the contemporary salon life of Budapest and for moving short stories. He published his first stories at the age of 19 and achieved his first great success with the play *Az ördög* (1907; The Devil, 1908). Although trained as a lawyer, he went into journalism and during World War I was a correspondent. Many of Molnár's plays, including *Liliom* (1909; English transl. 1927), *A hattyú* (1920; The Swan, 1929), and *A vörös malom* (1923; The Red Mill, 1932) were successfully played abroad, particularly in Austria, Germany, and the United States, where he made his home from 1940. Several plays were turned into films, but these translations and adaptations often emphasised the verbal beauty and romantic plots at the expense of their finely detailed characterisations and their often bitter cynicism. Some of Molnár's short stories, especially those collected in *Muzsika* (Music, 1908) are masterpieces; concise and moving, they look beneath the glittering facade of salon life to the problems of the poor and the underdog. Among his many novels, only *A Pál utcai fiúk* (1907; The Paul Street Boys, 1927) achieved much success. M. was willing to depict the victory of evil, egoism, and immorality, but these elements were offset by his light, amusing touch.

Monspart, Sarolta (1944)

Orienteer, sports administrator. Qualified teacher (1967); certificates as sports manager (1979), coach (1980) and master coach (1989). World champion in individual orienteering (1972). She was the first woman in Europe to run a marathon in under three hours (1972). National champion in cross-country skiing. Board member (1982–94) then vice-chair (1994–1996) of the International Orienteering Federation; board member of the Sports Aid Foundation (1988–); chair of the Miklós Wesselényi Public Sports Foundation (2000–). Awarded the Miksa Eszterházy Prize (2001).

Móricz, Zsigmond (1879–1942)

Novelist. A master of Hungarian, his style absorbing elements of old language and dialect, M. became one of the leaders of the progressive writers associated with the literary magazine *Nyugat*, which he co-edited between 1929 and 1933, later taking on the editorship of *Kelet Népe* (The People of the East) from 1940 until his death. He found his subjects in the life of the villages and country towns he knew, portraying figures such as the peasant fighting for his emancipation, as in *Sárarany* (Dirt and Gold, 1910) and *A boldog ember* (A Happy Man, 1935); the bankrupt way of life of the decaying provincial gentry, as in *Kivilágos kivirradtig* (Until Daybreak, 1924),

Úri muri (Gentlemen's Games, 1927) and *Rokonok* (Relatives, 1930); or the failure of those who work for the happiness of the people, as in *Fáklya* (The Torch, 1917). In M.'s world marriage and family life are fraught with bitter conflicts, but he also evokes pure, even idyllic, loves, as in *Pillangó* (Butterfly, 1925). He wrote one of the finest children's books in Hungarian, *Légy jó mindhalálig* (Be Faithful Unto Death, 1920), and also two monumental historical novels, the trilogy *Erdély* (Transylvania, 1922–35) and *Rózsa Sándor* (1940–42). The finely drawn characters of M.'s works are representative social types who are set on collision courses, their fierce energies being finally consumed by murderous passion; some novels thereby become veritable Dances of Death of the morbid and doomed elements of society.

Mosonyi, Mihály (1815–1870)

Composer. After 1842 he quickly became a leading figure in musical life, initially composing in a style influenced by Schumann. In 1859 he was converted to the cause of Hungarian national music. He was also an important music critic.

Munkácsy, Mihály (1844–1900)

Painter. The most significant Hungarian artist of critical realism, M. had been orphaned at a young age and was working as a joiner's apprentice in Arad (Oradea, Romania) when painter Elek Szamossy recognised the boy's artistic talent and adopted him in order to give him an education and proper training in the fine arts. M. travelled to Pest in 1863 to show his first paintings to Antal Ligeti and Mór Than, gaining their support and a scholarship to study with Karl Rahl in Vienna, then in Munich (1866), and finally with Ludwig Knaus in Düsseldorf. There he managed to throw off the hold of academicism and find his own style, working on *Ásító inas* (Yawning Apprentice, 1868–69) and, more particularly, *Siralomház* (The

Condemned Cell, 1869–72)—a picture awarded a Gold Medal at the 1870 Salon in Paris and sold for the then large sum of 2,000 thalers. This enabled him to move to Paris in 1872, where he settled permanently. For a while there was a stream of outstanding portraits and character studies, many based on childhood memories, which display a mastery in creating psychological tension: *Tépéscsinálók* (The Lint Makers, 1871), *Köpülő asszony* (Churning Woman, 1873), *Sátoros cigányok* (Tent-Dwelling Gypsies, 1873), *Éjjeli csavargók* (Nocturnal Tramps, 1872–73), and *Rőzsehordó nő* (Woman Carrying Faggot, 1873). By and large, however, these were not to the taste of fashionable society, with the exception of *Milton* (awarded a Gold Medal at the 1878 World Exhibition) and *Műteremben* (In the Studio, 1876). Though his final years were blighted by illness, M. was still able to bring off occasional triumphs, as in the 'trilogy' of *Krisztus Pilátus előtt* (Christ before Pilate, 1881), *Golgota* (1884), and *Ecce Homo!* (1896).

Nádas, Péter (1942)

Novelist, playwright. After studies in journalism and photography in the early 1960s, his early work was in short-story form. The novella *A Biblia* (The Bible, 1967), for instance, is about a young boy who, whilst living in an apparently ideal world of his own, confronts the terrors lurking both inside himself and the hell of the external reality of the early 1950s. The novel *Egy családregény vége* (The End of a Family Story, 1977) takes this further, with another boy exploring, through the figures of his father and grandfather, his family's mythical Jewish past and the termination imposed on it by brutal reality of the same era. In the monumental *Emlékiratok könyve* (The Book of Memories, 1986) N. adopts a Plutarchian model to track the lives of his protagonists in different times, presenting the emotional and ethical conflicts they experience when public and private spheres clash, and analysing the torments of creation and writing. In the essay *Az égi és a földi szerelemről* (On Heavenly and Earthly Love, 1991) N. places emphasis on love as the ultimate state of all human relations. The three one-act plays of *Nézőtér* (Auditorium, 1983), with their absurdist touches and cliché-free language, are experiments aimed at a reform of Hungarian drama.

Nagy, Imre (1933)

Modern pentathlete, sports administrator. Economics graduate (1955) and doctorate (1965). Won team gold (1960) and bronze (1964) at the Olympics, and team silvers at World Championships (1958, 1961, 1962). Captain of the Hungarian national modern pentathlon team (1970–76). Board member and secretary of the International Modern Pentathlon Federation (1976–92).

Nagy, László (1925–1978)

Poet, translator. In his poetry N. contrasted tragedy with defiant insolence, many of his images and forms being taken from heroic songs, peasant life or myths and applied to contemporary life, with the interest deriving primarily from his vibrant rhythms and the play between the imaginative, associative, and sound quality of the language. One of his major poems, *A csodafiú-szarvas* (The Young Miracle Stag, 1946), for instance, is based on the ancient Romanian–Hungarian myth of the miracle stag but retold from a modern Christian point of view and interwoven with elements of shamanism, traditional Hungarian folk culture and modern life, thereby transferring the values of ancient peasant communities and traditions into the world of contemporary civilisation. Nagy is also noted for his translations of Bulgarian heroic songs and the poems of García Lorca, Zbigniew Herbert, and Dylan Thomas.

Nagy, Tímea (1970)

Fencer. Diploma in special needs education (1996); coach's certificate (1991). In individual epée she became European (1995) and Olympic (2000) champion and is a winner of the

World Cup (1995); several times member of the bronze-winning team at the World Championships. She is a speech therapist. Awarded the Officers' Cross of the Order of Merit of the Republic of Hungary (2000).

Nébald, György (1956)

Fencer. Qualified lawyer (1979); coach's certificate (1979). Has won golds in individual sabre (1985–90) and four times in the team event at World Championships, and team gold (1988) and silver (1992) at the Olympics. Hungarian Fencer of the Year (1985, 1987–88, 1990). Recipient of the Golden Cross of Merit of the Republic of Hungary (1992) and the Ferenc Kemény Prize (1993).

Nemere, Zoltán (1942–2001)

Fencer. Qualified dentist (1971); coach's certificate (1965). Won golds in the epée team at the Olympics (1964, 1968) and World Championships (1970, 1971), as well as an individual gold at the latter (1965). Hungarian Fencer of the Year (1965). Lived in (West) Germany after 1972.

Németh, Angéla (1946)

Javelin thrower. PE instructor's certificate (1968). Olympic (1968) and European champion (1969) in javelin; the first woman to throw over 60 meters. Head of department (1990–).

Németh, Ferenc (1936)

Modern pentathlete, coach. Won both individual and team gold at the Olympics (1960). In the Hungarian Modern Pentathlon Federation he was fencing master of the modern pentathlon section (1961–63, 1968–74) then coach (1975) and later director (1981) of the fencing section, also serving as the Federation's secretary-general (1964–67). Since 1987 he has been a fencing master in Kuwait.

Németh, Imre (1917–1989)

Hammer thrower, sports manager. Olympic champion (1948) and set several world records in hammer. Director (1953–57) then director general (1957–80) of the Népstadion facility in

Budapest. Chairman of the Hungarian Athletics Federation (1964–74). Recipient of the International Olympic Committee's Olympic Order (1979).

Németh, László (1901–1975)

Novelist, playwright. His first published work appeared in *Nyugat*, but N. later severed all contacts with the periodical and from 1932 to 1936 published an entirely self-written review called *Tanú* (Witness). His prose, renewing the traditional form of realism, usually follows a straightforward, linear narrative, focused on decisive events in the life of the main character. His most popular novels—*Gyász* (Mourning, 1936), *Iszony* (Revulsion, 1947), *Égető Eszter* (1956), and *Irgalom* (Compassion, 1965)—portray women in the central role, depicting their reaction to certain situations. Similarly, N.'s major plays—*Villámfénynél* (Lightning, 1936), *VII. Gergely* (Gregory VII, 1939), *Galilei* (1954) and *A két Bolyai* (The Two Bolyais, 1961)—are based on the realist model of Ibsen, concentrating on the development of character, revelation of hidden motivations, and the disintegration of the inner being. Among his essays on medicine, philosophy, politics, culture and education, the most influential is *Kisebbségben* (In the Minority, 1939), which still defines the role and horizon of the Hungarian intelligentsia.

Németh, Miklós (1946)

Javelin thrower, coach, sports administrator. Olympic champion (1976) and world record-holder. Was selected for a World athletics team (1977). Hungarian Sportsman of the Year (1976). Javelin coach to the Italian national team and to the British women's Olympic champion Tessa Sanderson (1984–90). Vice-chairman of the Hungarian Athletics Federation (1991–93). Now manufactures and sells innovative sporting goods.

Neumann, John von [János] (1903–1957)
Mathematician. Although he attended lectures given by Kürschák and Fejér in Budapest, N. went to Berlin and Zurich to qualify as a chemical engineer (1926), meanwhile completing a doctoral dissertation on set theory for Budapest University. The next year he became Berlin's youngest-ever *Privatdozent* but in 1929 moved to Hamburg before leaving Europe for the USA in 1930, where he obtained a post at Princeton. During his relatively short life, N. made crucial contributions to a vast range of mathematically based fields, from set theory and function analysis to quantum mechanics, operations research and computer science, and founded the field of game theory. Amongst other distinctions, he was appointed to the US Atomic Energy Commission, was a President of the American Mathematical Society, and received the Fermi Prize. His published works include *Mathematische Grundlagen der Quantummechanik* (1932), *Continuous Geometry* (1936), *The Theory of Games and Economic Behavior* (with O. Morgenstern, 1944), *The Computer and the Brain* (1958), and *Theory of the Self-Reproducing Automata* (1966).

Nikisch, Arthur (1855–1922)
Conductor. Studied at the Vienna Conservatory. He became first conductor at Leipzig Opera in 1878. He spent four seasons in America as conductor to the Boston Symphony Orchestra (1889–93), returning to Europe to take up the post of director at Budapest Opera (1893–95). From then until his death he worked in Leipzig and London (1904–14).

Novák, Éva (1930)
Swimmer. Qualified physician (1953). Member of gold-winning Olympic 4 x 100 meters freestyle relay squad (1952), which set several world records. Married a Belgian journalist in 1952 and lived in Belgium from 1954 until returning to

Hungary in the 1990s. First Belgian woman athlete to be decorated with the Medal d'Or du Merit. Inducted into Hungarian Swimmers' Hall of Fame (1973).

Novák, Ilona (1925)

Swimmer, sports administrator. Certificates as PE instructor and coach (1963). Member of gold-winning Olympic 4x 100 meters freestyle relay squad (1952), which set several world records. Inducted into Hungarian Swimmers' Hall of Fame (1973). Vice-chair of the Hungarian Swimming Federation (1971–89); captain of the national swimming team (1971–72); board member of the European Swimming League (1975–94). Recipient of the International Olympic Committee's Olympic Order (1992).

Ónodi, Henrietta (1974)

Gymnast. Studies at the Incarnate World University (San Antonio, 1994–). Olympic gold in vaulting (1992), Olympic and World Championship silver in floor exercise (1992). Hungarian Woman Gymnast of the Year (1989–92). Set up a leisure/recreational centre. Awarded the Small Cross of the Order of Merit of the Republic of Hungary (1992).

Ormándy, Eugene [Jenő] (1899–1985)

Conductor. He was a pupil of Jenő Hubay at Budapest Academy of Music. Moving to the United States in 1921, he started conducting in 1924. He was invited to the Minneapolis Symphony Orchestra for five seasons from 1931. Between 1938 and 1973 he was leading conductor of the Philadelphia Orchestra.

Orowan, Egon (1901–1989)

Physicist, geophysicist. In successive semesters at Vienna, he studied physics, chemistry, mathematics and astronomy then, after a six-year gap, embarked on the mechanical and electrical engineering courses and (in 1928) solid-state physics. Working in the research laboratories of United Incandescent

Lamp between 1932 and 1936, working with Polanyi, he wrote a seminal paper on the plastic deformation of materials. In 1937 O. emigrated to England, obtaining post first at Birmingham then at Cambridge University, breaking all professional ties with Hungary. He moved to the USA in 1950, joining the Massachusetts Institute of Technology. His major achievement was to introduce dislocation theory, initially as an explanation for the plastic deformability of metals through faults in their crystal lattice structures, but subsequently its application to geophysics and processes of rock formation. His major works include *Die Zugfestigkeit von Glimmer und das Problem der technischen Festigkeit* (1933), *Zur Kristallplastizität* (1934), and *Continental Drift and the Origin of Mountains* (1964).

Ország, Lili (1926–1978)

Painter. After an early surrealist phase (1952–57), painting works like *Nő fal előtt* (Woman in Front of a Wall, 1956) which owed much to Magritte and Chirico, an extended study trip to Bulgaria yielded the so-called icon-pictures, such as *Bizánci ikon* (Byzantine Icon, 1958), after which O. moved closer to the constructivist legacy of Lajos Vajda. The main themes of her visual world were now records of memories of the past in the form of architectural ground-plans and wall surfaces, as in *Jeruzsálem fala* (The Wall of Jerusalem, 1966), or ancient hieroglyphs, like the stylised Hebrew letterings O. saw on gravestones in Prague's Jewish Cemetery (*Exodus*, 1963). Trips to Naples and Pompeii (1968–74) yielded further motifs, utilised in the summative series of 48 panel paintings that make up O.'s masterpiece, *Labyrinth* (1973–78), in which the geometric ordering, surface effects and emblems of earlier work are explored to render a sense of ancient myth and the timelessness of history and art.

Ottlik, Géza (1912–1990)

Novelist. O.'s first short stories, in the collection *Hamisjátékosok* (Con-men, 1941), apart from their structural economy and inventive poetical language, can only be considered as the re-workings of familiar themes from the literature of the 1930s. His first full-length novel, *Hajnali háztetők* (Rooftops at Dawn, 1943 and 1957), is also inconsistent: after the references of the frame story set up expectations of a relativised interpretation, the main story offers only simple, one-sided tableaux. O.' significance thus rests largely with his novel, *Iskola a határon* (School at the Frontier, 1959), a parable about the formation and the limits of individuality as shown through a story about the life of a group of boys under the strict régime of a cadet school. The narrator of the frame story dramatically counter-poses versions of key events at the school, as recorded and interpreted in a manuscript by an ex-classmate, with his own recollections and retrospective analysis of the motives and emotions that prompted the events, leading to the conclusion that lack of freedom destroys the values of individuality.

Papp, Bertalan (1913–1992)

Fencer. Graduated from the Szeged University (1936). Golds as member of sabre team at the Olympics (1948, 1952) and World Championships (1953, 1954). Board member of the Hungarian Fencing Federation (1958–74). Professional army officer (to 1975), attaining rank of lieutenant colonel. Professional stuntman (1975–90).

Papp, László (1926)

Boxer, master coach. Olympic champion at middleweight (1948) and light middleweight (1952, 1956); the first boxer to win three Olympic golds. From 1962 he boxed professionally at middleweight, becoming European champion, but was not allowed to fight for World title in 1964. Retired unbeaten after 29 bouts. Captain and later head coach of the national boxing team (1969–92 with breaks). Awarded World Boxing Council's belt as 'All-Time Best Amateur and Professional Middleweight' (1989). Inducted into the Hungarian Hall of Fame (1991). Recipient of the International Olympic Committee's Olympic Order (1982), an International Fair Play Prize (1992), the Silver Cross of Merit of the Republic of Hungary (1992), and the Ferenc Csík Prize (2001).

Parti, János (1932–1999)

Canoeist, coach, sports administrator. Qualified English translator (1955) and lawyer (1968); coach's certificate (1963). Won golds in single canoes over 1,000 meters at Olympics (1960) and World (1954) and European (1961) Championships as well as European golds over 10,000 metres (1957, 1959). Captain of the national team (1977–88). Board member of the Hungarian Kayaking and Canoeing Federation; member of the Coaches' Committee, chairman of the Disciplinary Committee. Youth Prize (1987).

Pataki, Ferenc (1917–1988)

Gymnast. Olympic champion in the floor exercise (1948). One of the first to incorporate acrobatic elements into his exercise. Instructor at National School for Acrobatics.

Pataky, Kálmán (1896–1964)

Tenor. Made his debut at the Budapest Opera, from 1926 he was engaged to the Vienna Staatsoper. He was fêted all over Europe, America and South America, principally for his Mozart and Verdi roles.

Pauer, Gyula (1941)

Sculptor. Specialised in decorative sculpture, P.'s first works in the late 60s were non-figurative. In 1970, he invented 'Pseudo' as both a method of surface preparation (a rumpled, 'fake' surface) and a creative principle (a metaphor for the relation between illusion and reality), which he thereafter explored in paintings, sculptures, conceptual works, environments,

'Pathosformels', film sets and stage designs. For a sculptors' symposium at Nagyatád in 1987, P. constructed *A Forest of Demonstration Boards*—131 'Pseudo' boards with slogans and inscriptions—which the authorities demolished the next day. Later that year, P., together with Zoltán Érmezei and János Rauschenberger, produced a large-scale landscape-cum-stage set entitled *Dráva Riverbank Pictures*, which they purported to be the work of female artist Puci P.É.R.Y. Now a professor at the Budapest Academy of Fine Arts, P. is a member of the Széchenyi Academy of Arts.

Perényi, Miklós (1948)

Cellist. An infant prodigy, he made his debut in 1957. Following studies at the Budapest Academy of Music, he was a pupil of Enrico Mainardi in Rome (1960–62) and later of Pablo Casals (1969–72). He now tours the world performing a repertoire ranging from Bach to contemporary music. He has been a teacher at the Budapest Academy of Music since 1974.

Peri, Peter [Péri, László] (1889–1967)

Sculptor. A stonemason by training, his drawing skills brought him into touch with Lajos Kassák's activist group. Strongly left-wing in his views, P. was touring Czechoslovakia with a theatrical group when Hungary's 1919 Commune fell. He moved first to Paris but was soon expelled and so moved to Berlin. There he began making his famous tinted reliefs in concrete (considered by some as the precursors of 'shaped' canvases, though others have suggested they existed only as ideas or wooden models at this stage), which were exhibited in a joint show with Moholy-Nagy at the Der Sturm gallery in 1922, which also published his similarly conceived linocuts (1922–23). Alongside continued communist activism, he worked for a while in Berlin's city planning office before emigrating to Britain in 1934, after Hitler's rise to power. After World War II he made realist graphics and statues, which won the praise of (left-wing) British critics such as Francis D. Klingender, Anthony Blunt and John Berger (who modelled the main character of his novel *A Painter of Our Time* on P.). His most characteristic works from that period were bronze statuettes of 'little people', figurative concrete reliefs that emphasised linear perspective, and almost three-dimensional murals for public building (mainly schools).

Petelei, István (1852–1910)

Writer, journalist. A highly master of shorter fictions, the pieces of P.'s short-story and novella collections, *Keresztek* (Crosses, 1882), *Az én utcám* (My Street, 1896), *Vidéki emberek* (Country People, 1898) and *Az élet* (Life, 1905) are characterised by their resignedly fatalistic but sharp observation of the little tragedies of everyday provincial life in his native Transylvania—death, the estrangement of married couples, the loneliness of old age, failures of modest plans, etc. Despite the fact that P. came from a well-to-do merchant family, they are informed by a passionate sympathy for the underdogs of society, with discernible influences from Transylvanian ballads and the Russian realists, especially Turgenev. Apart from a short novel, *A fülemüle* (The Nightingale, 1886), P.'s only other excursion into more extended prose was *Mezőségi út* (Journey to the Mezőség, 1884), a pioneering near-sociographic description of this Transylvanian region. The psychologism and atmospheric subtlety of his writing make Petelei a forerunner of twentieth-century Hungarian short stories.

Péter, Rózsa (1905–1977)

Mathematician. It was László Kalmár, a fellow student at Budapest University, who made the suggestion that she study the work of Gödel, which was eventually to lead to her enduring achievement, the concept of recursive functions, which P. introduced in 1932 at an international mathematics conference in Zurich. Excluded by Hungary's anti-Jewish laws from working in the subject during the 1930s and World War II, the only employment she could obtain in the immediate post-war period was as a teacher at a secondary school and then at a teachers' training college, until she secured a lecturing post at Budapest University, teaching the introductory maths course. In 1951, however, she published the definitive monograph, *Rekursive Funktionen*, and subsequently (1976) on their application to computer technology. A member of the Hungarian Academy of Sciences, Rózsa Péter was also a populariser of science; her book *Játék a végtelennel* (Playing with Infinity, 1944) reached several editions and was translated into several foreign languages.

Péterfy, Jenő (1850–1899)

Essayist, critic. Literary criticism and the history of Hungarian literature attracted some of the country's best minds, including Péterfy. He also wrote many essays on aesthetics, history and music, and was a editor of *Budapesti Szemle* (Budapest Review). With a vivid, allegorical style, P.'s interpretative approach, examining the links between novelists and their works, the character and life of the writers, shows the influence of Hegel. The most enduring of the essays are *A tragikumról* (On the Tragic, 1885), *A tragédiáról* (On Tragedy, 1887), *A költői igazságszolgáltatás a tragédiában* (Poetic Justice in Tragedy, 1891), and *Aiszkhülosz-tanulmány* (A Study of Aeschylus, 1898). During the final years of his life P. worked on a history of Greek literature, but failed to finish it before putting an end to his own life.

Petőfi, Sándor (1823–1849)

Poet. One of the greatest Hungarian lyrical poets, P. studied at eight different schools, briefly joined a group of strolling players, and flared in the centre of Hungary's public life during the Revolution before meeting his end during the War of

Independence as an enlisted soldier. Having had his first poem published in 1842, P., with support from Vörösmarty, became assistant editor of the literary periodical *Pesti Divatlap* (Pest Vogue). His first volume of poetry, *Versek*, appeared in the same year, instantly propelling him to fame, though many were scandalised by its tone. P. was inspired to write his best love poems, e.g. *Szeptember végén* (At the End of September, 1847) and *Minek nevezzelek?* (What Shall I Call You?, 1848) after falling in love with Júlia Szendrey in the autumn of 1846. Together with Jókai, he edited the magazine *Életképek* (Sketches of Life) from 1847, castigating the social conditions of his country, attacking the privileges of the nobles and the monarchy. His poems seared with political passion, one of them, *Nemzeti dal* (National Song), written on the eve of the revolution, becoming its anthem. P.'s poetry introduced a direct, unpretentious style and a clear, unornamented construction adapted from folk song. Of his epic poems, *János vitéz* (John the Valiant, 1845) is an entrancing fairy tale, whilst the comic *A helység kalapácsa* (The Hammer of the Village, 1845) and *Az apostol* (The Apostle, 1848) are the most popular. P. disappeared during the battle of Segesvár (Sigişoara, Romania), but the general public for a long time refused to believe his death.

Petrovics, Emil (1930)

Composer. A pupil of Ferenc Farkas, he has been a teacher at the Budapest Academy of Music since 1968, heading the composition department until 1995. His compositions, which include nine cantatas and the opera *C'est la guerre* (1962) combine Hungarian elements with late-Romanticist and jazz influences.

Petzval, József (1807–1891)

Engineer, mathematician. Although born in Upper Hungary and qualified as an engineer at the University of Pest, he spent all but two years of his working life in Vienna, where he became a lecturer at the Technical University and eventually a member of the Viennese Academy of Sciences. He is remembered primarily for working out the precise mathematics required to design the first camera lenses fast enough to be used for portraiture (exploited commercially by the Voigtländer company), as well as developing anastigmatic lenses and lens systems for use with high-powered light sources (e.g. film projectors). His published works include *Bericht über die Ergebnisse einiger dioptrischen Untersuchungen* (1843), *Eigenschaften einer guten Camera-Obscura* (1847), and *Über die neuen Landschaften als Fernrohrobjektiv* (1858).

Pézsa, Tibor (1935)

Fencer. Certificates as PE instructor (1959) and coach (1964). Won gold for individual sabre at the Olympic champion (1964) as well as individual silver (1966) and gold (1970) and team gold (1966) at World Championships. Hungarian Fencer of the Year (1964, 1966, 1970). Head coach to the national sabre team (1978–82).

Pilinszky, János (1921–1981)

Poet. His first poems were already being published by the time he enrolled to study law, literature, and art history at Budapest University in 1938, which he left before graduating. Enlisted into the army in 1944, P. was stationed in Harbach, Germany, where he witnessed the horrors of the concentration camps—an experience that profoundly shaped all his subsequent work. *Egy KZ-láger falára* (On the Wall of a KZ-Lager), one of the poems in his second collection, is considered a universal symbol of the tragedy of human existence as well as a memento to the innocents killed in the Holocaust. On being demobbed, P. became a co-editor of the literary magazine *Újhold* (New Moon) and had his first volume, *Trapéz és Korlát* (Trapeze and Parallel Bars, 1946), published; however, the advent of

Communist dictatorship forced him, like many others, into silence. In 1957, P. began contributing to the Catholic weekly *Új ember* (New Man). His subsequent slim volumes of poetry, *Harmadnapon* (On the Third Day, 1959), *Nagyvárosi ikonok* (Big City Icons, 1970), *Szálkák* (Splinters, 1972) and *Végkifejlet* (Denouement, 1974), along with the final collection of *Kráter* (Crater, 1975), a few short plays and prose texts such as *Beszélgetések Sheryl Suttonnal* (Conversations with Sheryl Sutton, 1977), bleak and enigmatic though they are in content, have exerted huge influence on younger generations through their integrity of vision.

Polanyi, Michael [Polányi, Mihály] (1891–1976)

Physical chemist, political scientist, philosopher. P. originally qualified at Budapest's medical school but was primarily interested in physics and theoretical chemistry, initially absorption thermodynamics. After a period in the research laboratories at the United Incandescent Lamp Co. (working e.g. on large-scale manufacture of krypton and, with Orowan, on crystal dislocation theory), he left Hungary for England, where he was professor of physical chemistry at the University of Manchester (1933–48), working mainly on reaction theory. Having become increasingly active as a political thinker and philosopher, P. then moved to the chair of social studies at the same university (1948–58) before ending his career as a Senior Research Fellow at Merton College, Oxford (1959–61). His many publications included *Atomic Reactions* (1933), *U.S.S.R. Economics* (1936), *Science, Faith and Society* (1946), and *Personal Knowledge* (1954).

Polgár, Judit (1976)

Chess player. Reached women's grandmaster and men's international master grades in 1988, men's international grandmaster and super-champion grades in 1991. Member of the gold-winning women's Chess Olympics team (1988, 1990) and the men's European Championship team (1989); first board on the men's Chess Olympics team (1994, 1996). Headed FIDE women's world rankings (1989–99). Winner of a number of men's international competitions and has recorded wins over Spassky (1992) and Karpov (1998) in doubles matches. Speaks five languages. Awarded a Chess 'Oscar' (1989).

Polgár, László (1946)

Chess coach, manager. Graduate in philosophy (1969) and child psychology (1973), and doctorate in education (1993); holds master coach's certificate (1993). Manager and head coach to his renowned chess-playing daughters, whom he educated according to his own methods. Has established a family chess museum and archive, including chess-related artworks. Author on chess.

Polgár, Zsófia (1974)

Chess player. Reached the grades of women's international master and men's FIDE master in 1988, men's international master and women's international grandmaster in 1989. Won men's World Junior Championship (1988); individual women's champion (1988, 1990) and a member of the silver-winning women's team (1994) at Chess Olympics. Silver medals in the women's World Rapid Chess Championship (1992) and the men's World Under 20s Championship (1994). Has been living in Israel since 1995. Studies art history and speaks six languages. Awarded a Chess 'Oscar' (1994).

Polgár, Zsuzsa (1969)

Chess player. Reached the grade of men's international master in 1984, women's (1987) then men's (1991) international grandmaster. Individual women's champion (1988, 1990) and a member of the silver-winning women's team champion (1994) at Chess Olympics. Gold medal in women's World Rapid Chess Championship (1992) and women's World Championship (1996). Has been living in the USA since 1994.

Speaks eight languages. Awarded two Chess 'Oscars' (1994 and 1996).

Pollack, Mihály (1773–1855)

Neo-classical architect. Whilst a mason's apprentice (to 1792) he studied at the Vienna Academy then spent 1794 in the Milan office of his architect half-brother, Leopoldo Pollach. After moving to Pest in 1798, he was one of the original appointees to the Pest Municipal Beautification Commission set up in 1808 by Archduke Joseph. Early works in Pest were the *Lutheran Church in Deák tér*; the *Municipal* (later German) *Theater* (1808–12); and a long series of neo-classical apartment buildings put up over the period up to 1839. His façade for the *County Hall at Székesfehérvár* (1807) was widely copied, and he later also designed the *County Hall at Szekszárd* (1828–36). Of many country mansions that P. designed, the most important were that built at Alcsút, some 40 km west of Buda, for Archduke Joseph (1819–29; destroyed during World War II), and the *Festetics Country House at Dég* (1815–19), between Siófok and Sárbogárd. His next major projects in Pest were the *Ludovika Military Academy* (1830–36), in the Orczy Garden off Üllői út (now home to the Natural History Museum), and the *Vigadó*, or Redoute Building (1826–32) on the Danube bank, later destroyed during the 1849 siege of Buda. P.'s masterpiece is the *National Museum* (1836–44). With the central rectangular block of its huge portico of eight Corinthian columns, set at the top of a broad flight of stairs, and its splendid interior spaces, the building is the most mature example of Hungarian neo-Classicist simplicity.

Pólya, György (1887–1985)

Mathematician. As a student at Budapest University he attended lectures in physics, philosophy, law and literature, as well as mathematics, then carried on postgraduate studies in

Vienna, Göttingen and Paris before gaining his doctorate at Budapest (1910) with a thesis on probability calculus. His other main fields were combinatorial analysis and graph theory. He became a teacher at the Zurich University in 1914. On emigrating to the USA, he became a professor at Stanford University in 1940, from where he retired in 1953, although he was famous for the lectures that he was still delivering in his nineties. With Gábor Szegő, P. co-authored the well-known *Aufgaben u. Lehresätze aus der Analyse* (1925; publ. as *Problems and Theorems in Analysis*, 1976) and a work on mathematical physics (*Isoperimetic Inequalities in Mathematical Physics*, 1951). Several of his books—most notably *Induction and Analogy in Mathematics* (1954) and *How to Solve It* (1954)—present new approaches to teaching mathematics and the merits of 'plausible reasoning', or the so-called 'heuristic' method. Other published works included *Inequalities* (with G. Hardy and J. Littlewood, 1934), *Applied Combinatorial Mathematics* (with E.F. Beckenbach, 1964), *Complex Variables* (with G. Latta, 1974), and *Collected Papers* (1974). Two prizes bear his name.

Popper, David (1843–1913)
Cellist. Prague-born and trained, he played in Vienna. From 1886 until his death he was a teacher at the Budapest Academy of Music. His virtuoso cello pieces are still widely played.

Portisch, Lajos (1937)
Chess player. Master coach's certificate (1984). Reached international grandmaster grade in 1961. Eight times in Candidate Tournament for World Championship. Seconded A. Karpov during his World Championship title match (1990). Hungarian Chess Player of the Year (1964, 1966–82, 1984–88). He is a writer on chess. Awarded Hungarian Sportsmanship Prize (1973).

Puskás, Ferenc (1927)
Footballer, coach. Member of the gold-winning Olympic team (1952), the team that beat England in the 1953 'match of the

century' at Wembley Stadium, and the 1954 World Cup silver-winning team. Was four times member of Hungarian league championship team, and top scorer in three seasons. Moved abroad after 1956, playing for Real Madrid (1958–67) and winning Spanish league championship with them six times; top scorer in four seasons. Capped 84 times for Hungary, 4 times for Spain. Selected for World XI (1963) and European XI (1965). Coached in several continents. He moved back to Hungary in the 1990s. Inducted into the FIFA Hall of Champions. Voted 'Sportsman of the Century' in a readers' ballot run by the daily *Nemzeti Sport* (2001). Recipient of the International Olympic Committee's Olympic Order (1995 and 1997) and the Ferenc Csík Prize (2001).

Puskás, Tivadar (1844–1893)

Electrical engineer. After school and university studies in Vienna and a brief stay in England, he emigrated to the USA. Though bankrupted by a business venture (in agriculture), through learning about the telegraphic service he had formulated the idea of an exchange—a concept that he patented in Brussels on his return to Europe. During a further two-year spell in America, he worked with Thomas Edison and also met Alexander Bell before going back to England then on to Paris, where he opened his first exchange in 1879. There he conceived the idea of transmitting performances at the Grand Opera by cable and, with younger brother Ferenc, he tried to implement this in Pest. After several failed attempts, the first successful relay of an opera performance (Erkel's *Bánk bán*) took place in 1892, and their *Telefonhírmondó* (Telephone Messenger) commenced regular transmissions in February 1893, functioning up till 1944.

Putty, Lia von [Putti, Amália] (1897–1931)

Actress. After a short career in Hungary, she became a popular starlet of the UFA Studio in Berlin, most notably in *Das Indische Grabmal: Der Tiger von Eschnapur* (dir. Joe May,

1921), before moving to Hollywood in the mid-1920s. There she starred in films by Murnau, Griffith and Michael Curtiz and, most famously, in *Variety* (dir. E. Dupont, 1925).

Rácz, István (1922–1991)

Electrical engineer. A graduate of Budapest Technical University, most of his career was spent as a lecturer, yet he was a influential figure within the Hungarian Academy of Sciences' Computer Technology and Automation Research Institute (SZTAKI) from its foundation (1964) onwards. The two-volume *Transiente Vorgänge in Wechselstrommaschinen* (co-authored with Pál K. Kovács, 1959) was for a long time a 'Bible' for engineers working with alternating-current machines, but R.'s main innovations were in the theory and practical applications of controlled electric drives, on which he he co-wrote a book with Otto Benedikt (*Különleges villamosgépek*, Special Electric Machines, 1957).

Radnóti, Miklós (1909–1944)

Poet, translator, essayist. R. took a degree in Hungarian and French at Szeged University but was unable to obtain a teaching post due to his Jewish origins. He invested the refined lyrical forms of Classical pastoral poetry with powerful Expressionist and Surrealist tonalities, influenced by his work translating German and French poets such as Apollinaire. This fusion had been perfected by the time of his fourth and fifth volumes *Újhold* (New Moon, 1935) and *Járkálj csak, halálraítélt!* (Just Walk On, Condemned to Die!, 1936), the poems which anticipate the impending barbarism and terror of a world in which all are threatened by the silence of death. After the German occupation of Hungary (1944), R. was taken to a labour camp in Serbia then, when the Nazis evacuated the country in October 1944, he and the rest of the squad were force-marched to north-western Hungary, where he and twenty-two others were shot and buried in a mass grave. On exhumation of his body after the war, a notebook of poems was found. Composed in the camps and on the march itself, *Levél a hitvesnek* (Letter to My Wife), *Hetedik ecloga* (Seventh Eclogue) and *Nyolcadik ecloga* (Eight Eclogue) speak of the horror of the situation, their emotional stream of personal confessions, dreams, wild desires, sober intellectual objectivity, and crude reality being structured in perfectly crafted classical verse measures.

Radványi, Géza (1907–1986)

Director. Younger brother of the writer Sándor Márai, R. began his career as a foreign correspondent in several West European countries before taking up film direction on returning to Hungary for a brief spell in the early 1940s. The easygoing style, transparent structure and suspense of his early films, especially *Európa nem válaszol* (Europe Doesn't Answer, 1940), marked him out as an unusual talent. *A beszélő köntös* (The Talking Robe, 1941), his adaptation of a Mikszáth novel, was the first partially colour Hungarian film. Leaving the country again, he shot a film in Rome in 1942 and did not return until, with Félix Máriássy as co-director, he made *Valahol Európában* (Somewhere in Europe, 1947), a film that was critically acclaimed as one of the key post-war European films. Emigrating again after the Communist take-over in Hungary, R.'s subsequent films included *Una donna senza nome* (1949, Rome) and *Mr. Bard* (1953, Paris; with Michel Simon in one of his greatest roles), as well as a number of lesser works in Germany (*Der Arzt von Stalingrad*, 1957; *Operation Caviar*, 1961; *Uncle Tom's Cabin*, 1965; *Der Kongreß amüsiert sich*, 1966). Re-established contacts with Hungary during the 1970s paved the way for *Circus Maximus* (1980)—a swan song conceptually and even stylistically reminiscent of *Somewhere in Europe*—to be shot there.

Rajcsányi, László (1907–1992)
Fencer, sports administrator. Qualified engineer (1930). Won gold as member of sabre team at the Olympics (1936, 1948, 1952) and World Championships (1937, 1951, 1953). Became chairman of the Hungarian Fencing Federation in October 1956; banned in 1957. Chairman of the Hungarian Athletics Club (1990–92).

Ránki, Dezső (1951)
Pianist. A pupil of Pál Kadosa at the Budapest Academy of Music, in 1969 he won the Schumann Competition in Zwickau and has toured Europe, Japan and the US. His repertoire is centred on the works of Schumann and the Romantic period.

Reiner, Fritz (Frigyes) (1888–1963)
Conductor. After studying at the Budapest Academy of Music, he became conductor of the city's Népopera (People's Opera, 1911–14) and then at the Dresden Opera (1914–21). He moved to the US in 1922. He was director of the Cincinnati Symphony Orchestra, the Pittsburgh Symphony and the New York Metropolitan Opera (1948–53); finally he led the Chicago Symphony Orchestra (1953–63). Among his pupils was Leonard Bernstein.

Rejtő, Ildikó (1937)
Fencer. Coach's certificate (1975). Won gold at the Olympics in individual and team foil (1964) and silvers in the team event (1960, 1968, 1972) and golds at World Championships in the individual (1963) and five times in the team event. Hungarian Sportswoman of the Year (1963–64) and Woman Fencer of the Year (1964, 1968–71).

Rényi, Alfréd (1921–1970)
Mathematician. His studies at Budapest University were interrupted by being detailed into forced labour during the war. Finally qualifying there, he gained his doctorate under Riesz at Szeged in 1945 then went on to Moscow to obtain a post-doctoral 'candidate's degree' in 1947 with a thesis relating to the

Goldbach conjecture in number theory. After returning to Hungary, he won a chair in probability calculus at Debrecen University in 1949 and the following year set up and became director of the Mathematics Research Institute of the Hungarian Academy of Sciences. An important populariser of his subject, he was the chief organiser of the 1st and 2nd Hungarian Mathematics Conferences (1950 and 1960). His published works included *Probabilistic Methods in Number Theory* (1958), *Sur les graphes aléatoires* (1962), *Dialógusok a matematikáról* (Dialogues on Mathematics, 1965), *Probability Theory* (1970).

Rerrich, Béla (1881–1932)

Architect, garden designer. After graduating from Budapest Technical University (1903), he studied garden design in Paris and England. In 1908 he became a teacher (and from 1923 director) at the Budapest Institute of Horticulture. Before World War I he entered many competitions and gained commissions to build apartment houses, schools and some villas, initially in a neo-Baroque and later, under the influence of the *Fiatalok* ('Young Ones'), a more overtly modern style. His six-storey *Apartment House for the Metropolitan Electric Board* on Hengermalom Street is one of the more valuable housing designs of the inter-war period. His masterpiece, however, was the arcaded brick scheme for *Dóm tér in Szeged* (1927–30) in collaboration with Ernő Foerk, the architect who completed the new Votive Church on the square. Influenced by the neo-Romanesque brickwork of Ragnar Østberg's Stockholm City Hall and the 'expressionist' brick architecture of Fritz Höger's Chile House in Hamburg, the façades that Rerrich devised for the three sides of the square accommodate the changing functions, not imposing any rigid symmetry, with the pillars of the arcades leading naturally to the two archway street entrances.

Reviczky, Gyula (1855–1889)

Poet, novelist, essayist. *Ifjúságom* (My Youth, 1883) and *Magány* (Solitude, 1889), his only two volumes of verse, contain poems of elegiac tone dealing primarily with his torment over what he felt to be his own ruined life and his sympathy for prostitutes and other social outcasts. In 1887 he became the drama critic of *Pesti Hírlap*. As an essayist he was concerned about humour, modern lyrical poetry and the questions of symbolism, and philosophically, determinism and Schopenhauer's pessimism. His novella *Apai örökség* (Paternal Heritage, 1884), depicting the downfall of a youth from the gentry class, influenced by Tolstoy and Turgenev, is largely autobiographical, as is the short story *Selyembogár* (Silk Beetle, 1889).

Richter, Gedeon (1872–1944)

Pharmacist. After qualifying, he travelled around Europe for four years obtaining experience before returning home to purchase his own pharmacy. On starting a side-business supplying batched products to other pharmacists in 1901 he soon found that demand was outstripping the limited facilities, leading him to set up a factory in the Kőbánya district of Budapest in 1907. Besides manufacturing the conventional preparations of the day, he set up an industrial laboratory and later a research facility to develop novel plant extracts and synthetic drugs (e.g. analgesics), as well. By the late thirties, it was briefly competitive with the efforts of companies abroad, for example, in manufacturing the first synthetic oestrogen and, with ten foreign affiliates, was Hungary's second largest exporter after Tungsram (as the United Incandescent Lamp Co. was known by then). Richter himself died at the hands of local Nazi militiamen following the German occupation of the country.

Richter, Hans (1843–1916)
Conductor. He studied in Vienna (1860–65) then was conductor at the National Theatre in Budapest (1871–75) and chief conductor of the Vienna Philharmonic Orchestra (1875–98). He lived in England between 1900 and 1910. As Wagner's close friend he conducted the opening performances of the 1876 Bayreuth Festival.

Riesz, Frigyes (1880–1956)
Mathematician. Beginning his university studies at Zurich, he graduated at Budapest then, after a year in Göttingen, obtained his doctorate at Budapest in 1902. A celebrated 1907 paper containing the Riesz–Fischer theorem made him one of the founders of a new branch of functional analysis, whilst in 1908 he introduced the concept of topological space. Having secured a post at Kolozsvár (Cluj, Romania) alongside Alfréd Haar, he moved to Szeged when the university was relocated to there after World War I. Building up an internationally renowned school in functional analysis, R. continued to do significant work on such areas as sub-harmonic functions, with his generalisation of the theory of potential in 1925, and his introduction of the concept of the vector lattice (known as Riesz space) in 1928. His publications include the fundamental monograph *Functional Analysis* (with B. Szőkefalvi Nagy, 1955).

Rimanóczy, Gyula (1903–1958)
Architect. With a family pedigree in architecture, he spent the first few years after graduating from Budapest Technical University (1925) travelling abroad (during which he acquired a taste for modern Italian neo-Classicism) and also working in the office of Gyula Wälder, who practised a neo-Baroque approach. A leading figure in the informal 'Little Robbers' group of young modernists, R. had considerable success in competitions. His first masterpiece in the capital was the

Franciscan Church of St. Anthony of Padua on Pasaréti tér, Buda (1931–34), which—along with Árkay's Church of the Holy Heart of Jesus in Városmajor park—is a pioneering example of the branch of Hungarian modernism that came under the influence of the Rome School. Other important works of this period included the *Szakáts Villa at No. 97 Pasaréti út* (1934) and the *Post Office Retirement Home on Cházár András utca* (1936–38), near the City Park in Pest's fourteenth district. On a larger scale are: the *Mester utca Polyclinic* (1938–41), in the Ferencváros (ninth) district; and the *Administration Building and Automatic Telephone Exchange for the Post Office and Telephone Company, Dob utca* (1937–40), in the Ferencváros district. During the war R. worked as one of Gedeon Gerlóczy's partners on the *National Sports Hall* (1941–43). In 1949 he joined IPARTERV, the state planning office for industrial construction, heading the team that designed the *Ganz Electricity Meter Factory at Gödöllő* (1949) and a *Tobacco Curing Plant at Nyíregyháza* (1950). From 1950 he was one of the leading architects in KÖZTI, the state planning office for public buildings, working under the constraints of the prevailing 'socialist realist' dogma on such projects as the *Mining School at Várpalota* (1950–51); the *Industrial Apprentices' School at Győr* (1951); and *Building R for Budapest Technical University* (1951–55). For the latter R. received the Kossuth Prize (1955), and he was one of the first architects to be awarded the Ybl Prize (1953). With István Janáky, he was picked as a supervisor in the first cycle (1953–56) of 'Master School' postgraduate training courses organised by the Union of Hungarian Architects.

Rippl-Rónai, József (1861–1927)

Painter and industrial designer. Having studied at the Munich Academy of Art and in Paris, he became an assistant to Mihály Munkácsy. After early salon pictures, his *Nő fehérpettyes ruhában* (Woman in White-Dotted Dress, 1889) already shows signs of an individual style. This marked the start of what he himself called his 'black period', a spell of over ten years that he spent at Neuilly, where he became a member of Les Nabis and made the acquaintance of Maillol. His modelling of contours, whether for interiors or figurative works, was indebted to Art Nouveau, the finest examples being *Kalitkás leány* (Girl with Birdcage, 1892) and *Öregasszony ibolyával* (Old Lady with Violets, 1895). His wife wove tapestries based on his designs. R.-R. returned home to an initially hostile reception, though he was commissioned to paint a window in the Ernst Museum. After a tour of Belgium and a trip to Moscow, he made his home at Kaposvár. The subjects of this period are small-town life and people, such as 'Uncle' Piacsek. His pictures became more colourful, including the employment of pastels, as in *Karácsony* (Christmas, 1903), *Szomorúság* (Sadness, 1903), and *Kövezik a kaposi főutcát* (Paving Kaposvár High Street, 1905). Nearly every picture shown at his 1906 exhibition was purchased, allowing him to buy a villa in Rome. His third period, which he himself referred to as his 'corn-cob' period, is reminiscent of pointillism, using virtually pure colours at full intensity and increasingly decorative surfaces, as in *Apám és Piacsek bácsi vörösbor mellett* (My Father and 'Uncle' Piacsek Over a Drop of Red Wine, 1907) and *Park aktokkal* (Park with Nudes, 1910). Later on he only produced portraits in pastels of writers like Zsigmond Móricz, Mihály Babits, and Lőrinc Szabó as well as numerous beautiful women.

Rózsa, Miklós (1907–1995)

Composer. He studied in Leipzig then lived in London (1935–40). Moving to Hollywood in 1940, he picked up Academy Awards for composition three times. He was appointed professor of composition at the University of Southern California in 1945.

Rózsa, Norbert (1972)
Swimmer. In breast-stroke won golds at World Championships over 100 meters (1991, 1994) and 200 meters (1994), then at the Olympics over 200 meters (1996). Briefly resident in Australia (1993–94).

Rózsavölgyi, István (1929)
Athlete, coach. Coach's certificate (1962). Middle- and long-distance runner. Won Olympic bronze at 1,500 meters (1960); set world records in several events. Running coach to the national modern pentathlon team (1962–64), then club coach (1964–84). Army officer.

Rózsavölgyi, Márk (1789–1848)
Composer. He built up a reputation as a strolling composer all over Hungary. In 1833 he moved to Pest. One of the last exponents of the *verbunkos* style, his dance pieces played a major part in the emergence of a distinctively Hungarian musical culture.

Rozsnyói, Sándor (1930)
Athlete. PE instructor's certificate (1953); qualified as teacher in Austria (1963). Won gold in 3,000 meters steeplechase (setting first official world record for the event) at European Championships (1954), silver at the Olympics (1956). Lived in Austria from 1957, then in Australia from 1965, working as a PE teacher and tennis coach.

Sándor, Károly (Csikar) (1928)
Footballer. Coach's certificate (1960). First division club player (1947–64), winning championship four times (1951, 1953, 1955, 1958), Hungarian Cup (1952), and Central European Cup (1955, 1963). He was capped 75 times for his country and picked for Europe XI (1964).

Schaár, Erzsébet (1908–1975)
Sculptor. The works of S.'s early career, like *Édesanyám* (My Mother, 1925), were realistic, classicising statues and reliefs, but from the late '50s they underwent a constant evolution towards great expressiveness, whether in group compositions

like the slim bronzes of *Kórus* (Choir, 1963) or the small terracotta bas-reliefs of *Halott katonák* (Dead Soldiers, 1965). Her interest then moved from pure plastic form to placing glass and bronze figurines within maquette-like spatial constructions, as in *Fal, Ablak, Boltívek, Kirakat* (Wall, Window, Arches, Shop-window, all in the '60s). In her late creative period S. completed several large-scale compositions of solitary figures, women and busts of prominent Hungarians framed timelessly in pseudo-architectural structures. Most memorable is the environment *Utca* (Street, 1974): the perishable materials (plaster, expanded foam) of the original construction not only provided a counterpoint to the 'archaic' figures, with their pitted surfaces, but also underscored the idea of transitoriness.

Schiff, András (1953)

Pianist. A pupil of Pál Kadosa at the Budapest Academy of Music. Based in London from 1979 then in Salzburg since 1986, his interpretations of Bach, Mozart, Schubert and Bartók have made him one of the world's most acclaimed artists as a soloist and in chamber music settings.

Schmitt, Pál (1942)

Fencer, sports diplomat. Graduate (1965) and doctorate (1992) in economics. Won Olympic golds with epée team (1968, 1972). Vice-chairman of the National Physical Education and Sport Office (1983–86); chairman of the Scientific Council on Physical Education and Sport (1983–86). Secretary-general of the Hungarian Olympic Committee (1983–89), and its chairman since 1989; member of the International Olympic Committee (1983–), sitting in its Executive Committee since 1991, as vice-president (1995–99) and chief of protocol (1999–). Ambassador of the Republic of Hungary to Spain (1993–97) and Switzerland (1999–). Awarded the Middle Cross of the Order of Merit of the Republic of Hungary (1992) and the Miksa Eszterházy Prize (2001).

Schoeffer, Nicolas [Schöffer, Miklós] (1912–1992)

Kinetic artist. After studies at the Budapest Academy of Fine Arts and the Paris École des Beaux-Arts, he took up permanent residence in France in 1936. A first show at the Galerie Breteau, Paris (1948) was followed by 15 statues of the *Spatiodynamik* (1949–53) series then a series of reliefs with perforated elements (1950–51). A 20m tower constructed for the Groupe Espace exhibition in Biot and a 50m audio-visual tower for a park in St. Cloud (1955) can be considered precursors of his cybernetic works, the first of which, *CYSP* (1956) was displayed at the Unité d'habitation, Marseille, accompanied by a performance from Maurice Béjart's ballet group. That same year, S. presented the prototype of a house with invisible dividing walls, later elaborated in his concept of the 'cybernetic city'. His subsequent theory of 'Luminodynamik' (1956–57) led to the *LUX I–XIII* series, whilst the ensuing *CHRONOS*-series (1960–62), which included a 'musiscope' (an early form of manually-controlled light-sound synthesiser) and a cybernetic tower in Liege, was based on the principles of 'relief sériel' and 'Chronodynamik'. The major work of that decade, however, was the huge *PRISM*, with its mirror walls. In 1968 S. designed a kinetic stage set for the Hamburg Opera and was awarded the Grand Prix at the Venice Bienniale. In the '70s, plans for a statue to be erected in the Défense district of Paris were followed by the car statue *SCAM* (1973) and culminated in *KYLDEX I*, a multimedia show at the Hamburg Opera. S.'s propounded his ideas in a number of books, the first of which was *La Ville cybernétique* (1968). S.'s native town of Kalocsa has erected one of his cybernetic towers and also founded a museum named after him.

'Sigmond, Elek (1873–1938)

Soil scientist. Graduating as a chemical engineer from Budapest Technical University, he went on to obtain a doctorate at Kolozsvár (Cluj) University for research on the enzymatic breakdown of starch and derivatives. After short spells of research in several laboratories in Europe and the USA, in 1908 'S. set up the Department of Agricultural Chemistry at Budapest Technical University, which he headed until his death. His main achievements stemmed from his studies on the chemical and physical structure of alkaline soils. His general system of soil classification, introduced in 1934, was the first to be based on genetic principles and to include all known soil types. He spent nearly four decades investigating the efficacy of organic and artificial fertilisers in soil amelioration, working out the regimens indicated for different soil test findings. A member of the Hungarian Academy of Sciences, his major published works included *Mezőgazdasági chemia* (Agricultural Chemistry, 1905), *Mezőgazdasági chemiai technológia* (Agricultural Chemical Engineering, 1919), *Hungarian Alkali Soils and the Method of Their Reclamation* (1927), *Általános talajtan*, 1934 (*The Principles of Soil Science*, 1938).

Solti, Sir Georg [György] (1912–1997)

Conductor. Was a student of Bartók, Dohnányi and Weiner. He tided through the Second World War in Switzerland. He was music director at the State Opera in Munich (1946–52) then director at the Frankfurt Opera House (1952–61). In London he was music director of the Covent Garden (1961–71). He became chief conductor to the Chicago Symphony Orchestra (1969–91). Wagner's *The Ring of the Nibelung* (1958–66) is outstanding among his recordings.

Starker, János (1924)

Cellist. He studied in Budapest but has resided in the US. Since 1948 he has been teacher at the Indiana University. Due to his solo career starting in 1958, he is now widely hailed as one of the greatest post-war cello virtuosos both in the concerto and chamber music repertoire, noted particularly as an interpreter of Kodály's *Cello Sonata*.

Stefanek, Gertrúd (1959)

Fencer. Qualification in catering. As a member of the foil team won silvers (1979, 1982, 1985) and gold (1987) at the World Championships and bronze at the Olympics (1980, 1988). Hungarian Woman Fencer of the Year (1982, 1983). Television hostess (1994); now a businesswoman in clothes retailing.

Steindl, Imre (1839–1902)

Historicist architect. After graduating from the Joseph Polytechnic in Buda, he studied a further six years at the Vienna Academy. From 1869 himself was teaching at his old school (by then the Budapest Technical University). His early designs for projects in Pest were in a neo-Renaissance style: the *New City Hall* at Nos. 62–64 Váci utca (1870–75) to accommodate the previously separate but newly unified council staffs of Buda, Pest and Old Buda; the newly founded *Veterinary School* in Rottenbiller utca (1881); and a new building for the *Technical University* on the Múzeum körút (1880–83). In 1883 S. won the international competition for *Parliament Building* on the Pest bank of the Danube with a design in the neo-Gothic style imposed on a Baroque groundplan and spatial organisation. Though originally supposed to be completed in time for the country's Millennial celebrations in 1896, the exterior was not finished until 1902 and the interior only in 1904. He retained a neo-Gothic style in the stone-decorated brick architecture of *St. Elizabeth Church in Rózsák tere* (1895) in Pest's seventh district (just round the corner from the Veterinary School). S. was also active as a meticulous restorer on many projects, including the mid-15th-century *Vajdahunyad Castle* (Hunedoara, Romania), the largely mid-18th-century *Inner City Parish Church of Pest* (now at the Pest foot of the Elizabeth Bridge), and the early 14th-century *Cathedral at Kassa* (Košice, Slovakia).

Straub F., Brunó (1914–1996)

Biochemist. Having begun to study medicine at Szeged, S. was persuaded by Albert Szent-Györgyi to swap disciplines and thus took his doctorate in chemistry. He researched into cell respiration at Szeged under Szent-Györgyi then for several years at Cambridge, England before returning to Szeged, where he isolated and purified malate and lactate dehydrogenases, two enzymes that play a vital role in biological oxidation and discovered a diaphorase that bears his name. Switching to muscle research, S. identified actin as the key contractile protein. After the war, he moved with Szent-Györgyi to Budapest Medical University and in 1948, when the latter left Hungary, took over his chair as well as directorship of the Biochemistry Institute (later the Institute of Enzymology) of the Hungarian Academy of Sciences, which he retained until he retired. One of the chief promoters of the Szeged Biology Centre for molecular biological research (1959), he headed it for seven years and was instrumental in fostering collaborations with laboratories elsewhere in Europe and America. Alongside duties as a teacher, administrator and author of major textbooks, he continued to pursue research, e.g. elucidating the role that ATP plays in the potassium pump of red blood cells and discovering the enzyme disulphide isomerase. His published works included *Biokémia* (Biochemistry, 1949), *Általános szervetlen és analitikai kémia* (General Inorganic and Analytical Chemistry, 1950), and *Enzimek, molekulák, életjelenségek* (Enzymes, Molecules, Phenomena of Life, 1966).

Svéd, Sándor (1906–1979)

Baritone. He studied in Italy. He made his debut in Budapest in 1927. He was engaged by the Vienna Staatsoper from 1936, then by New York's Metropolitan Opera (1940–50). He returned to Budapest (1950–56) before again working abroad. At the peak of his career he was particularly associated with his roles in Verdi's operas.

Szabó, Bence (1962)

Fencer, sports manager. Graduated from the College of Commerce and Catering (1987). At the Olympics won gold for individual sabre (1992) and both gold (1988) and silver (1992, 1996) in the team event; winner of the World Cup (1989), and at World Championships was twice with silver-winning team (1990, 1994). Hungarian Fencer of the Year (1989, 1992–93). Sports club chairman (1996–). Recipient of the Officers' Cross of the Order of Merit of the Republic of Hungary (1992).

Szabó, Dezső (1879–1945)

Poet, novelist and essay writer. A highly influential innovator of the modern Hungarian literary language, author of some powerful and bewildering novels and stories, Sz. alienated sympathy with his venomous—though often pertinent—attacks on almost everything. His work can be divided into two discrete sections. Having started as a polished essay writer, with a thorough grounding in world literature, he later became a militant public writer, strongly influenced by the major intellectual trends of his time, with politics as the dominant strand in his *oeuvre*. His sociographic *Az elsodort falu* (The Village That Was Swept Away, 1919) is a tableau of the disintegration of Hungarian society during the war, covering all the important problems of the time. He went on to write, amongst other things, a pamphlet novel, *Segítség!* (Help!, 1926), and the short story *Feltámadás Makucskán* (Resurrection at Makucska, 1931). Sz. was a forerunner of the Hungarian Populist, or agrarian, movement, but the most problematic aspect of his later work is its frequently overt anti-Semitism.

Szabó, Ferenc (1902–1969)

Composer. A pupil of Kodály at the Budapest Academy of Music (1923–26). Due to his involvement in the work of clandestine Communist party, he was forced to leave via Berlin for

Moscow in 1931. He returned to Hungary in 1945 as a political officer in the Red Army, playing a major role in shaping Stalinist cultural policy. His most important works are *Farkasok dala* (Song of the Wolves, 1930) and *Föltámadott a tenger* (The Sea Has Risen, 1955).

Szabó, István (1938)

Director. The lyricism of short films he made as a student signalled Sz.'s talent from the outset. His first feature films, *Álmodozások kora* (The Age of Daydreaming, 1964), *Apa* (Father, 1966) and *Szerelmesfilm* (Love Film, 1970), were portraits of members of his own generation coming of age in post-1956 Budapest and established him as one of the most interesting of a new wave of East European *auteurs* (the autobiographically based *Father*, in particular, garnered several festival awards). Always more than just a setting in Sz.'s films, Budapest was a 'protagonist' in its own right, as it were, in *Tűzoltó utca 25* (25 Fireman's Street, 1973) and *Budapesti mesék* (Budapest Tales, 1976). The suspense film *Bizalom* (Confidence, 1979) then the Oscar-winning *Mephisto* (1981), followed by *Redl ezredes* (Colonel Redl, 1984) and *Hanussen* (1988), marked a change in style, being studies of self-delusion and illustrations of the constraints that arise from compromise with corrupt authority. *A napfény íze* (Sunshine, 1999), Sz.'s most recent project, is the story of three generations of a Hungarian Jewish family, depicting the inherent conflicts between the desire to assimilate and to preserve one's identity.

Szabó, Lőrinc (1900–1957)

Poet. Sz. early on blended his intensely personal style, emotional yet intellectual, with the new perceptions of reality brought in by Expressionism, seen most clearly in his first volume, *Föld, erdő, isten* (Earth, Forest, God, 1922) but also in later poems such as *Mérget! Revolvert!* (Poison! Pistol!) from the volume *A sátán műremekei* (The Masterpieces of Satan, 1926). Subsequent disillusionment, on display in *Különbéke* (Private Truce, 1936), led him to philosophical musings on man's helplessness and his subordination to higher, external laws. Sz.'s seemingly objective style, its logical qualities and the realism implicit in his method increasingly inclined him towards narrative poetry. *Tücsökzene* (Cricket Music, 1947) comprises a loosely connected sequence of 352 extended sonnets that describe everyday episodes in a mode reminiscent of Proust. In his late poems, the rhymes are expressive of an outward order and composition but inner turbulence; they are given their own value, as the effect of the poem is centred on the ear, thus making it more musical, more vivid. These poems are no longer concerned with the disillusioned acceptance of the ruthless nature of life but rather with experiencing its joys. His last collection of 120 regular sonnets, *A huszonhatodik év* (The Twenty-sixth Year, 1957), inspired by Shakespeare, was dedicated to the memory of a 25-year love affair.

Szalay, Gyöngyi (1968)

Fencer. Certificates as coach (1990) and sports manager (1995). Seven golds at World Championships with epée team (1989, 1991–93, 1995, 1997, 1999) and a silver in the individual event (1995); individual bronze at the Olympics (1996).

Széchenyi, István (1791–1860)

Statesman, political writer. Sz. was called by Lajos Kossuth "the greatest Hungarian". As a result of his foreign travels, especially in England, he came to regard the introduction of economic reforms in Hungary as essential, expounding his aims in a series of influential political and economic works: *Lovakrul* (On Horses, 1828); *Hitel* (Credit, 1830), which emphasises the importance of curtailing legal restrictions on selling land or using it as collateral to raise liquid capital for investment in business enterprises as well as a host of other reforms, including the abolition of guilds, further clarified in *Világ, vagy is felvilágosító töredékek némi hiba s előítélet eligazítására* (Light, or Enlightening Fragments on Some Mistakes and Prejudices, 1831) and *Stádium* (The Stage We Are At, 1833), which argues for key legal reforms, including bringing in equality before the law. On the practical front, Sz. made the initial donation to set up the Hungarian Academy of Sciences, started the building (1842–48) of the first permanent bridge across the Danube between Buda and Pest (the Lánchíd, or Chain Bridge), instigated regulation of the lower Danube and Tisza rivers and the introduction of steam navigation on these waterways. *A Kelet népe* (The People of the East, 1841) attacked Lajos Kossuth for, as Sz. claimed, endangering Hungary's peaceful civic transition—misgivings that he felt were vindicated by Austria's suppression of the War of Independence in 1848–49. Having suffered a nervous breakdown in September 1848, Sz. withdrew to a mental hospital at Döbling, near Vienna. Having recovered, he penned a final important work, *Ein Blick auf den anonymen Rückblick* (1857), to pour scorn on the thesis of *Rückblick*, written anonymously by Alexander Bach, the Austrian Minister for Home Affairs, that the intervening decade of absolutism had been necessary and successful. The ensuing police harassment led Sz. to commit suicide. The mainly German-language diary that Sz. kept during his years at Döbling has been a continuing source of fascination since being decoded, translated and published during the 1930s.

Széchy, Tamás (1931)

Coach. Certificates as coach (1970) and master coach (1973). Set up the Central Sports School (1966); as head coach of its swimming section (1966–76), he developed coaching methods for age-group swimming. Captain (1976) and from 1977 head coach of the national swimming team, bringing on a dozen

Olympic and World champions. Recipient of State Prize (1988), the Officers' Cross of the Order of Merit of the Republic of Hungary (1992) and the Béla Bay Prize (2001).

Szegő, Gábor (1895–1985)

Mathematician. A student of Fejér and Eötvös. As an undergraduate at Budapest University, he was doing military service when he obtained his doctorate from Vienna. Due to his Jewish origins, unable to secure permanent employment in Hungary after the war, he moved to Berlin, but was similarly compelled to leave there for the USA, where György Pólya helped him to get a job. He held a chair at Stanford University between 1938 and 1953, where he established one of the world's top mathematical centres, with Pólya rejoining him there in 1940. The two prepared the well-known *Stanford Mathematical Problems Book.* As well as work on complex functions, harmonic functions and Fourier analysis, P. studied the problems of Toeplitz forms and, using these as a starting point, made major contributions to orthogonal polynomials. Amongst his published works are *Orthogonal Polynomials* (1939), *Toeplitz Forms and Their Applications* (with U. Grenander, 1955), and *Isoperimetric Inequalities in Mathematical Physics* (with G. Pólya, 1980).

Székely, Bertalan (1835–1910)

Painter. A master of almost virtuoso skill, best known for his vast canvases and frescoes on historical subjects but also an outstanding painter of portraits, genre pictures, and nudes. He studied at the Viennese Academy of Art with K. Rahl and P. J. Geiger then from 1859 at Munich with Karl von Piloty. That was where he completed his *Önarckép* (Self-Portrait, 1860), *II. Lajos holttestének megtalálása* (King Louis II's Body is Found After the Battle of Mohács, 1861) and *Dobozi Mihály és hitvese* (Mihály Dobozi and His Spouse, 1861), which prompted an invitation to return home. In 1863 he was given state funding to go to Munich again, where he painted *VII. Károly búcsúja* (The Deathbed Farewell of Emperor Charles VII) for the Bavarian National Museum. After a study trip paid for from the fee for that, Sz. completed *Egri nők* (The Women of Eger, 1867), probably his most famous work. A whole series of historical studies, such as *V. László és Czillei* (King Ladislas V and Czillei, 1870), along with many portraits, ensued during the following decade, after which fresco painting gained precedence. Amongst the latter were murals in the Pension Institute of the National Railway, the vestibule of the Royal Opera House, the cupola for the mausoleum of Ferenc Deák (all in Budapest) and the chapels of Pécs Cathedral. Meanwhile he set down his artistic credo in book form, in *Figurális rajz és festés elvei* (Principles of Figurative Drawing and Painting, 1877) and *Festészeti bonctan* (Anatomy for Painters, 1881). From 1871 he taught at the Royal Hungarian School of Design, Budapest, becoming its director in 1903 and then heading its second master school from 1905. His pupils included Jenő Gyárfás, Aladár Körösfői-Kriesch, Sándor Nagy, and János Vaszary.

Székely, Éva (1927)

Swimmer. Qualified pharmacist; certificates as coach (1965) and master coach (1969). Won Olympic gold in 200 meters breaststroke (1952) and set many world records. First Hungarian women's champion in butterfly over 100 and 200 meters (1949), she set first national records in freestyle over 1,500 meters (1950), in butterfly over 100 meters (1950) and 200 meters (1947), and in individual medley over 400 meters (1953). Inducted into the Hungarian Swimmers' Hall of Fame (1976) and the Hungarian Hall of Fame (1994). Coached her daughter, the swimmer Andrea Gyarmati. Writer on swimming.

Székely, Mihály (1901–1963)

Bass. He made his debut in 1923 in the Budapest Opera House and made many appearances in New York (1946–50) and at the Glyndebourne Festival (1957–62). Apart from his roles in the operas of Mozart, Verdi and Wagner, he was an incomparable interpreter of Boris Godunov (Mussorgski), Baron Ochs (Richard Strauss: *Der Rosenkavalier*) and Bartók's Duke Bluebeard.

Székely, Zoltán (1903–2001)

Violinist. A pupil of Hubay at the Budapest Academy of Music, he moved to the Netherlands in 1922. He often played with Bartók in sonata recitals and indeed gave the world premiere performance of Bartók's *Concerto for Violin No. 2* in Amsterdam (1939). Moved to the US in 1950. Since 1975 he has resided in Banff, Canada. In 1937 he formed the Hungarian String Quartet, which was renowned for its performances of the works of Beethoven and Bartók, leading it until his retirement in 1972.

Szekeres, Pál (1964)

Fencer. Graduated from the College for Foreign Trade (1997); coach's certificate (1990). Won an Olympic bronze with foil team (1988). Handicapped since a car accident, he has won golds for individual foil at European (1991, 1995, 1999) and World (1994) Paraplegic Championships and the Paralympics (1992, 1996). Board member of the Fencers' Interest Group (1991); vice-chairman of the Hungarian Sport Federation for the Handicapped (1992–); board member of the International Wheelchair Fencing Federation (1996–). Is deputy state secretary in the Ministry of Youth Affairs and Sport (1999–).

Szell, George [Széll, György] (1897–1970)

Conductor. He studied in Vienna. He worked in Berlin and Düsseldorf, then at the German Theatre at Prague (1929–37).

He moved to the US in 1939, working at the New York Metropolitan Opera (1942–46) before embarking on his long association with the Cleveland Orchestra as its chief conductor (1946–70).

Szentágothai, János (1912–1994)

Neurologist. Studying medicine under Lenhossék at Budapest, it was Sz. who provided the first physiological evidence to corroborate the principle of neuronal separation. He went on to clarify the innervation of eye movements and the role of the vestibular receptor organs in the inner ear, and was the first in Europe to utilise stereotaxic surgery in brain research. In 1946, he was given a chair at Pécs University, where he headed the Institute of Anatomy. Amongst highlights of his post-war research were the discovery of the monosynaptic masseter reflex (1948), and major contributions to the functional anatomy of the cerebellum and the medulla oblongata, being the first to adopt cybernetic approaches in the study of nerve centres and reflex mechanisms. An early pioneer of neuroendocrinology, his team showed that hormone production by the anterior lobes of the pituitary gland is regulated by peripheral endocrine organs indirectly via the hypothalamo-hypophyseal tract. A member of the Hungarian Academy of Sciences, he served as its president for several terms. His publications included *Atlas of Human Anatomy* (with Ferenc Kiss, 1946), *Die Rolle der einzelnen...* (1952), *The Cerebellum as a Neuronal Machine* (with J. C. Eccles and M. Ito, 1967), and *Funkcionális anatómia* (Functional Anatomy, 1971).

Szentgáli, Lajos (1932)

Athlete. Coach's certificate (1968). Middle-distance runner, representing the country 50 times. European champion over 800 meters (1954). Police lieutenant colonel at the Ministry of the Interior (1950–88).

Szent-Györgyi, Albert (1893–1996)

Biochemist. After qualifying in medicine at Budapest (1917), he began his research career in Cambridge, England, working with Hans Krebs on unravelling the biological oxidation pathway. He returned home in 1931 to take a chair at the University of Szeged, where he discovered the anti-scorbutic properties of a substance now called ascorbic acid, thereby identifying the chemical principle of Vitamin C. For this and the earlier work with Krebs he was awarded the Nobel Prize for Medicine in 1937. He went on to show that the formation of an actomyosin complex was the molecular basis of muscle contraction. In 1943–44 he played an active political role in trying to induce the Hungarian government to pull out of World War II. Leaving Hungary in 1947, he carried on his researches at Woods Hole in the USA, concentrating mainly on biophysical problems of electron transfer in living cells, including their role in cancer growth.

Szervánszky, Endre (1911–1977)

Composer. Completed the composition course at the Academy of Music in 1936. He was professor at the Academy from 1949 until his death. His compositions were initially influenced by Kodály then from the mid-fifties by Bartók, but after 1959 he took a leading role in assimilating more recent trends in European contemporary music. His most important works are the *Concerto József Attila emlékére* (Concerto in Memoriam Attila József, 1954) and *Hat zenekari darab* (Six Pieces for Orchestra, 1959).

Szigeti, József (1892–1973)

Violinist. He was a pupil of Hubay, in 1905 he started a solo career. Following an early virtuosic career, he achieved fame mainly as performer of sonatas and concertos. He was closely associated with Bartók, Busoni, Prokoviev and Stravinsky, among others, premiering many of their works for violin. He settled in the US in 1925, from 1951 he lived in Switzerland.

Szilard, Leo [Szilárd, Leó] (1898–1964)

Physicist. Though he began studying architecture at Budapest Technical University, he graduated as a chemist at Berlin Technische Hochschule before moving over to physics. Whilst there, in 1929 he wrote a seminal work on the correlation of intelligence, of information, and the 2. main theorem of the thermodynamics, which is today considered to be the starting point of information technology and brain research. He moved to England in 1933, then later to the USA. He believed in the possibilities of neutron multiplication, which he proved through experiments on uranium fission. He was one of the key scientists (with Einstein, Fermi and Wigner) who persuaded Roosevelt to authorise the establishment, in August 1942, of the Manhattan Project to develop the atomic bomb at Oak Ridge, Tennessee, where his own work was on the controlled release of nuclear energy. After the war, at Chicago, he devoted his energies to the biological applications of physics, thus helping to found the discipline of biophysics. Three volumes of his *Collected Works* have been published (1972, 1978 and 1987).

Szinyei Merse, Pál (1845–1920)

Painter. Introduced the principle of *plein-air* painting in Hungary without any prior knowledge of Manet or Monet. He studied first with Lajos Mezey at Nagyvárad (Oradea, Romania) before entering the Munich Academy of Art in 1864, where Karl von Piloty, of all people, encouraged him to paint outdoors. He freed himself from literary subjects by finding inspiration in everyday scenes, as in *Anya és gyermekei* (Mother and Children), *Vitorlás a starnbergi tavon* (Yacht on Lake Starnberg), etc. *Vadgesztenyefa* (Horse-Chestnut Tree, 1869) brought him close to Courbet's vision of nature, whilst his colour sketches show kinship with the Impressionists. *Gyermekek a domboldalon* (Children on a Hillside) is a precursor of his *Picnic in May* but in the meantime he had a spell back home before returning to Munich in 1872, where he again threw himself into capturing the direct experience of nature. Sz. M.'s most famous work, *Majális* (Picnic in May, 1873), is in fact a reworking of the popular French subject of the outdoors breakfast. He successfully harmonised light and shade with his discovery of the unifying effect of sunlight and atmosphere, the informality of the figures engendering a relaxed, fresh mood. However, this masterpiece attracted no attention at the time and, losing heart, Sz. M. turned to the Romanticism of Böcklin for a while before painting his other famous canvas, *Lilaruhás nő* (Woman in Violet, 1874). With lengthening gaps between works, *Pacsirta* (The Skylark, 1882) represented a retreat to humdrum naturalism. *Picnic in May* was a belated triumph when shown at the 1896 Millennial Exhibition, and the late works include the outstanding *Hóolvadás* (Snow Thawing, 1895).

Szirmai, Albert (1880–1967)

Operetta composer. He studied in Budapest, where his first works were staged. His operettas were more in the English than the Viennese style. After moving to New York in 1926 he continued composing at first but then went into business as a music publisher. His main work is *Mágnás Miska* (Max the Magnate, 1916).

Szokolay, Sándor (1931)

Composer. Completed the composition course under Ferenc Farkas. He is known principally as a composer of operas, exploring various genres and styles of expression. His main works are *Vérnász* (Blood Wedding, 1964), *Hamlet* (1968) and *Savitri* (1998).

Szondy, István (1925)

Modern pentathlete. Certificate as PE instructor (1950). At Olympics won gold in the team event and individual bronze (1952) and at World Championships a silver (1953) and bronze in the individual and two golds in the team event

(1954, 1955). Has lived in (West) Germany since 1956 and captained its national modern pentathlon team (1969–70).

Szőke, Katalin (1935)

Swimmer. Won golds at the Olympics in 100 meters freestyle (1952) and in the 4 x 100 meters freestyle relay (1952). Set several world records. Has lived in the USA since 1956.

Szőkefalvi-Nagy, Béla (1922–1998)

Mathematician. He studied at Szeged University, where his father, Gyula Sz.-N. (1887–1953), a distinguished geometrist was a lecturer at the local teacher's training college and succeeded his father there when the latter secured a university chair. His early interests included quantum mechanics, which led him to study the linear operators of Hilbert space, the subject of a monograph published in German in 1942 (later transl. as *Unitary Dilations of Hilbert Space Operators*). He was awarded the Kőnig Prize the same year, elected to the Hungarian Academy of Sciences in 1945, and became a professor at Szeged, in 1949, as successor to his mentor Frigyes Riesz, with whom he wrote the fundamental textbook *Functional Analysis* (1955), which was followed up by a further volume co-written with the Romanian C. Foias.

Szőllősy, András (1921)

Composer, musicologist. He studied initially with Kodály at the Budapest Academy of Music then with Goffredo Petrassi in Rome. After composing a considerable amount of music for films, he began to emerge as a composer for the concert hall in the late 1960s. As a musicologist he has made important contributions with his editions of Bartók's writings and his systematic cataloguing of Bartók's works (Sz. numbers). His main works are *Concerto no. III* (1969) and *Elegia* (1993).

Szőts, István (1912–1998)

Director. Strongly influenced by his Transylvanian roots, Sz. attracted early attention with a short film adaptation of a Hungarian folk ballad. With its blend of folkloristic subject and modern cinematic techniques, *Emberek a havason* (Men in the Alps, 1942), his first full-length feature film, was a unique artistic success amongst pre-war Hungarian films and picked up a major award at the 1942 Venice Film Festival. The next film, *Ének a búzamezőkről* (Song of the Cornfields, 1948), was banned by the Communist régime that was in sole power by the time it was ready for release. Branded an 'ideological enemy', Sz.'s work was limited to a few short films such as *Kövek, várak, emberek* (Stones, Castles, People, 1954) and *Melyiket a kilenc közül?* (Which of the Nine?, 1957). Choosing not to return to Hungary after the Venice Film Festival in autumn 1957, Sz. eventually settled in Austria, becoming a director of respected popular science and nature films. His artistic potential was thus largely frustrated, but his slender *oeuvre* was still a major influence on younger filmmakers such as Miklós Jancsó, Ferenc Kósa and István Gaál.

Szrogh, György (1915–1999)

Architect. On graduating from Budapest Technical University (1938), he worked in the office of Lajos Hidasi. After the war he refashioned two damaged villas on György Dózsa út in Pest, facing the City Park, into the *Headquarters of the Miners' Trade Union House* (1949) and also did much of the detailed design work for the team under Lajos Gádoros that planned the nearby *Office Building and Assembly Hall for MÉMOSZ*, the Hungarian Construction Workers' Trade Union (1947–49). From 1948 he worked first at LAKÓTERV, the state planning office for housing, then at KÖZTI, the state planning office for public buildings, which offered the chance to undertake study-trips to England and, on a UNESCO scholarship as an expert in school planning, to the USA and Mexico. Amongst his executed designs are: a *Movie Theater for Dunaújváros* (1951–53); *Csepel Police Station* (1953); *Medical Surgeries at Oroszlány and Komló* (1955); the *Nurses' Hostel at No. 8 Frankel Leó utca* (1957–59) on the

Buda embankment facing Margaret Island; *Schools at Dunavarsány and Mogyoród* (1958–61); the *Observatory and Institute of Astronomy Research Center on Piszkéstető* (1960–1964), in the Mátra Hills; a *Residential Home at Nagyszénás* (1962–65); *Salgótarján Community Centre* (1962–66); and the *Budapest Hotel on Szilágyi Erzsébet fasor* (1964–67), the distinctive cylindrical tower block at the terminal of the cogwheel railway up to Széchenyi Hill. In 1966 Sz. was appointed head of the Architecture Department at the Academy of Applied Arts, from which he retired officially in 1976 though continuing to teach there until 1983. His small-scale projects include: two *Artists' Summer Houses at Szigliget* (1972), by Lake Balaton; the Collection of History of Photographic Techniques in Vác (1985). Sz. was awarded a State Prize in 1968 and was twice a recipient of the Ybl Prize.

Tábori, László (1931)

Athlete. Medium- and long-distance runner. Fourth placed at Olympics over 1,500 meters (1956), having set world records at the distance (1955) and as a member of 4 x 1,500 meters relay squad (1953, 1954, 1955). Has lived in the USA since 1956, winning at the US Nationals.

Tarján, Rezső (1908–1978)

Computer engineer. A pioneer of computer and information technology, his career was dramatically broken by imprisonment for political reasons during the Rákosi era. On rehabilitation, he became head of Institute for Metrology and Instrumentation, established in 1955 and soon after (1957) of a new Cybernetics Research Group within the Hungarian Academy of Sciences, where he played a leading role in the design, construction and research application of the first Hungarian computer, the M3. Later he became a leading consultant to the National Technical Development Committee (OMFB), which played a significant role in the development of economy in Hungary.

Tarr, Béla (1955)

Director. Initially an amateur film-maker, his first projects were experiments in a blend of fiction and documentary that used non-professional actors. T.'s approach and style—tense, slow-moving fictions based on strongly literary texts—were crystallised in his work on *Őszi almanach* (Autumn Almanac, 1984) and *Kárhozat* (Damnation, 1988). *Sátántangó* (Satan's Tango, 1994), his biggest project to date, is striking not purely on account of its sheer seven and a half hours duration and apocalyptic feel (the climactic scene is a *danse macabre* in a village setting), but through the paradoxical effect that is generated by having the film's action run in real time. The subsequent *Werckmeister harmóniák* (Werckmeister Harmonies, 2001), a story about the closed community of a desolate provincial town who, having been cheated by a fraudulent mystic, prepare to vent their frustrations by lynching one of their number, has met with considerable acclaim at film festivals and art-house movies.

Tass, Olga (1929)

Gymnast, coach. Certificates as PE instructor (1951), coach (1978) and master coach (1981). Won Olympic gold as a member of the hand apparatus team (1956). Head coach to the French women's national gymnastics team then coached in Montreal (1961–68). Instructor at the Budapest College of Physical Education (1968–85); coached Hungarian women's national gymnastics team (1969), club coach (1969–74). International judge (1970–89). Now teaches the art of movement (1997–).

Teller, Edward [Ede] (1908)

After studying at the Budapest and the Karlburg Universities, he went to the US in 1928. He worked at the Washington University, then later at Berkley. He was in the team that developed the nuclear and later the hydrogen bombs. He elaborated the theory of thermonuclear reaction based on quantum mechanics. He proved that the normally repelling light weight nuclei could fuse at high temperatures. Over the last thirty years he has been a prominent advocate of high-technology defence systems within the USA. His publications include *Nuclear Structure* (with L. Eisenbud, 1958), *Energy* (1979), and *Better a Shield Than a Sword: Perspectives on Defence Technology* (1987).

Temes, Judit (1930)

Swimmer, sports administrator. Qualified physician (1955) and lawyer (1999). Won gold at Olympics as member of the 4 x 100 meters freestyle relay squad (1952). Set several world records. Member of the Hungarian Olympic Committee (1961–97); member of the Technical Committee of the European Swimming League (1976–94). She is now a patients' rights legal representative.

Than, Károly (1834–1908)

Chemist. Having studied medicine and pharmacy at Vienna, T. went on to learn chemistry under Bunsen at Heidelberg, obtaining his doctorate in Vienna (1859) and staying on there as a lecturer until he secured a chair at the University of Pest in 1862. He ventured into almost all the fields of chemistry, his name is still alive in two analytical measuring solutions. He was one of the first to use spectrum analysis. He urged for and partially designed the new, very modern chemistry building of the university. He was a member of the Hungarian Academy of Sciences, being vice chairman at one time. His major works are: *A qualitativ chemiai analysis elemei* (Elements of Qualitative Chemical Analysis, 1895), *Kisérleti chemia elemei* (Elements of Experimental Chemistry, 1898).

Tihanyi, Lajos (1885–1938)

Painter. A deaf-mute from the age of eleven, he was largely self-taught but received early training in the Nagybánya school, exemplified by *Nagybányai utcarészlet* (Street Detail—Nagybánya, 1908). A short trip to Paris in 1907 acquainted him with Cézanne's formal innovations and Fauvist colours—palpable influences on works like *Vörösinges fiú* (Boy in Red Shirt, 1909), *Erdős táj* (Wooded Landscape, 1911) and *Önarckép* (Self-Portrait, 1912). On his return to Budapest, T.'s works were included in the 1911 exhibition of a group of young artists known as The Eight. He then fell under the influence of German expressionism, notably Kokoschka's psychological sensitivity, as evidenced by *Kassák Lajos portréja* (Portrait of Lajos Kassák, 1918) or *Önarckép-tanulmány* (Study for a Self-Portrait, 1920). After the fall of the 1919 Commune, T. emigrated to Vienna, gaining a show at the Moderne Galerie as early as March 1920, which critics acclaimed, ranking him alongside Picasso, Munch and Kokoschka as a representative of modern expressionism, and which led to an exhibition of his collected works in Berlin in 1922. Moving to Paris the next year, he quickly struck up close friendships with many Hungarian expatriates as well as French avant-garde writers and musicians. Pictures from a 1925 Parisian one-man show were subsequently bought and exhibited across the world. Stylistically, his paintings of still-life subjects became increasingly abstract. After his death from meningitis, T.'s artistic estate passed to Brassaï, Jacques de la Fregonnière, and his cousin, Ervin Marton, but in the '60s was donated by them to the Hungarian state.

Tordasi, Ildikó (1951)

Fencer. Coach's certificate (1985). Olympic gold in individual foil (1976) and silver in team event (1972). Hungarian Woman Fencer of the Year (1973, 1976–77, 1979) and Sportswoman of the Year (1973, 1976). Club coach (1986–).

Török, Ferenc (1935)

Modern pentathlete, sports administrator. Qualified lawyer (1957); certificates as coach (1966) and master coach (1984). Golds at Olympics in individual (1964) and team (1968) events and four times in the team event at World

Championships. Captain of the Hungarian men's modern pentathlon team (1976–89, 1991–92). Board member of the Hungarian Olympic Committee (1989–97); chairman of the Hungarian Modern Pentathlon Federation (1990–96); vice-chairman of the International Modern Pentathlon Federation (1993–96). Member of Parliament (1990–94). Recipient of the Order of the Star of the Republic of Hungary (1988) and International Olympic Committee's Olympic Order (1995).

Török, Gyula (1938)

Boxer. Olympic flyweight champion (1960). Coach of the boxing section (1976–81). Second coach of the Hungarian national boxing team (1993–). Recipient of the 'For Budapest' Prize (1994).

Turán, Pál (1910–1976)

Mathematician. After studying with Lipót Fejér at Budapest University, where he obtained his doctorate in 1935, he worked as a private tutor then between 1938 and 1947—interrupted by conscription into a forced labour camp during World War II—as a secondary school teacher. Having secured a post as unsalaried lecturer at Budapest University in 1945, he spent several years teaching abroad (Copenhagen, Princeton), but from 1949 until his death held a chair at Budapest University. In addition to a significant pre-war contribution to extremal problems in graph theory, T. is remembered chiefly for developing the sum of powers method in 1949 to tackle the Riemann hypothesis of prime number theory, publishing some fifty papers on its wider applications. A member of the Hungarian Academy of Sciences, his major published work was *Az analízis új módszere és alkalmazása* (A New Method and Application of Analysis, 1953; expanded English, 1984).

Urányi, János (1924–1964)

Kayaker. Golds in kayak doubles at Olympics over 10,000 meters and 500 meters (1956), and at European (1957) and World (1958) Championships over 10,000 meters. Fire-fighter officer.

Uzsoky, Miklós (1925–1995)

Electrical engineer. An innovative engineer, he began his career in telecommunications, making a name for himself as designer of the first Hungarian television transmitter (1956) and obtaining several patents, before joining the Computer Technology and Automation Research Institute (SZTAKI) in the 1960s. There, he and his colleagues were the first in Hungary to implement a computer-aided design and manufacturing system for printed circuits.

Vágó, József (1877–1947)

Architect. On graduating from Budapest Technical University (1900), he gained initial practice working with Ignác Alpár, Zsigmond Quittner and Ödön Lechner, meanwhile making study trips to Italy and France, before opening his own office in partnership with his brother László (1902–11). Though they aspired to implement Lechner's ideas on national character, their buildings seem more obviously influenced by the Austrian Secession: the four-storey *Apartment House at Bartók Béla út* (1902–03), in Buda's Lágymányos (eleventh) district; the majolica-clad white *Arcade Bazaar on Dohány utca*, at the corner with Síp utca (1907–09), in Pest's seventh District; the *National Saloon*, an exhibition building in Pest's inner-city Erzsébet tér (1906–07, demolished 1959); the *Apartment House at No. 4 Gutenberg tér* (1906–07), in Pest's Józsefváros (eighth) district, which was built for the Hungarian Printers and Typesetters' Charity Fund; and, overlooking the City Park, at the corner of the Ajtósi Dürer sor and Dózsa út, the *Feld Theater* (1908–09), which was demolished in 1948 to make way for the widening of the latter into the 'parade' route of today. For accepting the post of president of the Directorate for Construction during the 1919 Hungarian Soviet Republic V. was obliged to emigrate to Switzerland after the revolution failed, later moving to Italy and finally France. Working under a pseudonym, he continued to produce designs for Hungary, including the three-story modern *Apartment House at No. 48 Városmajor utca* (1933–34), which since 1977 has housed the '*Nyugat*' Museum of Literature. His books are: *Városokon keresztül* (Passing Through, 1930) and *Budapest művészeti újjáépítése* (The Artistic Reconstruction of Budapest, 1936).

Vajda, János (1827–1897)

Poet, journalist. As a youth, V. joined a group of strolling players in 1845, then briefly became a tutor before moving to Pest in 1847, where he joined the 'Young Hungary' circle round Petőfi. After the War of Independence he was enlisted into the Austrian army. His first poems, mainly published in *Életképek* and later *Pesti Divatlap* under Petőfi's editorship, were strongly influenced both by the latter's popular tendency and by Romanticism. V. supported the Independence Party, and the two poems *Luzitán dal I.–II.* (Lusitanian Song, 1867) reject the Compromise of that year. His love poetry was inspired by Gina Kratochwill, for whom V. wrote the cycles *Sirámok* (Laments, 1854), *Szerelem átka* (The Curse of Love, 1855) and *Gina emléke* (The Memory of Gina, 1856), later revisited in the outstanding collection *Húsz év múlva* (Twenty Years Later, 1876). From the 1870s on V. turned to philosophical poetry, e.g. in *Végtelenség* (Infinity) and *A vaáli erdőben* (In the Forest of Vaal) from 1875, also writing two novels in verse: *Alfréd regénye* (The Novel of Alfred, 1875) and *Találkozások* (Encounters, 1877).

Vajda, Lajos (1908–1941)

Painter. One of the outstanding painters of the inter-war years, V.'s early schooling was obtained in the new Serbian state before his family moved to Szentendre in 1923. At the Budapest Academy of Fine Arts (1928–29) his teachers included István Csók and János Vaszary, and he banded with other 'young progressives'—amongst others, György Kepes,

Dezső Korniss and Béla Veszelszky—who in 1930 joined Lajos Kassák's *Munka* (Work) circle. Works of this period, such as *Lelátás a templomdombról* (View from Hilltop with Church, 1929), were in a Cézannesque cubist manner. During a spell in Paris (1930–34) V. became familiar with Russian constructivism and films, leading him into photographic collages. Back in Hungary, a 1936 letter to his wife, Júlia Vajda (Richter), mentions V.'s plan to work with Dezső Korniss on creating 'a unique, new, East-Central European form of art' based on Hungarian folk art, in emulation of what Bartók and Kodály had done in the field of music. Accordingly, he began to collect the motifs of folk and Eastern Orthodox art and architecture that were used in icon-based portraits like *Felmutató ikonos önarckép* (Icon Self-Portrait Pointing Upward, 1936) and works like *Szentendrei házak feszülettel* (Houses with Crucifix—Szentendre, 1937) or *Fej házzal* (Head with House, 1937), where the motifs are reduced to the overlapping or interwoven simple outlines that yield broad surfaces. In V.'s final years, the near-surreal associations of such objects gave way to an imagery that was more expressively visionary, as in *Maszk holddal* (Mask with Moon, 1938), and even frankly apocalyptic, as in *Szörny kék térben* (Monster in Blue Space, 1939), which reflects growing apprehensions about fascism.

Vasarely, Victor [Vásárhelyi, Győző] (1908–1997)

Visual artist. Having begun to study medicine in Budapest, V. quit in 1927 to enter art school. He started to take an interest in optical illusions in 1928–29 whilst frequenting Sándor Bortnyik's graphics workshop in Budapest. On moving to Paris in 1930, he made a living at first out of poster designs. The initial phases of V.'s painting career are represented by the works *Denfert* (1938–58), the early abstract-geometric *Belle-Isle* (1947–54), *Yellow Manifesto* (completed in 1955), *Crystal*

(1948–60), and *Black and White* (1955–63). V. became established internationally during the '60s as one of the acknowledged pioneers of variation and permutation art, kinetic art and op art in particular, gaining a Gold Medal at the Milan Triennial, and the Grand Prix at both Ljubljana and Sao Paulo in 1965. According to his utopian vision of 'planetary folklore', art—and algorithmic geometrical art in particular—should serve a social function, as evidenced, for instance, by the murals designed for the University of Caracas and his advocacy of replicated series (multiples) of works of art. In 1968 he had a successful show at Budapest's Exhibition Hall. V. donated two collections to Hungary, including pieces of his own and works by others (the Vasarely Museums in Pécs and Budapest), as well as founding a pedagogical museum at Gordes in France.

Vásárhelyi, Pál (1795–1846)

River engineer. Having graduated from the Institutum Geometrico-Hydrotechnicum of Pest University in 1816, V. primarily used his expertise as a surveyor, preparing hydrological maps of the Kőrös river system and the Danube (from 1826). He then became involved in projects to make the lower Danube navigable, working with István Széchenyi on a scheme to construct a towpath at the Iron Gates (1832–37). Meanwhile, he also carried out the surveying and planning work for his scheme (1833) to regulate the Tisza, making it navigable, and install reliable flood defences, though implementation of this started only after his death. V. was a member of the Hungarian Academy of Sciences.

Vaszary, János (1867–1939)

Painter. Extensively trained at the Budapest School of Drawing, under Bertalan Székely, then in Munich (1887–89) and finally at the Académie Julian in Paris, with spells in Rome and Paris to follow, V.'s painting career embraced many of the 'isms' of the early twentieth century, starting from a naturalism indebted both to Bastien-Lepage, as in *Pásztorsípot fúvó gyermek* (Child Blowing Shepherd's Pipes, 1896), and to Puvis de Chavannes, as in *Aranykor* (The Golden Age, 1900), then a period painting Hungarian peasant subjects under the influence of Adolf Fényes, and onwards through Art Nouveau, the impressionism of *Reggeli a szabadban* (Breakfast in the Open Air, 1907), the Cézanne-style cubism of *Fürdő fiúk* (Bathing Boys, 1911) and expressionist Fauvism. Working as a war correspondent during World War I, he painted pictures of great dramatic strength like *Vonuló foglyok* (Marching Prisoners of War, c. 1915) before entering successive 'blue' and 'black' periods during the early '20s, leading to a colourism of Japanese lightness in *Parkban* (In the Park, 1928). His versatile and dynamic personality made him a leading figure of the Hungarian avant-garde. As a liberal teacher at the Academy of Fine Arts from the mid '20s onwards, V. had a tremendous influence on students through the support he gave to new trends in art, in part through such books as *Természetlátás és képszerűség* (Seeing Nature and Pictorial Quality, 1922).

Vecsey, Ferenc (1893–1935)

Violinist. A child prodigy, he was a pupil of Hubay (1900–03). He made his debut in Berlin as early as 1903 and in the US in 1905. His virtuosity inspired Sibelius to compose the *Violin Concerto* for him (1903). He toured Europe giving concerts, moving to Italy in 1920.

Végh, Sándor (1912–1997)

Violinist. A pupil of Hubay at the Budapest Academy of Music (1929–31) and later also taught there (1941–46). He then took up residence in Switzerland and formed Végh String Quartet (1940–63). After moving to Salzburg he was the conductor of Camerata Academica there from 1978 until his death. His recordings of Mozart are the ideal realisations of the Central-European performing style.

Veress, Sándor (1907–1992)

Composer. A pupil of Kodály at the Budapest Academy of Music, he was to eventually become his successor as professor of composition (1943–49). In 1949 he took up residence in Bern, Switzerland. Although initially a Bartók–Kodály epigone, he strove to synthesise this with newer trends in Western-European music. His main works are *Szent Ágoston zsoltára az eretnekek ellen* (St. Augustine's Psalm Against Heresy, 1944), *Sinfonia Minneapolitana* (1953), and *Orbis tonorum* (1987).

Verőczi, Zsuzsa (1949)

Chess player. Coach's certificate (1972). Reached women's international grandmaster grades in 1978. Candidate for women's World Championship (1979). Silver medals as member of women's team at Chess Olympics (1969, 1972, 1978, 1980, 1986). Hungarian Woman Chess Player of the Year (1973, 1975, 1977, 1979–86). Member (1976–86) then vice-chair (1986–90) of the Women's Committee of the International Chess Federation. International referee (1995–); head referee at the Women's Chess Olympics (1996). Writer on chess.

Verrasztó, Zoltán (1956)

Swimmer. Qualified physician (1987); coach's certificate (1979). In 200 meters backstroke won gold at World (1975) and European (1977) Championships and silver at Olympics (1980). World record holder in 400 meters individual medley (1976). Hungarian Swimmer of the Year (1977).

Veszelszky, Béla (1905–1977)

Painter. Although an important body of graphic art dating back to the late 1920s (and resumed during his final years) survives, V.'s most characteristic works were paintings deeply shaped by the theories he developed in mid-life about the perception of life as imbedded in infinite space, and how the ineffable relations of that reality might be rendered on canvas by applied dots of pigment. Although the painted motifs may seem almost obliterated in the mass of dots on the canvas, V. insisted that he was depicting strictly what he saw—if anything, highlighting and not hiding his subjects. The resulting paintings can be grouped by subject and period, with *Köcsög, bögre* (Jug, Mug) representative of a 1953–64 still-life phase and *Filodendron és Kukorica* (Philodendron and Corn) from the ensuing 1965–73 phase; *Pusztaszeri úti panoráma* (Panorama—Pusztaszeri St.) from the 1957–65 landscapes; whilst a culminating series of portraits and self-portraits stems from 1967–76. A hole that he dug in his own garden during the '60s, as a kind of 'observatory', can be regarded as a forerunner of land art.

Vígh, László (1941)

Coach. Certificate as PE instructor. Head coach of the Ferencváros Athletics Club (1974–86); assistant professor at the Budapest College of Physical Education (1990). Coached gymnast Zoltán Magyar, working with him to develop new exercises for the pommel horse.

Villányi, Zsigmond (1950–1995)

Modern pentathlete, coach. Certificates as coach (1973, 1978) and master coach (1991). Won silvers at World Championships in individual (1971) and team (1971, 1974) events and in team event at Olympics (1972). Hungarian Modern Pentathlete of the Year (1971). Captain of the Hungarian modern pentathlon national junior team (1980–81); coach to the Italian national team (1982); fencing master to the Hungarian national team (1988–). Helped establish the women's section in this sport.

Vörösmarty, Mihály (1800–1855)

Poet. More than any other single figure, V. helped purify Hungarian language and literature from the previous dominance of Classical and German influences. As a schoolboy and later legal student, V. supported himself by private tutoring. He established his name in 1825 with *Zalán futása* (The Flight of Zalán), an epic poem in hexameters describing Árpád's defeat of Zalán to conquer Hungary, which, apart from its artistic merits, caught an upsurge in patriotic fervour by supplying a nation. He also wrote some minor epic poems, both on historical subjects—e.g. *Cserhalom* (1825) and *Eger* (1827)—and more romantic fantasies such as *Tündérvölgy* (Valley of Fairies, 1826), the incomplete *A Délsziget* (A Southern Isle, 1826), *Rom* (Ruin, 1830), and *A két szomszédvár* (Two Neighbouring Castles, 1831). Initially a regular contributor to Károly Kisfaludy's literary periodical *Aurora*, V. became full-time editor of the magazine *Tudományos Gyűjtemény* (Scholarly Anthology) in 1828 and the first member of the newly founded Hungarian Academy in 1830. V.'s early works for the stage were influenced by French and German romantic drama, but in 1830 he produced a symbolic fairy tale, *Csongor és Tünde* (Csongor and Tünde, 1831) and was later also an important Shakespeare translator (*Julius Caesar*, 1840; *King Lear*, 1853). V. is now valued, above all, for his shorter poems, among others, *Szózat* (Appeal, 1836), the second national poem after Kölcsey's *Himnusz*, *Gondolatok a könyvtárban* (Thoughts in a Library, 1844), and *Az emberek* (On Mankind, 1846), though some of the greatest lyrics, such as *A merengőhöz* (To a Day-Dreamer; 1843), were inspired by his wife, Laura Csajághy. Having been a parliamentary representative, V. was to go into temporary hiding after Hungary's defeat in the War of Independence (1849). Though still producing occasional splendid poems, such as *A vén cigány* (The Old Gipsy, 1854), he was unable to resume his earlier role in public life.

Wartha, Vince (1844–1914)
Chemist. W. began his studies in Budapest but qualified in Zurich and returned to Hungary only in 1870 to accept a new chair in chemical engineering at what was then still the Polytechnic School of Buda and became later dean and rector of the Technical University. A versatile scientist, his research topics ranged from the chemistry of wines and photochemistry to the raw materials for manufacturing gases and water chemistry (his method for determining water alkalinity still finds use today). However, he is remembered chiefly for his work during the early 1890s in developing the eosin red glaze employed by the Zsolnay factory in its ceramic wares. A member of the Hungarian Academy of Sciences, his published works include *Über eine einfache Methode zur Bestimmung der temporären Härte des Wassers* (1880), *Az anyagművesség* (Shaping Materials, 1905), and *Chemiai technológia* (Chemical Technology, 1906).

Weiner, Leó (1885–1960)
Composer, teacher. Although his career started with Bartók and Kodály, as a composer he preferred a light, French-influenced sensibility. He was appointed professor of chamber music (1923–60). Several generations of instrumental musicians and conductors remember him as their greatest master.

Weöres, Sándor (1913–1989)
Poet, translator, playwright. Recognised as a talented poet even as a young teenager, W. studied law then Hungarian language at Pécs University, where his doctoral dissertation on *A vers születése* (The Birth of the Poem, 1939) is significant both as a theoretical treatise and as a meditation of a poet. His reputation as a major innovator was secured before the war by *Hideg van* (It is Cold, 1934) and several further volumes, including *A teremtés dicsérete* (In Praise of Creation, 1938). It was consolidated further with subsequent collections: *Meduza* (Medusa, 1944), *A fogak tornáca* (The Colonnade of Teeth,

1947), *A hallgatás tornya* (The Tower of Silence, 1956), *Tűzkút* (Well of Fire, 1964) and *Merülő Saturnus* (Saturn Submerging, 1968). W. was attracted to a spiritual metaphysic which united ancient Chinese, Japanese, Christian and Mesopotamian myths, religions and epic tales. These themes often reflected his translation work, which was published in collected form as *A lélek idézése* (Conjuring Up the Soul, 1958). W.'s poetry ranges from tight syntactic construction to loosely woven free association, often yielding verse with the impact of a musical piece. In *Psyché* (1972) W. masqueraded as an imaginary poetess, Erzsébet Lónyai, creating for her a stylistically convincing *oeuvre* that is interwoven with authentic verses by overlooked poets at the turn of the eighteenth into the nineteenth century. A personal anthology of older Hungarian verse, *Három veréb, hat szemmel* (Three Sparrows, Six Eyes, 1977), was likewise an eye-opening antidote to the generally promoted canon.

Wigner, Eugene [Jenő] (1902–1995)

Physicist. Although qualifying as a chemist at Budapest University, he was more interested in quantum physics; his doctoral dissertation at Berlin is considered the first work in quantum chemistry. With J. Neumann he worked out the application of group-theory to quantum theory, but his major discovery—recognised in the joint award to him of the 1963 Nobel Prize for Physics—was the principle of basic symmetries. He played an important role, alongside Fermi and Szilard, in the construction of the first nuclear reactor, and as director of the Oak Ridge Laboratory was closely involved in the design of early high-performance plutonium reactors. His theoretical and practical works were very closely linked.

Wladár, Sándor (1963)

Swimmer. Won golds in backstroke at Olympics over 200 meters (1980) and at European Championships over 100 and 200 meters (1981), setting European records over 200 meters

(1980, 1981). Hungarian Swimmer of the Year (1979–83) and Sportsman of the Year (1981). Since 1995 manages a private veterinary clinic.

Ybl, Miklós (1814–1891)

Neo-Renaissance architect. Having been schooled in Vienna (1825–31), he worked for four years with Mihály Pollack then other architects before he continued his studies at the Munich Academy of Fine Arts (1840–42). Unable to gain admittance to the builders' guild himself until 20 years later, he entered into partnership with Pollack's son, Ágost, on several projects, including rebuilding the *Batthyányi Country House at Ikervár* (1844), just south of Sárvár in western Hungary, and the *Károlyi Mansion in Egyetem tér* in Pest's inner city (now the Petőfi Museum of Literature). He then worked on private commissions given by the Károlyi family, including the *Parish Church at Fót* (1845–55), which is still in Romanticist style, as was the *Unger House* at No. 7 on the Múzeum körút in Pest. In 1865 Y. was to build round the corner at No. 8 Bródy Sándor utca what functioned for a time as Parliament's *House of Representatives* (now the Italian Cultural Institute), having already designed palaces for Lajos Károlyi and György Festetics (1862) in what is now Pollack Mihály tér, behind the Museum. He later designed the *First Pest Savings Bank* building in what is now Károlyi Mihály utca (1868; restored in 2000). More outlying were the *Parish Church in Bakáts tér* (1867–79), in the Ferencváros (ninth) district, and the *Margaret Bath* on Margaret Island (1868–70; demolished 1950). His main neo-Renaissance works in the capital were built after 1870: *Main Custom's House* (1870–74) on the Danube embankment, at what is now the Pest foot of the Liberty Bridge (now the University for Economic Sciences); the *Várkert* (Castle Garden) *Bazaar* (1875–78) below Castle Hill and, across the road, the *Kioszk* (1875–78, today the Casino); the *'Danubius' Fountain*, with sculptures by Leo

Fessler, in Kálvin tér (1880–83; removed to its present site in Erzsébet tér in 1959). His masterpieces are the *Opera House* (1875–84; restored in 1893 and 1984) and *West Wing and Main Courtyard for the Royal Palace* (1880–91), later finished by Alajos Hauszmann (1891–98). In 1867 Y. took over from József Hild construction of the *Basilica of St. Stephen* in Pest's Lipótváros (fifth) district, but the collapse of its dome in 1868 (an event which prompted legislation in 1872 requiring builders to keep a construction log), forced its rebuilding in Y.'s neo-Renaissance style; it was finally completed by József Kauser, according to his own plans, in 1891–1905.

Zakariás, József (1924–1971)

Footballer. Member of the gold-winning Olympic team (1952), the team that beat England in the 1953 'match of the century' at Wembley Stadium, and the 1954 World Cup silver-winning team. Member of European Cup-winning club team (1953).

Zemplén, Győző (1879–1916)

Physicist. After graduating in Budapest he worked as an assistant to Loránd Eötvös before securing a chair at Budapest Technical University. His research was concentrated on the physics of shock waves in gases and the theory of discontinuous electro- and hydrodynamic phenomena. A member of the Hungarian Academy of Sciences, he was also very active in popularising science education. His publications included *Besondere Ausführungen über unstetige Bewegungen in Flüssigkeiten, Enzyklopädie der mathematischen Wissenschaften,* (1905), *Sur l'impossibilité des ondes de choc négatives dans les gas* (1905), *Über die Theorie der Stosswellen* (1912), *Az elektromosság és gyakorlati alkalmazásai* (Electricity and its Practical Applications, 1910).

Zipernowsky, Károly (1853–1942)

Electric engineer. After qualifying as a mechanical engineer from Budapest Technical University, Z. was asked by András Mechwart in 1878 to organise a new electric engineering division at the Ganz Works. Though he initially worked on direct-current generators, Z. became increasingly attracted to the merits of alternating current for electricity supply. With Miksa Déri he developed a self-inducing AC generator, and by the end of 1884, Ganz had installed over fifty of these to power small-scale local lighting systems. The main barrier to long-distance transmission of electric power was overcome by the transformer that Z., Déri and Bláthy patented in 1885. Z. left Ganz in 1893 to head a new department of high-tension electricity at Budapest Technical University.

Zsivótzky, Gyula (1937)

Hammer thrower. Certificate as PE instructor (1973). In hammer won silvers (1960, 1964) and gold (1968) at the Olympics. World record holder (1965, 1968) as the first to throw over 70 meters. Hungarian Sportsman of the Year (1965, 1968) and Throwing Athlete of the Year (1966, 1968). Board member (1993–99) and vice-chairman (1999–) of the Hungarian Athletics Federation. President of the Zselici Order (1995).

BIBLIOGRAPHY

Aczel, Richard. *National Character and European Identity in Hungarian Literature, 1772–1848*. Budapest: Nemzetközi Hungarológiai Központ, 1996.

Andrási, Gábor, Gábor Pataki, György Szücs, and András Zwickl. *The History of Hungarian Art in the Twentieth Century*. Budapest: Corvina, 1999.

Aristarco, Guido. *A filmművészetek története 1–2* (The History of Film Theories Vols. I–II). Budapest: Magyar Filmtudományi Intézet és Filmarchívum, 1962.

Miklós, D., V.T. Sós, and T. Szőnyi, eds. *Combinatorics: Paul Erdős Is Eighty.* Vols. I–II. Budapest: János Bolyai Mathematical Society, 1996.

Bakonyi, Tibor, and Mihály Kubinszky. *Lechner Ödön* (Ödön Lechner). Budapest: Corvina, 1981.

Balázs, Béla. *Der sichtbare Mensch, oder die Kultur des Films.* Wien–Leipzig: Deutsch-Österreichisches Verlag, 1924.

Balogh, Gyöngyi, and Jenő Király. *"Csak egy nap a világ..."—Magyar film 1929–1936* ("The World Lasts only a Day..."—Hungarian Film 1929–1936). Budapest: Magyar Filmintézet, 2000.

Bán, Ferenc. *Vallomások* (Confessions). Budapest: Kijárat Kiadó, 1997.

Bános, Tibor. *Kabos Gyula* (Gyula Kabos). Budapest: Athenaeum 2000, 2001.

Beke, László, and Zsuzsa Varga. *Kozma Lajos* (Lajos Kozma). Budapest: Akadémiai, 1968.

Bellman, Jonathan. *The 'Style Hongrois' in the Music of Western Europe*. Boston: Northeastern University Press, 1993.

Berend, Iván, and Miklós Szuhay. *A tőkés gazdaság története Magyarországon 1848–1944* (The History of Capitalist

Economy in Hungary 1848–1944). Budapest: Közgazdasági és Jogi Könyvkiadó, 1973.

Bierbauer, Virgil. *A magyar építészet története* (The History of Hungarian Architecture). Budapest: Magyar Szemle Társaság, 1937.

Biraghi, Marco, ed. *Béla Lajta ornamento e modernita.* Milan: Electa, 1999.

Biró, Yvette. *Miklós Jancsó.* Paris: Albatros, 1977.

Bónis, Ferenc, ed. *The Selected Writings of Zoltán Kodály.* London: Boosey & Hawkes, 1974.

Borsody, Stephen, ed. *The Hungarians: A Divided Nation.* New Haven: Yale Center for International and Area Studies, 1988.

Breuer, János. *A Guide to Kodály.* Budapest: Corvina, 1990.

Budapest 1869–1914. Modernité hongroise et peinture européenne. Dijon: Musée des Beaux-Arts, 1995.

Chalip, I., A. Johnson, and L. Stachura, eds. *National Sports Policies.* Westport, Connecticut: Greenwood Press, 1996.

Cook, Jeffrey. *Seeking Structure from Nature: The Organic Architecture of Hungary.* Basel: Birkhauser, 1996.

Császár, László. *Korai vas és vasbeton építészetünk* (Early Hungarian Iron and Reinforced Concrete Architecture). Budapest: Műszaki Könyvkiadó, 1978.

Czigány, Lóránt. *The Oxford History of Hungarian Literature: From the Earliest Times to the Present.* Oxford: Oxford University Press, 1984.

Dávid, Sándor. *Balczó.* Budapest: Sport Könyvkiadó, 1973.

Dercsényi, Balázs. *Árkay Aladár* (Aladár Árkay). Budapest: Akadémiai–Zrínyi, 1976.

Déry, Attila, and Ferenc Merényi. *A magyar építészet 1867–1945* (Hungarian Architecture 1867–1945). Budapest: Urbino Kft.–Mediart Grafikai Stúdió Kft., 2000.

Dévényi, Sándor. *Vallomások* (Confessions). Budapest: Kijárat, 1996.

Dobszay, László. *A History of Hungarian Music.* Budapest: Corvina, 1993.

Domokos, Mária. *János Lavotta.* Budapest: Mágus, 2001.

Eösze, László. *Kodály: His Life and Work.* Budapest: Corvina, 1962.

Éri, Gyöngyi, and Zsuzsa Jobbágyi. *A Golden Age: Art and Society in Hungary 1896–1914.* London: Barbican Art Gallery and Miami: Center for the Fine Arts, 1990.

Fábián, Gáspár. *Nagy magyar építőművészek* (Great Hungarian Master Builders). Budapest: the author, 1936.

Faller, Gusztáv, Béla Kun, and László Zsámboki, eds. *A magyar bányászat évezredes története* (The Millennial History of Hungarian Mining). Budapest: Országos Magyar Bányászati és Kohászati Egyesület, 1997.

Farkas, Zoltán. *Benedek Istvánffy.* Budapest: Mágus, 2000.

Ferkai, András. *Buda építészete a két világháború között* (The Architecture of Buda between the Wars). Budapest: Magyar Tudományos Akadémia, Művészettörténeti Kutató Intézet, 1995.

————, ed. *Jánossy György építőművész 1923–1998* (György Jánossy Architect. 1929–1998). Budapest: 6 BT K., 2001.

————, ed. *Pest építészete a két világháború között* (The Architecture of Pest between the Wars). Budapest: Modern Építészetért Építészettörténeti és Műemlékvédelmi Kht., 2001.

Filep, László. *A tudományok királynője: a matematika fejlődése* (The Queen of Sciences: the Development of Mathematics). Budapest: Typotex & Bessenyei, 1997.

Fitz, Péter, ed. *Kortárs magyar művészeti lexikon 1–3* (Encyclopedia of Contemporary Hungarian Arts. Vols. I–III). Budapest: Enciklopédia, 1999–2001.

Fleischmann, Wolfgang Bernard, ed. *Encyclopaedia of World Literature in the 20th Century.* New York: Frederick Ungar, 1969.

Földesi, Gyöngyi Szabó. *Magyar olimpikonok önmagukról és a sportról* (Hungarian Olympic Athletes on Themselves and on Sport). Budapest: Közgazdasági és Jogi Könyvkiadó, 1984.

Földesi, Gyöngyi Szabó. *Félamatőrök, félprofik* (Half Amateurs, Half Professionals). Budapest, MOB, 1999.

———, ed. *A magyar sport szellemi körképe* (A Mental Panorama of Hungarian Sport). Budapest, OTSH–MOB, 1996.

Frank, János. *Makovecz Imre* (Imre Makovecz). Budapest: Corvina, 1980.

Fülep, Lajos. *Magyar művészet* (Hungarian Art). Budapest: Athenaeum, 1923.

Gábor, Eszter. *A CIAM magyar csoportja (1928–1938)* (The Hungarian CIAM-Group (1928–1938)). Budapest: Akadémiai, 1972.

Gati, Charles. *Hungary and the Soviet Bloc.* Durham: Duke University Press, 1986.

Genthon, István. *Az új magyar festőművészet története 1800-tól napjainkig* (The History of New Hungarian Painting from 1800 to the Present Day). Budapest: Magyar Szemle Társaság, 1935.

Gerle, János, Attila Kovács, and Imre Makovecz. *A századforduló magyar építészete* (The Hungarian Architecture of the Turn of the Century). Budapest: Szépirodalmi Könyvkiadó: Bonex, 1990.

———, ed. *Makovecz Imre műhelye* (The Imre Makovecz Studio). Budapest: Mundus, 1996.

———, ed. *Tíz éves a Kós Károly Egyesülés* (Tenth Anniversary of the Károly Kós Union). Budapest: Kós Károly Alapítvány, 1999.

Gillespie, Harold, ed. *Romantic Drama.* Amsterdam: John Benjamins, 1994.

Gillispie, C. C., ed. *Bibliography of Scientific Biography.* New York: Scribner, 1970.

Glatz, Ferenc. *Gróf Klebelsberg Kunó válogatott beszédei és írásai (1917–1932)* (Selected Speeches and Writings of Count Kunó Klebelsberg 1917–1932). Budapest: Európa Könyvkiadó, 1990.

Goulding, Daniel J., ed. *Post New Wave Cinema in the Soviet Union and Eastern Europe.* Bloomington & Indianapolis: Indiana University Press, 1989.

Görög, Sándor. *Kémiai Tudományok Osztálya: múlt, jelen, jövő* (The Department of Chemical Sciences: Past, Present and Future). Lecture, 2001.

Griffiths, Paul. *György Ligeti.* London: Robson Books, 1983.

Guttmann, Allen. *From Ritual to Record. The Nature of Modern Sport.* New York: Columbia University, 1978.

Gyertyán, Ervin. *Mit ér a film, ha magyar?* (What Are Hungarian Films Worth?). Budapest: MOKÉP–MMA, 1994.

Györgyi, Dénes, Dezső Hültl, and Lajos Kozma, eds. *Új Magyar Építőművészet I–II.* (New Hungarian Architecture. Vols. I II). Budapest: Budai István, 1938.

Halász, Péter. *György Kurtág.* Budapest: Mágus, 1999.

Hanák, Péter, and Zsuzsa Szentgyörgyi, eds. *Gazdaság és társadalom a dualizmus korában (Egy modernizációs korszak)* (Economy and Society in the Austro–Hungarian Monarchy: an Era of Modernisation), *Magyar Tudomány* 10–11 (1989).

Hencsei, Pál, Tibor Ivanics, Ferenc Takács, and Dezső Vad. *Magyarok az olimpiai játékokon* (Hungarians at the Olympic Games). Budapest: IQ Press Lapkiadó, 2000.

Hoensch, Jörg K. *A History of Modern Hungary 1867–1994.* London–New York: Longman, 1996.

Hoffman, Paul. *The Man Who Loved Only Numbers.* London: Fourth Estate, 1998.

Ivanics, Tibor, Bende István. *Athén, Atlanta. Az olimpiai játékok magyar érmesei 1896–1996* (From Athens to Atlanta: Hungarian Olympic Medal Winners 1896–1996). Budapest: Magyar Olimpiai Bizottság and Magyar Testnevelési és Sportmúzeum, 1996.

Janos, Andrew C. *The Politics of Backwardness in Hungary 1825–1945.* Princeton, N.J.: Princeton University Press, 1982.

Jánszky, Béla. *A magyar formatörekvések története építészetünkben 1894–1914* (History of Hungarian Architectural Design 1894–1914). Budapest: Kőművesmesterek Országos Szövetsége, 1929.

Jeancolas, Jean-Pierre. *Allegro barbaro—Miklós, István, Zoltán et les autres.* Budapest: Corvina, 1989.

———. *L'oeil hongrois—Quatre décennies de cinéma a Budapest 1963–2000.* Budapest: Magyar Filmunió, 2001.

Jékely, Zsolt, and Alajos Sódor. *Budapest építészete a XX. században* (The Architecture of Budapest in the 20th Century). Budapest: Műszaki Könyvkiadó, 1980.

Jurcsik, Károly. *Vallomások* (Confessions). Budapest: Kijárat, 1998.

Kállai, Ernő. *Új magyar piktúra 1900–1925* (New Hungarian Painting 1900–1925). Budapest: Gondolat, 1990.

Kathy, Imre. *Medgyaszay István* (István Medgyaszay). Budapest: Akadémiai, 1979.

Kerényi, József. *Vallomások* (Confessions). Budapest: Kijárat, 1998.

Kincses, Károly. *Fotográfusok Made in Hungary—Aki elment, aki maradt.* Kecskemét: Magyar Fotográfiai Múzeum, 1998.

Kiss, Elemér. *Mathematical Gems from the Bolyai Chests.* Budapest: Akadémiai Kiadó & Typotex, 1999.

Klaniczay, Tibor, ed. *A History of Hungarian Literature.* Budapest: Corvina, 1983.

Knoll, Hans, ed. *Die zweite Öffentlichkeit. Kunst in Ungarn im 20. Jahrhundert.* Dresden: Verlag der Kunst, 1999.

Komárik, Dénes. *Feszl Frigyes 1821–1884* (Frigyes Feszl 1821–1884). Budapest: Akadémiai Kiadó, 1993.

Kontha, Sándor, ed. *Magyar Művészet 1919–1945, I–II.* (Hungarian Art. 1919–1945. Vols. I–II). Budapest: Akadémiai, 1985.

Kontler, László. *Millennium in Central Europe: A History of Hungary.* Budapest: Atlantisz, 1999.

Koós, Judit. *Kozma Lajos munkássága* (The Work of Lajos Kozma). Budapest: Akadémiai, 1975.

Kozák, Péter, ed. *Ki kicsoda a magyar sportéletben I–III.* (Who is Who in Hungarian Sport. Vols. I–III). Budapest: Babits Kiadó, 1995.

Kő, András. *Papp Laci.* Budapest: Sport Könyvkiadó, 1973.

Kőháti, Zsolt. *Tovamozduló ember tovamozduló világban—A magyar némafilm története 1896–1930* (People Moving On in a World That Is Moving On—A History of Hungarian Silent Film, 1896–1930). Budapest: Magyar Filmintézet, 1996.

Kőrös, Endre. *Honnan és merre? A hazai kémiakutatás fél évszázada: emlékezés a múltra* (Where Has It Come from and Where Is It Going? A Half Century of Chemical Research in Hungary: Remembering the Past). Lecture, 2001.

Kroó, György. *A Guide to Bartók.* Budapest: Corvina, 1974.

————. *Ungarische Musik—gestern und heute.* Budapest: Corvina, 1980.

Kubinszky, Mihály. *Györgyi Dénes* (Dénes Györgyi). Budapest: Akadémiai, 1974.

————, ed. *Modern építészeti lexikon* (Encyclopedia of Modern Architecture). Budapest: Műszaki Kiadó, 1978.

Kulcsár-Szabó, Ernő, and Mihály Szegedy-Maszák, eds. *Epoche–Text–Modalität.* Tübingen: Niemeyer, 1999.

Laki, Peter, ed. *Bartók and His World.* Princeton: Princeton University Press, 1995.

László Kósa, ed. *A Companion to Hungarian Studies.* Budapest: Akadémiai Kiadó, 1999.

Lendvai, Paul. *Die Ungarn. Ein Jahrtausend Sieger in Niederlagen.* München: C. Bertelsmann, 1999.

Lexikon bedeutender Mathematiker. Leipzig: Bibliographisches Institut, 1990.

Lichtmann, Tamás, ed. *Angezogen und abgestossen: Juden in der ungarischen Literatur.* Frankfurt am Main: Peter Lang, 1999.

Liehm, Mira, and Antonin J. Liehm. *The Most Important Art: East European Film After 1945.* Berkeley, Los Angeles and London: University of California Press, 1977.

Lipták, Irén. *Farkasdy Zoltán* (Zoltán Farkasdy). Budapest: Képzőművészeti Alap Kiadóvállalata, 1975.

Liszt, Ferenc. *A czigányokról és a czigány zenéről Magyarországon* (About Gypsies and Gypsy Music in Hungary). Pest: Heckenast, 1861.

Lőrinczi, Zsuzsa, and Mihály Vargha, eds. *Architecture in Budapest from the Turn-of-the-Century to the Present.* Budapest: 6B Építész Bt., 1997.

Lukacs, John. *Budapest 1900. A Historical Portrait of a City and Its Culture.* New York: Weidenfeld and Nicolson, 1988.

Lukács, László, and György Szepesi. *A magyar olimpiai aranyérmek története (1896–1992)* (History of the Hungarian Olympic Gold Medals, 1896–1992). Budapest: Sport Könyvkiadó, 1992.

Lyka, Károly. *Nemzeti romantika. Magyar művészet 1850–1867* (National Romanticism. Hungarian Art 1850–1867). Budapest: Singer & Wolfner, 1942.

Macrae, Norman. *John von Neumann.* New York: Pantheon Books, 1992.

Magyar organikus építészet. Velencei Építészeti Biennálé, Hungarian Organic Architecture. La Biennale di Venezia 1991. Budapest: Novotrans Kft., 1991.

Magyar, Bálint. *A magyar némafilm története 1–2* (A History of Hungarian Silent Films. Vols. I–II). Budapest: Filmtudományi Intézet, 1961.

Major, Máté. *Breuer Marcell* (Marcel Breuer). Budapest: Akadémiai, 1970.

———, and Judit Osskó, eds. *Új építészet, új társadalom, 1945–1978* (New Architecture, New Society, 1945–1978). Budapest: Corvina, 1981.

Makai, Adam, ed. *In Quest of the Miracle Stag: the Poetry of Hungary.* Chicago–Budapest: Atlantis–Centaur and M. Szivárvány–Corvina, 1996.

Mansbach, Steven A. *Standing in the Tempest: Painters of the Hungarian Avant-Garde 1908–1930*. Cambridge, Mass. & London: MIT Press, 1991.

Marx, George. *The Voice of the Martians*. Budapest: Akadémiai Kiadó, 1997.

Marx, György. *A marslakók érkezése* (The Voice of the Martians). Budapest: Akadémiai Kiadó, 1994.

Mendöl, Zsuzsa. *Málnai Béla* (Béla Málnai). Budapest: Akadémiai, 1974.

Merényi, Ferenc. *A magyar építészet 1867–1967* (The Hungarian Architecture 1867–1967). Budapest: Műszaki Könyvkiadó, 1969.

Mezei, Ottó. *Molnár Farkas* (Farkas Molnár). Budapest: Akadémiai, 1987.

———, ed. *A Bauhaus* (The Bauhaus). Budapest: Gondolat, 1975.

Mihailich, Győző, and Győző Haviár. *A vasbetonépítés kezdete és első létesítményei Magyarországon* (The Debut of the Reinforced Concrete Construction and the First Works In Hungary). Budapest: Akadémiai, 1966.

Móra, László, and István Próder. *A magyar kémia és vegyipar kronológiája*. Budapest–Piliscsaba: MATI, 1997.

Moravánszky, Ákos. *Építészet az Osztrák–Magyar Monarchiában 1867–1918* (Architecture in the Austro–Hungarian Monarchy 1867–1918). Budapest: Corvina, 1988.

———. *Versengő látomások. Esztétikai újítás és társadalmi program az Osztrák–Magyar Monarchia építészetében 1867–1918* (Competing Visions. Esthetic Innovation and Social Program in the Austro-Hungarian Monarchy 1867–1918). Budapest: Vince, 1998.

Nádori, László, ed. *Sportlexikon I–II.* (Sport Encyclopaedia. Vols. I–II). Budapest: Sport Könyvkiadó, 1985.

Nagy, Elemér. *Az építő Kós Károly* (The Architect Károly Kós). Budapest–Kolozsvár: Balassi–Kós Károly Alapítvány–Polis, 1995.

Nagy, Ferenc, ed. *Magyar tudóslexikon* (Encyclopedia of Hungarian Scientists). Budapest: OMIKK, 1997.

Nagy, Ferenc, ed. *Magyarok a természettudomány és technika történetében* (Hungarians in the History of Natural Sciences and Technology). Budapest: OMIKK, 1986.

Nagy, Gergely, and Károly Szelényi. *Kertvárosunk, a Wekerle* (Wekerle, Our Garden City). Budapest–Veszprém: Magyar Képek Kft.–F. Szelényi Ház BT, 1994.

Nemes, Károly. *Sodrásban. A magyar film 25 éve, 1945–1970* (In the Current—25 Years of Hungarian Cinema, 1945–1970). Budapest: Magvető Kiadó, 1972.

Nemeskürty, István. *A képpé varázsolt idő* (Time Magicked into Pictures). Budapest: Magvető Kiadó, 1983.

————. *A magyar film története (1912–1963)* (The History of Hungarian Film (1912–1963)). Budapest: Gondolat, 1965.

————. *Word and Image: History of the Hungarian Cinema.* Budapest: Corvina, 1968.

Németh, Amadé. *Ferenc Erkel: Sein Leben und Wirken.* Budapest: Corvina, 1979.

Németh, Lajos. *Modern Hungarian Art.* Budapest: Corvina, 1972.

————, ed. *Magyar művészet 1890–1919* (Hungarian Art 1890–1919). Budapest: Akadémiai Kiadó, 1981.

Palmer, Alan. *The Lands Between: A History of East-Central Europe since the Congress of Vienna.* New York: Macmillan, 1970.

Pámer, Nóra. *Magyar építészet a két világháború között* (Hungarian Architecture between the Wars). Budapest: Műszaki Könyvkiadó, 1986.

Passuth, Krisztina. *A Nyolcak festészete* (Painting of The Eight). Budapest: Corvina, 1967.

————. *Avantgarde kapcsolatok Prágától Budapestig 1907–1930* (Avant-garde Connections from Prague to Budapest 1907–1930). Budapest: Balassi, 1998.

————. *Moholy Nagy László* (László Moholy Nagy). Budapest: Corvina, 1982.

Perneczky, Géza, ed. *Kortársak szemével. Írások a magyar művészetről 1896–1945* (Through the Eyes of Contemporaries: Writings on Hungarian Art 1896–1945). Budapest: Corvina, 1967.

Petrie, Graham. *History Must Answer to Man: The Contemporary Hungarian Cinema.* Budapest: Corvina, 1978.

Polgár, László. *Nevelj zsenit* (Bring up a Genious). Budapest: Interart, 1989.

Porter, Roy, and Mikulas Teich, eds. *Romanticism in National Context.* Cambridge, UK: Cambridge University Press, 1988.

Prakfalvi, Endre. *Szocreál, Budapest építészete 1945 és 1959 között* (Socialist Realism, The Architecture of Budapest between 1945 and 1959). Budapest: Budapest Főváros Önkormányzata Főpolgármesteri Hivatala, 1999.

———, and Virág Hajdú, eds. *Architecture and Design in Hungary 1945–1959.* Vols. I–II. Budapest: OMvH Magyar Építészeti Múzeum, 1996.

Pungor, Ernő. *Telecommunication in Central and Eastern Europe—an Impetus for Economic Growth and Regional Integration.* Lecture, 1992.

Pynsent, Robert B., ed. *The Everyman Companion to East European Literatures.* London: J. M. Dent, 1983.

Rácz, Zoltán. *Borsos József és Debrecen korai modern építészete* (József Borsos and the Early Modern Architecture of Debrecen). Debrecen: Csokonai, 1990.

Rados, Jenő. *Hild József, Pest nagy építőjének életműve* (The Oeuvre of József Hild, the Great Builder of Pest). Budapest: Akadémiai, 1958.

———. *Magyar Építészettörténet* (History of Hungarian Architecture). Budapest: Műszaki Könyvkiadó, 1971.

Rappai, Zsuzsa, ed. *Finta József műhelye* (The József Finta Studio). Budapest: Mundus, 1997.

Rhodes, Richard. *The Making of the Atomic Bomb.* New York: Simon&Schuster, 1986.

Romsics, Ignác. *Hungary in the Twentieth Century.* Budapest: Corvina–Osiris, 1999.

Sárosi, Bálint. *Gypsy music.* Budapest: Corvina, 1978.

Schéry, Gábor, ed. *Évek, művek, alkotók, Ybl Miklós-díjasok és műveik 1953–1994* (Years, Works, Artists. Architects Awarded with the Miklós Ybl-Prize between 1953–1994). Budapest: Építésügyi Tájékoztatási Központ Kft., 1995.

Sietz, Barbara, ed. *Zeitgenössische Kunst aus Ungarn, Installation Malerei Skulptur Videokunst.* München: Matthes & Seitz, 1999.

Siklóssy, László. *Hogyan épült Budapest? (1870–1930)* (How Budapest Was Constructed? 1870–1930). Budapest: Fővárosi Közmunkák Tanácsa, 1931.

Sisa, József, and Dora Wiebenson, eds. *Magyarország építészetének története* (The History of Hungary's Architecture). Budapest: Vince Kiadó, 1998.

————. *The Architecture of Historic Hungary.* Cambridge, Mass. and London: MIT Press, 1998.

Somfai, László. *Béla Bartók: Composition, concepts, and autograph sources.* Berkeley: University of California Press, 1996.

Stäckel, Paul. *Wolfgang und Johann Bolyai: Geometrische Untersuchungen.* Leipzig, 1913.

Stade, George, ed. *European Writers: The Twentieth Century*, vol. 10. New York: Scribner, 1990.

Suchoff, Benjamin, ed. *Béla Bartók Essays.* London: Faber & Faber, 1976.

Sugar, Peter F., Péter Hanák, and Tibor Frank, eds. *A History of Hungary.* Bloomington–Indianapolis: Indiana University Press, 1990.

Szabadi, Judit. *A magyar szecesszió művészete* (The Art of the Hungarian Secession). Budapest: Corvina, 1979.

Szabadváry, Ferenc. *A magyar kémia művelődéstörténete* (The Cultural History of Hungarian Chemistry). Budapest: Mundus, 1998.

Szabó, Júlia. *A magyar aktivizmus művészete, 1915–1927* (The Art of Hungarian Activism, 1915–1927). Budapest: Corvina, 1981.

————. *A XIX. század festészete Magyarországon* (Nineteenth-Century Hungarian Painting). Budapest: Corvina, 1985.

Szabolcsi, Bence. *A Concise History of Hungarian Music.* Budapest: Corvina, 1964.

Székely, Éva. *Sírni csak a győztesnek szabad.—Jöttem, láttam... vesztettem?—Megúsztam.* (Only the Winner May Cry.—I've Come, Seen... Lost?—I Have Survived). Budapest: Új Mandátum Könyvkiadó, 1992.

Szénássy, Barna. *History of Mathematics in Hungary until the 20th Century.* Budapest: Akadémiai Kiadó, 1992.

Szendrei, János, ed. *Mathematics in Hungary.* Budapest: Bolyai János Matematikai Társaság, 1996.

Szendrői, Jenő, ed. *Magyar építészet 1945–1995* (Hungarian Architecture 1945–1995). Budapest, 1972.

Szent-Györgyi, Albert. "Az iskolai ifjúság testnevelése" (Physical Education of School Children), *Testnevelés* 12 (1930): 928–933.

Szentgyörgyi, Zsuzsa. "Kérdések és álkérdések—Adalékok a magyar számítástechnikai kutatás történetéhez" (Questions and Pseudo-Questions—Contributions to a History of Computer Technological Research in Hungary), *Magyar Tudomány* 9 (1989): 711–718.

————. "A Short History of Computing in Hungary," *IEEE Annals of the History of Computing* 3 (1999): 49–57.

Szentiványi, Tibor. "A számítástechnika kezdetei Magyarországon" (The Beginnings of Computer Technology in Hungary), *Természet Világa* 6–8 (1994): 250–255, 312–316, 362–366.

Szrogh, György. *Vallomások* (Confessions). Budapest: Kijárat, 1999.

Szűcs, Júlia P. *A római iskola* (The School of Rome). Budapest: Corvina, 1987.

Takács, Ferenc. *A modern olimpiai játékok* (Modern Olympic Games). Budapest: Útmutató Kiadó, 1999.

———. *Az olimpiák* (Olympic Games). Budapest: Útmutató Kiadó, 1996.

———. *Száz év olimpia* (Hundred Years of Olympics). Budapest: IQ Press Lapkiadó Kft., 2000.

Tallián, Tibor. *Béla Bartók.* Budapest: Corvina, 1988.

Tamás, Pál, ed. *BIT-korszak—Fejezetek a magyar számítástechnika történetéből* (The BIT Era—Chapters from the History of Hungarian Computer Technology). Budapest: MTA Politikai Tudományok Intézete and MTA Társadalmi Konfliktusok Kutató Intézete, 1992.

Toop, Richard. *György Ligeti.* London: Phaidon Press, 1999.

Török, Ferenc. *Vallomások* (Confessions). Budapest: Kijárat, 1996.

Turjányi-Papp, Melinda. *Szrogh György* (György Szrogh). Budapest: Képzőművészeti Alap Kiadóvállalata, 1976.

Umminger, Walter. *A sport krónikája* (Chronicle of Sport). Budapest: Officina Nova, 1990.

Vadász, György. *Vallomások* (Confessions). Budapest: Kijárat, 1997.

Vámos, Ferenc. *Lajta Béla* (Béla Lajta). Budapest: Akadémiai, 1970.

Vámos, Tibor: Informatika (Information Technology), *Természet Világa,* II. (Special Number) 2000.

———. "Kutatások a kibernetika és az automatizálás közös területein" (Research in the Common Areas of Cybernetics and Automation), *Magyar Tudomány* 9 (1966): 556–565.

———. "A számítástechnika az Akadémián" (Computer Technology at the Academy of Sciences), *Magyar Tudomány* 4 (1971): 235–241.

Vargha, Mihály, ed. *Magyar építészet! Architektur in Ungarn 1989–1999* (Architecture in Hungary. 1989–1999). Budapest, 1999.

Wagner, Francis S. *Hungarian Contributions to World Civilization.* Center Square, PA: Alpha Publications, 1977.

Walker, Alan. *Franz Liszt.* Vols. I–III. New York: Knopf, 1983.

Weibel, Peter. *Jenseits von Kunst.* Vienna: Passagen Verlag, 1997.

Ybl, Ervin. *Ybl Miklós* (Miklós Ybl). Budapest: Képzőművészeti Alap Kiadóvállalata, 1956.

Young, Percy M. *Zoltán Kodály: a Hungarian Musician.* London: Benn, 1964.

Zádor, Anna. *Pollack Mihály 1773–1855* (Mihály Pollack 1773–1855). Budapest: Akadémiai, 1960.

———, ed. *Magyar művészet 1800–1945* (Hungarian Art 1800–1945). Budapest: Képzőművészeti Alap, 1958.

Zsugán, István. *Szubjektív magyar filmtörténet* (A Subjective History of Hungarian Films). Budapest: Osiris Kiadó, 1998.

LIST OF ILLUSTRATIONS

CONTRIBUTORS

Beke, László

Art historian. Chief curator of the 19th- and 20th-century collections at the National Gallery, Budapest (1988–95), professor for interdisciplinary studies of the Hungarian Academy of Fine Arts (1990–94), then director-general of the Műcsarnok/Kunsthalle, Budapest. Since 2000 he has been director of the Research Institute for Art History of the Hungarian Academy of Sciences. His publications include *Schaár Erzsébet* (The Sculptor Erzsébet Schaár, 1973); *Moholy-Nagy László munkássága* (The Work of László Moholy-Nagy, 1980); *Jovánovics* (The Sculptor György Jovánovics, 1980); *Caspar David Friedrich* (1982); *Dóra Maurer, Arbeiten/Munkák/Works 1970–1993* (1994, with Dieter Ronte); *Művészet/elmélet* (Art/Theory, 1994); *Médium/elmélet* (Media/Theory, 1998).

Bikácsy, Gergely

Film and literature critic. His publications include *Andrzej Wajda* (1975), *Pajkos nő az árnyas utcán* (Wanton Woman in a Shady Lane, 1987), *Bolond Pierrot moziba megy – A francia film 50 éve* (1939–1990) (Pierrot le Fou Goes to the Movies: 50 years of French Cinema, 1939–1990, 1991), *Buñuel-napló* (Buñuel Diary, 1998), *"Minden út Róma"* ("All Roads are Rome", 2002).

Filep, László

Mathematician, lecturing on algebra and the history of mathematics at Nyíregyháza College since 1973. He is a member of the International Commission on the History of Mathematics and of the Hungarian Academy of Sciences' Interdisciplinary

Committee for the History of Science. His publications include *A számírás története* (The History of Numeration, 1982); *Játékelmélet* (Theory of Games, 1985); *A tudományok királynője, a matematika fejlődése* (The Development of Mathematics, Queen of Sciences, 1997).

Földesi-Szabó, Gyöngyi

Professor of sociology in the School of Physical Education and Sports Sciences at Semmelweis University, Budapest, with doctorates in sociology and in physical education. She served as vice president of the International Committee for Sociology of Sport (1984–92), was associate editor of the *International Review for the Sociology of Sport* (1988–96), and is president of the National Confederation of the Hungarian Leisure Sports Federations. She is the author and editor of several books.

Frank, Tibor

Professor of history and former director (1994–2001) of the School of English and American Studies at Loránd Eötvös University, Budapest. He was a Fulbright visiting professor at the University of California, Santa Barbara, and the University of Nevada, Reno (1987–91) and, recently, visiting professor at Columbia University, New York. His recent books include *Ethnicity, Propaganda, Myth-Making* (1999); *From Habsburg Agent to Victorian Scholar: G. G. Zerffi 1820–1892* (2000) and *Genius in Exile* (2002). He was co-editor, with the late Peter F. Sugar and Péter Hanák, of *A History of Hungary* (1990).

Kovács, György

Electrical engineer. After working as a research engineer for a precision-mechanics company that produced medium-wave, short-wave and VHF radio stations, broadband microwave sets and radio locators, he became technical manager and later manager in chief of the Budapest Electro-acoustics Works

(BEAG). Since 1991 he has been managing director and joint proprietor of Transcoder Ltd. He is the author of several important feasibility studies and popularising articles on technology.

Marx, György

Physicist. Member of the Hungarian Academy of Sciences, he was professor in the Department of Atomic Physics, Loránd Eötvös University, Budapest, and a visiting professor at, amongst others, the universities of Stanford, Vienna, Nanjing, and Kyoto until his recent death in 2002. He was a fellow of the International Astronautical Academy, Academia Europaea, the New York Academy of Sciences, the American Physical Society, and the Institute of Physics. He has published several hundred papers and about 30 books, including *Kvantummechanika* (Quantum Mechanics, 1957); *Túl az atomfizikán* (Beyond Atomic Physics, 1960); *A természet játékai* (The Games Nature Plays, 1984); *Atommagközelben* (Up Close to the Atomic Nucleus, 1996); *The Voice of the Martians* (1997).

Matskási, István

Biologist. Gaining his PhD degree in parasitology, he was research officer in the Hungarian Natural History Museum (1965–74) before entering the Veterinary Institute of the Hungarian Academy of Sciences, later working in the Academy's Central Office. He became director-general of the Hungarian Natural History Museum in 1986. He is president of the Society of Hungarian Museologists and editor in chief of several scientific journals. He has published seventy scientific papers and numerous popularising articles on science.

Szabadváry, Ferenc

Chemical engineer, science historian, member of the Hungarian Academy of Sciences. He has taught in the Department of General and Analytical Chemistry at Budapest Technical University since 1948. He was director of the

Hungarian Technological Museum and editor in chief of the *Review of Technical History*. His publications include *Lavoisier és kora* (Lavoisier and His Age, 1968); *A kémia története Magyarországon* (The History of Chemistry in Hungary, 1972); *A magyar kémia művelődéstörténete* (A Cultural History of Hungarian Chemistry, 1998).

Szegedy-Maszák, Mihály

Professor of Central Eurasian Studies and Comparative Literature at Indiana University, Bloomington, and of Comparative Literature at Loránd Eötvös University, Budapest. A member of the Hungarian Academy of Sciences, vice-president of the International Comparative Literature Association, editor in chief of the journal *Hungarian Studies*. He is the author of numerous articles and books, including: *Világkép és stílus* (Ethos and Style, 1980); *"A regény, amint írja önmagát"* ("The Novel as it Writes Itself", 1980); *Kubla kán és Pickwick úr* (Kubla Khan and Mr Pickwick, 1982); *Márai Sándor* (1991); *Ottlik Géza* (1994); *Irodalmi kánonok* (Literary Canons, 1998) and *Literary Canons: National and International* (2001). Co-author of *Epoche—Text—Modalität* (1999) and *National Heritage—National Canon* (2001).

Szentgyörgyi, Zsuzsa

Electrical engineer. A research worker at the Research Institute for Computing and Automation (SZTAKI) of the Hungarian Academy of Sciences, in the mid-eighties she headed the Ministry of Industry's R&D program. For many years she was deputy secretary-general and co-president of the Hungarian Society for Measurement and Automation (MATE). Since 1991 she has been a freelance writer and editor in the fields of science and technology. She is author of some 60 scientific papers and 500 popularising articles, and editor or co-author of five books.

Vámossy, Ferenc

Emeritus professor of architectural history and monuments at Budapest Technical University, he gained his doctorate from the Hungarian Academy of Sciences. Earlier he taught at the College of Applied Arts and in the master school for young architects. He was awarded the Ybl Prize in 1984. Among his numerous books and other publications are *A XX. század építészete* (Architecture of the Twentieth Century, 1961) and *Korunk építészete* (Architecture of Our Times, 1974).

Vargha, Magda

Librarian at the Astronomical Research Institute of the Hungarian Academy of Sciences from 1965 to 2000, her main scholarly interest is the collection and publication of historical sources in the field of Hungarian astronomy. Her publications as editor or co-editor include: *Briefe Franz Xaver von Zachs in seine Vaterland* (1984, with Peter Brosche); *Correspondence de Ferenc Weiss Astronome Hongrois du 18e siècle*, 2 vols. (1990 and 1992); and *The Role of Miklós Konkoly Thege in the History of Astronomy in Hungary* (1992, with László Patkós and Imre Tóth).

Name Index

Volumes Published in

"Atlantic Studies on Society in Change"

No. 1 — *Tolerance and Movements of Religious Dissent in Eastern Europe.* Edited by Béla K. Király. 1977.

No. 2 — *The Habsburg Empire in World War I.* Edited by R. A. Kann. 1978

No. 3 — *The Mutual Effects of the Islamic and Judeo-Christian Worlds: The East European Pattern.* Edited by A. Ascher, T. Halasi-Kun, B. K. Király. 1979.

No. 4 — *Before Watergate: Problems of Corruption in American Society.* Edited by A. S. Eisenstadt, A. Hoogenboom, H. L. Trefousse. 1979.

No. 5 — *East Central European Perceptions of Early America.* Edited by B. K. Király and G. Barány. 1977.

No. 6 — *The Hungarian Revolution of 1956 in Retrospect.* Edited by B. K. Király and Paul Jonas. 1978.

No. 7 — *Brooklyn U.S.A.: Fourth Largest City in America.* Edited by Rita S. Miller. 1979.

No. 8 — *Prime Minister Gyula Andrássy's Influence on Habsburg Foreign Policy.* János Decsy. 1979.

No. 9 — *The Great Impeacher: A Political Biography of James M. Ashley.* Robert F. Horowitz. 1979.

No. 10, Vol. I* — *Special Topics and Generalizations on the Eighteenth and Nineteenth Century.* Edited by Béla K. Király and Gunther E. Rothenberg. 1979.

No. 11, Vol. II — *East Central European Society and War in the Pre-Revolutionary 18th Century.* Edited by Gunther E. Rothenberg, Béla K. Király, and Peter F. Sugar. 1982.

* Vols. no. I through XXXVII refer to the series *War and Society in East Central Europe*

No. 12 Vol. III *From Hunyadi to Rákóczi: War and Society in Late Medieval and Early Modern Hungary.* Edited by János M. Bak and Béla K. Király. 1982.

No. 13 Vol. IV *East Central European Society and War in the Era of Revolutions: 1775-1856.* Edited by B. K. Király. 1984.

No. 14 Vol. V *Essays on World War I: Origins and Prisoners of War.* Edited by Samuel R. Williamson, Jr. and Peter Pastor. 1983.

No. 15 Vol. VI *Essays on World War I: Total War and Peacemaking, A Case Study on Trianon.* Edited by B. K. Király, Peter Pastor, and Ivan Sanders. 1982.

No. 16 Vol. VII *Army, Aristocracy, Monarchy: War, Society and Government in Austria, 1618-1780.* Edited by Thomas M. Barker. 1982.

No. 17 Vol. VIII *The First Serbian Uprising 1804-1813.* Edited by Wayne S. Vucinich. 1982.

No. 18 Vol. IX *Czechoslovak Policy and the Hungarian Minority 1945-1948.* Kálmán Janics. Edited by Stephen Borsody. 1982.

No. 19 Vol. X *At the Brink of War and Peace: The Tito-Stalin Split in a Historic Perspective.* Edited by Wayne S. Vucinich. 1982.

No. 20 *Inflation Through the Ages: Economic, Social, Psychological and Historical Aspects.* Edited by Edward Marcus and Nathan Schmuckler. 1981.

No. 21 *Germany and America: Essays on Problems of International Relations and Immigration.* Edited by Hans L. Trefousse. 1980.

No. 22 *Brooklyn College: The First Half Century.* Murray M. Horowitz. 1981.

No. 23 *A New Deal for the World: Eleanor Roosevelt and American Foreign Policy.* Jason Berger. 1981.

No. 24 *The Legacy of Jewish Migration: 1881 and Its Impact.* Edited by David Berger. 1982.

No. 25 *The Road to Bellapais: Cypriot Exodus to Northern Cyprus.* Pierre Oberling. 1982.

No. 26 *New Hungarian Peasants: An East Central European Experience with Collectivization.* Edited by Marida Hollos and Béla C. Maday. 1983.

No. 27 *Germans in America: Aspects of German-American Relations in the Nineteenth Century.* Edited by Allen McCormick. 1983.

No. 28 *A Question of Empire: Leopold I and the War of Spanish Succession, 1701-1705.* Linda and Marsha Frey. 1983.

No. 29 *The Beginning of Cyrillic Printing — Cracow, 1491. From the Orthodox Past in Poland.* Szczepan K. Zimmer. Edited by Ludwik Krzyżanowski and Irene Nagurski. 1983.

No. 29a *A Grand Ecole for the Grand Corps: The Recruitment and Training of the French Administration.* Thomas R. Osborne. 1983.

No. 30 Vol. XI *The First War between Socialist States: The Hungarian Revolution of 1956 and Its Impact.* Edited by Béla K. Király, Barbara Lotze, Nandor Dreisziger. 1984.

No. 31 Vol. XII *The Effects of World War I, The Uprooted: Hungarian Refugees and Their Impact on Hungary's Domestic Politics.* István Mócsy. 1983.

No. 32 Vol. XIII *The Effects of World War I: The Class War after the Great War: The Rise of Communist Parties in East Central Europe, 1918-1921.* Edited by Ivo Banac. 1983.

No. 33 Vol. XIV *The Crucial Decade: East Central European Society and National Defense, 1859-1870.* Edited by Béla K. Király. 1984.

No. 35 Vol. XVI *Effects of World War I: War Communism in Hungary, 1919.* György Péteri. 1984.

No. 36 Vol. XVII *Insurrections, Wars, and the Eastern Crisis in the 1870s.* Edited by B. K. Király and Gale Stokes. 1985.

No. 37 Vol. XVIII *East Central European Society and the Balkan Wars, 1912-1913.* Edited by B. K. Király and Dimitrije Djordjevic. 1986.

No. 38 Vol. XIX *East Central European Society in World War I.* Edited by B. K. Király and N. F. Dreisziger, Assistant Editor Albert A. Nofi. 1985.

No. 39 Vol. XX *Revolutions and Interventions in Hungary and Its Neighbor States, 1918-1919.* Edited by Peter Pastor. 1988.

No. 41 Vol. XXII *Essays on East Central European Society and War, 1740-1920.* Edited by Stephen Fischer-Galati and Béla K. Király. 1988.

No. 42 Vol. XXIII *East Central European Maritime Commerce and Naval Policies, 1789-1913.* Edited by Apostolos E. Vacalopoulos, Constantinos D. Svolopoulos, and Béla K. Király. 1988.

No. 43 Vol. XXIV *Selections, Social Origins, Education and Training of East Central European Officers Corps.* Edited by Béla K. Király and Walter Scott Dillard. 1988.

No. 44 Vol. XXV *East Central European War Leaders: Civilian and Military.* Edited by Béla K. Király and Albert Nofi. 1988.

No. 46 *Germany's International Monetary Policy and the European Monetary System.* Hugo Kaufmann. 1985.

No. 47 *Iran Since the Revolution—Internal Dynamics, Regional Conflicts and the Superpowers.* Edited by Barry M. Rosen. 1985.

No. 48 Vol. XXVII *The Press During the Hungarian Revolution of 1848-1849.* Domokos Kosáry. 1986.

No. 49 *The Spanish Inquisition and the Inquisitional Mind.* Edited by Angel Alcala. 1987.

No. 50 *Catholics, the State and the European Radical Right, 1919-1945.* Edited by Richard Wolff and Jorg K. Hoensch. 1987.

No. 51 Vol.XXVIII *The Boer War and Military Reforms.* Jay Stone and Erwin A. Schmidl. 1987.

No. 52 *Baron Joseph Eötvös, A Literary Biography.* Steven B. Várdy. 1987.

No. 53 *Towards the Renaissance of Puerto Rican Studies: Ethnic and Area Studies in University Education.* Maria Sanchez and Antonio M. Stevens. 1987.

No. 54 *The Brazilian Diamonds in Contracts, Contraband and Capital.* Harry Bernstein. 1987.

No. 55 *Christians, Jews and Other Worlds: Patterns of Conflict and Accommodation.* Edited by Philip F. Gallagher. 1988.

No. 56 Vol. XXVI *The Fall of the Medieval Kingdom of Hungary: Mohács 1526, Buda 1541.* Géza Perjés. 1989.

No. 73 *Jadwiga of Anjou and the Rise of East Central Europe.* Oscar Halecki. Edited by Thaddeus V. Gromada. A joint publication with the Polish Institute of Arts and Sciences of America, New York. 1991.

No. 74 Vol. XXIX *Hungarian Economy and Society during World War Two.* Edited by György Lengyel. 1993.

No. 75 *The Life of a Communist Revolutionary, Béla Kun.* György Borsányi. 1993.

No. 76 *Yugoslavia: The Process of Disintegration.* Laslo Sekelj. 1993.

No. 77 Vol. XXX *Wartime American Plans for a New Hungary. Documents from the U.S. Department of State, 1942-1944.* Edited by Ignác Romsics. 1992.

No. 78 Vol. XXXI *Planning for War against Russia and Serbia. Austro-Hungarian and German Military Strategies, 1871-1914.* Graydon A. Tunstall, Jr. 1993.

No. 79 *American Effects on Hungarian Imagination and Political Thought, 1559-1848.* Géza Závodszky. 1995.

No. 80 Vol. XXXII *Trianon and East Central Europe: Antecedents and Repercussions.* Edited by Béla K. Király and László Veszprémy. 1995.

No. 81 *Hungarians and Their Neighbors in Modern Times, 1867-1950.* Edited by Ferenc Glatz. 1995.

No. 82 *István Bethlen: A Great Conservative Statesman of Hungary, 1874-1946.* Ignác Romsics. 1995.

No. 83 Vol. XXXIII *20th Century Hungary and the Great Powers.* Edited by Ignác Romsics. 1995.

No. 84 *Lawful Revolution in Hungary, 1989-1994.* Edited by Béla K. Király. András Bozóki Associate Editor. 1995.

No. 85 *The Demography of Contemporary Hungarian Society.* Edited by Pál Péter Tóth and Emil Valkovics. 1996.

No. 86 *Budapest, A History from Its Beginnings to 1996.* Edited By András Gerő and János Poór. 1996.

No. 87 *The Dominant Ideas of the Nineteenth Century and Their Impact on the State.* Volume 1. *Diagnosis.* József Eötvös.

No. 101 *Jewish Budapest.* Edited by Géza Komoróczy. 1999.

No. 102 *Evolution of the Hungarian Economy, 1848-1998.* Vol. I. *One-and-a-Half Centuries of Semi-Successful Modernization, 1848-1989.* Edited by Iván T. Berend and Tamás Csató. 2001.

No. 103 *Evolution of the Hungarian Economy, 1848-1998.* Vol. II. *Paying the Bill for Goulash-Communism.* János Kornai. 2000.

No. 104 *Evolution of the Hungarian Economy, 1848-2000.* Vol. III. *Hungary: from Transition to Integration.* Edited by György Csáki and Gábor Karsai. 2001.

No. 105 *From Habsburg Agent to Victorian Scholar: G. G. Zerffi (1820-1892).* Tibor Frank. 2000.

No. 106 *A History of Transylvania from the Beginning to 1919.* Vol. I. Edited by Zoltán Szász and Béla Köpeczi. 2001.

No. 107 *A History of Transylvania from the Beginning to 1919.* Vol. II. Edited by Zoltán Szász and Béla Köpeczi. 2001.

No. 108 *A History of Transylvania from the Beginning to 1919.* Vol. III. Edited by Zoltán Szász and Béla Köpeczi. 2002.

No. 109 *Hungary: Governments and Politics, 1848–2000.* Edited by Mária Ormos and Béla K. Király. 2001.

No. 110 *Hungarians in the Voivodina, 1918–1947.* Enikő A. Sajti. 2002.

No. 111 *Hungarian Arts and Sciences, 1848–2000.* Edited by László Somlyódy and Nóra Somlyódy. 2002.

No. 112 *Hungary and International Politics in 1848–1849.* Edited by Domokos Kosáry. 2002.

No. 113 *Social History of Hungary from the Reform Era to the End of the Twentieth Century.* Edited by Gábor Gyáni, György Kövér and Tibor Valuch. 2002.

No. 114 Vol.XXXVII *A Millennium of Hungarian Military History.* Edited by László Veszprémy and Béla K. Király. 2002.

No. 115 *Hungarian Relics. A History of the Battle Banners of the 1848-49 Hungarian Revolution and War of Independence.* Jenő Györkei and Györgyi Cs. Kottra. 2000.

No. 116 *From Totalitarian to Democratic Hungary. Evolution and Transformation, 1990-2000*. Edited by Mária Schmidt and László Gy. Tóth. 2000.

No. 117 *A History of Eastern Europe since the Middle Ages*. Emil Niederhauser. 2003.

No. 118 *The Ideas of the Hungarian Revolution, Suppressed and Victorious, 1956–1999*. Edited by Lee W. Congdon and Béla K. Király. 2002.